Water Science and Application 3

Land Surface Hydrology, Meteorology, and Climate

Observations and Modeling

Venkataraman Lakshmi
John Albertson
John Schaake
Editors

American Geophysical Union
Washington, DC

Land Surface Hydrology, Meteorology, and Climate: Observations and Modeling
Water Science and Application 3

Library of Congress Cataloging-in-Publication Data
Land surface hydrology, meteorology, and climate: observations and modeling/
Venkataraman Lakshmi, John Albertson, John Schaake, editors.
p.cm -- (Water science and application ; 3)
Included bibliographical references (p.)
ISBN 0-87590-352-5
1. Hydrology--Mathematical models. 2. Meteorology--Mathematical models. 3. Climatology--Mathematical models. I. Lakshmi, Venkataraman, 1966-. II. Albertson, John (John D.). III. Schaake, John. IV. Series

ISBN 0-87590-352-5
ISSN 1526-758X

2000 Florida Avenue, N.W.
Washington, DC 20009

Printed in the United States of America.

CONTENTS

CONTENTS

PREFACE

Land surface hydrology integrates various physical, chemical and biological processes that occur above, on, and below the surface of the Earth. As a result, it is critical to accurately account for land surface processes within predictive models of hydrology, meteorology, and climate.

One of our main difficulties, however, concerns the broad range of spatial and temporal scales that characterize land surface hydrological processes. For example, we determine infiltration by pore scale physics, while soil hydraulic conductivity remains a field scale property. Photosynthesis, respiration, and transpiration occur at the leaf scale. Runoff is a catchment scale process, and the variability of groundwater storage is a regional scale issue. Turbulence in land-atmosphere exchanges of heat, moisture, and momentum occur on the order of seconds to minutes, while variations in land surface and air temperatures occur much more gradually: on the order of hours. The persistence of floods and droughts is seasonal to annual, and so is the effect of El Nino on regional hydrology. Long-term climate effects occur much more slowly, on the order of years to decades.

Studies at each scale are vital to describing the processes just mentioned. There is also a pressing need to move information across these spatial and temporal scales to describe their aggregate effects at scales different from those by which we describe individual processes. In this respect, one of the most important issues in hydrological studies concerns process representation by parameterization and the assimilation of observations into prediction models. In recent times, spatially distributed data through remote sensing has played a major role in model development, parameter estimation, and model validation. However, in situ measurements remain both the main vehicle for theoretical advances and the only tool for validating and calibrating remote sensing tools. Even in the best of cases, with satellite remote sensing limited by inadequate, intermittent temporal coverage, we must turn to model dynamics to complete such observations. We also understand that the tools that integrate observations and models (i.e., data assimilation) are in their infancy.

In response to this situation we offer this book - a unique representation of land surface hydrologic processes, observations, modeling, and the techniques now emerging that will help us to assimilate observations into models that improve our predictive capability in hydrology, meteorology, and climate.

This book is divided into three sections, each discussing a general area of importance: Observations, Modeling, and Integration of Observations and Modeling. The book is designed to bring together under one roof the current state of the science used by experimentalists, modelers, and those working to bridge the two through comparison/validation studies and assimilation of observations into models.

This book will be a valuable reference for working scientists in hydrology, meteorology, and climate; scientists leading graduate seminars in the area of Earth-atmosphere interaction; and graduate students seeking a thorough introduction to the current literature.

Venkataraman Lakshmi
University of South Carolina

Introduction

Venkataraman Lakshmi

Department of Geological Sciences, University of South Carolina, Columbia, South Carolina

INTRODUCTION

Hydrology is the study of processes that involve movement of water from the land surface to the subsurface and inversely from the subsurface to the surface and atmosphere. In order to understand the various components of the hydrological cycle, we study the processes involved by observations, modeling, and a synthesis of observations and modeling. With these three "modes" of study, we can characterize the magnitudes, timing, and the spatial and temporal scales associated with each component of the hydrological cycle in terms of their effects on ecological, biological and physiological processes. Although we commonly answer questions that arise at this juncture (e.g., regarding the recurrent frequency and magnitude of floods or the seasonal variability of soil moisture) by evaluating individual components of the hydrological cycle, we must also evaluate them simultaneously. The magnitudes of the various components should exist in a "balance" such that no mass of water is created or destroyed.

Texts on hydrology usually describe the physical processes involved in water and energy transport in the context of land surface models. As an observational science, however, most hydrological theories have originated from observations. It is thus incumbent upon us to explain appropriate physical equations in relation to the observations that corroborate them.

In order to improve our knowledge of the parameterization of physical processes we have recently developed the field of data assimilation, which encompasses the need to relate observations with models. This book originates here: to inspire an understanding of hydrology as an integrated knowledge of observations, models and the equations that describe them.

Land Surface Hydrology, Meteorology, and Climate: Observations and Modeling
Water Science and Application 3

A Description of Land Surface Hydrology

Hydrology is based on a set of equations for water and energy balance of the land surface. These equations capture the essence of the various processes that occur between the atmosphere and the water table. So as to place this book in its proper perspective, I have listed then explained these equations briefly below. Various papers in this book will discuss in greater detail aspects of these equations.

The equations describing the water balance in the soil-vegetation column (Figure 1) for five layers follow:

$$Z_1 \frac{\partial \theta_1}{\partial t} = P - R - E - f_1 T - q_{1,2}$$

$$Z_2 \frac{\partial \theta_2}{\partial t} = q_{1,2} - I_2 - f_2 T - q_{2,3}$$

$$Z_3 \frac{\partial \theta_3}{\partial t} = q_{2,3} - I_3 - f_3 T - q_{3,4}$$

$$Z_4 \frac{\partial \theta_4}{\partial t} = q_{3,4} - I_4 - f_4 T - q_{4,5}$$

$$Z_5 \frac{\partial \theta_5}{\partial t} = q_{4,5} - I_5 - f_5 T - q_{5,wt}$$

In the above set of equations, θ_1, θ_2, θ_3, θ_4 and θ_5 are the volumetric soil moistures of the layers with thickness Z_1, Z_2, Z_3, Z_4 and Z_5, respectively); P is the precipitation, E is the bare soil evaporation, R is the surface runoff, T is the transpiration, and f_1, f_2, f_3, f_4 and f_5 denote the fraction (sum of all fractions equal unity) of the transpiration from the various layers; $q_{1,2}$ is the moisture flux from layer 1 to layer 2, $q_{2,3}$, $q_{3,4}$, $q_{4,5}$ are fluxes from layer 2 to 3, 3 to 4 and 4 to 5, and $q_{5,wt}$ is the moisture flux from layer 5 to the water table minus the capillary rise from the water table (water table is considered to lie below layer 5). The interflow/baseflow

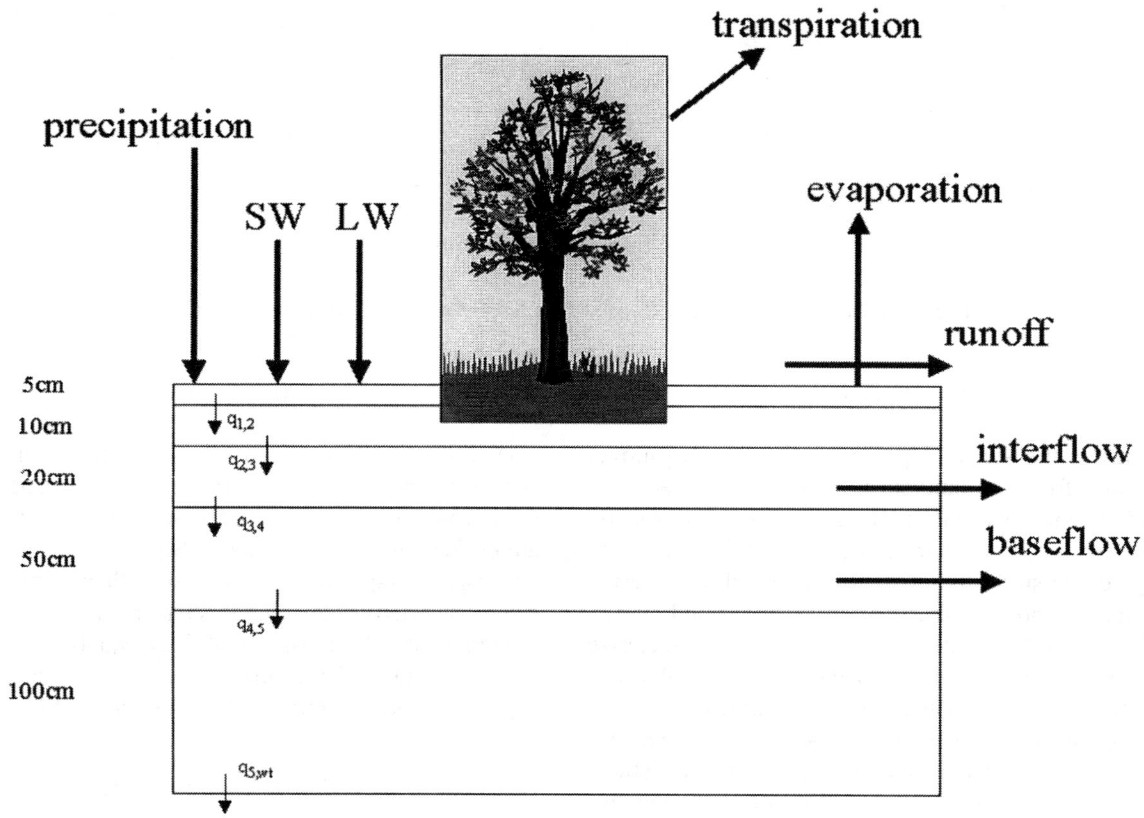

Figure 1. Representation of the soil-vegetation column.

from layers 2, 3, 4 and 5 are denoted by I_2, I_3, I_4 and I_5, respectively. The moisture flow between the layers is modeled using the Richard's equation accounting for the gravity advection and the moisture gradient. The bare soil evaporation and the vegetation transpiration are estimated using the supply and demand principle (i.e., if there is enough moisture to satisfy the potential value, the evaporation and transpiration occur at the potential rate, else they occur at a rate limited by the amount of available soil moisture).

The fraction f_I of transpiration from layer i is determine by the fraction of roots in the layer. These layers are 5, 10, 20, 50 and 100cm thick, respectively, going from the surface towards the water table.

The exchange fluxes from one layer to the next layer are given by

$$q = -K(\theta)\frac{\partial h}{\partial z}$$

where the vertical direction is denoted by z, q is the exchange flux, $K(\theta)$ is the unsaturated hydraulic conductivity, which is dependent on the soil moisture of the layer and h, is the hydraulic head, which is a sum of the pressure head and the gravitational head. This above expression can be used for computation of infiltration (following precipitation) as well as exfiltration (evaporation). Infiltration and exfiltration are calculated using the mid-height soil moisture of the top layer θ_1 and the surface soil moisture, which is θ_s (saturated) in the case of infiltration (the surface is completely wet) and θ_r (residual) in the case of evaporation (the surface is completely dry).

The parameterization of $K(\theta)$ is dependent on soil type. There are various empirical relationships that can be used to quantify this relation. Prior knowledge of the type of soil in the study area is helpful in assigning appropriate values for soil hydraulic properties such as saturated hydraulic conductivity, bubbling pressure, variation of hydraulic conductivity with soil moisture and the maximum and minimum values of soil moisture.

The energy balance equation for the land surface can be written as

$$R_{sd}(1-\alpha)+R_{ld}-\varepsilon\sigma T_s^4-\frac{\rho_w L}{\gamma r_{av}}(e_s(T_s)-e_a)-\frac{\rho C_p}{r_{ah}}(T_s-T_a)-\frac{\kappa}{D}(T_s-T_d)=0$$

where R_{sd}, R_{ld} are the incoming shortwave and longwave radiation, respectively. α, ε and σ are the albedo, emissivity and the Stefan-Boltzmann's constant, respectively. ET is the evapotranspiration flux $ET = E + T$; and T_s, T_a and T_d are the surface temperature, air temperature and the 10cm soil temperature, respectively; $e_s(T_s)$ and e_a are the saturated vapor pressure at surface temperature T_s and actual vapor pressure of the air, respectively. ρ, C_p, γ, ρ_w, are the density, specific heat and psychrometric constant of air and density of water, respectively; r_{av} and r_{ah} are the aerodynamic resistances to vapor and heat and r_c is the canopy resistance. κ and D are the thermal conductivity and the diurnal damping depth of the soil. The aerodynamic resistances to vapor (r_{av}) and heat (r_{ah}) are taken as equal to each other and are evaluated using established methods.

Requirements for the Study of Land Surface Hydrology and the Content of this Book

In order to study land surface hydrology, one must obtain comprehensive observations of land surface and atmospheric variables. These variables include those that serve as input to land surface models, viz., wind speed, vapor pressure and air temperature at a reference height, longwave and shortwave radiation and precipitation (among the meteorological variables); leaf area index/canopy density and stomatal resistance (among the biophysical variables). The variables that serve for validation include surface temperature, soil moisture, heat and moisture fluxes and streamflows. These variables (validation variables) are computed within a model that can then be compared to their corresponding observations. Quite often, the models can simulate at a higher spatial and temporal resolution than the observations.

Part 1 features papers on observations of atmospheric and land surface variables of the hydrological cycle. Eichinger discusses water vapor profiles in the atmosphere while Katul examines the strengths of the sources and sinks of water vapor, carbon and heat in a canopy volume. Cuenca studies an aspect crucial for parameterizing unsaturated hydraulic conductivity: observation of the variability in space (with depth) and time of soil moisture and soil hydraulic properties. Kustas compares the synergistic use of satellite data and observations from weather stations for computing latent and sensible heat fluxes with a detailed model-based computation, and demonstrates the adequacy of observations (ground and satellite) in calculation of heat fluxes.

Part 2 is devoted to advances in land surface modeling. The papers in this section include Bastidas' and Duan's parameter estimations; Koster's comparison of various GCM-generated land water budgets; Chen's development and application of land surface components in mesoscale models; and Maurer's comparison of water budgets from a mesoscale model with corresponding values obtained from a macroscale hydrological model. It is important to note that hydrological models are becoming increasingly complex, especially in mesoscale and GCM frameworks. These models have a number of associated parameters with only a limited amount with a physical basis. All such models, however, vary in time and space, and estimation is essential to obtaining accurate predictions from them. In addition to model development and placement in the proper context (mesoscale or GCM), we must also reconcile differences between the outputs of different models. Analysis of various elements of the water and energy budget to pinpoint areas of weaknesses in the models we use is essential, especially in regard to implications for weather forecasts (mesoscale models), climate change, and global warming predictions (GCM).

Part 3 presents papers that feature integration of the above two areas, viz., observations and modeling. Integrating observations and modeling should improve our understanding of land surface processes if such integration involves innovative methodology. In this respect, Woods discusses treating spatial variability in the design of a field experiment and compares results from a satellite-based model with field data. In studying a perceived increase in the accuracy of model simulations on incorporating field scale variability, Mohr demonstrates the need for carefully designed field experiments. Knorr studies assimilation of satellite data into a biosphere-hydrology model so as to increase the accuracy of simulations. Lawford outlines in detail, with some scientific results, studies from the ongoing GCIP (GEWEX: Global Energy and Water Cycle, Continental Scale International Project), which serves as test sites for many land surface and atmospheric models. Finally, Piechota presents a demonstration of the connection between forecasts of the El Nino Southern Oscillation as key for prediction of river discharge with implications for agriculture and power supply.

SUMMARY

The spectrum of content in this book has been specifically organized to invoke interest in hydrological modeling from different scientific angles. Land surface modeling is no longer a tool just for hydrologists but a valuable component in various land-atmosphere-climate research issues that require integration of the two procedures before us: observation and modeling. In this introduction, of course, we recognize the wide extent of land surface hydrology and mention its broad edges. By providing a flavor of some of the topics involved, we wish to spark a sense of curiosity in the interested reader.

Section 1

OBSERVATIONS

Lidar Measurements of the Dimensionless Humidity Gradient in the Unstable Atmospheric Surface Layer

William E. Eichinger

Iowa Institute for Hydraulic Research, Department of Civil and Environmental Engineering, University of Iowa, Iowa City, Iowa

Marc B. Parlange

Department of Geography and Environmental Engineering, Center for Environmental and Applied Fluid Mechanics, Johns Hopkins University, Baltimore, Maryland

Gabriel G. Katul

School of the Environment, Duke University, Durham, North Carolina

The description of velocity and scalar statistics in the atmospheric surface layer are often based on Monin and Obukhov [1954] similarity theory. Kader and Yaglom [1990], in a reexamination of the theory, proposed the existence of separate dynamic, dynamic-convective and free convective sublayers in the unstable atmospheric surface layer. This has led to a number of experimental studies on the formulation of the similarity functions for dimensionless profiles of velocity and temperature variances, scalar variance dissipation and dissipation of turbulent kinetic energy. There has not been a study of the mean humidity gradient given the common difficulties of measuring accurate mean water vapor profiles in the lower atmosphere. Raman-Lidar measurements of mean water vapor profiles in the atmospheric surface layer are analyzed and the existence of the three sublayers is demonstrated. The data support the assumption that the similarity function for water vapor is similar to that for temperature in the dynamic and the dynamic-convective sublayers. In practice the traditional "Businger-Dyer" formulation for the specific humidity dimensionless gradient is sufficiently accurate for routine applications of Monin-Obukhov similarity theory.

1. INTRODUCTION

The general form of Monin-Obukhov similarity for the dimensionless gradient of the mean specific humidity (q) in the Atmospheric Surface Layer (ASL) may be written as:

$$\varphi(\zeta) = -\frac{\rho k_v (z-d)}{Q_*} \frac{dq}{dz}. \qquad (1)$$

Land Surface Hydrology, Meteorology, and Climate: Observations and Modeling
Water Science and Application Volume 3, pages 7-13

where z is the height above the surface, d is the displacement height, Q_* [$= E / (\rho u_*)$] is the turbulent moisture scale, u_* [$=(\tau_o/\rho)^{1/2}$] is the friction velocity, τ_o is the surface shear stress, E is the evaporation rate, ρ is the density of the air, k_v [=0.4] is the analog of von Karman's constant for water vapor. The symbol φ(ζ) represents the dimensionless ASL humidity gradient which is a similarity function of the dimensionless parameter ζ = -(z-d)/L, where L is the Obukhov length defined as

$$L = \frac{-u_*^3}{k(g/T_a)\ (H_v/\rho c_p\ )} \tag{2}$$

where g is the acceleration due to gravity, T_a is the air temperature near the ground, $H_v = (H + 0.61\ T_a c_p E)$ is the virtual sensible heat flux, H is the sensible heat flux, and c_p is the specific heat at constant pressure. To estimate surface fluxes, the integrated version of (1) is typically used in practice,

$$q(z_1) - q(z_2) = \frac{E}{k_v \rho u_*}\left[\ln\left(\frac{z_2 - d_0}{z_1 - d_0}\right) + \Psi_v(\zeta_2) - \Psi_v(\zeta_1)\right] \tag{3}$$

where z_1 and z_2 are two heights within the surface layer,

$$\Psi_v(\zeta) = \int_{z_{0v}/L}^{\zeta} \frac{[1-\varphi(x)]}{x}\,dx \tag{4}$$

is the integrated scalar profile function and z_{ov} is the water vapor roughness length [*Brutsaert*, 1982].

For some 30 years there has been a general agreement that the Businger-Dyer parameterization

$$\varphi(\zeta) = (1+16\zeta)^{-1/2} \tag{5}$$

represented the existing data [*Businger et al.*, 1971; *Dyer*, 1974]. Thus the form of the similarity functions for mean gradient profiles (under unstable atmospheric stability) had become a relatively closed issue in boundary layer meteorology.

The issue was opened up again when Kader and Yaglom [1990] presented evidence for the existence of 3 sublayers (dynamic, dynamic-convective, free-convective) in the atmospheric surface layer. The concept of the separate sublayers in the ASL, discussed in detail by Kader and Yaglom, was based on the application of directional dimensional analysis theories of the atmosphere developed by Zilitinkevich [1971] and Betchov and Yaglom [1971] as well as the availability of new data. Kader and Yaglom divided the surface layer into three sections based upon the processes responsible for the creation of turbulent kinetic energy. In the dynamic sublayer, where the predominate motion is in the horizontal and in which turbulent kinetic energy is created primarily by mechanical forces, u_* is the appropriate velocity scale along with a horizontal length scale (l_x) to be used in characterization of the flow. In the free convective sublayer buoyant forces are primarily responsible for the creation of turbulent kinetic energy, so that the convective velocity, w_*, is the appropriate velocity scale along with a vertical length scale (l_z). In the dynamic-convective sublayer both mechanical and buoyant forces create turbulent kinetic energy, thus both of the length and velocity scales must be used to characterize the flow field. The data presented by Kader and Yaglom focused on momentum and sensible heat transfer from which they derived the mean similarity functions for momentum φ_m and sensible heat φ_h. The work of Kader and Yaglom supports the concept of different sublayers given the different behavior of the φ values for the z/L ranges for each of the three sublayers. Based upon the mean temperature profile data and the overall fit of φ_h by Kader and Yaglom, Brutsaert [1996] proposed an interpolation formula

$$\varphi_h(\zeta) = (0.33 + 0.057\zeta^{0.78})/(0.33 + \zeta^{0.78}) \tag{6}$$

which allows for an analytical integration of equation 5. Brutsaert also presented a version of φ_m, based on Kader and Yaglom, which was tested by Parlange and Brutsaert [1993] and Parlange and Katul [1995] in two separate field experiments. The new form of φ_m was found to match a wider range of wind profile data as compared to the Businger-Dyer form, especially in the free convective limit. Other analyzes of field data which were inspired, in part, by Kader and Yaglom include: a) studies of the dissipation rate of temperature variance [*Kiely et al.*, 1996], b) the standard deviation of the temperature [*Albertson et al.*, 1995] and c) the dissipation rate of turbulent kinetic energy [*Albertson et al.*, 1997].

A similarity between latent heat and sensible heat transfer is often assumed in atmospheric transport, such that $\varphi = \varphi_h$, though there is evidence [*Warhaft*, 1976; *Katul and Parlange*, 1994] that this is not necessarily true. Less information is available concerning the behavior of φ. One of the few studies is due to Pruitt et al. [1973]. Part of the reason that latent heat transfer is assumed to follow the same parameterization as sensible heat transfer is the difficulty in obtaining

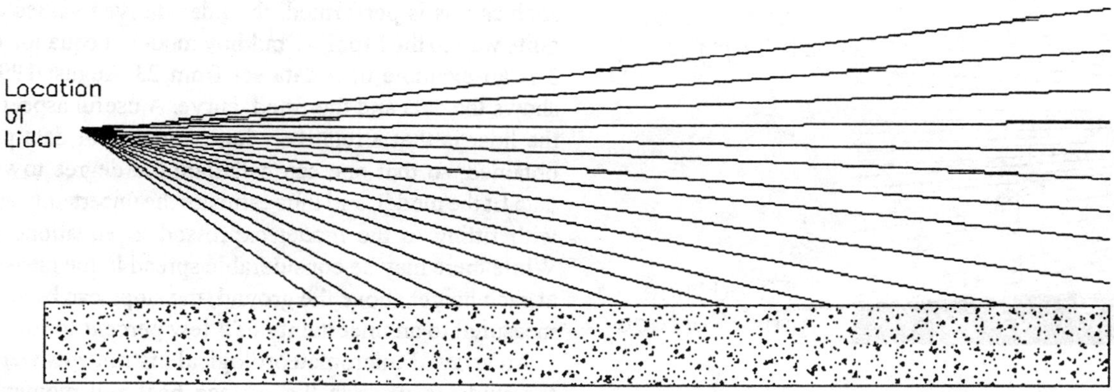

Figure 1. A conceptual drawing showing the Davis site and how different lines of sight from the lidar are combined to make the water vapor concentrations map. The water vapor concentration is determined every 1.5 meters along each of the lines shown. The lines of sight in actual practice are only 0.15 to 0.25 degrees apart. The drawing is exaggerated in the vertical direction for clarity. The lidar lines of sight are actually nearly parallel to the surface with the vertical distance between lines of sight on the order of 5 to 10 centimeters.

measurements of φ. The measurement of humidity profiles in the lower atmosphere is notoriously difficult since the differences between fixed sensors, even with extensive calibration, can be the same order of magnitude as the actual humidity differences themselves near the land surface. Further, the presence of a number of sensors in close proximity may affect the results. A Raman lidar allows a non-intrusive measurement of the humidity profile throughout the lower atmosphere using the same instrument [*Parlange et al.*, 1999]. The objective of this paper is to obtain accurate and detailed information on φ and investigate its behavior through the different ASL sublayers.

2. EXPERIMENT

The field experiment was carried out at the Campbell Tract of the University of California, Davis in August, 1991. The site consisted of an irrigated bare field of yolo light clay soil. The scanning, solar-blind, water-Raman lidar used in this experiment was built at Los Alamos and is based upon the technique pioneered by Cooney [1970] and Melfi et al. [1969]. Details of this instrument are described by Eichinger et al. [1994]. Other work using the lidar measurements of water vapor over this particular research site include a multifractal analysis of water vapor [*Pinzon et al.*, 1995] and the growth of the water vapor blanket over the field [*Parlange et al.*, 1999]. The use of a Raman lidar enables high spatial resolution (1.5 m) data elements and the application of scanning techniques. Pulses are summed to increase the signal-to-noise ratio along each line of sight. The number of pulses considered to have the minimum acceptable signal-to-noise ratio is dependent upon the desired range and laser energy per pulse, and varies from five to twenty-five. This lidar system has been shown to be capable of making measurements with an accuracy of 3 to 5 percent when operated in this fashion.

The lidar system can be programmed to sample along a number of lines of sight in order to execute a RHI (Range-Height-Indicator) scan. These scans are made by taking measurements using multiple laser pulses at various elevation angles at constant azimuth. Water vapor concentrations are measured every 1.5 meters along each line of sight. Figure 1 is a graphical representation of the lines of sight that comprise the scan. This effectively generates a map of the water concentration in a vertical slice of the atmosphere along the azimuth direction. An RHI scan over an 16 degree field of view (4 degrees down to 12 degrees up) with 0.25 degree angle increments takes about 50 s to complete.

Measurements of the time averaged humidity gradient are historically difficult to make because they require non-interfering measurements of the humidity profile at various heights inside the ABL surface layer. A lidar RHI scan provides a large number of measurements at many heights over distances of several hundred meters. The concentrations at all heights in a given range interval are used in the analysis as shown in figure 2. The data were plotted as water vapor concentration with height above the ground and were examined prior to processing to estimate the height of the inner region of the local boundary layer over the field. A height of 75 percent of this estimate was used as the highest point in processing to minimize effects at the upper boundary [see *Parlange et al.*, 1999]. While this ensures that the measurements are well inside the surface layer, this limits the number of measurements at large z/L. When spatial averaging

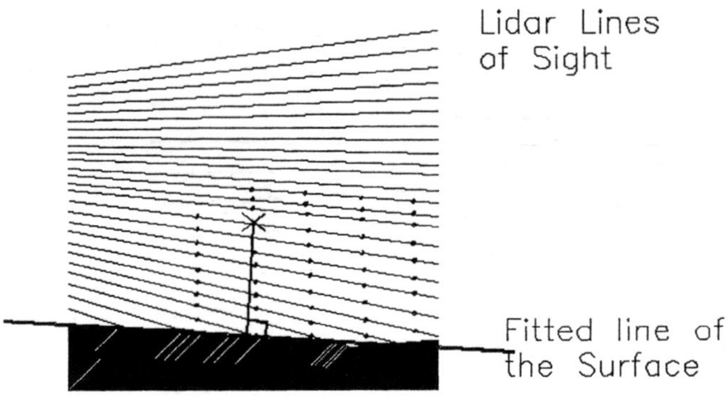

Figure 2. A conceptual drawing of a 50 meter region and all of the lidar lines of sight within it. The location of a line approximating the surface is determined, and the distance from each measured value to this line along a perpendicular to the line is calculated. All of the measured values of water vapor concentration are used in the calculation of the slope shown in figure 3.

such as this is performed, the lidar-derived values converge quite well to the Monin-Obukhov model of equation 4. Figure 3 is an example of a data set from 23 August 1991 which shows the data and the fitted curve. A useful aspect of using the lidar is that a relatively large number of data points are obtained, so that one can calculate the degree to which the data fit the model, and thus estimate the uncertainty associated with fitting to the model expressed in equations 1 and 4. While there may be considerable spread in the measurements at each height above the ground, the slope can be statistically measured to an uncertainty of a few percent for most cases..

An array of micrometeorological instruments were used in the study to provide the surface heat and momentum flux values to estimate L. The evaporation rate was obtained using a six meter diameter weighing lysimeter [*Parlange and Katul*, 1992]. A Campbell Scientific Inc. eddy-correlation system (krypton hygrometer and one-dimensional sonic anemometer with a fine wire thermocouple) was used to check the lysimeter estimate of E and to determine H. A three-

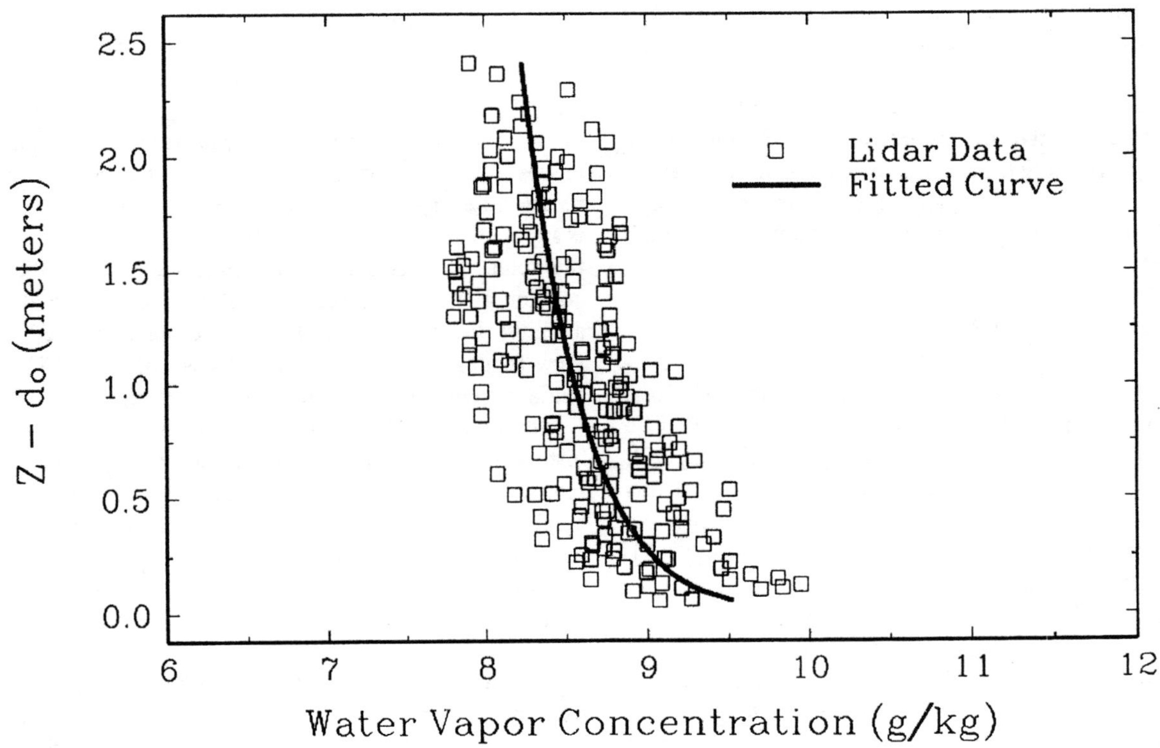

Figure 3. An example of the profile of water vapor concentration with height above the ground from 23 August, 1991. Also shown is the specific humidity profile from equation 7.

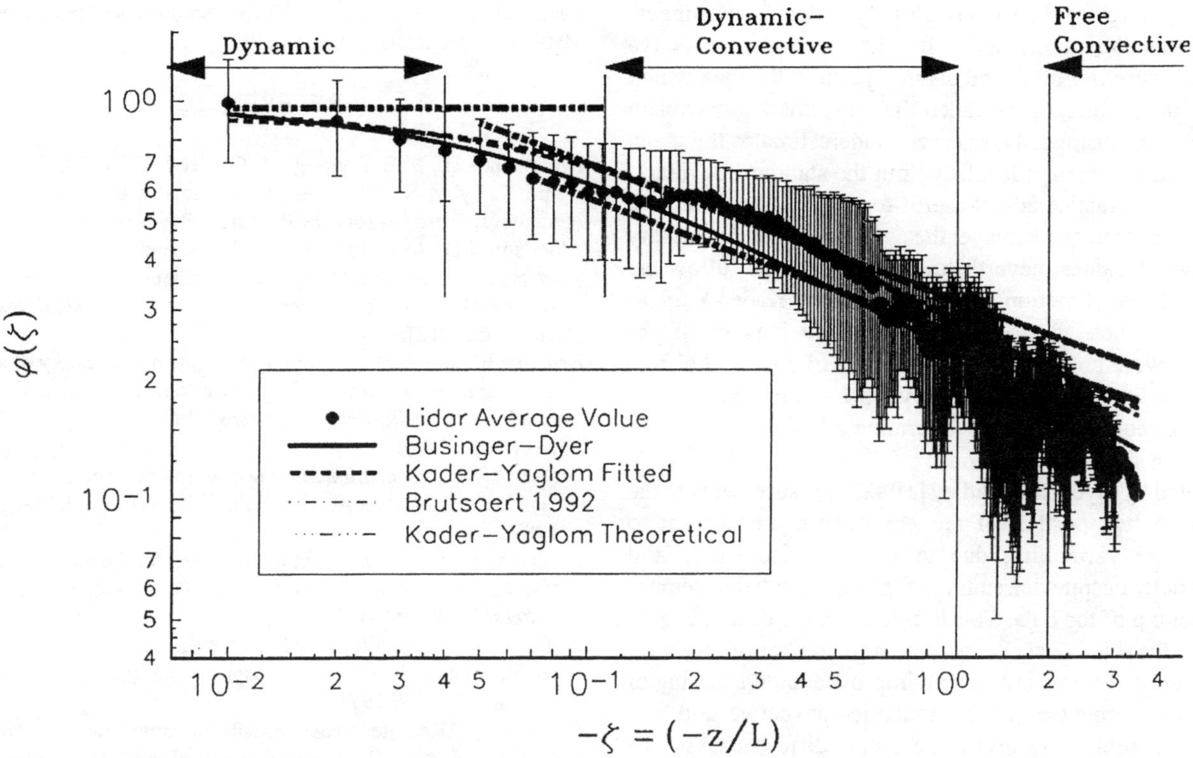

Figure 4. Lidar-derived φ_v values with the three sublayer ranges of Kader and Yaglom indicated. Comparison of lidar-derived stability correction functions to the traditional Businger-Dyer parameterization and the Kader and Yaglom parameterization. The error bars represent the standard deviations of the measurements within the interval. The small error bars on the largest values of ζ are indicative of a limited number of measurements and not necessarily decreased uncertainty.

dimensional sonic anemometer was used to obtain u_*. Soil heat flux plates and a net radiometer were used to complete the measurements of the surface energy balance and to serve as an overall check on the measurements.

3. RESULTS

The water vapor concentration data were fitted using a least squares technique to a second order polynomial in ln(z-d) in a manner similar to Hogstrom [1988],

$$q = q_0 + A\,[\ln(z-d)] + B\,[\ln(z-d)]^2 \,. \tag{7}$$

The gradients were determined from the derivatives of equation 7 to obtain the dq/dz values. Simultaneous values of u_*, L, and E were obtained from the meteorological instruments. These were used to calculate the values of $\varphi(\zeta)$ with

$$\varphi(\zeta) = -\frac{[k_v u_*(z-d_0)\rho]}{E}\frac{dq}{dz} \tag{8}$$

Lidar measurements of $\varphi(\zeta)$ were made and the results averaged over a wide range of values of ζ. Values were binned over ζ intervals of 0.01 and the mean and standard deviation of all the values in the interval calculated. Because of the very large number of measurements (greater than 41,000), an uncertainty can be calculated in each interval of ζ as shown in Figure 4. The range of each of the sublayers identified by Kader and Yaglom is marked in Figure 4a with a transition section separating each interval.

The φ_v values are plotted in Figure 4 along with the Businger-Dyer function, the Kader-Yaglom sublayer models and the smooth Brutsaert function based on the Kader-Yaglom sublayer models. In the dynamic layer, the φ data clearly show a fall-off in comparison to the constant value predicted by the

Kader and Yaglom theory though they match the Businger-Dyer parameterization well. In the dynamic-convective sublayer, there is a clear and distinct jump in the data which matches the φ_h function of Kader-Yaglom remarkably well on average. The Businger-Dyer curve underestimates the mean φ observations, though it falls within the standard deviation limits and in practice is still useful for application purposes. In the free convective sublayer the data are limited in quantity and range of values, nevertheless, they appear to follow the Businger-Dyer function better than the Kader-Yaglom function. Which is interesting since the Businger-Dyer functions were not based on a great deal of data in the free-convective limit. In the region beyond $\zeta = 2$ more measurements are required to better determine all of the scalar and momentum similarity functions.

As noted by Dyer and Bradley [1982], measurements in the free convective portion of the atmosphere are subject to considerable variability, due to lack of stationarity and homogeneity despite detrending of data and spatial or temporal averaging of the data. This is evident in the data in Figure 3 despite the relative accuracy of a large number of measurements. While the data indicate an improved understanding of the physical processes in the dynamic-convective and free convective sublayers, given the small differences in the various similarity functions in comparison to the inherent uncertainties, changing from the Businger-Dyer parameterization is, in practice, probably unwarranted.

As an aside, an additional result from the measurement of the φ values is the determination of k_v, the analog of von Karman's constant for latent heat. This value is normally taken as 0.40 and was used in Figure 4. Using the raw data and fitting the twelve values of φ below ζ=0.11 to a second order polynomial in ln(ζ), an asymptotic value for the y intercept can be determined. The intercept value is the inverse of k_v which was found to be 0.37 ± 0.02. Given the scatter in the data, using a number for k_v other than 0.4 is probably unwarranted.

4. CONCLUSIONS

The theoretical justification given by Kader and Yaglom for the existence of three distinct sublayers in the atmospheric surface layer is supported by the extensive field data which are presented in this study. This is the strongest indication to date of the existence of these sublayers experimentally. Free convection conditions, which distinguish the third sublayer, require further measurement. Nevertheless, the existence of a third sublayer for large |z/L| is supported by the existing data.

Acknowledgments. This research has been supported and financed, in part, by the National Science Foundation, (NSF-ATM-9726270 and NSF-ATM-9818485) and by the New Mexico Bureau of Reclamation. The authors gratefully acknowledge the help of Mike Mata in the preparation of the field.

REFERENCES

Albertson, J.D., M.B. Parlange, G.G. Katul, C.R. Chu, H. Stricker, and S. Tyler, Sensible heat flux estimates using flux variance methods, *Water Resources Research*, 31(4):969-974, 1995.

Albertson, J.D., M.B. Parlange, G. Kiely, and W.E. Eichinger, The average dissipation rate of turbulent kinetic energy in the neutral and unstable atmospheric surface layer, *J. Geophysical Research-Atm.*, 102(D12):13,423-13,432, 1997.

Betchov, R., and A.M. Yaglom, Comments on the theory of similarity as applied to turbulence in an unstably stratified fluid, *Izv. Akad. Nauk SSSR. Ser. Fiz. Atmosf. i Okeana* 7, p. 1270-1279, 1971.

Brutsaert, W., *Evaporation into the atmosphere: Theory history and applications*, Kluwer Academic Publishers., Norwell MA, pp.299, 1982.

Brutsaert, W., Stability correction functions for the mean wind speed and temperature in the unstable surface layer, *Geophys. Res. Letters*, 19(5), 469-472, 1996.

Businger, J.A., J.C. Wyngaard, Y. Izumi, and E.F. Bradley, Flux-profile relationships in the atmospheric surface layer, *J. Atm. Sci.*, vol. 28, 181-189, 1971.

Cooney, J., Remote measurements of atmospheric water-vapor profiles using the Raman component of laser backscatter, *Appl. Meteor.* 9(1), 182-184, 1970.

Dyer, A.J., A review of flux-profile relationships, *Bound. Lay. Met.*, 7, 363-372, 1974.

Dyer, A.J., and E.F. Bradley, An alternative analysis of flux-gradient relationships at the 1976 ITCE, *Bound. Lay. Met.*, 22, 3-19, 1982.

Eichinger, W.E., D.I. Cooper, D.E. Hof, D.B. Holtkamp, R.R. Karl, C.R. Quick, and J.J. Tiee, Development of a scanning solar-blind, water Raman lidar, *Appl. Opt.*, 33(18), 3923-3932, 1994.

Hogstrom, U., Non-dimensional wind and temperature profiles in the atmospheric surface layer: A re-evaluation, *Bound. Lay. Met.* 42, 55-78, 1988.

Kader, B.A., and M.Y. Yaglom, Mean fields and fluctuation moments in unstably stratified turbulent boundary layers, *J. Fluid Mech.*, 212, 637-662, 1990.

Katul, G.G. and M.B. Parlange., On the active role of temperature in surface layer turbulence, *Journal of the Atmospheric Sciences*, 51(15),2181-2195, 1994.

Kiely, G., J.D. Albertson, M.B. Parlange, and W.E. Eichinger, Scaling the average dissipation rate of temperature variance in the atmospheric surface layer, *Boundary Layer Meteorology*, 77:267-284, 1996.

Melfi, S.H., J.D. Lawrence, and M.P. McCormick, Observations of Raman scattering by water-vapor in the atmosphere, *Appl. Phys. Lett.*, 15(9), 295-297, 1969.

Monin, A.S., and A.M. Obukov, Basic laws of turbulent mixing in the ground layer of the atmosphere, *Trudy Geofiz. Inst. Akad. Nauk SSSR* 24(151), 163-167, 1954.

Parlange, M.B. and W. Brutsaert., Regional shear stress of broken forest from radiosonde wind profiles in the unstable surface layer, *Boundary Layer Meteorology*, 64, 355-368, 1993.

Parlange, M.B., and G.G. Katul, Estimation of the diurnal variation of potential evaporation from a wet bare soil surface, *J. Hydr.*, 132, 91-106, 1992.

Parlange, M. B., and G. G. Katul., Watershed scale shear stress from tethersonde wind profile measurements under near neutral and unstable atmospheric stability, *Water Resources Research*, 31(4):961-968, 1995.

Parlange, M.B., J.D. Albertson, W.E. Eichinger, and A.T. Cahill, Evaporation: Use of fast response turbulence sensors, raman lidar, and passive remote sensing, in: *Vadose Zone hydrology: Cutting across disciplines* (Parlange, M.B. and Hopmans, J.W. Editors), Oxford University Press, 260-278, 1999.

Pinzon, J.E., C.E. Puente, M.B. Parlange, and W. Eichinger, A multifractal analysis of lidar measured water vapour, *Boundary Layer Meteorology*, 76:323-347, 1995.

Pruitt, W.O., D.L. Morgan, and F.J. Lourence, Momentum and mass transfers in the surface boundary layer, *Quart. J. Roy. Meteorol. Soc.*, 99:370-386, 1973.

Warhaft, Z., Heat and moisture flux in the stratified boundary layer, *Quart. J. Roy. Met. Soc.* 102, 703-707, 1976.

Zilitinkevich, S.S., Turbulence and diffusion in free convection, *Izv. Akad. Nauk SSSR. Ser. Fiz. Atmosf. i Okeana* 7, 1263-1269, 1971.

William E. Eichinger, Iowa Institute for Hydraulic Research, Department of Civil and Environmental Engineering, University of Iowa, Iowa City, IA 52242

Marc B. Parlange, Department of Geography and Environmental Engineering, Center for Environmental and Applied Fluid Mechanics, Johns Hopkins University, Baltimore, MD 21218

Gabriel G. Katul, School for the Environment, Duke University, Durham, NC 27706

Time Difference Methods for Monitoring Regional Scale Heat Fluxes with Remote Sensing

William P. Kustas

USDA-ARS Hydrology and Remote Sensing Lab, Beltsville, Maryland

George R. Diak

CIMSS Space Science and Engineering Center, University of Wisconsin, Madison, Wisconsin

John M. Norman

Department of Soil Science, University of Wisconsin, Madison, Wisconsin

Satellite-based surface radiometric temperatures, $T_R(\phi)$ at viewing angle ϕ, can serve as a key boundary condition, defining the degree to which available surface net radiation is partitioned between sensible and latent heat fluxes. However, traditional single-source flux-gradient approaches have not provided reliable estimates, mainly due to significant differences existing between $T_R(\phi)$ and the so-called 'aerodynamic' temperature (T_O) of the surface. Unfortunately, T_O is a construct that cannot be measured and many of the factors affecting $T_R(\phi)$ are not well correlated to T_O. For regional scale applications, there is the additional problem of defining meteorological conditions, primarily air temperature (T_A) and wind speed, for each satellite pixel. These limitations have been recently addressed through an energy closure modeling scheme, Atmosphere-Land-Exchange-Inverse (ALEXI). ALEXI uses the growth of the atmospheric boundary layer (ABL) coupled to a two-source flux-gradient modeling scheme with temporal changes in $T_R(\phi)$ observations. The two-source modeling scheme, which accounts for soil and vegetation contributions to the mass and energy exchanges, addresses the main factors affecting the $T_R(\phi) - T_O$ relationship, while the use of temporal changes in $T_R(\phi)$ and ABL height significantly reduces measurement errors associated with satellite observations and eliminates the need for T_A observations, respectively. An alternative approach, the Dual-Temperature-Difference (DTD), uses the two-source modeling scheme with time rate of change in $T_R(\phi)$ and T_A observations in a simple formulation that is computationally efficient for regional applications. By using a

Land Surface Hydrology, Meteorology, and Climate: Observations and Modeling
Water Science and Application Volume 3, pages 15-29

time difference in T_A, errors associated with using local shelter level observations for representing a region are reduced. The utility of the DTD approach is evaluated at the local scale with ground-based $T_R(\phi)$ observations driven by regional weather station data in a semiarid region. In addition, a comparison of regional scale heat fluxes between the more rigorous ALEXI model and the simple DTD method using satellite data over the Southern Great Plains is presented.

INTRODUCTION

One of the inherent difficulties with using satellite remote sensing of surface radiometric temperature to infer fluxes on a regional scale arises because the surface sensible heat flux is proportional to the temperature difference between the surface and the air. Except in areas of low vegetation cover, the magnitude of this temperature difference is often smaller than the uncertainties associated with measurements of the surface radiometric temperature and in defining screen-level air temperature on a regional basis.

The most direct method for relating temperature to surface sensible heat flux (H) uses the temperature difference between the surface and the overlying air, divided by the aerodynamic resistance:

$$H = \rho C_P \frac{T_O - T_A}{R_A} \quad (1)$$

where T_O is the aerodynamic temperature of the surface, T_A is the near-surface air temperature, ρ is the density of air, C_P is the heat capacity of air, and R_A is the aerodynamic resistance computed from the surface roughness, displacement height and wind speed (Huband and Monteith, 1986). Surface brightness temperature measurements from satellites offer the possibility of mapping surface heat fluxes on a regional scale if the surface brightness temperature can be reliably related to the surface aerodynamic temperature.

In a recent review on monitoring evaporation with remote sensing, Kustas and Norman (1996) conclude that obtaining reliable estimates of the heat fluxes using radiometric temperature, $T_R(\phi)$, derived from remote brightness temperature measurements at viewing angle ϕ has been hampered by several factors. These include correcting the remotely-sensed brightness temperatures for atmospheric and emissivity effects, sensor calibration issues and the nonuniqueness of the aerodynamic-radiometric temperature relationship due to viewing angle and vegetation characteristics (e.g., Kustas, 1990; Vining and Blad, 1992; Mahrt et al., 1997).

The use of Eq. (1) has proven troublesome leading some (e.g. Hall et al., 1992) to conclude that uncertainties in atmospheric corrections, surface emissivity, instrument calibrations, radiometric-aerodynamic temperature relationship, and air temperature measurements will accumulate, significantly degrading the accuracy of T_O - T_A required in computing H. Uncertainties of 1-2 K in the absolute accuracy of the radiometric temperature alone might appear to support this skeptical position, and clearly near-surface air temperature measurements will never be available on the spatial scale of remote brightness temperature measurements. Attempts at using satellite remote sensing for estimating T_A at regional scales indicates an uncertainty of 3-4 K (Goward et al, 1994; Prince et al., 1998). The cumulative effect of errors in $T_R(\phi)$ and T_A estimation severely reduces the reliability of Eq. (1) with regional satellite data, and suggests operational application of Eq. (1) might be limited to areas where surface-air temperature differences $\gtrsim$ 5 K.

Sensitivity analyses conducted by Zhan et al. (1996), Kustas and Norman (1997) and Anderson et al. (1997) show that models using Eq. (1) at a single time step will produce errors > 50% in sensible heat flux predictions when absolute accuracy in the surface and air temperature estimates is 10%. This percentage uncertainty translates to a 2-4 K error in either $T_R(\phi)$ or T_A, which is the likely variation one would observe for regional applications (e.g., Gao et al., 1998). Variations in wind speed have much less impact on computed fluxes and appear significant only when the uncertainty is greater than 50% (Anderson et al., 1997; Kustas and Norman, 1997.)

An important conceptual step in improving the procedure for estimating surface fluxes was using the time rate of change of $T_R(\phi)$ from a geostationary satellite such as GOES (Geosynchronous Operational Environmental Satellite) with an atmospheric boundary layer model (Wetzel et al., 1984). By using time rate of change of $T_R(\phi)$, one reduces the need for absolute accuracy in satellite calibration, and atmospheric and emissivity corrections, all significant challenges . Diak and Stewart (1989) and Diak (1990) implemented this approach with a method for partitioning the available energy into latent, LE, and sensible heat, H, flux by using the rate of rise of $T_R(\phi)$ from GOES and atmospheric boundary layer (ABL) growth. Diak and Whipple (1993) refined the model by including a procedure to account for effects of horizontal and vertical temperature advection and vertical motions above the ABL.

In a related approach, the Two-Source Time-Integrated Model (TSTIM) of Anderson et al. (1997) provides a practical algorithm for using a combination of satellite data, synoptic weather data and ancillary information to map surface energy flux components. From the TSTIM framework, the Atmospheric-Land Exchange Inverse (ALEXI) model for the diagnosis of the land-surface energy balance on a continental scale has been recently developed and uses the time-difference of brightness temperatures (measured from GOES) , remotely-sensed vegetation cover, plus operationally available atmospheric and land-use information (Mecikalski et al., 1999). The ALEXI approach builds on the earlier work with the Two-Source Model, TSM, (Norman et al., 1995) by using remote brightness temperature observations at two times in the morning hours and ABL growth for energy balance closure. The methodology eliminates the need for a measurement of near-surface air temperature and is relatively insensitive to uncertainties in surface thermal emissivity and atmospheric corrections of GOES brightness temperature measurements. Anderson et al. (1997) and Mecikalski et al. (1999) have shown that surface fluxes retrieved from the ALEXI approach compare well with ground-based measurements. The ALEXI approach is a practical means for obtaining operational estimates of surface fluxes over continental scales with a 5 - 10 km pixel resolution. Mecikalski et al. (1999) detail the required inputs to ALEXI for large-scale applications and how they were produced in that continental-scale study.

Implementing ALEXI in a real-time mode relies on a mesoscale forecast model (the CIMSS Regional Assimilation System; CRAS, Diak et al. 1998) to provide the required atmospheric and surface meteorological fields. This reduces the time-consumptive analysis of surface meteorological reports at multiple times over large regions. Secondly, an operational real-time surface temperature product produced by the National Oceanic and Atmospheric Administration (NOAA) from geostationary satellite data (Hayden and Wade, 1996) has been incorporated into ALEXI. This hourly product is already atmospherically corrected. As a result of these procedures, the number of job steps and resulting execution time for the ALEXI flux diagnosis has been significantly reduced.

An alternative approach that is computationally very fast and can be applied without the need for a mesoscale forecasting model nor an atmospheric boundary layer growth algorithm has been proposed (Norman et al., 2000). The equations of Anderson et al. (1997) are used to form a dual difference of radiometric and air temperatures so that an estimate of sensible heat flux can be obtained from temporal measurements of surface radiometric temperature, air temperature and wind speed. The utility of the new method (called the Dual-Temperature-Difference, DTD, approach) has been tested with data from several different field sites covering a wide range of environmental conditions. Case studies where weather station observations of T_A and wind speed, u, 50-100 km from the site are used and compared to heat fluxes predicted by the TSM approach, which uses absolute $T_R(\phi)$ - T_A values (cf. Eq. 1), indicate that the DTD technique is not strongly affected by biases in the meteorological data. The results of Norman et al. (2000) support the potential of this simple and computationally very efficient technique for regional applications.

In this chapter we present an overview of the two schemes. The DTD method is tested and compared to the original TSM at the local or field scale with ground-based remote sensing observations from a semiarid region using both local and regionally available meteorological data. Then the output of ALEXI and DTD methods are compared using satellite data collected over the Southern Great Plains. Discrepancies between TSM and DTD output versus flux observations at the field scale are investigated using both local and regional shelter level air temperature and wind speed data. In addition, differences between ALEXI and DTD flux predictions with regional satellite data are analyzed and discussed.

MODEL OVERVIEW

In both ALEXI and DTD approaches, the TSM serves as the framework for determining soil and canopy temperatures from $T_R(\phi)$ and partitioning the available energy (net radiation less soil heat flux) between the soil and vegetation. Appendix A provides an overview of the algorithms in the TSM scheme. The main advantage of the TSM approach is that the model distinguishes the soil and canopy contributions to the radiative and aerodynamic temperature, explicitly accounting for view angle effects of the $T_R(\phi)$ observations. The TSM formulations are a significant advancement over single-source approaches that substitute $T_R(\phi)$ for T_O in Eq. (1) and hence require empirical excess resistance parameterizations (e.g., Stewart et al., 1994); this results in large uncertainty in heat flux predictions for noncalibrated surfaces (Kustas et al., 1996).

ALEXI *diagnoses* the air temperatures at the two observation times (at a height of about 50 m, through an energy continuity approach, matching fluxes in the surface layer and those calculated using the rise of the ABL height and the lapse rate in the ABL at this 50 m interface). The *time-integral* of the fluxes over the two observation times is the important output of ALEXI. The use of *temporal changes* of radiometric temperatures (rather than absolute temperatures) mitigates errors that can be caused by uncertainties in the estimation of atmospheric corrections to brightness temperatures and estimation of the surface emissivity (Wetzel et al. 1984; Diak 1990; Diak and Whipple, 1993; Anderson et al. 1997; Mecikalski et al. 1999). A detailed mathematical description of ALEXI can be found in Anderson et al. (1997) and Mecikalski et al. (1999).

Briefly, a simple slab model describing atmospheric boundary layer dynamics constrains the upper boundary condition in air temperature (T_A) used in the TSM component (cf. Eqs. A.12 and A.13). McNaughton and Spriggs (1986) give a simplified conservation equation relating the rise in height (z) and potential temperature (θ_M) of the mixed layer to the time-integrated sensible heating from the surface and the entrainment flux at the top:

$$\int_{t_1}^{t_2} H(t)dt = \rho C_P(z_2\theta_{M2} - z_1\theta_{M1}) - \rho C_P \int_{z_1}^{z_2} \theta_S(z)dz. \qquad (2)$$

where t_1 and t_2 (and subscript i = 1 and 2) represent two periods in the morning hours, from shortly after sunrise to just before local noon. θ_S (z) is the potential temperature profile above the mixed layer or the capping inversion and can be estimated from an early morning sounding or from output of the analysis component of a forecast system (Diak et al., 1998). The relationship between potential and air temperature is $\theta \approx T_A(100/P)^{0.286}$ (Brutsaert, 1982) where P is the atmospheric pressure (kPa). The growth of the mixed layer from z_1 to z_2 is determined by the sensible heat flux supplied during this interval and the strength of the capping inversion θ_S (z) above z_1 derived from the early morning sounding. This simple boundary layer model assumes a uniform potential temperature in the mixed-layer with gradients locally needed at the top and bottom of the slab. Moreover, the flux divergence of sensible heat is neglected, usually a reasonable assumption (e.g., Peters-Lidard and Davis, 2000).

Adopting a linear function for $H(t)$ during the morning period and substituting into Eq. (2) yields a time-integrated heat flux equation for the atmospheric boundary layer of the form,

$$H_i = \frac{2\rho C_P t_i}{t_2^2 - t_1^2}\left[\left(\frac{100}{P}\right)^{.286}\left(z_2 T_{A,2} - z_1 T_{A,1}\right) - \int_{z_1}^{z_2} \theta_S(z)dz\right] \qquad (3)$$

Under clear sky conditions, observations suggest assuming a linear form for $H(t)$ during the morning period is reasonable (Tennekes, 1973; Stull, 1988; Mecikalski et al., 1999). By combining Eqs. (A.12) - (A.13) from TSM for two time periods, Anderson et al. (1997) obtain for the surface component

$$T_{R,i}(\phi) - T_{A,i} = f(\phi)\frac{H_{C,i}\, R_{A,i}}{\rho C_p} + (1 - f(\phi))\frac{(H_i - H_{C,i})\,(R_{A,i} + R_{S,i})}{\rho C_p} \qquad (4)$$

where $f(\phi)$ is the fraction of the radiometer's field of view that is occupied by vegetative cover at zenith viewing angle ϕ, $H_{C,i}$ is the sensible heat flux from the vegetative canopy at time i, H_i is the total sensible heat flux above the canopy arising from both vegetation and soil, $R_{A,i}$ is the aerodynamic resistance to heat transport above the canopy, and $R_{S,i}$ is the resistance to heat transport of the air layer immediately above the soil (see Appendix A for resistance formulations). The quantity $f(\phi)$ can be estimated from canopy architecture and view angle; for a random or clumped canopy with a spherical leaf angle distribution and leaf area index, F, Eq. (A.2) is applied. The sensible heat from the canopy, $H_{C,i}$, is estimated from the net radiation divergence of the vegetative canopy ($\Delta R_{N,i}$) using

$$H_{C,i} = \Delta R_{N,i}\left[1 - \alpha_{PT}\, f_g \frac{s}{s+\gamma}\right] \qquad (5)$$

where α_{PT} is the Priestly-Taylor coefficient ~ 1.3 (Priestly and Taylor, 1972), f_g is the fraction of the vegetative canopy that is green, s is the slope of the saturation vapor pressure versus temperature curve, and γ is the psychrometer constant. The estimate from Eq. (5) can be modified if it produces unrealistic evaporative fluxes, such as condensation at the soil surface or $LE_{S,i} < 0$ when solving Eq. (A.6)-(A.7). This can occur when vegetation is under stress resulting in a high $T_{R,i}(\phi)$ and H_i; since $H_{S,i} = H_i - H_{C,i}$, Eq. (5) will compute a relatively small estimate of $H_{C,i}$ resulting in a high $H_{S,i}$, and by invoking energy conservation via Eq. (A.19), namely $LE_{S,i} = R_{N,S,i} - H_{S,i} - G_i$, could force $LE_{S,i} < 0$ (see Norman et al., 1995).

The net radiation divergence equation used by Norman et al. (1995) and modified by Anderson et al (1997) for solar zenith angle assumes an exponential decay of R_N through the canopy layer (Ross, 1981). This is a reasonable parameterization for moderate to high canopy covers (i.e., $F \gtrsim 2$) but not reliable for sparse canopies leading to significant underestimates of $\Delta R_{N,i}$ because of inadequate treatment of thermal emission from the soil. If the equation presented by Norman et al. (1995) is used for the net solar radiation and net thermal radiation calculated by the approach of Kustas and Norman (1999), $\Delta R_{N,i}$ estimates are more accurate (see Eqs A.9a and A.9b).

The six unknowns in Eqs. (3)-(5) are H_1, H_2, $H_{C,1}$, $H_{C,2}$, $T_{A,1}$ and $T_{A,2}$. With $T_{R,i}(\phi)$ observations at the t_1 and t_2, an early morning sounding providing $\theta_S(z)$, and with z_1 and z_2 determined by $T_{A,1}$ and $T_{A,2}$, the surface and boundary layer components of the model yield six equations for the six unknowns. A solution is obtained through iteration until the sensible heat flux estimates from both Eq. (3) and (4) converge. Anderson et al. (1997) provide further details concerning the solution sequence used in the ALEXI model.

Optimal times t_1 and t_2 for the observations have been determined in a sensitivity study by Anderson et al. (1997) and are nominally 1.5 and 5.5 hours after local sunrise. Sky conditions for ALEXI evaluations need to be mostly clear at the two observation times, so that the surface brightness temperature measurements are not contaminated by clouds. The use of measurements taken in the mid-morning hours, however, increases the chances of clear skies, as well as mitigating the effects of advection in the ABL (Wetzel et al., 1984), which can be detrimental to ALEXI flux evaluations.

The DTD algorithm is derived by Norman et al. (2000) starting from Eq. (4). Equation (4) assumes the vegetative canopy and soil are in parallel (i.e., heat or scalar fluxes, H and LE, from the soil and vegetation do not interact), usually a good assumption (Norman et al,. 1995). However, there are instances where the parallel assumption will yield acceptable estimates of the total heat fluxes H and LE, but the partitioning of available energy between H and LE for the soil and canopy is not reliable since heat cannot be advected from the soil to the vegetation (Kustas and Norman, 1999). In the DTD method, Eq. (4) is applied at two times. The first time is chosen when all the fluxes are small and temperatures are similar, typically about one hour after sunrise (similar to the optimal time t_1 derived by Anderson et al., 1997). To distinguish this derivation from ALEXI we define this initial time $i = 0$. The second time can actually be any hour during the day. Applying Eq. (4) at two times (0,i) and subtracting the equations yields

$$H_i = \rho C_p \left[\frac{\left(T_{R,i}(\phi) - T_{R,0}(\phi)\right) - \left(T_{A,i} - T_{A,0}\right)}{(1 - f(\phi))(R_{A,i} + R_{S,i})} \right] + H_{C,i} \left[1 - \frac{f(\phi)}{1 - f(\phi)} \frac{R_{A,i}}{R_{A,i} + R_{S,i}} \right] + \left(H_0 - H_{C,0}\right) \left[\frac{R_{A,0} + R_{S,0}}{R_{A,i} + R_{S,i}} \right] + H_{C,0} \left[\frac{f(\phi)}{1 - f(\phi)} \frac{R_{A,0}}{R_{A,i} + R_{S,i}} \right] \quad (6)$$

The last two terms on the right side of Eq. (6) involving H_0 and $H_{C,0}$ are usually negligible, because of the choice of the first time (i=0). Shortly after sunrise the sensible heat flux from the soil, $H_{S,0}$ (= H_0 - $H_{C,0}$) is clearly negligible; this is fortunate because there is no direct means for estimating H_0. The last term involving $H_{C,0}$ usually is small but can be evaluated so is kept in the equation. The simplified equation for H_i is

$$H_i \cong \rho C_p \left[\frac{\left(T_{R,i}(\phi) - T_{R,0}(\phi)\right) - \left(T_{A,i} - T_{A,0}\right)}{(1 - f(\phi))(R_{A,i} + R_{S,i})} \right] + H_{C,i} \left[1 - \frac{f(\phi)}{1 - f(\phi)} \frac{R_{A,i}}{R_{A,i} + R_{S,i}} \right] + H_{C,0} \left[\frac{f(\phi)}{1 - f(\phi)} \frac{R_{A,0}}{R_{A,i} + R_{S,i}} \right] \quad (7)$$

Equation (7) represents a relatively simple result with the advantage that any offset between measurements of $T_{R,i}(\phi)$ and $T_{A,i}$ cancels out of the temperature term. Given measurements or estimates of net radiation ($R_{N,i}$) and soil heat conduction flux (G_i) at time i (see Appendix A), the latent heat flux, LE_i, can be calculated from the energy budget:

$$LE_i = R_{N,i} - G_i - H_i \quad (8)$$

THE DATA

The utility of the DTD approach was first evaluated at the field scale with ground-based remote sensing data from the Monsoon '90 Experiment conducted within the USDA-ARS Walnut Gulch Experimental Watershed (Kustas and Goodrich, 1994). The main study period was during the monsoon season in the desert southwest when the vegetation is most active (highest green biomass) and soil moisture is highly variable due to localized and frequent precipitation events. A Meteorological-Flux (METFLUX) station was located within the Lucky Hills (LH) subwatershed, a shrub-dominated site with heterogeneous vegetative cover ~25 % and significant variability in vegetation height and architecture. During the main field campaign, continuous measurements of canopy and soil temperatures were made at this site using a similar measurement design deployed by Nichols (1992). Reliable field-scale composite temperatures with the continuous infrared observations were determined using a set of transect data collected periodically (see Flerchinger et al., 1998). Shrub height was nominally 0.5 m and $F \approx 0.5$. Continuous surface energy balance components were computed using the variance technique for estimating sensible heat flux H and measurements of R_N and G for computing LE as a residual (Kustas et al., 1994). This technique was calibrated with periodic measurements of H and LE made using a one-dimensional eddy covariance technique (Stannard et al., 1994). Comparisons between eddy covariance and variance techniques by Kustas et al. (1994) showed differences were 20%, on average, which is typically found between different measurement techniques (e.g., Dugas et al., 1991; Nie et al., 1992) Continuous measurements of wind speed and air temperatures were made at a nominal height of 4 m. Twenty-minute data were recorded; hourly-averaged data were used in the analysis. The impact of using weather station data 100 km away from the site with DTD and TSM schemes was evaluated by using a Tucson AZMET station (see WEB site http://128.196.42.70/azmet/.html and station name TUSC). This station is ~ 120 km away and nearly 800 m lower in elevation.

To illustrate the disparity in shelter level meteorological conditions, u and T_A from the METFLUX station at the LH site are compared to the observations from the TUSC station for the period starting at i = 0 or ~0630 MST (Mountain Standard Time) with R_N ~ 25 W m^{-2} and ending at ~1730 MST with R_N ~ 100 W m^{-2} (Figure 1). There is a strong bias in T_A ; however, the bias varies both with time of day (bias is less in the morning than the afternoon) and between days (cf. Figures 1a and 1b), which is due in part to differences in cloud cover and other local climate conditions affected by the land cover/land use type surrounding the weather stations.

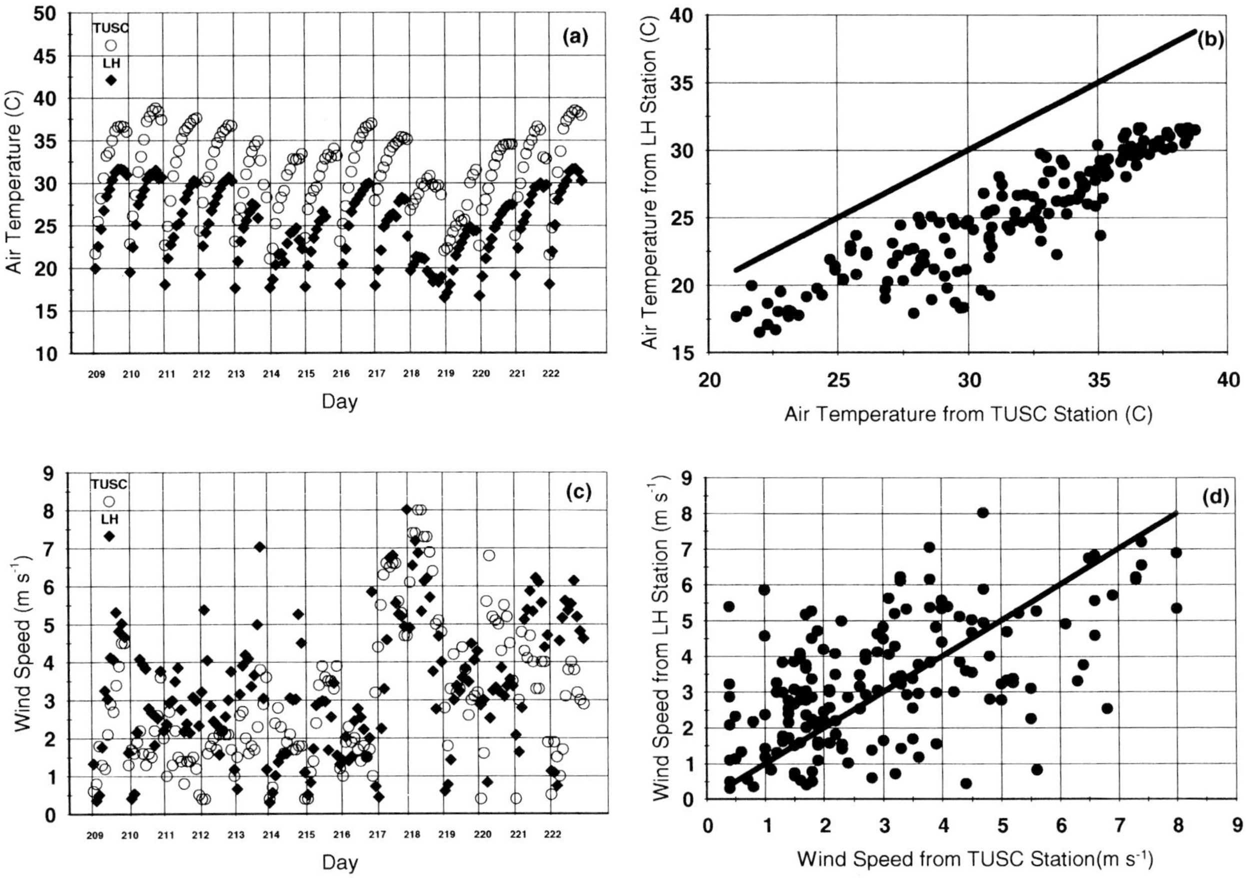

Figure 1. Comparison of air temperature (a) and (b) and wind speed (c) and (d) observations from the Lucky Hills (LH) station near the METFLUX site within the Walnut Gulch Experimental Watershed and a weather station in Tucson (TUSC) approximately 120 km away and 800 m lower in elevation.

This changing air temperature bias with time between stations will instill a bias in heat flux predictions with the DTD method (see Eq. 7). For example, average values of $T_{A,i} - T_{A,0}$ using $i = 1330$ MST were ≈ 9 K for the LH station while ≈ 11 K for the TUSC station. There is considerably more scatter in u with the best agreement between stations occurring under high wind conditions (cf. Figure 1c and 1d). This comparison illustrates the type of differences one is likely to observe when applying standard weather station observations to pixel data from a regional satellite image.

Lastly, to illustrate the application of the DTD approach to a large region and to compare with the more complex ALEXI algorithm, GOES and AVHRR (Advanced Very High Resolution Radiometer) satellite observations are combined with surface synoptic data on June 12, 1995; the same date as was chosen by Mecikalski et al. (1999). The AVHRR data were used to estimate a vegetative-cover fraction. The GOES thermal images were obtained 1.5 and 5.5 hours after local sunrise. For the DTD approach, the satellite data are combined with near-surface air temperature and wind speed from the surface weather station nearest the pixel of interest (see below).

MODEL RESULTS

The first set of results compare the original TSM approach using absolute differences with the DTD technique, using both meteorological data at the LH site and the observations from the TUSC station. The comparison between predicted and observed H and LE using the LH site meteorological data indicates both the TSM and DTD approaches yield similar results (Figures 2a and 2b). The Root-Mean-Square-Difference (RMSD) statistic (Willmott, 1982) is approximately 30 and 55 W m^{-2} for H and LE, respectively, using TSM while RMSD values are slightly higher, namely ~ 40 and 60 W m^{-2} for H and LE, respectively, with the DTD

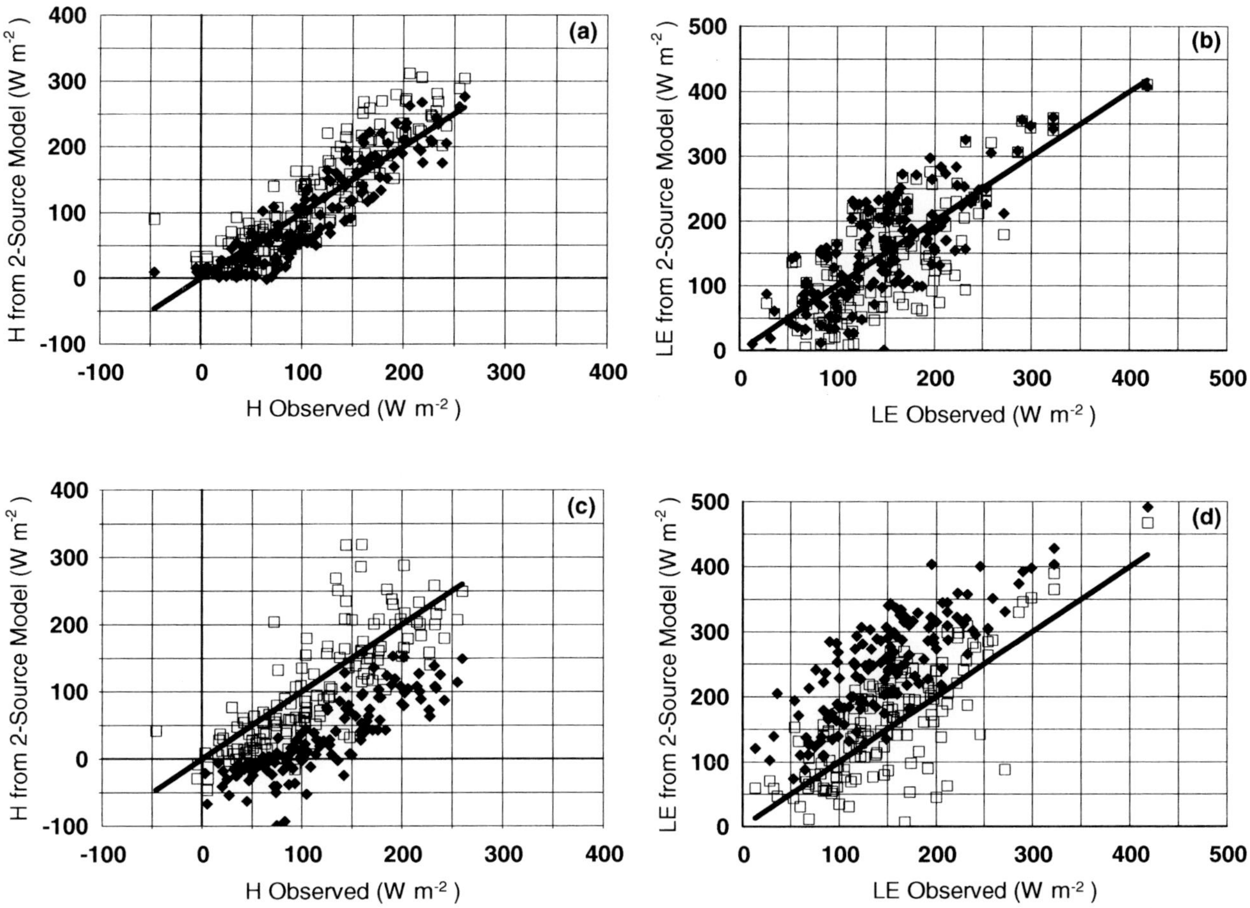

Figure 2. Predicted Sensible (H) and latent heat flux (LE) from the TSM (◆) and DTD (□) approaches versus observed values from the LH METFLUX site using the local meteorological data (LH station), (a) and (b), and meteorological data from the Tuscon (TUSC) weather station, (c) and (d). The lines represent perfect agreement with the observations.

approach. The average bias with both the DTD and TSM predictions are comparable (≈ 10 W m^{-2}) with the DTD method overestimating and TSM underestimating H. When the TUSC meteorological data are used, the discrepancies between predicted and observed H and LE increase dramatically for TSM, where the RMSD for H and LE are 104 and 108 W m^{-2}, respectively. This rise in RMSD is largely due to a significant bias illustrated in Figures 2c and 2d between the H and LE predictions and the flux observations (≈ 95 W m^{-2}). Using the DTD approach, RMSD increases slightly for H to approximately 50 W m^{-2}, but essentially remains the same for LE with RMSD ≈ 57 W m^{-2}. There does appear to be increasing scatter between predicted and measured heat fluxes when using the TUSC data; the average bias remains at approximately 10 W m^{-2}, but with a tendency (except for a few outliers) for the DTD method to underestimate H. This change in bias is mainly caused by the tendency for higher values of $T_{A,i}$ - $T_{A,0}$ using the TUSC station (Fig. 1b). However, the overall consistency in the heat flux predictions using data from either station clearly indicates a significant advantage in using the DTD method with synoptic weather data.

To illustrate a regional application of the DTD approach, GOES and AVHRR satellite observations were combined with surface synoptic data for the case study day of 12 June 1995, the same day chosen by Mecikalski et al. (1999). The domain investigated was divided into 10-km by 10-km grid cells, with 223 cells east-to-west and 201 in the meridional direction, a total of 44,823 cells.

AVHRR visible and near-infrared measurements were used to calculate the Normalized Difference Vegetation Index (NDVI) for the region, and these NDVI data were subsequently utilized to estimate fraction vegetation cover via the method of Carlson et al. (1995); these fractional cover values are shown for the regional domain in Plate 1a. Hourly GOES thermal brightness temperature measurements for the region were cloud screened and subsequently linearly time-interpolated to 1.5 and 5.5 hours after local sunrise. These top-of-atmosphere brightness temperatures were then atmospherically corrected using the method of Price (1983) and emissivity corrected (assuming emissivities of 0.94 and 0.98 for soil and vegetation respectively) using a procedure described by Norman et al. (1990). The resulting estimates of time differences of surface radiometric temperatures ($\Delta T_R = T_{R,i} - T_{R,0}$, Eq. 7) are shown for the domain in Plate 1b. A strong inverse relationship between the fraction cover values (Fig. 3a) and the ΔT_R values is evidenced.

Near-surface fields of air temperature differences ($\Delta T_A = T_{A,i} - T_{A,0}$) were generated at hourly intervals by objectively analyzing hourly synoptic data and performing a similar time interpolation to 1.5 and 5.5 hours past local sunrise. Wind speeds at the two observation times were produced by *averaging* the three closest hours to the observation time; this averaging procedure has proved advantageous to flux observations using the full ALEXI model (Anderson et al., 1997), as hourly wind observations (generally only five minute averages) tend to be noisy.

For regional applications using satellite data, net solar radiation plus downwelling and upwelling longwave radiation components are summed to estimate the required total net radiation for the DTD method. The required net solar radiation values over the domain were constructed by combining an estimate of clear-air incident solar radiation (Diak et al., 1996) with an albedo of the surface determined from land cover type and fraction vegetation cover in a formulation described in Campbell and Norman (1998). Downwelling clear-sky longwave radiation was estimated using an empirical formulation developed by Prata (1998), the most accurate of several similar parameterizations that were tested against pyrgeometer data. Upwelling longwave radiation was calculated dynamically within the DTD iteration process using the derived temperatures of the soil surface and vegetation canopy, the component emissivities detailed above, and fraction vegetation cover to produce a *hemispherical* emissivity for the soil/vegetation system based on an empirical fit to simulations from a detailed plant-environmental model Cupid (Norman and Campbell, 1983) described in detail by Norman et al. (1990) (see also Kustas et al., 2000). The procedures for determining the net shortwave and net longwave balance is similar to the description in Appendix A (see Kustas and Norman, 1999).

The application of the DTD method (Eq. 7) implies that ΔT_A at the locations where radiometric temperatures are taken can be represented either by measurements taken at synoptic weather stations, or for this regional application, objective analyses of such air temperature data. Because synoptic weather stations tend to be located near airports and populated areas, air temperature differences at these locations may be systematically larger than at surrounding rural areas, which are generally more vegetated. Positive biases in air temperature differences from such measurements will result in an *underestimate* of sensible heat flux from the DTD method (see Eq. 7). This was observed in the field scale application of DTD with the weather station data from Tucson (TUSC station) having ≈ 2 K larger ΔT_A value than the local LH station around midday.

To investigate this possibility, output from the ALEXI model (which *diagnoses* low-level air temperatures, their time differences and the relationship of air temperatures to surface fluxes) was used to explore the relationship between ΔT_A and ΔT_R and determine if systematic biases in ΔT_A from synoptic weather station reports could be identified and compensated for. A graph of ΔT_A versus ΔT_R from ALEXI model output on 12 June for the regional domain is shown in Figure 3 (a total of about 20,000 points).

Clearly ΔT_A and ΔT_R are correlated with a mean slope of about 0.5 ($\Delta T_A = 0.5\ \Delta T_R$) and few values of ΔT_A exceeding ΔT_R (Fig. 3). As a first-order correction to the air temperature differences in the DTD method, we therefore check for $\Delta T_A > 0.75\ \Delta T_R$, and if true set $\Delta T_A = 0.75\ \Delta T_R$. In extreme cases where $\Delta T_R > 20$ K, ΔT_A is limited to 0.5 ΔT_R. The result of these two empirical adjustments is a lowering of extreme ΔT_A values, a resulting 10% increase in the domain-average sensible heat flux (about 20 W m^{-2}), as well as an improved agreement between the DTD and ALEXI methods.

Latent heating results for 5.5 hours after local sunrise for the domain are shown in Plate 1c and 1d from the DTD and ALEXI schemes, respectively. Areas that are white in this figure were those identified as cloudy by screening procedures, and thus were not evaluated in either method. The DTD method displays very similar spatial features compared to the ALEXI output, although even with the empirical adjustments detailed above, domain-average sensible heating results are still about 20 W m^{-2} lower than with ALEXI. Patterns in latent heat fluxes strongly resemble features in the fraction cover and ΔT_R fields of Plate 1a and 1b. Regions of sparse cover showing large temperature excursions in Texas and New Mexico have been assigned low values of latent heating. Soils are largely bare across the corn belt this early in the growing season and also show very low evaporation rates. The DTD method is very encouraging in its ability to duplicate the results from ALEXI, a much more complicated and data-intensive parameterization. Computer processing time for the domain shown in Plate 1 for the

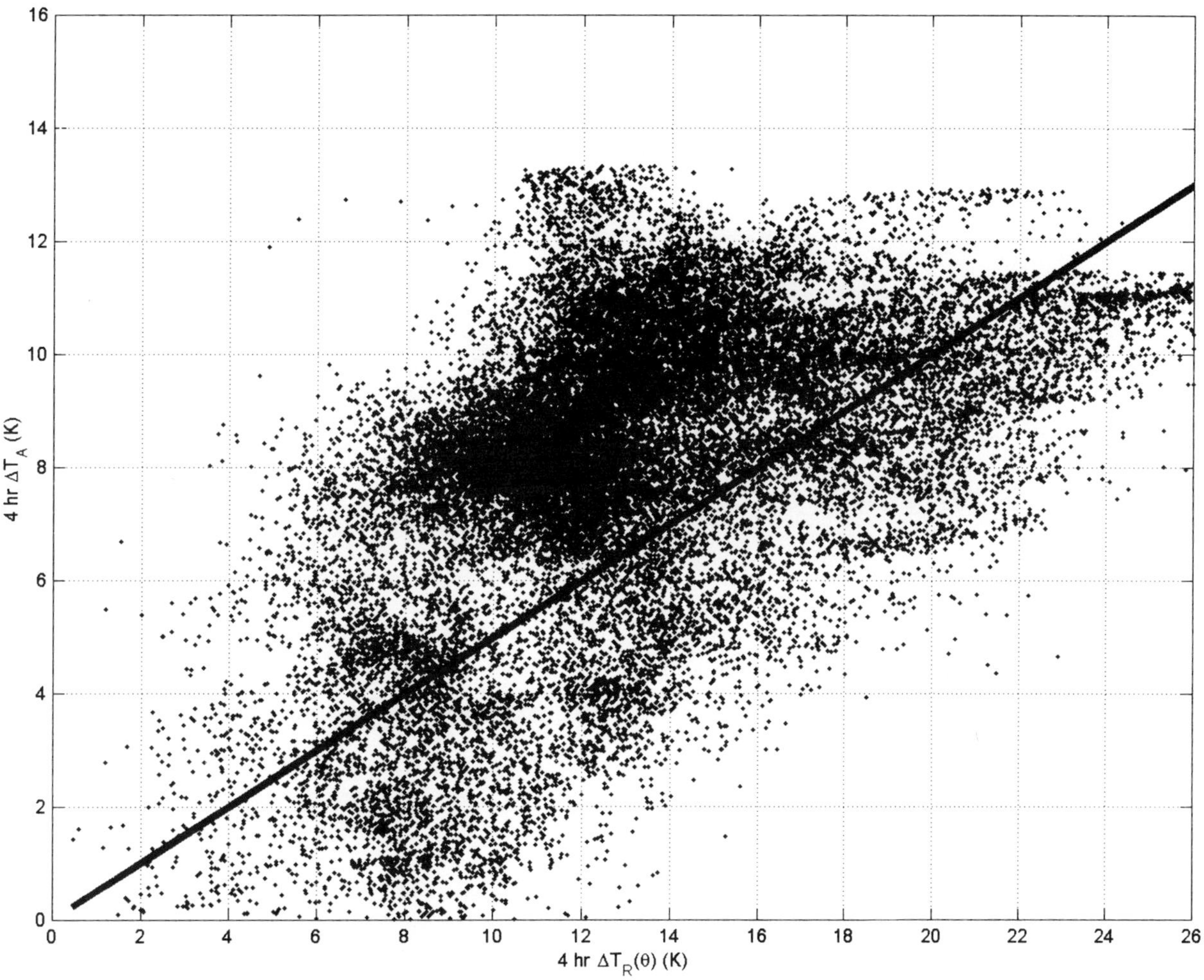

Figure 3. A comparison of time difference in air temperature, ΔT_A (= $T_{A,i}$ - $T_{A,0}$) from the ALEXI model output versus time difference in surface temperature ΔT_R(=$T_{R,i}$ - $T_{R,0}$) derived from GOES for the regional domain (a total of about 31,500 points). The line represents ΔT_A= 0.5ΔT_R.

ALEXI model was about 35 minutes, while the DTD scheme required only about one minute of processing time on the same UNIX workstation.

CONCLUSIONS

The utility of the DTD method at the field scale is supported by the ground-based data collected from a semiarid region. Weather station data from ~100 km away and 800 m lower in elevation from this site caused large discrepancies with the TSM approach, which relies on absolute temperature differences, while the performance of the DTD method did not significantly degrade. Regional application of DTD using satellite data and synoptic weather stations gave satisfactory heat flux predictions compared to ALEXI, a much more complicated and data-intensive set of algorithms simulating atmospheric boundary layer growth for achieving energy balance closure (Mecikalski et al., 1999).

Correction to weather station air temperature data may have to be applied in practice, however, because the DTD method cannot account for differences in time-rate-of-change in air temperature caused by local land cover/land use conditions. This error is due to the fact that synoptic weather stations tend to be located near airports and populated areas resulting in systematically larger temporal changes in air temperature than what would be observed in surrounding rural areas containing generally more vegetation. Positive biases in air temperature differences from such measurements will result in an *underestimate* of sensible heat flux from the DTD method, which is supported by both the field scale and regional scale

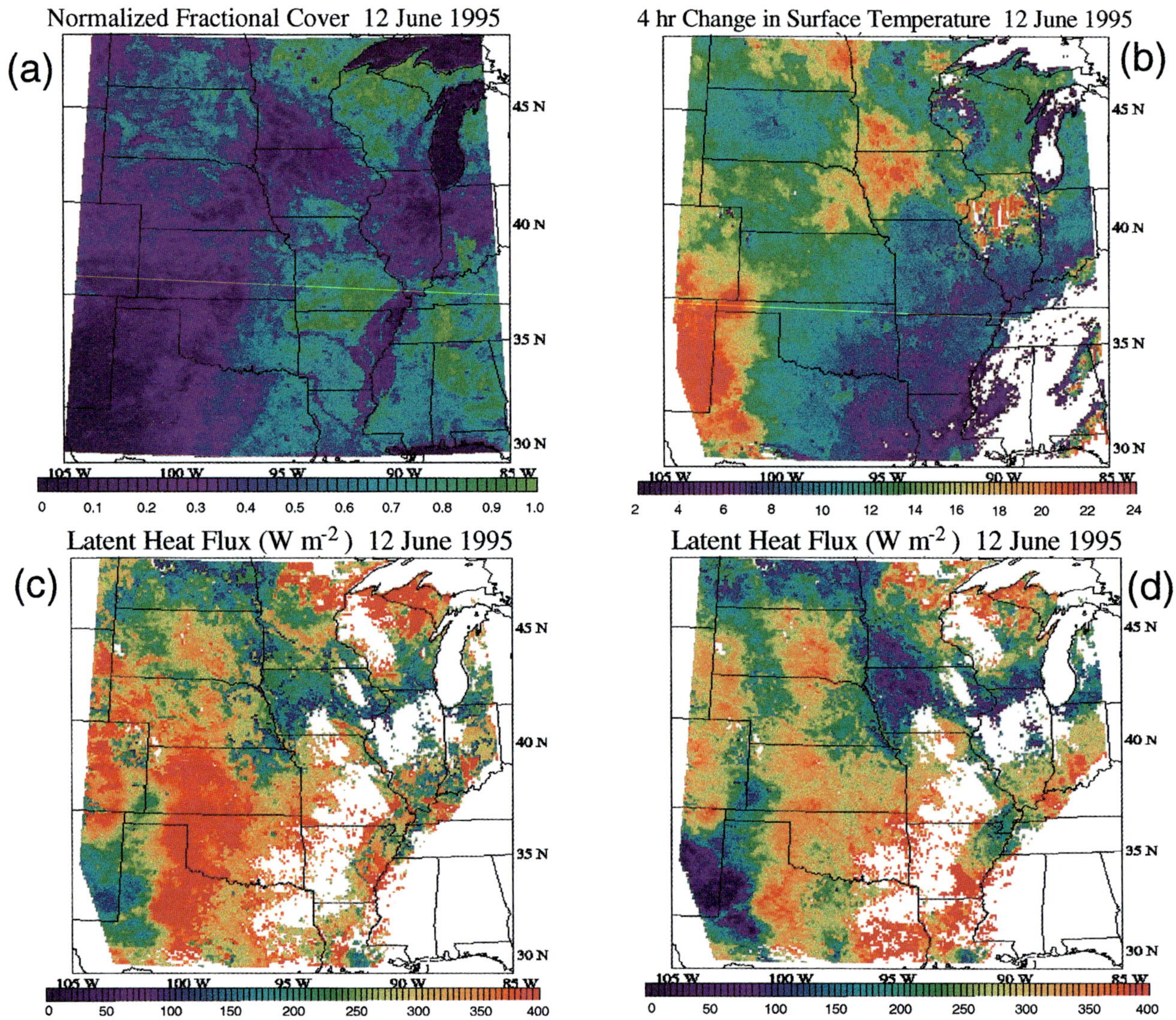

Plate 1. Estimated fractional vegetation cover map (a) via the method of Carlson et al. (1995) using NDVI product from NOAA-AVHRR for the region. A map (b) of time differences in surface radiometric temperatures $\Delta T_R (= T_{Ri} - T_{R0})$ for the domain using GOES brightness temperatures estimated at approximately 1.5 and 5.5 hrs after local sunrise. Maps of latent heat flux for the domain estimated by (c) DTD method and (d) ALEXI.

studies. Fortunately, a relatively simple correction reduces the bias in heat fluxes to around 20 W m^{-2}, which is well within model and measurement uncertainties.

Although this correction is not entirely general, it was derived from observations having a wide variety of surface conditions, including fractional vegetation cover ranging from 0 to ~1 and climatic regimes ranging from the hot and dry desert southwest to the cool and wet northern coniferous forests (Plate 1). Therefore, this scheme for adjusting time-rate-of-change in air temperature has wider applicability than what is normally expected of empirically-based corrections. The utility of this simple correction needs further evaluation, particularly as affected by different seasonal conditions.

APPENDIX A

Overview of TSM model

With the use of a single emissivity to represent the combined soil and vegetation, the ensemble directional radiometric temperature, $T_R(\phi)$, is related to the fraction of the radiometer view occupied by soil versus vegetation expressed as

$$T_R(\phi) = \left[f(\phi) T_C^n + (1 - f(\phi) T_S^n \right]^{1/n} \tag{A.1}$$

where T_C and T_S are the thermodynamic temperatures of the vegetation canopy and soil surface, respectively, and are assumed to represent spatially weighted averages of the sunlit and shaded portions of the canopy and soil, respectively, and $n \sim 4$ (Becker and Li, 1990). The fraction of the field of view of the infrared radiometer occupied by canopy, $f(\phi)$, depends upon the view zenith angle, ϕ, canopy type and fraction of vegetative cover, f_C. $f(\phi)$ can be estimated from canopy architecture and view angle; but for many vegetated surfaces, which are either random or clumped, a spherical leaf angle distribution can be assumed and with an estimate of leaf area index, F,

$$f(\phi) = 1 - \exp\left(\frac{-0.5\, \Omega(\phi)\, F}{COS\, \phi} \right) \tag{A.2}$$

where $\Omega(\phi)$ is unity for random canopies and less than one for clumped canopies (Kucharik et al.,1999; Kustas and Norman, 1999).

The use of $T_R(\phi)$ in a convective heat flux equation frequently involves the controversial assumption that $T_R(\phi)$ is equivalent to the so-called "aerodynamic temperature", T_O, of the surface. T_O is the temperature that satisfies the bulk transport expression having the form (cf. Eq. 1)

$$H = \rho C_P \frac{T_O - T_A}{R_A} \tag{A.3}$$

where H is the sensible heat flux (W m^{-2}), ρC_P is the volumetric heat capacity of air (J m^{-3} K^{-1}), T_A is the air temperature at some reference height above the surface (K) and R_A is the resistance to heat transport (s m^{-1}), which has the following form in the surface layer (Brutsaert, 1982):

$$R_A = \frac{\left[ln\left(\frac{z_U - d_O}{z_{OM}} \right) - \Psi_M \right] \left[ln\left(\frac{z_T - d_O}{z_{OH}} \right) - \Psi_H \right]}{k^2 u} \tag{A.4}$$

In this equation d_O is the displacement height (m) , u is the wind speed (m s^{-1}) measured at height z_U (m), k is von Karman's constant ($\approx$ 0.4), z_T is the height (m) of the T_A measurement, Ψ_M and Ψ_H are the Monin-Obukhov stability functions for momentum and heat, respectively, and are functions of the variable $(z - d_O)/L$ (see Brutsaert, 1982) where $L = -u_*^3 / [k(g/T_A)(H_V / \rho C_P)]$ is the Monin-Obukhov length (m), u_* is the friction velocity (m s^{-1}), g is the acceleration of gravity (9.8 m s^{-2}), $H_V = (H + 0.61 T_A C_P E)$ is the virtual sensible heat flux (W m^{-2}), and E is the rate of surface evaporation (kg m^{-2} s^{-1}). The roughness parameter z_{OM} is the local roughness length (m) for momentum transport and z_{OH} is the local roughness length (m) for heat transport. T_O cannot be measured, so it is often replaced with an observation of $T_R(\phi)$ in Eq. (A.3).

The net energy balance of the soil-canopy system is given by (neglecting photosynthesis)

$$R_N = H + LE + G \tag{A.5}$$

The system of equations for computing fluxes from the soil and canopy components, denoted by subscripts s and c , respectively, are listed below and will be used by all versions of the model. The energy budgets for the soil and vegetation are given by

$$R_{N,S} = H_S + LE_S + G \tag{A.6}$$

$$R_{N,C} = H_C + LE_C \tag{A.7}$$

with $R_N = R_{N,S} + R_{N,C}$. Similar to Eq. (A.1) for estimating the contribution of soil and canopy temperatures to the observed radiometric temperature, Eq. (A.8a) and (A.8b) are used for partitioning net radiation, R_N, between the soil and vegetation in order to properly weight the contributions of sensible, H, and latent heat flux, LE, from the soil and vegetation, and estimate the soil heat flux, G:

$$R_{N,S} = R_N \exp\left(\frac{-\kappa F}{\sqrt{2COS(\beta_S)}} \right) \tag{A.8a}$$

$$R_{N,C} = R_N\left[1-\exp\left(\frac{-\kappa F}{\sqrt{2COS(\beta_S)}}\right)\right] \quad (A.8b)$$

Equations (A.8a) and (A.8b) are modifications from the original Norman et al. (1995) formulations (where $\sqrt{2COS(\beta_S)} = 1$ and $\kappa = 0.45$) proposed by Anderson et al. (1997) based on simulations with a detailed soil-plant-atmosphere model, Cupid (Norman and Campbell, 1983) where the net radiation divergence is found to be a function of the solar zenith angle β_S. The value of $\kappa > 0.45$ for sparse canopies because the quantity $\kappa/\sqrt{2COS(\beta_S)}$ will have a value of $\approx$ 0.6 to 0.8 (Campbell and Norman, 1998; Kustas and Norman, 1999).

A more physically-based algorithm for estimating the divergence of R_N has been constructed by Kustas and Norman (1999) requiring incident solar radiation observations and formulations for the transmission of direct and diffuse shortwave radiation and for the transmission of long-wave radiation through the canopy. Since the reflection and absorption of radiation in the visible and near-infrared wavelengths is markedly different for vegetation and soils, the visible and near-infrared reflectances of the soil and vegetation are evaluated separately before combining to give an overall shortwave albedo. The equations for estimating the transmission and reflection of direct and diffuse shortwave radiation are described in Chapter 15 of Campbell and Norman (1998); hence the net shortwave radiation balance for the soil ($S_{N,S}$) and canopy ($S_{N,C}$) are computed separately from the net long-wave radiation balance for the soil ($L_{N,S}$) and canopy ($L_{N,C}$). The long-wave balance for the soil-vegetation-atmosphere system is derived by calculating diffuse radiation transmission through the canopy (Ross , 1975). A simpler formulation of the net long-wave radiation balance than described in Ross (1975) was derived where a single exponential equation is used for estimating the transmission for both the soil and canopy,

$$L_{N,S} = \exp(-\kappa_L F)L_{SKY} + [1-\exp(-\kappa_L F)]L_C - L_S \quad (A.9a)$$

$$L_{N,C} = [1-\exp(-\kappa_L F)][L_{SKY} - L_S - 2L_C] \quad (A.9b)$$

where the extinction coefficient for diffuse radiation depends on leaf area index, and if $F \lesssim 0.5$, $\kappa_L \approx 0.95$ (Campbell and Norman, 1997); L_C, L_S and L_{sky} are the long-wave emissions from the canopy, soil and sky, respectively. L_C, L_S and L_{sky} are computed from the Stefan-Boltzmann equation using canopy temperature, soil temperature and shelter level air temperature and vapor pressure (Brutsaert, 1982). Thus Eqs. (A.8a) and (A.8b) are replaced by visible and near-infrared radiation penetration equations from Chapter 15 of Campbell and Norman (1998) combined with Eqs. (A.9a) and (A.9b) above (i.e., $R_{N,S} = S_{N,S} + L_{N,S}$ and $R_{N,C} = S_{N,C} + L_{N,C}$).

For computing G, the original formulation from Norman et al. (1995) was simply

$$G = c_G R_{N,S} \quad (A.10)$$

where the value of $c_G \approx 0.35$ (Choudhury et al., 1987). However c_G is constant only for several hours around solar noon (Kustas and Daughtry, 1990). Friedl (1996) included the effects of a temporally varying c_G by multiplying Eq. (A.10) by $COS(\beta_S)$. Another approach developed by Kustas et al. (1998b) is based on time differences with the local solar noon quantified by the following nondimensional time parameter, t_N,

$$t_N = \frac{|t_i - t_{SN}|}{t_{SN}} \quad (A.11)$$

where t_i is the time nominally $\pm$ 5 hours of the local time of solar noon, t_{SN}, and the $|\;|$ represents the absolute value of the difference. Using experimental data to compute $G/R_{N,S}$ or c_G as a function of time t_i, an empirical function was fit between G/R_{NS} and t_N. The results indicated that a constant $G/R_{N,S}$ could be used for $t_N < 0.3$ (i.e., several hours around solar noon) and linear least squares regression equation between $G/R_{N,S}$ and t_N was needed for $t_N > 0.3$. Neither Eq. (A.11) nor the approach suggested by Friedl (1996), however, considers the fact that G and R_N are not in phase, and hence the temporal change in the ratio $G/R_{N,S}$ will not be the same between the morning and afternoon. In addition, these approaches also do not account for possible variations in c_G due to soil moisture conditions (e.g., Friedl, 1996).

With $H = H_S + H_C$ and with the soil and vegetation taken in "parallel" (i.e., the resistance network provides for no interaction between the soil and vegetation), the heat fluxes from the soil and vegetation are computed by

$$H_S = \rho C_P \frac{T_S - T_A}{R_A + R_S} \quad (A.12)$$

$$H_C = \rho C_P \frac{T_C - T_A}{R_A} \quad (A.13)$$

With H_C and H_S taken in "series" (i.e., the resistance network allows for interaction between the soil and vegetation) yields:

$$H_S = \rho C_P \frac{T_S - T_{AC}}{R_S} \quad (A.14)$$

$$H_C = \rho C_P \frac{T_C - T_{AC}}{R_X} \qquad (A.15)$$

where T_{AC} is related to T_O in Eq. (A.3), namely,

$$H = \frac{T_{AC} - T_A}{R_{AO}} \qquad (A.16)$$

See Figures 1 and 11 in Norman et al. (1995) illustrating the "parallel" and "series" resistance network, respectively. R_S is the resistance to heat flow in the boundary layer immediately above the soil surface and is estimated from an empirical expression developed by Sauer et al. (1995) from extensive studies of this soil-surface resistance in a wind tunnel and beneath a corn canopy. R_X is the total boundary layer resistance of the complete canopy of leaves (see Appendix A in Norman et al., 1995) estimated with the wind speed in the canopy air space computed from the equations of Goudriaan (1977). R_A is estimated via Eq. (A.4) with local d_O and z_{OM} estimated as a fraction of canopy height, h_C, (i.e., $d_O \approx 0.65\ h_C$; $z_{OM} \approx 0.13\ h_C$; see Brutsaert, 1982) and z_{OH} is estimated as a fraction of z_{OM} as postulated by Garratt and Hicks (1973), namely $z_{OH} \approx z_{OM}/7$ or $kB^{-1} \approx 2$. A more sophisticated approach based on canopy structure has been proposed by McNaughton and Van den Hurk (1995). R_{AO} is computed from Eq. (A.4) with $z_{OH} = z_{OM}$. T_{AC} is the momentum aerodynamic temperature and only approximates the temperature in the canopy air space (see Appendix A in Norman et al., 1995).

Although soil-surface resistances depend on many factors (Sauer and Norman, 1995), a reasonable, simplified equation has been developed where

$$R_S = \frac{1}{a + bu_S} \qquad (A.17)$$

In Eq. (A.17), $a \approx 0.004$ m s^{-1} is the free convective velocity "constant", $b \approx 0.012$, and u_S is the wind speed at a height above the soil surface where the effect of the soil surface roughness is minimal; typically 0.05 to 0.2 m. The coefficients in Eq. (A.17) depend on turbulent length scale in the canopy, soil-surface roughness and turbulence intensity in the canopy and are discussed by Sauer et al. (1995). The numerical value for the coefficient a was taken from data presented in Sauer (1993) as the mean intercept of plots of soil surface transfer coefficients versus wind speed in the canopy. For the smooth aluminum plates used by Sauer (1993) the value of the parameter b was measured to be 0.007. The value of 0.012 for b used in Eq. (A.17) was estimated from a combination of wind tunnel data for surfaces of various roughnesses and the field data on smooth plates to represent the more typical roughness that soil surfaces have.

Eq. (A.17) has been recently modified by Kustas and Norman (1999) based on results from Kondo and Ishida's (1997) and Sauer's (1993) experimental results

$$R_S = \frac{1}{c(T_S - T_C)^{1/3} + bu_S} \qquad (A.18)$$

where $c \approx 0.0025$. This allows for the free convective velocity to vary with surface-canopy temperature differences as predicted by the model.

Finally, for $LE = LE_S + LE_C$ the fluxes are estimated by the following expressions:

$$LE_S = R_{N,S} - G - H_S \qquad (A.19)$$

$$LE_C = \alpha_{PT} f_G \frac{s}{s+\gamma} R_{N,C} \qquad (A.20)$$

The Priestley-Taylor parameter, α_{PT}, is set equal to 1.26 (Priestley and Taylor, 1972) for the green part of the canopy, s is the slope of the saturation vapor pressure-temperature curve at T_C (Pa K^{-1}) and γ is the psychrometric constant (≈ 66 Pa K^{-1}). The fraction of F that is "green" or actively transpiring, f_G, may be obtained from knowledge of the phenology of the vegetation. If no information is available for estimating f_G, then it is assumed to equal unity.

Equation (A.20) only provides an initial calculation of LE_C, and it can be overridden if the temperature difference between the soil-canopy system and the atmosphere is large causing erroneous flux estimates, such as negative LE_S or condensation during the daytime period. This can occur when the Priestley-Taylor approximation in Eq. (A.20) leads to an underestimate of T_C and hence an overestimate of T_S in order to satisfy Eq. (1), which then causes an overestimate of H_S leading to $LE_S < 0$ computed by Eq. (A.19). Therefore an iteration procedure will compute LE_C values below estimates given by Eq. (A.20) until values of T_C and T_S satisfying Eq. (A.1) leads to LE_C and $LE_S > 0$. Further details concerning model convergence issues for the energy budgets of the soil and vegetation in later iterations and the justification for the Priestley-Taylor assumption used in Eq. (A.20) are given in Norman et al. (1995) and Kustas and Norman (1999).

Acknowledgments. Funding for the University of Wisconsin was provided by NASA grant NAG5-9436 and USDA Cooperative Agreement 58-1270-7-008 with the assistance of the University of Wisconsin Agricultural Experimental Station. Funding from NASA Interdisciplinary Research Program in Earth Sciences (NASA reference number IDP-88-086) and USDA ARS Beltsville Area funds provided the necessary support for the Monsoon '90 Experiment.

REFERENCES

Anderson, M.C., Norman, J.M., Diak, G.R., Kustas, W.P., Mecikalski, J.R., A two-source time-integrated model forestimating surface fluxes from thermal infrared satellite observations, *Remote Sens. Environ.*, 60, 195-216, 1997.

Becker, F., and Z.-L. Li, Temperature-independent spectral indices in thermal infrared bands. *Remote Sens. Environ.*, 32, 17-33, 1997.

Brutsaert, W., *Evaporation Into The Atmosphere.* D. Reidel Publ. Co., Dordrecht, Holland, 299 pp., 1982.

Campbell, G.S., and Norman, J.M., *An Introduction to Environmental Biophysics.* Springer-Verlag New York, Inc. 286 pp., 1998.

Carlson, T. N., Capehart, W .J., Gillies, R.R., A new look at the simplified method for remote sensing of daily evapotranspiration, *Remote Sensing Environ.*, 54, 161-167, 1995.

Diak, G. R., 1990. Evaluation of heat flux, moisture flux and aerodynamic roughness at the land surface from knowledge of the PBL height and satellite-derived skin temperatures, *Agric. For. Meteorol.*, 52, 181-198, 1990.

Diak, G. R., Anderson, M.C., Bland, W.L., Mecikalski, J.M., Norman, J. M., and Aune, R.M., Agricultural management decision aids driven by real-time satellite data.,*Bull.Am.Meteorol.Soc.*, 79, 1345-1355, 1998.

Diak, G. R., Bland, W.L., and Mecikalski, J.R., A note on first estimates of surface insolation from GOES-8 visible satellite data. *Ag. For. Meteorol.*, 82, 219-226, 1996.

Diak, G. R., and Stewart, T.R., Assessment of surface turbulent fluxes using geostationary satellite surface skin temperatures and a mixed layer planetary boundary layer scheme. *J. Geophys. Res.*, 94, 6357-6373, 1989.

Diak, G. R., Whipple, M. A., Improvements to models and methods for evaluating the land-surface energy balance and "effective" roughness using radiosonde reports and satellite-measured "skin" temperatures, *Agric. For. Meteorol.*, 63, 189-218, 1993.

Gao, W., Coulter, R. L., Lesht, B. M., Qiu, J., and Wesely, M. L., Estimating clear-sky regional surface fluxes in the southern great plains atmospheric radiation measurement site with ground measurements and satellite observations. *J. Appl. Meteorol.*, 37, 5-22, 1998.

Dugas, W.A., Fritschen, L.J., Gay, L.W., Held, A.A., Matthias, A.D., Reichosky, D.C., Steduto, P., Steiner, J.L., Bowen ratio, eddy correlation, and portable chamber measurements of sensible and latent heat flux over irrigated spring wheat, *Agric. For. Meteorol.*, 56,1-20, 1991.

Flerchinger, G.N., Kustas, W.P., Weltz, M.A., Simulating surface energy fluxes and radiometric surface temperatures for two arid vegetation communities using the SHAW model, *J. Appl. Meteorol.*, 37, 449-460, 1998.

Friedl, M.A., Relationships among remotely sensed data, surface energy balance, and area-averaged fluxes over partially vegetated land surfaces. *J. Appl. Meteorol.* 35: 2091-2103, 1996.

Garratt, J.R., Hicks, B.B., Momentum, heat and water vapor transfer to and from natural and artificial surfaces, *Quart. J. Roy. Meteorol. Soc.*, 99, 680-687, 1973.

Goudriaan, J., *Crop Micrometeorology: A Simulation Study.* Center for Agric. Publ. and Doc., Wageningen, the Netherlands., 249 pp., 1977.

Goward, S.N., Waring, R,H., Dye, D.G., Yang, J.,Ecological remote sensing at OTTER: Satellite macroscale observations, *Ecol. Appl.*, 4, 322-343, 1994.

Hall, F. G., Huemmrich, K. F., Geotz, S. J., Sellers, P. J., Nickerson, J. E., Satellite remote sensing of surface energy balance: success, failures and unresolved issues in FIFE, *J. Geophys. Res.* 97(D17), 19,061-19,090, 1992.

Hayden, C. M., and Wade, G.S., Derived product imagery from GOES-8. *J. Appl. Meteorol.*, 2, 153-162, 1996.

Huband, N. D. S., Monteith, J. L., Radiative surface temperature and energy balance of a wheat canopy. Part I: Comparison of radiative and aerodynamic canopy temperature, *Bound.-Layer Meteorol.* 36, 1-17, 1986.

Kondo, J., and Ishida, S., Sensible heat flux from the earth's surface under natural convective conditions. *J. Atmos.Sci.*, 54, 498-509, 1997.

Kucharik, C.J., Norman, J.M., Gower, S.T., Characterizing the radiation regime of nonrandom forest canopies: Theory, measurements, modeling and a simplified approach. *Tree Physiol.*, 19, 695-706, 1999.

Kustas, W.P., Estimates of evapotranspiration with a one-and two-layer model of heat transfer over partial canopy cover. *J. Appl. Meteorol.* 29, 704-715, 1990.

Kustas, W.P., Blanford, J.H., Stannard, D.I., Daughtry, C.S.T., Nichols, W.D., Weltz, M.A., Local energy flux estimates for unstable conditions using variance data in semiarid rangelands, *Water Resour. Res.*, 30, 1351-1361, 1994.

Kustas, W. P., Daughtry, C.S.T., Estimation of the soil heat flux/net radiation ratio from spectral data, *Agric.For. Meteorol.*, 49, 205-223, 1990.

Kustas, W. P., Goodrich, D.C., Preface, MONSOON'90 Multidisciplinary Experiment, *Water Resour. Res.*, 30(5), 1211-1225, 1994.

Kustas, W. P., Humes, K.S., Norman J.M., Moran, M.S., Single- and dual-source modeling of surface energy fluxes with radiometric surface temperature, *J. Appl. Meteorol.*, 35, 110-121, 1996.

Kustas, W.P., Norman, J.M., Use of remote sensing for evapotranspiration monitoring over land surfaces, *Hydrological Science J.* 41, 495-516, 1996.

Kustas, W.P., Norman, J.M., A two-source approach for estimating turbulent fluxes using multiple angle thermal infrared observations, *Water Resour. Res.*, 33, 1495-1508, 1997.

Kustas, W.P., Norman, J.M., Evaluation of soil and vegetation heat flux predictions using a simple two-source model with radiometric temperatures for a partial canopy cover, *Agric. For. Meteorol.*, 94, 13-29, 1999.

Kustas, W.P., Norman, J.M., Schmugge, T.J., Anderson, M.C., Mapping surface energy fluxes with radiometric temperature. In: *Thermal Remote Sensing in Land Surface Processes*. Eds. D.A. Quattrochi and J.C. Luvall. Ann Arbor Press, Chelsea, MI., 2000.

Kustas, W.P., Zhan, X., Schmugge, T.J., Combining optical and microwave remote sensing for mapping energy fluxes in a semiarid watershed, *Remote Sens. Environ.*,64, 116-131, 1998.

Mahrt, L., Sun, J., MacPherson, J.I., Jensen,N.O.,Desjardins,

R.L., Formulation of surface heat flux: Application to BOREAS, *J. Geophys. Res.*, 102(D24), 29,641-29,649, 1997.

McNaughton, K.G., Van den Hurk, B.J.J.M., A 'Lagrangian' revision of the resistors in the two-layer model for calculating the energy budget of a plant canopy, *Boundary-Layer Meteorol.*, 74, 262-288, 1995.

McNaughton, K. J., and T. W. Spriggs, 1986: A mixed-layer model for regional evaporation. *Boundary Layer Meteorol.*, 74, 243-262.

Mecikalski, J.R., Diak, G.R., Anderson M.C., Norman, J.M., Estimating fluxes on continental scales using remotely-sensed data in an atmospheric-land exchange model, *J. Appl. Meteorol.*, 38, 1352-1369, 1999.

Nichols, W.D., Energy budgets and resistances to energy transport in sparsely vegetated rangeland, *Agric. For. Meteorol.*, 50, 221-247, 1992.

Nie, D., Kanemasu, E.T., Fritschen, L.J., Weaver, H.L., Smith, E.A., Verma, S.B., Field, R.T., Kustas, W.P., Stewart, J.B., An intercomparison of surface energy flux measurement systems used during FIFE 1987, *J. Geophys. Res.*, 97(D17), 18715-18724, 1992

Norman, J.M., Campbell, G.S., Application of a plant- environment model to problems in irrigation. In (D.I. Hillel, Editor) *Advances in Irrigation*, p. 155-188, Vol. II. Academic Press, New York, 1983.

Norman, J. M., Chen, J. -L., and Goel, N. S., Thermal emissivity and infrared temperature dependence of plant canopy architecture and view angle. *Proc. Tenth Annu. Int. Geoscience Remote Sensing Symp.*, Vol. 1., 755-756, IEEE, Piscataway, N.J., 1990.

Norman, J. M., Kustas, W. P., Humes, K. S., A two-source approach for estimating soil and vegetation energy fluxes from observations of directional radiometric surface temperature, *Agric. For. Meteorol.*, 77, 263-293, 1995.

Norman, J.M., Kustas, W.P., Prueger, J.H., and Diak, G. R., Surface flux estimation using radiometric temperature: A double-difference method to minimize measurement errors. *Water Resour. Res.* In Press, 2000.

Peters-Lidard, C.D., and Davis, L.H., Regional flux estimation in a convective boundary layer using a conservation approach. *J. Hydromet.*, 1, 170-182, 2000.

Prata, A. J., A new long-wave formula for estimating downward clear-sky radiation at the surface. *Q. J. Roy. Meteorol Soc.*, 122, 1127-1151, 1998.

Price, J. C., Estimating surface temperatures from satellite thermal infrared data. A simple formulation for the atmospheric effect. *Remote. Sens. Environ.*, 13, 353-361, 1983.

Priestly, C.H.B., Taylor, R.J., On the assessment of surface heat flux and evaporation using large-scale parameters, *Monthly Weath. Rev.*, 100, 81-92, 1972.

Prince , S.D., Goetz, S.J., Dubayah, R.O., Czajkowski, K.P., and Thawley, M., Inference of surface and air temperature, atmospheric precipitable water and vapor pressure deficit using Advanced Very high-Resolution Radiometer satellite observations: comparison with field observations, *J. Hydrol.*, 212-213, 230-249, 1998.

Ross, J., The radiation regime and architecture of plants. In *Tasks for Vegetation Sciences 3* (H. Lieth, Series Ed.), Dr. W. Junk, The Hague, Netherlands, 1981.

Ross, J., Radiative transfer in plant communities. In *Vegetation and the Atmosphere* (J.L. Monteith, Editor). Academic Press, London. pp 13-55, 1975.

Stannard, D.I., Blanford, J.H., Kustas, W.P., Nichols W.D., Amer, S.A.,Schmugge T.J., Weltz, M.A., Interpretation of surface- flux measurements in heterogeneous terrain during the MONSOON 90 experiment, *Water Resour. Res.*, 30(5), 1227-1239, 1994.

Stewart, J.B., Kustas, W.P., Humes, K.S., Nichols, W.D., Moran, M.S., De Bruin, A.A.R., Sensible heat flux-radiometric surface temperature relationship for eight semiarid areas, *J. Appl. Meteorol.*, 33, 1110-1117, 1994.

Sauer, T.J., 1993. Sensible and latent heat exchange at the soil surface beneath a maize canopy. *Ph. D. Thesis*, University of Wisconsin, Madison, WI. 292 pp.

Sauer, T.J., and Norman, J.M., 1995. Simulated canopy microclimate using estimated below-canopy soil surface transfer coefficients. *Agric. For. Meteorol.* 75: 135-160.

Sauer, T.J., Norman, J.M., Tanner, C.B., Wilson, T.B., Measurement of heat and vapor transfer at the soil surface beneath a maize canopy using source plates, *Agric. For. Meteorol.*, 75, 161-189, 1995.

Stull, R.B., *An Introduction to Boundary layer Meteorology*, Kluwer Academic Pub., Norwell, MA, U.S.A., 666 pp., 1988.

Tennekes, H., A model for the dynamics of the inversion above a convective boundary layer. *J. Atmos.Sci.*,30, 558-567, 1973.

Vining, R.C., Blad, B.L. Estimation of sensible heat flux from remotely sensed canopy temperatures, *J. Geophys. Res.*, 97, D17, 18951-18954, 1992.

Wetzel, P. J., Atlas, D., Woodward, R,. Determining soil moisture from geosynchronous satellite infrared data: A feasibility study, *J. Clim. Appl. Meteorol.*, 23, 375-391, 1984.

Willmott, C.J., Some comments on the evaluation of model performance, *Bull. Am. Meteorol. Soc.*, 11, 1309-1313, 1982.

Zhan, X., Kustas, W. P., Humes, K. S., An intercomparison study on models of sensible heat flux over partial canopy surfaces with remotely sensed surface temperature, *Remote Sens. Environ.*, 58, 242-256, 1996.

George R. Diak, CIMSS Space Science and Engineering Center, University of Wisconsin-Madison, 1225 West Dayton St., Madison WI 53706

William P. Kustas, USDA-ARS Hydrology and Remote Sensing Laboratory, Bldg. 007, Rm. 104, BARC-West, Beltsville MD 20705

John M. Norman, Department of Soil Science, University of Wisconsin-Madison, 1525 Observatory Drive, Madison WI 53706

Inferring Scalar Sources and Sinks Within Canopies Using Forward and Inverse Methods

Gabriel G. Katul, Chun-Ta Lai, Mario Siqueira and Karina Schäfer

School of the Environment, Duke University

John D. Albertson

Department of Environmental Sciences, University of Virginia

Karen H. Wesson, David Ellsworth[1], and Ram Oren

School of the Environment, Duke University

It is becoming clear that the description of water exchange over vegetated surfaces can benefit from a simultaneous consideration of heat and CO_2 exchanges, as all these exchange processes are intertwined at the most fundamental level. Over the last two decades, several approaches have been developed to infer scalar sources and sinks within canopies without resorting to gradient-diffusion theory. This study investigates recent developments in multi-layer methods to compute distributions and strengths of scalar sources and sinks within the canopy volume. Two types of model formulations are considered: 1) forward methods which require vertical foliage distribution along with canopy radiative, physiological, biochemical, and drag properties and 2) inverse methods which require measured mean scalar concentration distribution within the canopy. These approaches are able to reproduce measured turbulent fluxes above the canopy without relying on empirical relationship between turbulent scalar fluxes and mean concentration gradients. However, both approaches share the need for accurate description of the second moments of the velocity statistics inside the canopy.

1. INTRODUCTION

Over the past decade, much research effort has been dedicated to understanding and predicting land-surface fluxes from vegetated surfaces. The recognition that vegetation is not simply a source or sink of matter but can modify its microclimate via complex turbulent exchange processes transferred the study of land-atmosphere interaction from a disciplinary research to a multidisciplinary research theme combining expertise from surface hydrology, micrometeorology, and physiological ecology. Research efforts were stimulated by the need to quantify how vegetated surfaces affect their local microclimate (i.e. intercepting radiation, atten-

[1]Brookhaven National Laboratory, Upton, New York.

Land Surface Hydrology, Meteorology, and Climate: Observations and Modeling
Water Science and Application Volume 3, pages 31-45

uating wind, and acting as a source or sink for mass and energy), which in turn influences the exchanges of carbon dioxide, water vapor, and many biogenic compounds [*Baldocchi*, 1989]. In particular, interest in Net Ecosystem Exchange (NEE) of carbon, commonly quantified by the net vertical turbulent flux of CO_2 near the canopy-atmosphere interface, has received significant attention in the United States and Europe as evidenced by the emergence of two continental flux-monitoring networks: *EuroFlux* and *AmeriFlux* [*Kaiser*, 1997]. Such attention is not surprising given the modulating role of the biosphere on global atmospheric CO_2 concentration [*Wofsy et al.*, 1993] and on the general spatial and temporal characteristics of the water cycle. Early attempts to quantify mass and energy exchanges between the biosphere and atmosphere made use of "single layer" models, now called the "big-leaf" approach. This approach considers the entire canopy as a single transferring surface with all the transfer dynamics "compressed" into empirical bulk transfer coefficients (e.g. aerodynamic resistance, bulk stomatal resistance, etc..). One widely used example of such formulation is the Penman-Monteith equation [*Penman*, 1948; *Monteith*, 1965; *Brutsaert*, 1982; *McNaughton and Spriggs*, 1989; *Stuart*, 1989]. Here we focus on multi-layer canopy theories [*Finnigan and Raupach*, 1987; *Raupach*, 1988], which explicitly consider the non-uniform vertical structure of the canopy and can resolve the subsequent feed backs of such non-uniformity on the microclimate. The objective of this chapter is to present and evaluate recent advances in methods to estimate scalar source-sink distribution within the canopy volume using multi-layer theories. Multi-layer theories are commonly classified as either of the "forward" or "inverse" inferences of scalar source and sink distributions within the canopy volume depending on the types of measurements performed [*Raupach*, 1989a,b]. In forward approaches, the source strength and location is specified such that knowledge of the turbulent transport mechanics permits the estimation of down-wind mean scalar concentration and flux distribution. The inverse problem utilizes measured mean scalar concentration distribution downwind from the source in concert with knowledge of the turbulent transport mechanics to infer the source-sink strength distribution. We present recent advances in both forward and inverse approaches to canopy transport processes and evaluate their performance with flux and concentration measurements in a uniform even-aged pine forest.

2. MODEL FORMULATION AND PERFORMANCE

2.1. Forward Methods

In forward methods, the source strength and distribution are calculated from the physiological and biophysical properties of the foliage, which must be assumed or, preferably, measured. Hence, the definition of forward methods in this context is modified somewhat so that physiological, biophysical, and radiative properties of the canopy are specified rather than the source strength distribution.

2.1.1. Model Development. In multi-layer approaches, the canopy height h is divided into N layers, each characterized by a thickness Δz and a rigid leaf area density $a(z)$ with $LAI=\int_{z=0}^{h} a(z)\, dz$, where LAI is the leaf area index. For each layer, the one-dimensional scalar continuity equation for a planar homogeneous turbulent flow, after proper time and horizontal averaging, is given by

$$\frac{\partial \bar{C}}{\partial t} + \frac{\partial F_c}{\partial z} = S_c \qquad (1)$$

where $\bar{C}$ is the mean scalar concentration of a scalar entity (i.e. H_2O, CO_2, and air temperature T_a), F_c is the mean vertical flux of the scalar entity (e.g. F_{CO2}, F_{H2O}, and F_T are the CO_2, H_2O, and sensible heat turbulent fluxes at height z, respectively), and S_c is the mean vegetation source strength (i.e. sink implies $S_c \leq 0$) at time t and height z above the ground surface. All mean quantities are subject to both time and horizontal averaging as described by *Raupach and Shaw*, [1982]. The scalar continuity equation in (1) is used to compute one of the unknown series ($\bar{C}$, F_c, S_c), provided additional information is available. One approach to establish additional equations is to consider the interdependency between $\bar{C}$ and S_c via Lagrangian dispersion theory. *Raupach*, [1988; 1989a,b] introduced the dispersion matrix (D_{ij}) that relates mean concentration difference between a given level inside the canopy and a reference level above the canopy ($z_R \geq h$) to the scalar source strength by

$$\begin{aligned} \bar{C}(z_i) - \bar{C}(z_R) &= \sum_{j=1}^{N} S_{cj} D_{ij} \Delta z_j \\ D_{ij} &= \frac{c_i - c_r}{s \Delta z_j} \end{aligned} \qquad (2)$$

where i and j are the indices for concentration and source location, respectively, Δz is the discrete layer thickness within the canopy, c_i is the concentration at a layer i above a reference concentration resulting from a unit source s placed at layer $j(= 1, 2, \cdots N)$ inside the canopy, and, as before, N is the number of layers within the canopy volume. The $c_i - c_r$ is calculated from the velocity statistics inside the canopy by following the trajectory of an ensemble of fluid parcels released uniformly from s placed at the jth layer. The random walk algorithm of *Thomson* [1987] is commonly used to trace these particle trajectories. Note that the difference between the lower and upper case $C's$ is due to the difference between the hypothetical unit source effects between a source and concentration level (lower case) and the actual concentration (upper case). In order to use the Lagrangian dispersion algorithm of *Thomson* [1987], the vertical velocity standard deviation σ_w and the Lagrangian integral time scale (T_L) must be described. The estimation of can be achieved via higher-order turbulent closure models (e.g. *Katul and Albertson*, 1998; *Katul and Chang*, 1999; or simplifying analytical solutions to them as in *Massman and Weil*, 1999) or Large Eddy Simulations (LES) techniques (e.g.*Albertson and Parlange*, 1999). The T_L is assumed constant inside the vegetation as discussed in *Raupach*, [1988]. Even after combining (1) and (2), the number of equations are still not sufficient to allow mathematical closure. A third equation can be derived by assuming that the transfer of mass and heat from the leaf surface is governed by molecular (or Fickian) diffusion so that is given by

$$S_c = -\rho_a a(z) \frac{\bar{C}(z) - \bar{C}_i(z)}{r_b + r_s} \tag{3}$$

where ρ_a is the mean air density, $\bar{C}_i$ is the mean intercellular scalar concentration at height z, r_b is the boundary layer resistance, and r_s is the stomatal resistance. Equations (1), (2), and (3) permit a complete mathematical description of $\bar{C}$, S_c and F_c if $\bar{C}_i$, r_s , and r_b are known or parameterized. A variant on the above approach was developed by *Meyers and Paw U*, [1987] who proposed a third-order closure model that retains (1) and (3) but replaces (2) with a set of prognostic equations derived from Eulerian closure principles. The closure employed in the Meyers-Paw U model is the so-called "quasi-Gaussian" approximation in which fourth moments are related to second moments via Gaussian approximations. As discussed in *Raupach*, [1988], such approximations neglect the diffusive terms in fourth moments and appear to result in unrealistic oscillations in model results (as anticipated earlier by *Deardorff*, 1978) that are not supported by measurements inside vegetation (*Shaw and Seigner*, 1987). Also, closure at third-order does not appear to provide increased accuracy over second-order closure models (e.g. *Katul and Albertson*, 1998) suggesting that increased complexity does not always translate into increased accuracy in such models. Much of the present research in forward methods (e.g. the *CANVEG* model of *Baldocchi and Meyers*, 1997 is aimed at predicting $\bar{C}$, S_c and F_c through coupling radiative, physiological, and biochemical properties of the foliage with (1) to (3) to arrive at a complete description of $\bar{C}$, S_c, F_c, $\bar{C}_i$, r_s , and r_b. Next we briefly describe the canonical form of the parameterizations for the unknowns r_s, r_b, and $\bar{C}_i$ without resorting to detailed explicit formulations. The leaf boundary layer resistance, r_b, is computed from a flat plate analogy. Monteith and Unsworth (1990), Schuepp (1993), and Campbell and Norman (1998) derived explicit equations for heat and mass boundary layer transfer subject to forced and free convection. After an appropriate characteristic leaf length scale (l_d) is specified for a particular vegetation type (e.g. needle diameter for pine forests), the leaf boundary layer resistance is given by

$$r_b = \frac{l_d}{d_m \times Sh} \tag{4}$$

where d_m is the molecular diffusivity of the scalar entity, and Sh is the Sherwood number, which requires the mean velocity inside the canopy. The mean velocity variation inside the canopy can be determined via higher-order closure models (*Wilson and Shaw*, 1977; *Wilson*, 1988; *Meyers and Paw U*, 1986; *Katul and Albertson*, 1998) or LES techniques (*Shaw and Schumann*, 1992; *Shen and Leclerc*, 1997; *Dyer et al.*, 1997). Other simplifications such as the *Massman and Weil*, [1999] approach to the mean momentum equation are also viable alternative since modeling mean velocity inside the canopy requires less sophisticated closure formulations when contrasted with σ_w. The parameterization of $\bar{C}_i$ and r_s is much more complex and requires knowledge of the enzymatic biochemistry of carbon assimilation in leaves (*Williams et al.*, 1996). The stomatal conductance r_s^{-1} is a function of photosynthesis A, relative humidity (rh) and mean CO_2 concentration at the leaf surface ($\bar{C}_s$), and is given by (*Collatz et al.*, 1991) as

$$g_s(z) = m \frac{A(z) \times rh(z)}{\bar{C}_s(z)} + b \tag{5}$$

where m and b are parameters that vary with vegetation type. The biochemical model of *Farquhar et al.* [1980] can be used to couple A with internal CO_2 concentration (i.e. the net mathematical outcome is a relationship between F_c and $\bar{C}_i$ for CO_2). For water vapor and heat, the leaf energy balance provides the necessary equation to couple leaf surface temperature (T_{sl}), surface water vapor concentration, and the sensible and latent heat fluxes. Also, because the *Farquhar et al.*, [1980] model requires photosynthetically active radiation (PAR) to estimate the biochemical kinetic constants at all levels within the canopy, a radiation attenuation model (e.g. *Campbell and Norman*, 1998) is needed. Gas exchange experiments, commonly performed using porometry techniques, can be employed to determine independently the parameters of the Collatz model for g_s (e.g. m and b) and the Farquhar photosynthesis model. Note that $\bar{C}_i$, rh and T_{sl} are all required to calculate g_s; therefore, all three scalars (H_2O, CO_2 and T_a) must be simultaneously considered at each canopy layer. Also, the modeled T_{sl} at different levels within the canopy are needed to estimate plant respiration which adjusts the *Farquhar et al.*, [1980] modeled A.

2.1.2. Experiment. As an illustration of a *CANVEG* type model, we applied the above forward method to model sources and sinks of CO_2, H_2O and T_a in a pine forest. The model inputs are mean air temperature, mean air relative humidity, mean CO_2 concentration, mean wind speed, and mean PAR every 30 minutes above the canopy. The model parameters include leaf area density distribution, the Collatz model parameters (m and b) and the Farquhar photosynthesis model parameters, the radiative properties of the canopy (for use in the Campbell and Norman radiation attenuation model) and the drag coefficient (C_d) of the foliage to use in a second-order closure model to estimate $\sigma_w(z)$. The data set used in this model validation was collected in April 2 – 11, 1999 at the Blackwood Division of the Duke Forest near Durham, North Carolina (36°2′N, 79°8′W, elevation = 163 m). This site is a uniformly planted loblolly pine (*Pinus taeda* L.) forest that extends 300 to 600 m in the east-west direction and 1000 m in the north-south direction. The mean canopy height was 14.5 m (±0.5 m) at the time of the experiment. The CO_2, H_2O, and sensible heat fluxes above the canopy (z/h=1.11) were measured by a conventional eddy-covariance system comprised of a Licor-6262 CO_2/H_2O infrared gas analyzer (LI-COR, Lincoln, NE, USA) and a Campbell Scientific triaxial sonic anemometer (CSAT3, Campbell Scientific, Logan, UT, USA). The infrared gas analyzer was housed in an enclosure 4.5 m from the inlet cup, which is positioned under the eddy covariance system. The sampling flow rate for the gas analyzer is 9 L min^{-1}, sufficient to maintain turbulent flow in the tubing. A krypton hygrometer (KH2O, Campbell Scientific) was positioned with the CSAT3 to measure water vapor concentration fluctuations and assess the magnitude of the tube attenuation and time lag between vertical velocity and infrared gas analyzer measured scalar concentration fluctuations as discussed in *Katul et al.*, [1997a,b]. The analog signals from these instruments were sampled at 10 Hz using a Campbell Scientific 21*X* data logger with all the digitized signals transferred to a portable computer via an optically isolated *RS232* interface for future processing. All 10 Hz raw measurements processing was performed using the procedures described in *Katul et al.*, [1997a,b] with scalar covariance computed after maximizing the cross correlation between vertical velocity fluctuations and scalar concentration fluctuations for each 30 minute run. Other measurement and processing corrections are described in *Katul et al.*, [1997a,b]. The three flux measurements above the canopy (i.e F_T, F_{CO2}, and F_{H2O}) provide indirect validation of the model performance for scalars strongly controlled by stomata (e.g. H_2O and CO_2) and scalars strongly controlled by aerodynamic transfer processes (e.g. temperature). In addition to the eddy covariance flux measurements above the canopy, a HMP32C T/RH Vaisala probe (Campbell Scientific) was positioned at the same height as the CSAT3 to measure the mean air temperature and mean relative humidity. A *Q7* Fritchen type net radiometer and a LI-190SA quantum sensor (LI-COR) were installed to measure net radiation (R_n) and PAR, respectively. The R_n measurements were also used to assess the energy closure in the measurements. The mean energy budget closure for this experiment, estimated by regressing $F_T + F_{H2O}$ on $R_n - G_s - H_{st}$, was ±15%. Here, G_s is the soil heat flux estimated at 10% of R_n from Campbell and Norman's radiation attenuation calculation, and H_{st} is the heat storage flux in the canopy volume, determined from modeled air temperature profiles within the canopy. All the meteorological variables were sampled every second and averaged every 30 minutes using a *21X* Campbell Scientific datalogger. A multi-port system was installed to measure the CO_2 concentration inside the canopy at 10 levels (0.1 m, 0.75 m, 1.5 m, 3.5 m, 5.5 m, 7.5 m, 9.5 m, 11.5 m, 13.5 m and 15.5 m). Each level was sampled for 1 minute (45 s sampling and 15 s purging) at the beginning, the middle, and the end of a

30 minute sampling duration. The mean CO_2 concentration profile measurements inside the canopy are also used to assess the model performance at multiple depths within the canopy. The leaf-level physiological measurements needed to parameterize the Collatz stomatal conductance and the Farquhar photosynthesis models were collected by a portable infra-red gas analyzer system for CO_2 and H_2O (CIRAS-1, PP-Systems). The system is operated in open flow mode with a 5.5 cm long leaf chamber and an integrated gas CO_2 supply system as described in *Ellsworth*, [1999]. The chamber was modified with an attached Peltier cooling system to maintain chamber temperature near ambient atmospheric temperature. The gas exchange measurements were collected for upper canopy foliage at 11 – 12 m height, accessed with a system of towers and vertically telescoping mobile lifts. The shoot silhouette area index, a value analogous to the leaf area index (LAI), was measured vertically in increments of 1 m by a pair of LICOR LAI 2000 plant canopy analyzers on April 2, 1999 (Figure 1). The findings and resulting physiological parameters from these gas exchange measurements are presented in *Lai et al.*, [2000]. Linear interpolation was used to estimate the leaf area density at other depths within the canopy. The total LAI for the stand was 2.17 m^2 m^{-2} at the time of the experiment. A unique feature about the Duke Forest experiment is the availability of long-term sap-flux measurements collected on a 30-minute time step over the past 3 years. Sap flux (in $gm\ H_2O\ m^2\ xylem\ s^{-1}$) within individual trees was measured with constant-heat probes described by *Granier* [1985, 1987]. Sap flux densities of individual trees are scaled to the stand level using respective sapwood area per unit ground area in the stand multiplied by the average sap flux densities corrected for a reduced flux rate within the inner xylem (*Phillips et al.*, 1996; *Oren et al.*, 1998). Such measurements permit detailed assessments of the relative contribution of individual species (e.g. pine trees) to the over-all water flux from the forest ecosystem.

2.1.3. Discussion and Model Validation. The vertical variation in normalized leaf area density as well as the measured and estimated σ_w by the Wilson and Shaw (1977) model within the canopy are shown in Figure 1. The Wilson-Shaw model solves the coupled mean momentum, Reynold stress, and the three velocity variance equations simultaneously using measured $a(z)$, a specified C_d=0.2 (typical for pine needles), and a prescribed mixing length scale as discussed in *Katul and Albertson*, [1998] and *Katul and Chang*, [1999]. The closure constants in the Wilson-Shaw model are estimated to

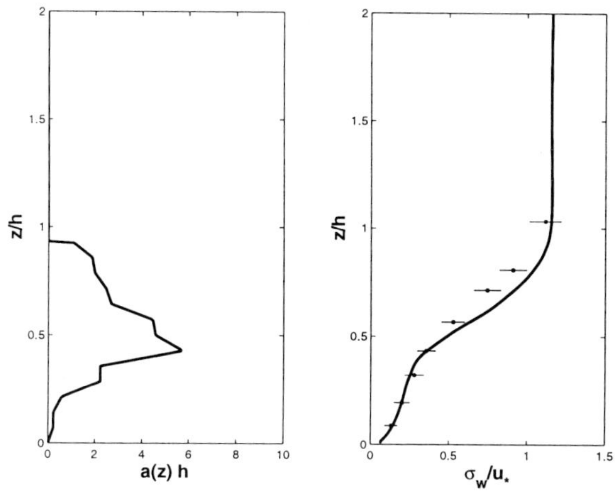

Figure 1. Comparisons between measured (dots) and second-order closure modeled σ_w (solid) as a function of height. The horizontal bars represent one standard deviation. The measured leaf area density ($a(z)$), normalized by the canopy height h is also shown. The measured and modeled σ_w are normalized by the friction velocity (u_*) above the canopy and the height from the forest floor z is normalized by h.

match the well-established Monin-Obukhov similarity functions in the atmospheric surface layer. The computed $\sigma_w(z)$ was used to estimate in equation (2) using *Thomson*'s, [1987] algorithm after *Baldocchi*, [1992]. The modeled time-depth evolution of mean sources and sinks for CO_2, H_2O, and T_a within the canopy and comparisons between measured and modeled F_T, F_{H2O}, and F_{CO2} are shown in Plate 1. To assess how well such forward methods reproduce the canopy environment, the time-depth comparison between measured and modeled mean CO_2 concentration is presented in Plate 2. From Plate 1, it is evident that much of the modeled CO_2 and H_2O sources and sinks are co-located with maximum height (shown as a horizontal solid line); however, daytime heating occurs in the upper fifth of the canopy volume. The source - sink patterns shown in Plate 1 demonstrate dissimilarity in zero-plane displacement heights for CO_2 and heat. The zero-plane displacement is commonly computed from the centroid of the source (or sink) profile as described in *Jackson*, [1981] and *Shaw and Pereira*, [1982]. Additionally, the dissimilarities in heat sources and sinks within the canopy volume suggest why flux-variance and flux-gradient similarity relationships for heat and water vapor, derived within the canopy sublayer, do not converge (*Padro*, 1993; *Katul et al.*, 1995; 1996). Another implication of the heat source being concentrated in the upper layers of the canopy layers (vis--vis water vapor) is its co-location

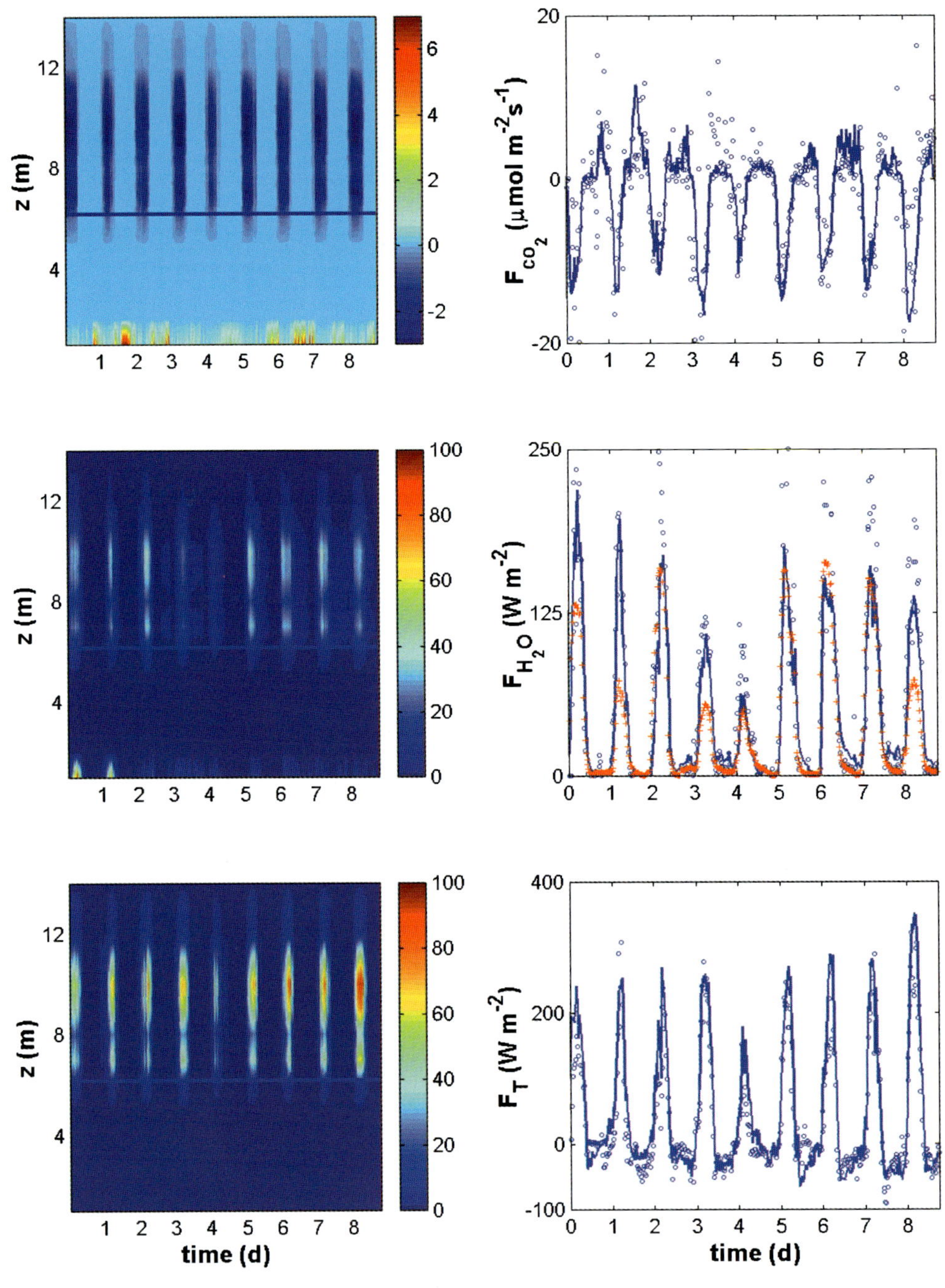

Plate 1. Results from forward model calculations and comparisons with field measurements: **Left Panels**: Time-depth variations of modeled scalar sources and sinks for CO_2 ($\mu mol\ m^{-3}s^{-1}$), H_2O (Wm^{-3}), and heat (Wm^{-3}) (top to bottom, respectively). The horizontal lines correspond to the height of maximum leaf area density. Right panels: Comparisons between eddy-covariance measured (open circle) and modeled scalar fluxes at z/h=1.11 for CO_2 ($\mu mol\ m^{-2}s^{-1}$), H_2O (Wm^{-2}), and sensible heat F_T (Wm^{-2}) (top to bottom, respectively). For F_{H2O}, the sap-flux measurements are also shown (red circles).

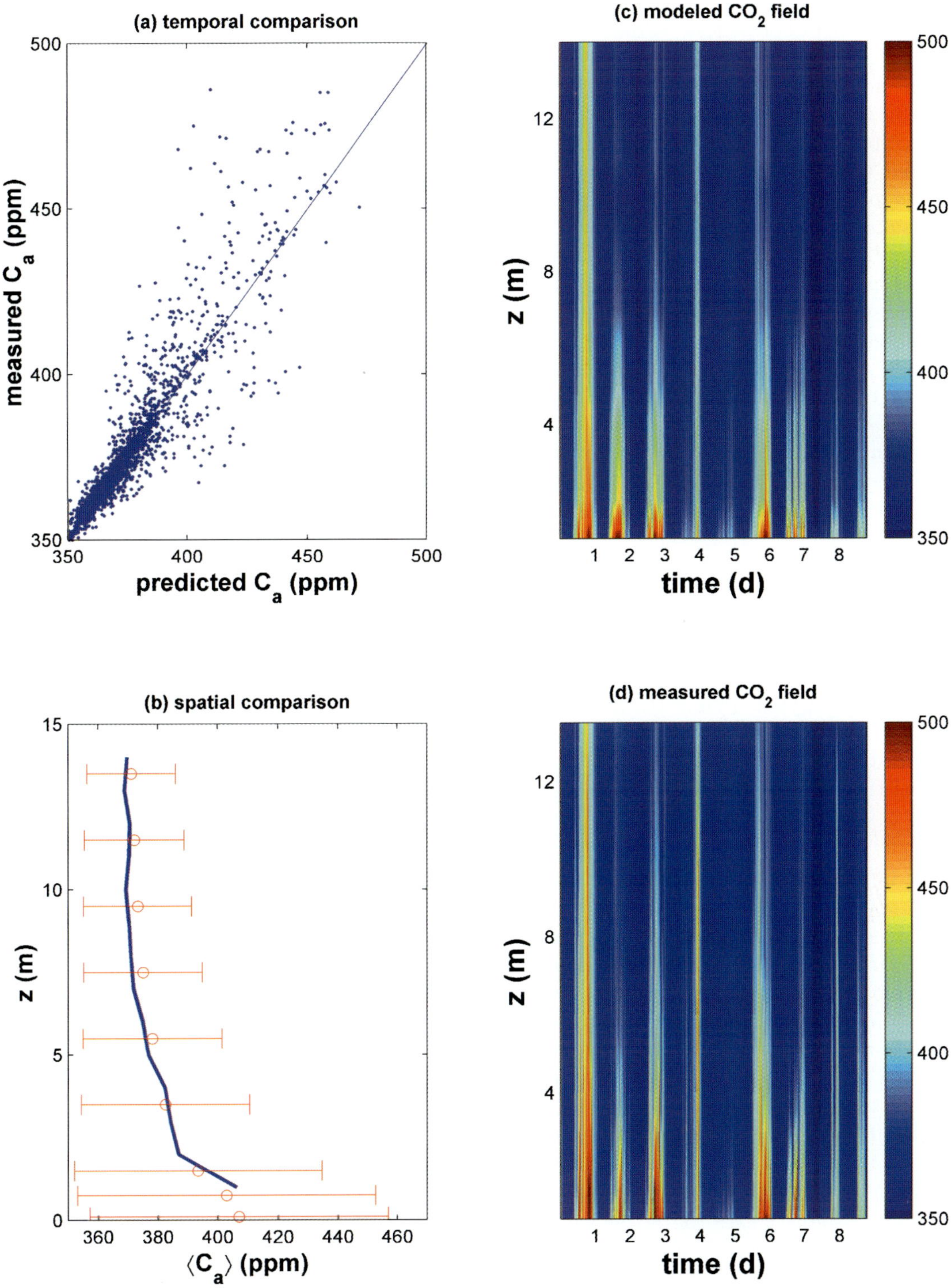

Plate 2. Results from forward model calculations and comparisons with CO_2 concentration field measurements. **Top Left panel**: Measured versus modeled 30 minute mean CO2 ($\mu mol\ mol^{-1}$) for all layers for the duration of the experiment. The 1 : 1 line is also shown. **Bottom Left Panel**: Variations of ensemble measured and modeled mean CO_2 ($\mu mol\ mol^{-1}$) for all the layers. The bars represent one standard deviation. **Top and Bottom Right**: Time-depth variation of measured (bottom) and modeled (top) mean CO_2 ($\mu mol\ mol^{-1}$) for the duration of the experiment.

with the penetration depth of "active" eddies whose size is $\sim \frac{h}{2}$. As shown by *Raupach et al.*, [1996] and *Katul et al.*, [1998], such "active" eddies are responsible for much of atmosphere-biosphere mass exchange. Given that the heat source is completely immersed within such eddies (when compared to water vapor), there is more efficient heat transport compared to water vapor or CO_2 whose source is concentrated in the middle canopy layers. In fact, many field experiments have demonstrated that heat is transported more efficiently than water vapor above a wide range of vegetated surfaces (*Katul et al.*, 1995). The close agreement between measured and modeled fluxes as well as mean CO_2 concentration (see Plate 2) suggests that the modeled sources and sinks as well as the dispersion calculations are reasonable. For water vapor calculations, differences between measured F_{H2O} using eddy-covariance and sapflux and modeled F_{H2O} were within 20% (Table 1). The model calculations of water vapor fluxes systematically underestimated the eddy covariance measurements, but systematically overestimated the sapflux measurements (see Table 1). The fact that eddy covariance measured and modeled water fluxes exceed the sapflux measurements (reflecting pine trees only) is not surprising given that the understory contribution as well as the ground evaporation are neglected. Generally, the modeled ground fluxes are rather uncertain given the large uncertainty in the flow statistics near the forest floor.

2.2. *Inverse Models*

The interest in inverse models is driven, in part, by the fact that $\bar{C}(z)$ can be more easily measured within the canopy than can the actual sources and sinks, or fluxes. Since sources (or sinks) and fluxes are directly related by a scalar concentration budget equation, the problem is in inferring S_c from the readily measured $\bar{C}(z)$ and modeled flow statistics. Early attempts to estimate S_c from $\bar{C}(z)$ relied on the so-called *K-theory*, which relates F_c to the vertical gradient of $\bar{C}(z)$ using eddy diffusivity $K_m(z)$. Combining this approximation with (1) under steady-state conditions, it can be shown that $S_c(z) = -\frac{\partial}{\partial z}[K_m(z)\frac{\partial \bar{C}(z)}{\partial z}]$ which depends on $\bar{C}(z)$ and $K_m(z)$ only, where the latter variable is strictly dependent on the flow statistics. However, for turbulent transport within vegetation, a local imbalance between production and dissipation commonly exists in the scalar flux- budget, which can lead to large *counter-gradient* fluxes and the ultimate failure of *K-theory*. The limitations of *K-theory* are now well recognized (e.g. *Raupach*, 1988; *Wilson*, 1989) and have been documented by many field experiments (e.g. *Denmead and Bradley*, 1985) and laboratory studies (e.g. *Coppin et al.*, 1986). Over the past two decades, two basic approaches have emerged to circumvent the limitations of *K-theory* for inverse methods: Lagrangian dispersion models (e.g. localized near field theory or LNF) and higher-order Eulerian closure models.

2.2.1. *The Localized Near Field (LNF) Theory.*

Over the last decade, the LNF became the first popular alternative to *K-theory* and has been successfully applied over a wide range of vegetation types (*Raupach et al.*, 1992; *Denmead and Raupach*, 1993; *Denmead*, 1995; *Katul et al.*, 1997a; *Massman and Weil*, 1999; *Leuning et al.*, 1999). Based on the LNF theory, the difference in the mean scalar concentration $\bar{C}(z)$ at any height z and a reference value $\bar{C}_R$ measured above the canopy at ($z_R > h$) is calculated by super-imposing near field (C_n) and far field (C_f) contributions:

$$\bar{C}(z) - \bar{C}_R = C_n + C_f \tag{6}$$

As shown in *Raupach*, [1989a,b], the near field contribution is computed via a kernel function:

$$\begin{aligned} C_n &= \int_0^{\infty} \frac{S_c(z_o)}{\sigma_w(z_o)} \left(f_1(z_o) + f_2(z_o)\right) dz_o \\ f_1(z_o) &= k_n \left[\frac{z - z_o}{\sigma_w(z_o) T_L(z_o)}\right] \\ f_2(z_o) &= k_n \left[\frac{z + z_o}{\sigma_w(z_o) T_L(z_o)}\right] \end{aligned} \tag{7}$$

where z_o is a dummy variable. An analytical approximation for the kernel function k_n was derived by *Raupach*, [1989a] and is given by:

$$k_n(\xi) = -0.39894 \log\left(1 - e^{-|\xi|}\right) - 0.15623\, e^{-|\xi|} \tag{8}$$

The far field contribution is calculated using results from the near field and a gradient-diffusion relationship given by:

$$C_f(z) = \bar{C}_R(z_R) - C_n(z) + \int_z^{z_R} \frac{F_c(z)}{\sigma_w(z) T_L(z)} dz \tag{9}$$

The Lagrangian time scale can be estimated from:

$$\frac{T_L \times u_*}{h} = \beta \tag{10}$$

where u_* is the friction velocity at the canopy top and β is a constant ($\sim 0.1 - 0.3$). With these formulations for the near and far field concentrations, and with direct concentration measurements within the canopy, it

Table 1. Regression statistics for the comparisons between modeled and measured flux variables. The regression slope (A), the intercept (B), the coefficient of determination (R^2), and the root-mean squared error ($RMSE$) are shown. For this comparison, the modeled flux is the independent variable, and N_p is the number of points in the regression analysis.

Variable	N_p	A	B	R^2	RMSE	Flux Measurements
F_{H2O} (Wm^{-2})	421	1.20	2.26	0.86	29.4	Eddy Covariance System
F_{H2O} (Wm^{-2})	421	0.87	2.25	0.76	29.4	Sapflux
F_T (Wm^{-2})	421	0.80	3.92	0.87	40.5	Eddy Covariance System
F_{CO2} ($\mu mol\ m^{-2}s^{-1}$)	421	0.85	0.16	0.56	4.6	Eddy Covariance System

is possible to estimate the sources and sinks via (2). However, to avoid numerical instability in the source profile calculations, redundant concentration measurements are necessary (i.e. the number of concentration measurements must exceed the number of source layers). As shown by *Raupach* [1989a], such redundancy reduces the source inference problem to a regression problem with the source strengths calculated by a least-squares error approach given by

$$\begin{aligned}\sum_{k=1}^{m} A_{jk} S_k &= B_j \\ A_{ij} &= \sum_{i=1}^{n} D_{ij}\Delta z_j D_{ik}\Delta z_k \\ B_j &= \sum_{l=1}^{n} \left(\bar{C}_l - \bar{C}_R\right) D_{lj}\Delta z_j \end{aligned} \quad (11)$$

Once A_{jk} and B_k are determined from the measured concentration and modeled dispersion matrix, the estimation of the source strength can be readily achieved. The LNF approach does not allow for non-zero vertical velocity skewness, strong inhomogeneity in vertical source strength variation and mean horizontal velocity variation within the canopy. These limitations can be relaxed in practice if one invokes a second-order Eulerian closure "inverse" model (EUL), as proposed by Katul and Albertson (1999) and described next.

2.2.2. Eulerian (EUL) Methods. Applying time and horizontal averaging, the steady state scalar conservation equation in (1) for planar homogeneous high Reynolds and Peclet numbers flow (neglecting molecular diffusion) can be written as (*Finnigan*, 1985; *Raupach*, 1988):

$$\frac{\partial \bar{C}}{\partial t} + \frac{\partial \overline{w'c'}}{\partial z} = S_c \quad (12)$$

where, primes denote fluctuations from time averages; w' is the vertical velocity fluctuation and $F_c = \overline{w'c'}$ is the scalar vertical flux. The corresponding time and horizontally averaged conservation equation for the vertical scalar flux budget is:

$$\begin{aligned}\frac{\partial \overline{w'c'}}{\partial t} &= 0 = -\overline{w'w'}\frac{\partial \bar{C}}{\partial z} - \frac{\partial \overline{w'w'c'}}{\partial z} - \overline{c'\frac{\partial p'}{\partial z}} \\ & \qquad\qquad P_1 \qquad + \qquad P_2 \qquad + \quad P_3\end{aligned} \quad (13)$$

In equation (13), buoyancy, scalar drag, and waving source production were neglected (*Finnigan*, [1985]). The terms P_2 and P_3 are unknowns requiring closure approximations. The flux transport term (P_2) is modeled after *Meyers and Paw U*, [1987] and the dissipation term (P_3) is modeled after *Finnigan*, [1985]. These approximations are:

$$\begin{aligned}\overline{w'w'c'} &= \frac{\tau}{C_8}\left[\overline{w'^3}\frac{\partial \bar{C}}{\partial z} + \overline{w'c'}\frac{\partial \overline{w'w'}}{\partial z} - 2\overline{w'w'}\frac{\partial \overline{w'c'}}{\partial z}\right] \\ \overline{c'\frac{\partial p'}{\partial z}} &= C_4\frac{\overline{w'c'}}{\tau}\end{aligned} \quad (14)$$

In these approximations, C_4 and C_8 are closure constants and τ is a Eulerian time scale given by:

$$\tau = \frac{q^2}{\bar{\epsilon}} \quad (15)$$

where $q = \left(\overline{u_i'u_i'}\right)^{\frac{1}{2}}$ is a characteristic turbulent velocity, $\bar{\epsilon}$ is the mean rate of viscous dissipation, and u_1, u_2 and u_3 are the velocity components in the x (or x_1), y (or x_2), and z (or x_3) directions, respectively, with x_1 aligned along the mean wind direction. The relaxation time is strictly dependent on the velocity statistics and can be computed from momentum higher-order closure models (*Meyers and Paw U*, 1986; *Katul and Albertson*, 1998, 1999; *Katul and Chang*, 1999) or LES techniques. We note that if $P_2 \ll P_3$, then the flux-transport term can be neglected and *K-theory* is recovered with

$$\overline{w'c'} = -\frac{\tau\sigma_w^2}{C_4}\frac{\partial \bar{C}}{\partial z} \tag{16}$$

Hence, it is common to attribute the failure of *K-theory* to the large flux transport term within the canopy volume. To some extend, this failure in the Eulerian framework mirrors the large near-field contribution in the Lagrangian framework. In fact, for some canopy layers, P_2 is the only term balancing the dissipation (or term P_3) and can be much larger than P_1. Upon combining these closure approximations with the scalar budget in (13), a second order ordinary differential equation (ODE) can be derived to describe the variations of the scalar turbulent flux with height (*Katul and Albertson*, 1999):

$$A_1(z)\frac{\partial^2 \overline{w'c'}}{\partial z^2} + A_2(z)\frac{\partial \overline{w'c'}}{\partial z} + A_3(z)\overline{w'c'} = A_4(z) \tag{17}$$

where:

$$\begin{aligned}
A_1(z) &= \frac{2\tau}{C_8}\overline{w'w'} \\
A_2(z) &= \frac{\tau}{C_8}\frac{\partial \overline{w'w'}}{\partial z} + 2\frac{\partial}{\partial z}\left(\frac{\tau}{C_8}\overline{w'w'}\right) \\
A_3(z) &= \frac{\partial}{\partial z}\left(\frac{\tau}{C_8}\frac{\partial \overline{w'w'}}{\partial z}\right) - \frac{C_4}{\tau} \\
A_4(z) &= \overline{w'w'}\frac{\partial \bar{C}}{\partial z} - \frac{\partial}{\partial z}\left(\frac{\tau}{C_8}\overline{w'w'w'}\frac{\partial \bar{C}}{\partial z}\right) \\
&\quad - \frac{\tau}{C_8}\overline{w'w'w'}\frac{\partial^2 \bar{C}}{\partial z^2}
\end{aligned} \tag{18}$$

The measured mean concentration profile is used only to calculate its gradient and curvature in A_4, the non-homogeneous term of the ODE. The flow statistics $\overline{w'w'}$, $\overline{w'w'w'}$, and τ can be estimated from momentum transport second order closure principles such as the model proposed by *Wilson and Shaw*, [1977] or *Wilson*, [1988] (see *Katul and Albertson*, 1998). The EUL model of *Katul and Albertson*, [1999] can incorporate non-Gaussian statistics (e.g., nonzero $\overline{w'w'w'}$) and advective scalar transport, two mechanisms neglected in the LNF approach. However, the EUL model calculations of S_c and F_c are sensitive to measurement errors in $\bar{C}$ (at least when compared to LNF) due to the lack of any redundancy in the inversion for S_c from measured $\bar{C}$ (*Katul and Albertson*, 1999; *Siqueira et al.*, 1999).

2.2.3. Hybrid (HEL) Methods. Recently, a hybrid Eulerian Lagrangian (HEL) method was proposed and tested by *Siqueira et al.*, [1999] that combines many of the advantages of both EUL and LNF. The method adopts a second-order closure model, similar to the one described above, to estimate the elements of D_{ij} but computes S_c from the "robust" regression algorithm of *Raupach*, [1989a,b] given by (11). The second order ODE describing the concentration profile from a prescribed unit source (and hence the flux as in (1)) is given by:

$$B_1(z)\frac{\partial^2 \bar{C}}{\partial z^2} + B_2(z)\frac{\partial \bar{C}}{\partial z} = B_3(z) \tag{19}$$

where:

$$\begin{aligned}
B_1(z) &= \frac{\tau}{C_8}\overline{w'w'w'} \\
B_2(z) &= -\overline{w'w'} + \frac{\partial}{\partial z}\left(\frac{\tau}{C_8}\overline{w'w'w'}\right) \\
B_3(z) &= -\frac{\partial}{\partial z}\left(\frac{\tau}{C_8}\left[\overline{w'c'}\frac{\partial \overline{w'w'}}{\partial z} + \overline{w'w'}\frac{\partial \overline{w'c'}}{\partial z}\right]\right) \\
&\quad + C_4\frac{\overline{w'c'}}{\tau}
\end{aligned} \tag{20}$$

In (19), $\overline{w'c'}$ is calculated by simply integrating the known unit source placed at one layer. In essence, the elements of D_{ij} are computed by: 1) positioning a unit source at a layer located at node j; 2) integrating this source profile to obtain the $\overline{w'c'}$ profile; and 3) solving the ODE in (19) for the concentration at all nodes $i(=1,2,\cdots,N$ resulting from the unit source placed at node j, where $N$is the number of layers within and above the canopy for which concentration measurements are available. This procedure is repeated for $j=1,\cdots,m$ to obtain all the elements of the D_{ij} matrix needed in (11). The scheme embeds the robustness of the LNF regression inversion in a more physically sound forward calculation of D_{ij}, which includes the effects of $\overline{w'w'w'}$ on D_{ij} through the coefficients B_1 and B_2. We note that *Thomson's*, [1987] algorithm does not uniquely define D_{ij} in forward methods (described earlier) and the above approaches can also be used to establish a relationship between S_c and $\bar{C}$.

2.2.4. Discussion and Model Validation. In order to compare all three methods with a "benchmark" data set for which the scalar source strength and location as well as the flow statistics are known with minimal error, we consider the laboratory experiment of *Coppin et al.*, [1986]. In this pioneering wind tunnel experiment, a planar heat source is positioned at $\frac{z}{h} = 0.8$ within a set of vertical cylinders spaced in a regular diamond array with roughness concentration of 0.23. The planar heat source is generated by an array of horizontal

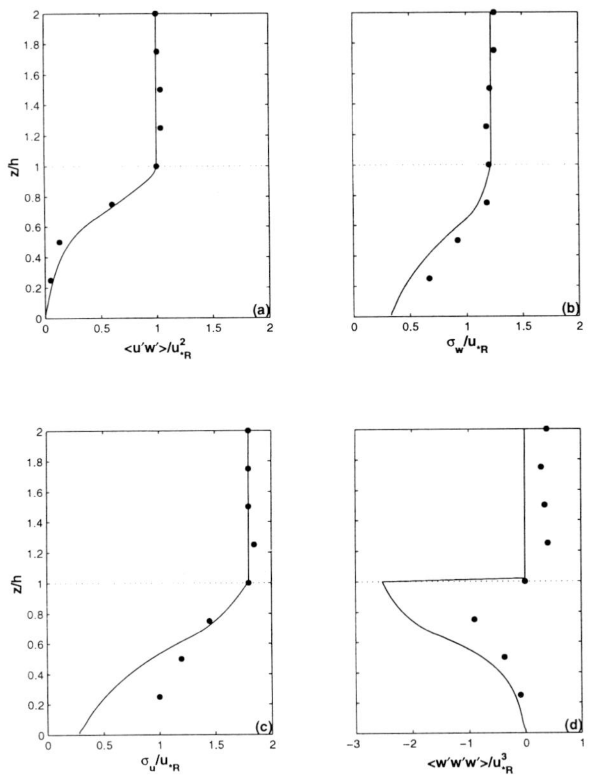

Figure 2. Comparison between measured (circles) and modeled (solid line) velocity statistics (Reynolds stress $\overline{w'u'}$, vertical velocity standard deviation σ_w, longitudinal velocity standard deviation σ_u, and third-moment of vertical velocity $\overline{w'w'w'}$) by *Wilson*'s, [1988]) closure model for the Strips wind tunnel experiment. All velocity statistics are normalized by the friction velocity (u_{*R}) at the $z/h = 1$.

wires placed between the cylinders. In this experiment, mean temperature profiles in concert with sensible heat flux and key velocity statistics are measured. Because EUL requires a relaxation time scale, σ_w, and $\overline{w'w'w'}$ for scalar transport calculations, we matched one of the published second-order closure models (*Wilson and Shaw*, [1977]) to the velocity statistics reported in *Coppin et al.* [1986] and *Raupach* [1988] to obtain these flow variables. The comparison between modeled and measured Reynolds stress ($\overline{w'u'}$), vertical velocity standard deviation (σ_w), longitudinal velocity standard deviation (σ_u), and vertical velocity third moment ($\overline{w'w'w'}$) are shown in Figure 2. The calculated velocity statistics are sufficiently close to the modeled values thus allowing the use of modeled τ, σ_w, and $\overline{w'w'w'}$ in scalar transport. Using the measured temperature profile in *Raupach*, [1989] and the modeled velocity statistics in Figure 2, the heat sources and fluxes are computed using all three methods and compared to the measurements in Figure 3. We computed the LNF and HEL sources and fluxes using 4 and 8 layers to assess how sensitive the model results are to the number of layers. In the LNF calculations, the T_L profile used is identical to *Raupach*, [1989a]. The LNF source location is sensitive to the number of layers, naturally with 8 layers better capturing the source location than 4 layers. However, with 8 layers, spurious sources in the lower-canopy layers exist. All in all, the HEL model calculations identified the source position slightly better than LNF for the same number of source layers. When

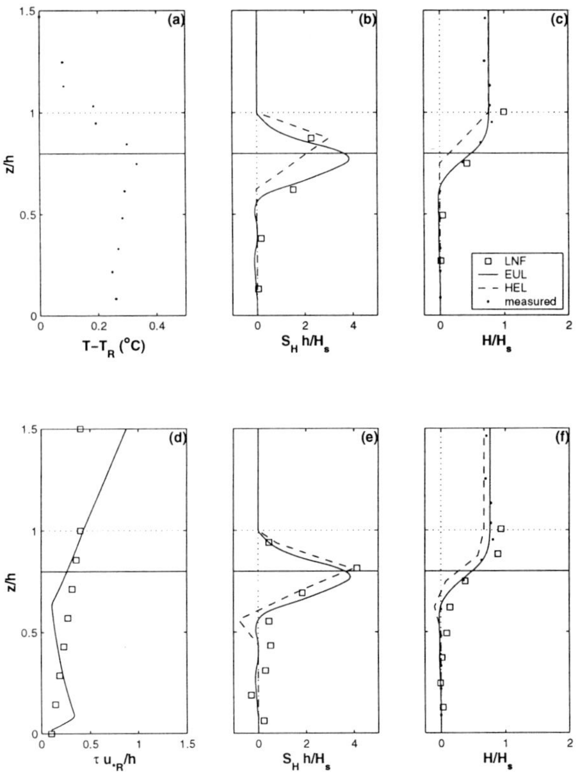

Figure 3. Comparison between measured and modeled sources and fluxes by EUL (solid line), LNF (squares), and HEL (dashed line) for the wind tunnel experiment. For LNF and HEL, the top panels are computations performed with 4 layers and the bottom panels are for computations performed using 8 layers. The EUL is shown in both for reference. Panel (a) Measured temperature profile from *Raupach* [1989a]. Panels (b) and (e) Source comparisons by the three methods normalized by the source strength H_s and canopy height h. The source location is at z/h=0.8 (horizontal solid line). Panels (c) and (f) Normalized flux comparisons for the three methods with *Coppin et al.*[1986] sensible heat flux measurements (dots). Panel (d) is the relaxation time scale τ (line) and the Lagrangian integral time scale T_L (squares) both normalized by $\frac{u_{*R}}{h}$. Subscript R refers to the variable value at $z/h = 1$.

compared to HEL (dashed line), the EUL calculations (solid line) appear to smooth out random fluctuations in the source strength. In all model calculations, the maximum source location was well reproduced for the LNF and HEL eight-layer calculations and EUL. It is evident from Figure 2 that the modeled fluxes by the three methods agree well with *Coppin et al.* [1986] sensible heat flux measurements (H_s). Additionally, the agreement between LNF and HEL flux calculations with measurements is robust to the number of layers. For the EUL method, the sources are sensitive to how the temperature gradients (and curvatures) are estimated from the measurements. To alleviate such sensitivity, we experimented with three interpolation schemes that permit numerical and analytical estimation of the temperature gradient from a prescribed discrete temperature profile. These schemes include: 1) Fitting the variations in temperature to z via an analytical function whose parameters are determined from nonlinear regression analysis, 2) Cubic spline interpolation, and 3) performing local linear regression on consecutive measurement levels to estimate a local temperature gradient (using only three points) and curvature (using only five points). We found that the local regression method is more suitable for such inverse calculations because it best reproduces the observed sources and fluxes (analysis presented in *Siqueira et al.*, [1999] but not reproduced here). This method was used in the EUL calculations shown above and is adopted throughout this study. The good agreement noted in Figure 3 for EUL and HEL was achieved after the closure constants and were altered from 9.9 and 9 typical for canopy flows (*Meyers and Paw U*, 1986) to 3.3 and 27.0. It must be emphasized that the Coppin et al. (1986) wind tunnel experiment does not resemble entirely a canopy flow experiment. In fact, the first and second moments of the velocity field are large at the base of the roughness elements (see *Raupach*, 1988). The fact that the turbulence is not damped at the base of the roughness elements, typical of canopy flows, suggests that the *Coppin et al.*, [1986] experiment is a combination of both canopy and boundary layer flows. It is perhaps for this reason that the closure constants had to be modified in order to reproduce the sensible heat flux measurements. An additional uncertainty that impacts the choice of the closure constants is the estimate of τ from the modeled dissipation by closure models. The dissipation rate was not measured in the Coppin experiment (and in fact is rarely measured), and hence, no direct validation of is possible.

An analogous experiment to the Coppin wind tunnel study was conducted at the Duke Forest field site (described earlier) using six sonic anemometers and 8 thermocouples. Near neutral runs were used in order to assess the attenuation of the sensible heat fluxes within the canopy volume. The difference between this experiment and the Coppin wind tunnel experiment is the complex heat source distribution (see, e.g., Plate 1). Unlike the wind tunnel experiment, the source distribution here is not known; however, the heat flux distribution within the canopy could be measured by the sonic anemometers. Hence, these flux comparisons are used to investigate the performance of the three models (EUL, HEL, and LNF). The modeled and measured velocity statistics relevant to this comparison are discussed in *Siqueira et al.*, [1999]. The measured sensible heat flux statistics were constructed from the ensemble of 30 neutral runs along with one standard deviation. For each of the 30 runs, the measured mean temperature profile and the modeled flow statistics were used to compute the flux distribution within the canopy by the three inverse methods. The ensemble measured and modeled sensible heat fluxes are compared along with one standard deviation of the measured heat flux. For reference, the ensemble temperature profile for all 30 runs is shown along with one standard deviation. As evidenced in Figure 4 is that all three models over attenuated the heat flux inside the canopy. Furthermore, none of the models reproduced well the counter gradient heat flux in the lower canopy layers. The measured and modeled sensible heat flux variations by the three approaches at each of the 6 layers and for all 30 runs are shown in Figure 5. It is evident that the LNF method overestimated the eddy-covariance measured sensible heat flux near the canopy top more than HEL and EUL.

3. CONCLUSIONS

This chapter described recent developments in forward and inverse methods to infer distributions and strengths of scalar sources and sinks of water vapor, carbon, and heat within the canopy volume. Both approaches require detailed turbulent transport mechanics formulation within the canopy volume. Recent developments in higher order closure principles or Large Eddy Simulation techniques provide reasonably accurate description formulation for these velocity statistics. Both forward and inverse approaches considered in this study do not resort to empirical relationship between turbulent scalar fluxes and mean concentration gradients. Having demonstrated the ability of forward models to describe the canopy microclimate, such approaches can now be used to investigate the effect of variations in physiological or biophysical properties

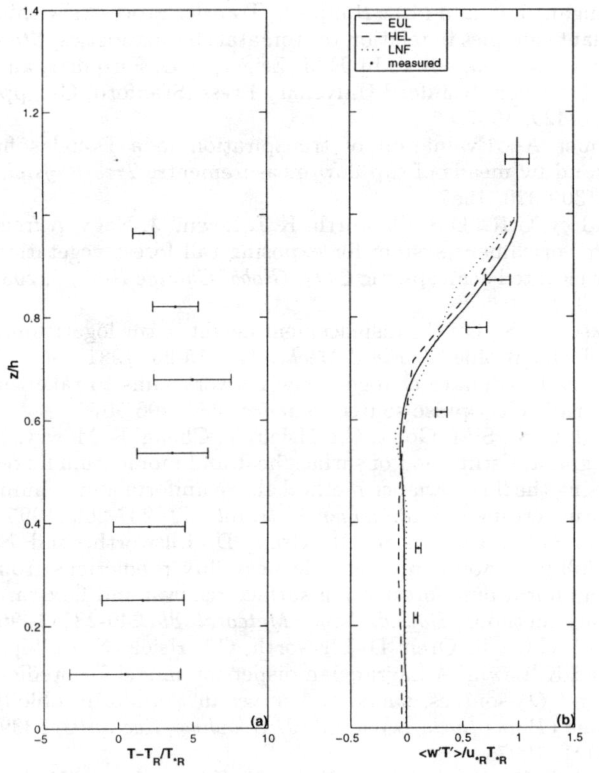

Figure 4. Ensemble vertical distribution for the Pine Forest heat dispersion experiment. (a) normalized temperature difference, and (b) normalized sensible heat flux. The normalizing variables are u_*R and h for velocity and length, and T_{*R} for temperature, where $T_{*R} = \frac{\overline{w'T'}}{u_{*R}}$ and subscript R refers to the variable value at $z/h = 1$.

on biosphere-atmosphere water and carbon exchange. For example, changes or adjustments in foliage physiological properties or distribution under elevated atmospheric CO_2 can be readily implemented in such a forward modeling framework to assess how much the net ecosystem carbon exchange and water vapor flux will be altered in response to such atmospheric CO_2 perturbation. Certain physiological adjustments are currently studied through the Free Air CO_2 Experiments at Duke Forest and elsewhere (*Ellsworth et al.*, 1995; *Hendrey et al.*, 1999). Other immediate applications for both forward and inverse approaches is guiding the development and assessment of simpler models, such as single layer models, which are currently used in many hydrologic, meso-scale, and climatic models. In this way, the multi-layer "forward" approach described and demonstrated here can be used to develop and refine empirical bulk transfer coefficients needed in big-leaf models. The comparisons between the three inverse methods with published wind tunnel measurements and recent field measurements at Duke Forest highlighted some strengths and weaknesses in each method. Given that the inverse problem itself is ill posed, we envision the application of all three methods for routine source inference to be more beneficial than recommending a particular method. Given that these methods differ in their basic assumptions, agreement amongst all three methods adds confidence in the modeled source strength and simplifications for a particular scalar concentration profile. Similarly, disagreement between them highlights the magnitude (and location) of the uncertainty in the modeled source distribution.

Acknowledgments. Support for this project was provided by the National Science Foundation (*NSFBIR12333* and *NSFEAR9903471*), the Department of Energy (DOE) through the FACE FACTS project and through the National Institute of Global Environmental Change (NIGEC) through the Southeast Regional Center at the University of Alabama, Tuscaloosa *DEFC03090ER61010*).

The third author is supported by the Conselho Nacional de Desenvolvimento Cientifico e Tecnologico (CNPq) of Brazil.

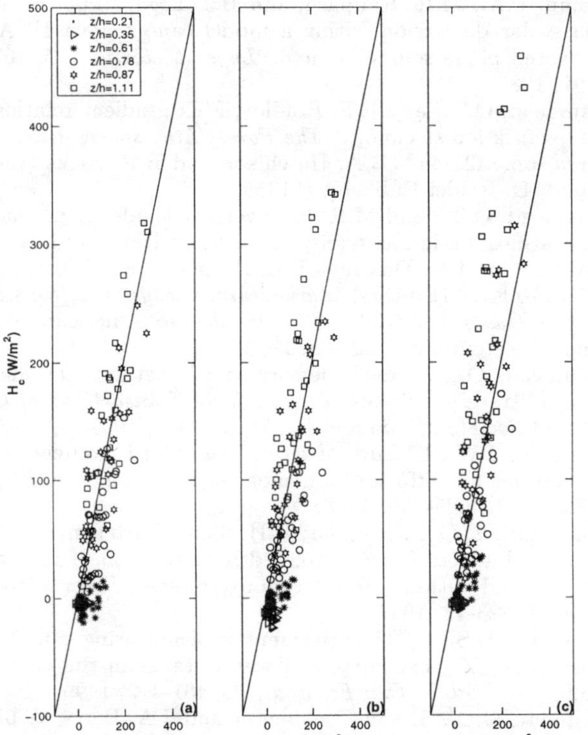

Figure 5. Comparison between measured (H_m) and modeled (H_c) sensible heat fluxes for the Pine Forest heat dispersion experiment under near-neutral conditions. Different symbols are for different z/h values. Panel (a) for EUL, Panel (b) for HEL, and Panel (c) for LNF.

REFERENCES

Albertson, J.D, and M.B. Parlange, Surface length scales and shear stress: Implications for land-atmosphere interaction over complex terrain, *Water Resour Res., 35*, 2121-2132, 1999.

Baldocchi, D.D., Turbulent transfer in a Deciduous Forest, *Tree Phys., 5*, 357-377, 1989.

Baldocchi, D.D., Measuring and modelling carbon dioxide and water vapour exchange over a temperate broad-leaved forest during the 1995 summer drought, *Plant Cell Environ., 20*, 1108-1122, 1997.

Baldocchi, D.D., A Lagrangian random-walk model for simulating water vapor, CO_2 and sensible heat flux densities and scalar profiles over and within a soybean canopy, *Bound. Layer Meteorol., 61*, 113-144, 1992.

Brutsaert, W., *Evaporation into the atmosphere: Theory, history, and applications*, pp. 209-230, Kluwer Academic Publishers, Dordrecht, The Netherlands, 1982.

Campbell, G.S. and J.M. Norman., *An Introduction to Environmental Biophysics*, pp. 148-259, Springer-Verlag, N. Y., 1998.

Collatz, G.J., J.T. Ball, C. Grivet, and J.A. Berry, Physiological and environmental regulation of stomatal conductance, photosynthesis and transpiration: A model that includes a laminar boundary layer, *Agric. For. Meteorol., 54*, 107-136, 1991.

Coppin, P.A., M.R. Raupach, and B.J. Legg, Experiments on scalar dispersion within a model canopy Part II: An elevated plane source, *Bound. Layer Meteorol., 35*, 167-191, 1986.

Deanmead, O.T., and E.F. Bradley, Flux gradient relationships in a forest canopy, *The Forest-Atmosphere Interaction*, pp. 421-442, B.A. Hutchison and B.B. Hicks (Editors), D. Reidel Publishing, 1985.

Deanmead, O.T., and M.R. Raupach, Methods for measuring atmospheric gas transport in agricultural and forest system, in: J.M. Duxbury, L.A. Harper, A.R. Mosier, and D.E. Rolston (Editors), *Agricultural Ecosystem Effects on Trace Gases and Global Climate Change*, American Society of Agronomy, Madison, 1993.

Deanmead, O.T., Novel meteorological methods for measuring trace gas fluxes. *Philosophical Transactions of the Royal Society of London A, 351*, 383-396, 1995.

Deardorff, J.W., Closure of second and third moment rate equations for diffusion in homogeneous turbulence, *Phys. Fluids, 21*, 525-530, 1978.

Dyer, M.J., E. G. Patton, and R.H. Shaw, Turbulent kinetic energy budgets from a large eddy simulation of air flow above and within a forest canopy, *Bound. Layer Meteorol., 84*, 23-43, 1997.

Ellsworth, D.S., 'CO2 enrichment in a maturing pine forest: Are CO_2 exchange and water status in the canopy affected?', *Plant Cell Environ., 22*, 461-472, 1999.

Farquhar, G.D., S. von Caemmerer and J.A. Berry, A biochemical model of photosynthetic CO_2 assimilation in leaves of C3 species, *Planta, 149*, 78-90, 1980.

Finnigan, J.J., Turbulent transport in plant canopies, *The Forest-Atmosphere Interactions*, edited by B.A. Hutchinson and B.B. Hicks, pp 443-480, D. Reidel, Norwell, Mass, 1985.

Finnigan, J.J., and M.R. Raupach, Transfer processes within plant canopies in relation to stomatal characteristics, *Stomatal Function*, edited by E.M. Zeiger, G.D. Farquhar, and I.R. Cowan, Stanford University Press, Stanford, CA, pp. 385-429, 1987.

Granier A., Evaluation of transpiration in a Douglas fir stand by means of sap flow measurements, *Tree Physiol., 3*, 309-320, 1987.

Hendrey, G.R., D.S. Ellsworth, K.F. Lewin, J. Nagy, A free-air enrichment system for exposing tall forest vegetation to elevated atmospheric CO_2, *Global Change Biol., 5*, 293-309, 1999.

Jackson, P.S., On the displacement height in the logarithmic velocity profile, *J. Fluid Mech., 111*, 15-25, 1981.

Kaiser, J., Climate change - new network aims to take the world's CO_2 pulse source, *Science, 281*, 506-507.

Katul, G.G., S.M. Goltz, C.I. Hsieh, Y. Cheng, F. Mowry, J. Sigmon, Estimation of surface heat and momentum fluxes using the flux-variance method above uniform and nonuniform terrain *Bound. Layer Meteorol., 74*, 237-260, 1995.

Katul G.G., C.I. Hsieh, R. Oren, D. Ellsworth, and N. Phillips, Latent and sensible heat flux predictions from a uniform pine forest using surface renewal and flux variance methods, *Bound. Layer Meteorol, 80*, 249-282, 1996.

Katul, G.G., R. Oren, D. Ellsworth, C.I. Hsieh, N. Phillips, and K. Lewin, A Lagrangian dispersion model for predicting CO_2 sources, sinks, and fluxes in a uniform loblolly pine (*Pinus taeda L.*) stand, *J. Geophys. Res., 102*, 9309-9321, 1997a.

Katul, G.G., C.I. Hsieh, G. Kuhn, D. Ellsworth, and D. Nie, Turbulent eddy motion at the forest-atmosphere interface, *J. Geophys. Res., 102*, 13,409-13,421, 1997b.

Katul, G.G. and J.D. Albertson, An investigation of higher-order closure models for a forested canopy, *Bound. Layer Meteorol., 89*, 47-74, 1998.

Katul, G.G. and J.D. Albertson, Modeling CO_2 sources, sinks and fluxes within a forest canopy, *J. Geophys. Res., 104*, 6081-6091, 1999.

Katul, G.G. and W.-H. Chang, Principal length scales in second-order closure models for canopy turbulence, *J. Appl. Meteorol., 38*, 1631-1643, 1999.

Lai, C.T., G.G. Katul, D. Ellsworth, and R. Oren, Modelling vegetation-atmosphere CO_2 exchange by a coupled Eulerian-Lagrangian approach, *Bound. Layer Meteorol., 95*, 91-122, 2000.

Leuning, R., O.T. Denmead, A. Miyata, J. Kim, Source/sink distributions of heat, water vapour, carbon dioxide and methane in a rice canopy estimated using Lagrangian dispersion analysis, submitted to the *Agric. Forest Meteorol*, 1999.

Massman, W.J., J.C. Weil, An analytical one dimensional second-order closure model of turbulence statistics and the Lagrangian time scale within and above plant canopies of arbitrary structure, *Bound. Layer Meteorol., 91*, 81-107, 1999.

McNaughton, K.G., and T.W. Spriggs, An evaluation of the Priestley and Taylor equation and the complementary relationship using results from a mixed layer model of the convective boundary layer, in *Estimation of Areal Evapotranspiration*, IAHS Publi. No. 177, 89-116, 1989.

Meyers, T., and K.T. Paw U, Testing of a higher-order closure model for modeling airflow within and above plant canopies, *Bound. Layer Meteorol., 37*, 297-311, 1986.

Meyers, T., and K.T. Paw U, Modeling the plant canopy micrometeorology with higher order closure principles, *Agric. For. Meteorol., 41*, 143-163, 1987.

Monteith, J.L., Evaporation and environment, *Symp. Soc. Exp. Biol., 19*, 205-234, 1965.

Monteith, J.L., and M.H. Unsworth, *Principles of Environmental Physics*, Edward Arnold, London, 1990.

Oren R., N. Phillips, G. Katul, B.E. Ewers, and D.E. Pataki, Scaling xylem sap flux and soil water balance and calculating variance: a method for partitioning water flux in forests, *Ann. Sci. For., 55*, 191-216, 1998.

Padro, J., An investigation of flux-variance methods and universal functions applied to a 3 land-use types in unstable conditions, *Bound. Layer Meteorol., 66*, 413-425, 1993.

Penman, H.L., Natural Evaporation from Open Water, Bare Soil and Grass, *Proc. R. Soc. London, Ser. A.*, 193, 120-146, 1948.

Phillips, N., and R. Oren, A comparison of daily representations of canopy conductance based on two conditional time-averaging methods and the dependence of daily conductance on environmental factors, *Ann. Sci. For., 55*, 217-235, 1998.

Raupach, M.R., and R.H. Shaw, Averaging procedures for flow within vegetation canopies, *Bound. Layer Meteorol., 22*, 79-90, 1982.

Raupach, M.R., Canopy transport processes, in *Flow and Transport in the Natural Environment*, ed. W.L. Steffen, and O.T. Denmead, pp.95-127, Springer-Verlag, New York, 1988.

Raupach, M.R., Applying Lagrangian fluid mechanics to infer scalar source distributions from concentration profiles in plant canopies, *Agric. Forest Meteorol., 47*, 85-108, 1989a.

Raupach, M.R., A practical Lagrangian method for relating scalar concentrations to source distributions in vegetation canopies, *Q.J.R. Meteorol. Soc., 115*, 609-632, 1989b.

Raupach, M.R., O.T. Denmead, and F.X. Dunin, Challenges in linking atmospheric CO_2 concentrations to fluxes at local and regional scales, *Aust. J. Bot., 40*, 697-716, 1992.

Schuepp, P.H., Tansley Review No. 59: Leaf Boundary Layers, *New Phytologist, 125*, 477-507, 1993.

Shaw, R.H. and I. Seginer, Calculation of velocity skewness in real and artificial canopies, *Bound. Layer Meteorol., 39*, 315-332, 1987.

Shaw, R.H. and A.R. Pereira, Aerodynamic roughness of a plant canopy: A Numerical Experiment, *Agric. Meteorol., 26*, 51-65, 1982.

Shaw, R.H., and U. Schumann, Large-eddy simulations of turbulent flow above and within a forest, *Bound. Layer Meteorol.,61*, 47-64, 1992.

Shen, S.H., and M.Y. Leclerc, Modelling the turbulence structure in the canopy layer, *Agric. For. Meteorol.*, 87, 3-25, 1997.

Siqueira, M., C.-T. Lai, G. Katul, Estimating CO_2 sources, sinks, and fluxes in a forest canopy using Lagrangian, Eulerian, and hybrid inverse models, *J. Geophys. Res.*, in press, 2000.

Stuart, J.B., On the use of the Penman-Monteith equation for determining areal evapotranspiration, in *Estimation of Areal Evapotranspiration*, IAHS Publi. No. 177, 3-12, 1989.

Thomson, D.J., Criteria for the selection of stochastic models of particle trajectories in turbulent flows, *J. Fluid Mech., 180*, 529-556, 1987.

Williams, M., E.B. Rastetter, D.N. Fernades, M.L. Goulden, S.C. Wofsy, G.R. Shaver, J.M. Meillo, J.W. Munger, S.M. Fan, and K.J. Nadelhoffer, Modeling the soil-plant-atmosphere continuum in a *Quercus-Acer* stand at Harvard Forest: the regulation of stomatal conductance by light, nitrogen and soil/plant hydraulic properties, *Plant, Cell and Environment, 19*, 911-927, 1996.

Wilson, J.D., A second order closure model for flow through vegetation, *Bound. Layer Meteorol., 42*, 371-392, 1988.

Wilson, J.D., Turbulent transport within the plant canopy, in *Estimation of Areal Evapotranspiration*, IAHS Publ. no. 177, pp.43-80, 1989.

Wilson, N.R., and R.H. Shaw, A higher order closure model for canopy flow, *J. Appl. Meteorol., 16*, 1198-1205, 1977.

Wofsy, S.C., M.L. Goulden, J.W. Munger, S.M. Fan, P.S. Bakwin, B.C. Daube, S.L. Bassow, and F.A. Bazzaz, Net exchange of CO_2 in a mid-latitude forest, *Science, 260*, 1314-1317, 1993.

Katul, G.G., C.-T.Lai, M. Siqueira, K.V.R. Schäfer, K.H. Wesson, David Ellsworth, and R. Oren, School of the Environment, Box 90328, Duke University, Durham, NC 27708-0328,

(e-mails: gaby@duke.edu; cl9@duke.edu; mbs4@duke.edu; karina@duke.edu; khw7f@duke.edu; ellswort@bnl.gov, and ramoren@duke.edu)

Albertson, J.D., Department of Environmental Sciences, University of Virginia, Charlottesville, VA 22903, (e-mail: jdalbertson@virginia.edu)

Ground-Based Soil Moisture and Soil Hydraulic Property Observations in Regional Scale Experiments

Richard H. Cuenca and Shaun F. Kelly

Hydrologic Science Team, Department of Bioresource Engineering, Oregon State University

Determination of the variation in depth and time of soil moisture content and soil hydraulic properties is critical for quantification of environmental conditions and for accurate simulation of soil-vegetation-atmosphere transfer (SVAT) processes. This applies to local models of the vadose zone, models of the diurnal energy balance and atmospheric boundary layer, mesoscale weather models and climate models. Typically the soil hydraulic properties of concern are the soil water retention function and unsaturated hydraulic conductivity function. Soil moisture content and soil hydraulic property measurements have been applied in large-scale hydrologic-atmospheric-remote sensing experiments to denote the environmental conditions and for simulation model initialization and verification. Examples of data from the HAPEX-MOBILHY, HAPEX-Sahel, BOREAS and CASES experiments are presented. The evolution of instrumentation and sampling protocol applied is discussed.

INTRODUCTION

Measurements of soil moisture content and soil hydraulic properties were incorporated into the first regional scale hydrologic-atmospheric science-remote sensing experiment, HAPEX-MOBILHY (MOdelisation de BILan HYdrique) in France in the mid-1980's (André et al., 1988; Goutorbe et al., 1989; Cuenca and Noilhan, 1991). The methods employed and instrumentation applied have evolved along with other experimental techniques. The same questions arise with respect to soil moisture content and soil hydraulic properties as with measurements of other parameters: a) what is the optimum sampling frequency in time and space, and b) what instrumentation will be cost effective and of adequate accuracy. The answers to these questions depend on the environment in which the measurements are to be made, the anticipated application of the data, and the evolution of instrument technology. This chapter will address these questions within the context of a number of solutions applied to various regional scale experiments over a span of 15 years.

Land Surface Hydrology, Meteorology, and Climate: Observations and Modeling
Water Science and Application Volume 3, pages 47-61

RATIONAL FOR AND APPLICATION OF SOIL MOISTURE CONTENT MEASUREMENTS

State Variable of Soil-Water System - Model Initialization and Verification

The measurement of soil water content requires calibration of the sensor employed rather than assumptions about the physical system. Since soil moisture content measurements are generally based on point sampling, assumptions as to the representativeness of the sampling site are required. But no assumptions are required for the measurements themselves, other than the general fact that the sensor signal can be directly correlated to soil moisture content. (The difficulties in calibration of various types of sensor systems will be described later in the article.) The soil moisture content measurements, assuming correct calibration, can therefore be used as state variables for initialization and verification of simulation models that include a soil moisture parameter. The measured soil water content in combination with soil hydraulic properties is also used to describe the environmental conditions that may affect other measurement systems (e.g. evaporative flux from a vegetative canopy).

Dynamics of the Soil-Water-Atmosphere-Plant System

The dynamics of the soil water-atmosphere-plant (SWAP) system for one-dimensional (vertical) soil water movement can be explained by Richards equation (Richards, 1931), which for an isothermal system can be expressed as (Kutílek and Nielsen, 1994),

$$\frac{\partial \theta}{\partial t} = \frac{\partial}{\partial z}\left[K(h)\left(\frac{\partial h}{\partial z}\right)\right] + \frac{\partial K(h)}{\partial z} - S \quad (1)$$

where θ = volumetric soil water content
t = time
z = depth taken as positive downwards
$K(h)$= hydraulic conductivity as a function of h
h = soil water pressure relative to atmospheric pressure ($h \le 0$)
S = soil water uptake by roots within the soil column

If the soil is wetting or drying, θ will be dependent upon h and the change in soil moisture content can be expressed as,

$$\frac{\partial \theta}{\partial t} = \frac{d\theta}{dh}\left(\frac{\partial h}{\partial t}\right) \quad (2)$$

which leads to the capacitance form of the Richards equation (Kutílek and Nielsen, 1994),

$$C(h)\frac{\partial h}{\partial t} = \frac{\partial}{\partial z}\left[K(h)\left(\frac{\partial h}{\partial z}\right)\right] + \frac{\partial K(h)}{\partial z} - S \quad (3)$$

where $C(h)$= specific capacity of the soil.

$$C(h) = \frac{d\theta}{dh} \quad (4)$$

Alternatively, using the following expression for the change in soil water pressure with depth,

$$\frac{\partial h}{\partial z} = \frac{dh}{d\theta}\left(\frac{\partial \theta}{\partial z}\right) \quad (5)$$

leads to the diffusivity form of Richards equation (Kutílek and Nielsen, 1994),

$$\frac{\partial \theta}{\partial t} = \frac{\partial}{\partial z}\left[D(\theta)\frac{\partial \theta}{\partial z}\right] + \frac{dK(\theta)}{d\theta}\left(\frac{\partial \theta}{\partial z}\right) - S \quad (6)$$

where $D(\theta)$ = soil water diffusivity as a function of θ.

$$D(\theta) = K(\theta)\frac{\partial h}{\partial \theta} \quad (7)$$

$K(\theta)$ = hydraulic conductivity as a function of θ.

Various parameters in Eqs. (1) through (7) are monitored or measured in large-scale field experiments to define the state and dynamics of the SWAP system. Measurements of soil water content, θ, versus time, t, can be used to define the left-hand side of Eqs. (1) and (6). Such measurements can be made periodically, as was common using the neutron probe, or continuously which is possible using probes based on principles that measure the dielectric constant of the soil water matrix. Measurements for changes in θ versus time have been the most common type of measurement applied in large-scale experiments. Probes that measure soil water potential (e.g. heat dissipation probes, gypsum blocks, thermocouple psychrometers) or soil water tension (e.g. tensiometers for relatively moist regimes) can be used to quantify the h versus t relationship given in the left-hand side of Eq. (3).

Data for the hydraulic conductivity, expressed as either $K(h)$ or $K(\theta)$, have been made in the laboratory using core samples or *in situ* using typically the tension infiltrometer for surface measurements and the Guelph permeameter for measurement at depth. Likewise the soil water retention function ($dh/d\theta$) can be measured using core samples in the laboratory or *in situ* for sites instrumented with concurrent measurements of h and θ at the same depth levels. *In situ* measurements are generally felt to be more representative of field conditions and not affected by limited sample size or edge effects of soil cores. Combinations of *in situ* and lab analysis of core samples are preferred.

Evidence of Impact of Soil Moisture Content in Simulation and Actual Systems

Soil moisture content of the surface layer, related to soil evaporation, or throughout the depth of the root profile, related to plant transpiration, plays a key role in partitioning net radiation into the sensible, latent and soil heat fluxes. Correct assessment of the diurnal energy balance and other related processes, such as evolution of the atmospheric boundary layer, therefore depends on accurate specification of soil moisture conditions, and ultimately soil hydraulic properties. Ek and Cuenca (1994) demonstrated the serious impact of the variation of soil hydraulic properties, originally based on soil texture, on the simulation of the daytime atmospheric boundary layer (ABL). Cuenca et al. (1996) followed to show the surprising impact and possibly systematic error caused by application of different soil hydraulic property parameterizations in surface energy balance and ABL simulations. Betts et al. (1996, 1998a, 1998b) have indicated the serious effects of errors in land surface parameterization in global forecast and climate models. Systematic biases resulting from these errors limit the accuracy of medium and seasonal forecasts and the ability to correctly model the present-day climate (Betts et al., 1999). Improvements in the representation of land surface boundary conditions have led to clear improvements in the accuracy of mesoscale and global forecast models (Beljaars et al., 1996). For these reasons ground-based soil moisture and soil hydraulic property assessment have been incorporated into numerous large-scale hydrologic and atmospheric experiments.

EVOLUTION OF INSTRUMENTATION AND APPLICATIONS - CASE STUDIES

The experiments described in this section are those for which the authors have personal experience, normally through tens or hundreds of hours of field work installing and monitoring sampling sites. The experiments cover the evolution of large-scale hydrologic-atmospheric-remote sensing experiments from HAPEX-MOBILHY (France, 1985-86) to CASES (Cooperative Atmosphere-Surface Exchange Study) in Kansas which is ongoing at the time of writing (André et al., 1988; LeMone et al., 2000). Other experiments, notably FIFE [First ISLSCP (International Satellite Land Surface Climatology Project) Field Experiment] (Sellers et al., 1992) and EFEDA [ECHIVAL (European International Project on Climate and Hydrological Interactions between Vegetation, Atmosphere and Land Surfaces) Field Experiment in a Desertification-threatened Area] (Bolle et al., 1993), also occurred during this time period but for the most part used technology covered in the other experiments described in this section. Results from these other experiments will be mentioned, but not to the detail of those experiments with which the authors are personally familiar.

HAPEX-MOBILHY

HAPEX-MOBILHY (in spite of the title of the FIFE experiment which was conducted one year later) was the first large scale hydrologic-atmospheric-remote sensing experiment. It was conducted in the Landes region of France and headquartered at the Centre National des Réchérches Météorologiques (CNRM) in Toulouse, France (André et al., 1988). The project area of the experiment encompassed a region of 100 km by 100 km which included 12 energy balance sites, 2 of which had multiple atmospheric flux measuring systems (Central Site at Lubbon and Southern Site at Castelnau). All energy balance sites included soil moisture monitoring measurements plus there was one additional soil monitoring site not equipped with an energy balance station.

In many respects HAPEX-MOBILHY represented an experiment in how to conduct large-scale experiments. The justification for the experiment, as well as its leadership, came from the atmospheric sciences. Considering the atmospheric science orientation of the experiment, credit has to be given to experiment leaders (J.-C. André) for incorporating the soil moisture component. The standard soil moisture monitoring instrumentation at the time was the neutron probe and the French-designed Sodar probe was used at all sites. Other experimental probes based on application of the dielectric properties of the soil water matrix were tested in HAPEX-MOBILHY. Complications involving inadequate contact between the sensor and soil precluded satisfactory results with these probes.

The governing equation for soil moisture content as a function of the neutron probe measurement is given as,

$$\theta_v = a_1 + b_1\left(\frac{C_{NM}}{C_{Std}}\right) \tag{8}$$

where θ_v = volumetric soil water content (percent)
a_1 = intercept constant
b_1 = slope constant
C_{NM} = neutron meter count
C_{Std} = standard count

The standard count is normally taken with the neutron probe on top of the transport case or inserted into an access tube that is placed in a water drum. The intercept and slope constants are determined by regression analysis on soil samples for which θ_v is determined using standard gravimetric sampling methods. The bulk density is required to convert the gravimetric soil water content to a volume basis. Therefore a separate calibration equation is required for each soil layer in which there is a change in bulk density (see Cuenca and Noilhan, 1991 for examples). Experience has shown that the intercept constant in Eq. (8) is more prone to error than the slope constant (Vandervaere et al., 1994). Therefore accuracy is improved when neutron meter data are used to compute the change in soil water content (which cancels the intercept constant), as is required in water balance calculations over the soil profile (Cuenca, 1988; Carrijo and Cuenca, 1992).

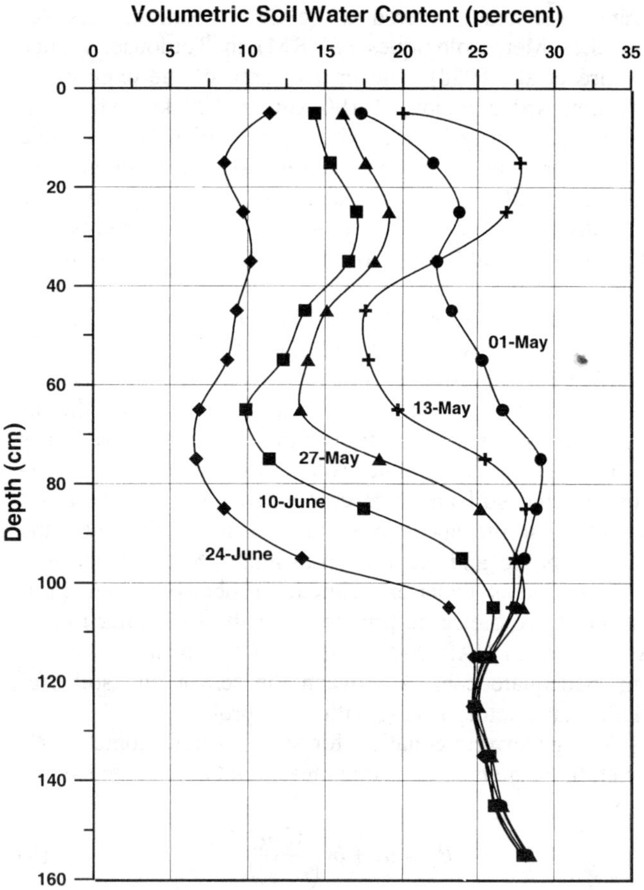

Figure 1. Volumetric soil water content as a function of depth at the Central Site (Lubbon, France) for oat vegetative cover for five dates during HAPEX-MOBILHY. Note response to rainfall event proceeding 13 May and drying trend indicated for all other dates.

The radius of influence of soil water content as measured by the neutron probe is given by (Burman et al., 1983),

$$R_i = (15.0)\sqrt[3]{\frac{100}{\theta_v}} \tag{9}$$

where R_i = radius of influence (cm). Measurements made with the neutron probe at depths less than 20 to 40 cm (i.e. less than the radius of influence), depending on soil moisture content, require adjustment to account for the escape of neutrons close to the surface. This adjustment is normally made using a nonlinear calibration equation for this layer (see Cuenca et al., 1997 for an example) or using the iterative adjustment technique of Parkes and Siam (1979).

The protocol for soil moisture measurements was similar to that applied for monitoring by neutron probe in crop water use experiments at the time, with some slight modifications. Representative sites where monitored at each location using a total of four aluminum access tubes with measurements made every 10 cm down the profile starting at 5 cm from the surface. In those sites with row crops, two tubes were placed in the crop row and two tubes were placed in middle of the furrow. Sites were monitored on a weekly basis. No concerted effort was made to incorporate soil water tension, soil water characteristic, or soil hydraulic conductivity data into the data set. Some follow-up work to characterize soil water retention and hydraulic conductivity functions at each site was performed the year following the HAPEX-MOBILHY field campaign. This was not as effective as it would have been to make soil water tension and other *in situ* measurements during the time of the experiment when soil water content was monitored.

Nonetheless, soil moisture data from HAPEX-MOBILHY were useful and have been applied to the purposes for which they were collected. Sample data shown in Fig. 1 indicate the change in profile soil water content over time at the Lubbon Central Site. Profile data to 155 cm depth are indicated approximately every 14 days with volumetric soil moisture content ranging from a high of about 28 to a low of 7 percent. There are a number of notable features in this figure that are somewhat typical of neutron probe data from other experiments. One is the unambiguous stability of the neutron probe data demonstrated by overlapping values of soil moisture content at depths below 120 cm. This is a typical trace of neutron probe data that show no change in soil moisture content. The profile for 01 May represents generally moist conditions while the 13 May profile indicates soil water uptake by roots below 35 cm depth and re-wetting due to rainfall above 35 cm depth. The traces for 27 May, 10 and 24 June indicate a classical depletion of soil water content by roots which increases in depth with time. (Later data not shown indicate a maximum depletion of soil water content to approximately 5 percent at 75 cm depth.)

Data like those indicated in Fig. 1 were used to integrate the change in soil moisture content over time which was used to compute evapotranspiration using water balance techniques. These results were compared with evapotranspiration measured by the energy balance stations at various sites (Cuenca and Noilhan, 1991). The soil water data were also invaluable in defining the time of decline of the water table at the Central Site at Lubbon (later data not shown in Fig. 1). This site housed a weighing lysimeter which was a unique instrument for the experiment. The water table level affected the lysimeter operation and its decline was not anticipated.

HAPEX-Sahel

HAPEX-Sahel, conducted in Niger in 1992, was a large-scale hydrologic-atmospheric-remote sensing experiment carried out in an arid tropical zone (Goutorbe et al., 1994). It was conducted in the Sahel region of Africa which is the zone of sharp rainfall gradient extending south from the Sahara Desert to more tropical conditions. It followed FIFE conducted in Kansas in 1987 and EFEDA conducted in Spain in 1991. Although the objectives of HAPEX-Sahel were similar to those of HAPEX-MOBILHY and involved some of the same teams, additional emphasis was placed on obtaining remote sensing imagery and ground-based measurements that could be related to remotely sensed parameters. The soil moisture - soil hydraulic property program in HAPEX-Sahel involved eight teams from African, European and American research laboratories and was by far the most intensive effort in this subject area coupled to a large-scale experiment up to that time. HAPEX-Sahel offered the opportunity to work in an experiment with flux measurements over various vegetative canopies, aircraft- and satellite-borne sensors, and ground-based support measurements coupled with difficult logistics and working conditions.

The HAPEX-Sahel project covered an area of approximately 100 km by 100 km within which were three so-called supersites that were intensively instrumented. The entire area was covered by a dense network of precipitation gauges that were analyzed in conjunction with a precipitation radar at Niamey as part of the EPSAT-Niger (Estimation des Pluies par Satellite, expérience Niger) project (Lebel et al., 1997; Lebel and Le Barbé, 1997). The neutron probe was employed as the standard soil moisture measurement instrument, much as was done in HAPEX-MOBILHY, FIFE and EFEDA. However, much more attention was applied to *in situ* measurements of soil water tension and soil hydraulic properties using tensiometers, tension infiltrometers and soil core techniques. In addition, HAPEX-Sahel saw the deployment of instrumentation that had basically been used on an experimental basis in previous experiments.

A full description of soil moisture and soil hydraulic property instrumentation deployed at each supersite is given in Cuenca et al. (1997a). Initial sampling over a fixed grid using either gravimetric methods or neutron probe was carried out at two of the supersites to determine approximately 10 long-term monitoring sites per vegetative cover for each supersite. The 10 sites were selected to cover the range of dry to wet soil moisture conditions within each vegetative cover following the experience of Vachaud et al. (1985). Each supersite included three to four vegetative covers, i.e. millet, so-called tiger bush, grassland, bush and fallow, which were individually monitored for energy balance and soil moisture conditions (Goutorbe et al. 1994).

The protocol for monitoring using the neutron probe was basically as described for HAPEX-MOBILHY except that measurements were more frequent, up to three times per week or every other day, particularly during the rainy period and dry-down immediately following rainfall events. Figure 2 depicts the soil moisture integrated over a profile depth of 340 cm as a function of time for four non-irrigated vegetative covers. The intensive observation period (IOP) of the experiment was designed to capture the effects of increased vegetation growth, and evapotranspiration, due to heavy seasonal rainfall and the subsequent dry-down with concurrent declines in soil moisture and evapotranspiration at the end of the rainy season. Figure 2 is an intensive data set from the Central Site East for the millet (sandy skirt) and tiger bush sites. Similar to other graphics of wet and dry sites (Cuenca et al., 1997a), the wet and dry millet sites are shown to parallel each other over the period of record. This consistency between the wet and dry sites is in agreement with the work of Vachaud et al. (1985) mentioned earlier. The period of record, which both precedes and runs past the IOP, is seen to capture the large seasonal shift in soil moisture content. It should be emphasized that the millet crop in this region is hand-sown, with the sowing date dependent upon rainfall and the farmer's experience. The resulting vegetative canopy is uneven and sparse by most production agriculture standards.

Tiger bush is a unique vegetation cover found in this region of the Sahel. It is made up of serpentine vegetation with alternating bare soil strips. The vegetation bands are 10 to 30 m wide and 100 to 300 m long. They are interspersed with bare soil bands that are 30 to 100 m wide. The bare soil bands are virtually cemented soil layers in which there is extremely limited infiltration. Consequently one expects very limited changes in soil moisture content in the bare soil sites. The runoff from the bare soil sites causes the vegetation strips to see in excess of 100 percent infiltration of rainfall. These conditions are clearly shown in Fig. 2 in which there is no perceptible change in soil moisture content in the bare soil and an enormous range of soil moisture conditions in the vegetated section.

HAPEX-Sahel saw one of the earlier, successful applications of time domain reflectometry (TDR) instrumentation. The dielectric constant of a media is the ratio of the velocity with which an electromagnetic pulse will travel along a waveguide in the media to the speed through a vacuum (speed of light). A pulse propagating along a waveguide surrounded by air is characterized by a dielectric constant of 1, while the same waveguide surrounded by water will propagate approximately 78 to 80 times slower because the dielectric constant of water is much higher (78 to 80). The dielectric constant of dry soil is approximately 4 to 5 and is generally independent of the bulk soil ionic conductivity that affects the amplitude of the signal but not the propagation time. TDR methods depend on changes in the dielectric constant of the soil water matrix which therefore varies from about 4 to 5 for dry soil to approximately 80 in saturated conditions. The intrinsic dielectric constant of soils in the dry state varies due to different parent materials and bulk density. The dielectric constant of the soil is can be modeled as a weighted summation of the dielectric constants of the soil constituents (air, water, mineral) (Roth et al., 1990). Subsequent development of TDR instrumentation has allowed for the device to be used to measure surface layer soil moisture content, i.e. from 0 to 15 cm depth, down to depths of 1 or 2 m with the installed TDR rods segmented into five or more depth layers. (See additional description for BOREAS and CASES experiments.)

The Trase TDR instrument was used for surface-layer soil moisture measurement at the Central Site East. Calibration was carried out using a gamma probe for soil moisture content on a mass basis and bulk density in the following two-step calibration for volumetric soil water content,

$$\theta_c = 0.013 + 0.614(\theta_T) \tag{10}$$

where θ_c = calibrated volumetric soil water content (percent)

θ_T = volumetric soil water content given by Topp et al. (1980)

$$\varepsilon = 3.03 + 9.3(\theta_T) + 146(\theta_T)^2 - 76.3(\theta_T)^3 \tag{11}$$

where ε = dielectric constant.

Equation (11) is one of the earlier equations for extracting the volumetric soil water content from the dielectric constant based on the development work of Topp et al. (1980) for TDR devices. Additional calibration equations for other TDR devices used in this experiment are found in Cuenca et al. (1997a).

HAPEX-Sahel also saw deployment of a capacitance probe, the so-called surface capacitance insertion probe (SCIP) developed by the Institute of Hydrology (Dean et al., 1987). The SCIP was calibrated using gravimetric analysis of soil cores in the procedure described by Robinson and Dean (1993). The final calibration equation was of the form,

$$\theta_v = \frac{a_2}{(SCIP_c)^2} + b_2 \tag{12}$$

where $SCIP_c$ = capacitance probe reading

a_2, b_2 = regression constants

Sample data taken using this probe is shown in Cuenca et al., (1997a). The calibration requirements for the capacitance probe for each new soil type have limited its application at this time.

A determined effort to incorporate measurements of soil hydraulic properties, including soil water retention and unsaturated hydraulic conductivity functions, was made in HAPEX-Sahel at each supersite. Work was led by personnel from the Institute of Hydrology (United Kingdom – Southern Supersite), Agricultural University of Wageningen (The Netherlands – Central Site West), and Laboratoire d'etude des Transferts en Hydrologie et Environnement (LTHE) (France – Central Site East). A total of on the order of 25 personnel from these and other institutions were involved in soils related field measurements making this probably the largest such effort in a large-scale experiment. Soil water tension measurements made using either mercury (Hg) or hypodermic (Tensimeter) tensiometers were generally coordinated with soil moisture monitoring sites and depths. The combined measurements allowed for determination of *in situ* soil water characteristic curves over the low range of tensions over which tensiometers are valid (i.e. up to about 800 cm tension). In addition to the tensiometer measurements, widespread measurements were made to determine the unsaturated hydraulic conductivity function through infiltration experiments using tension infiltrometers. (See additional description for the BOREAS experiment.).

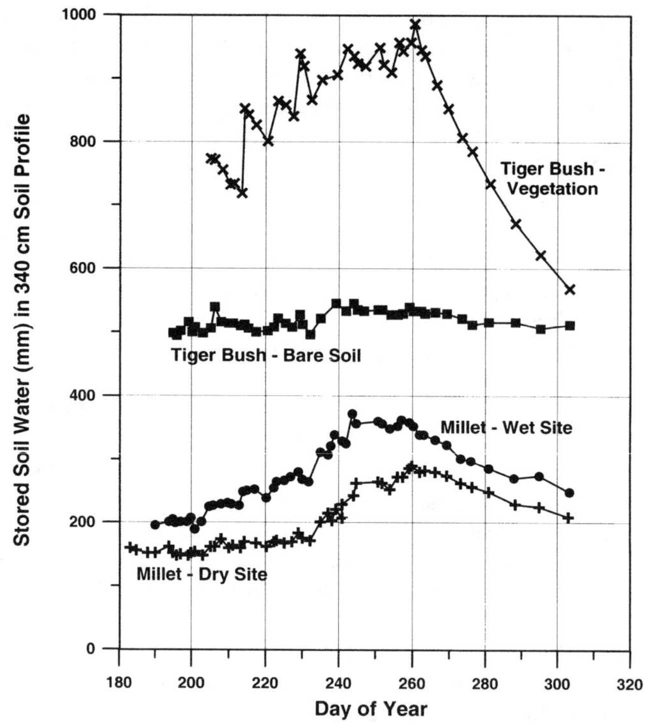

Figure 2. Variation of stored soil water content versus time over a soil profile of 340 cm for four vegetative conditions at the Central Site East in HAPEX-Sahel. Note relative consistency of wet and dry millet sites and extreme contrast in vegetation strip and bare soil in tiger bush. Data compliments of S. Galle, currently at Laboratoire d'etude des Transferts en Hydrologie et Environnement (LTHE), Grenoble, France.

BOREAS

Soil Water Content

The Boreal Ecosystem-Atmospheric Study (BOREAS) experiment was conducted in Saskatchewan and Manitoba provinces of Canada in 1994 with follow-up measurements made in 1996 (Sellers et al., 1995). The boreal ecosystem is a forested system made up predominantly of black spruce, alder, jack pine and aspen species with various moss and lichen ground covers. A significant proportion of the landscape is made up of fens which are low-lying areas, saturated during the wet season, which support tremendous vegetation growth. This is a massive ecosystem found in northern latitudes in Canada, Europe and the former Soviet Union.

Budgetary constraints limited the amount of personnel that could be committed to soil moisture monitoring and measurement of soil hydraulic properties vis-à-vis the HAPEX-Sahel experiment. A limited number of representative sites were therefore monitored for soil moisture content within each vegetation canopy. This typically numbered five to six sites that were selected along a radial line from the site flux tower so as to avoid any bias in sampling (see Cuenca et al., 1997b for examples). Sampling was done on an alternate day basis in the 1994 Intensive Observation Periods (IOP) using the neutron probe with a protocol similar to that applied in HAPEX-Sahel. Towards the end of the last IOP, a few selected sites were monitored using manual TDR instrumentation with segmented rods. All of the data sets presented in Cuenca et al. (1997b) were taken using the neutron probe. These data sets indicate the stability and responsiveness of neutron probe measurements to changes in soil moisture content as well as the means for two-dimensional representation of soil moisture distribution along a transect.

Following the 1994 IOP's, it was decided to monitor some sites for longer-term, seasonal changes in soil moisture content. This work was performed by the Hydrologic Science Team (HST) of Oregon State University (OSU). Two sites monitored were the old black spruce (OBS) in the Northern Study Area (NSA) and the old aspen (OA) site in the southern study area (SSA). These sites were monitored with transects comprised of 8 segmented TDR rods yielding soil moisture content from 0 to 15 cm, 15 to 30 cm, 30 to 60 cm, 60 to 90 cm, and 90 to 120 cm depth layers. The TDR meter, or reflectometer, used in the BOREAS experiment was developed by MoisturePoint (MP-917). It allows utilization of segmented probes to obtain a soil water profile from a single probe, with minimal disturbance of the surrounding soil. Automated electronic equipment to query the TDR transect along with one or two tipping bucket raingauges and micro-logger gave nearly continuous recordings of soil moisture content and the response of the soil profile to precipitation.

Two other sites were monitored using CS615 (Campbell Scientific, Inc.) reflectometer probes installed horizontally in a profile at 15, 30, 60 and 90 cm depths. The sites instrumented were the young jack pine (YJP) and old jack pine (OJP), both in the NSA. A tipping bucket raingauge was installed next to each profile.

The water content using the CS615 is derived from the effect of a changing dielectric constant on the propagation velocity of electromagnetic waves along a waveguide. This sensor is similar to TDR except it provides a frequency of output instead of calculating the actual round trip travel time of the electromagnetic pulse. The frequency of period output signal of the CS615 is related to the volumetric water content through an empirically derived third degree polynomial typical of most mineral soils. The calibration equation can be optimized for each particular soil through the last term in the third degree polynomial (zero order term). The soil volume measured is a cylinder approximately 30 cm long with a radius of 6 to 7 cm surrounding the waveguide weighted toward the regions closest to the stainless steel probes. Typically, these probes provide a stable soil moisture value and multiple readings to obtain an average moisture content are not required in contrast to the typical TDR measurement.

Figure 3 indicates sample data from the two sites monitored using the CS615 probes. These sites were monitored on a 15 min frequency resulting in an almost continuous record of soil moisture content and response to precipitation. Figure 3a indicates the response at the YJP site that has trees on the order of 2 m tall with limited root development. The 15 cm depth layer is seen to respond most rapidly to precipitation events shown by the bar graph at the bottom of the plot, with deeper layers responding more gradually. There is little variation in soil moisture content in the 90 cm depth layer until the rainfall event on DOY 198 which penetrates all depth layers. Following this rain, there is a long, relatively dry period with gradual soil moisture depletion by drainage and root extraction until just before the end of the record shown on DOY 220.

Figure 3b shows the contrasting data plot for the OJP site made up of mature trees on the order of 10 m tall with a well-developed root system. There is a similar response in the two shallow-most layers to rainfall at this site as seen at the YJP site. However the two deeper layers at 60 and 90 cm show little or no change in soil moisture content until the relatively large rainfall event just preceeding the end of record on DOY 220. The assumption is that the more developed root zone at the OJP site was able to intercept the rainfall and eventually use it in the transpiration process before the wetting front could penetrate the deeper layers. These two sites, separated by some 30 km and sub-

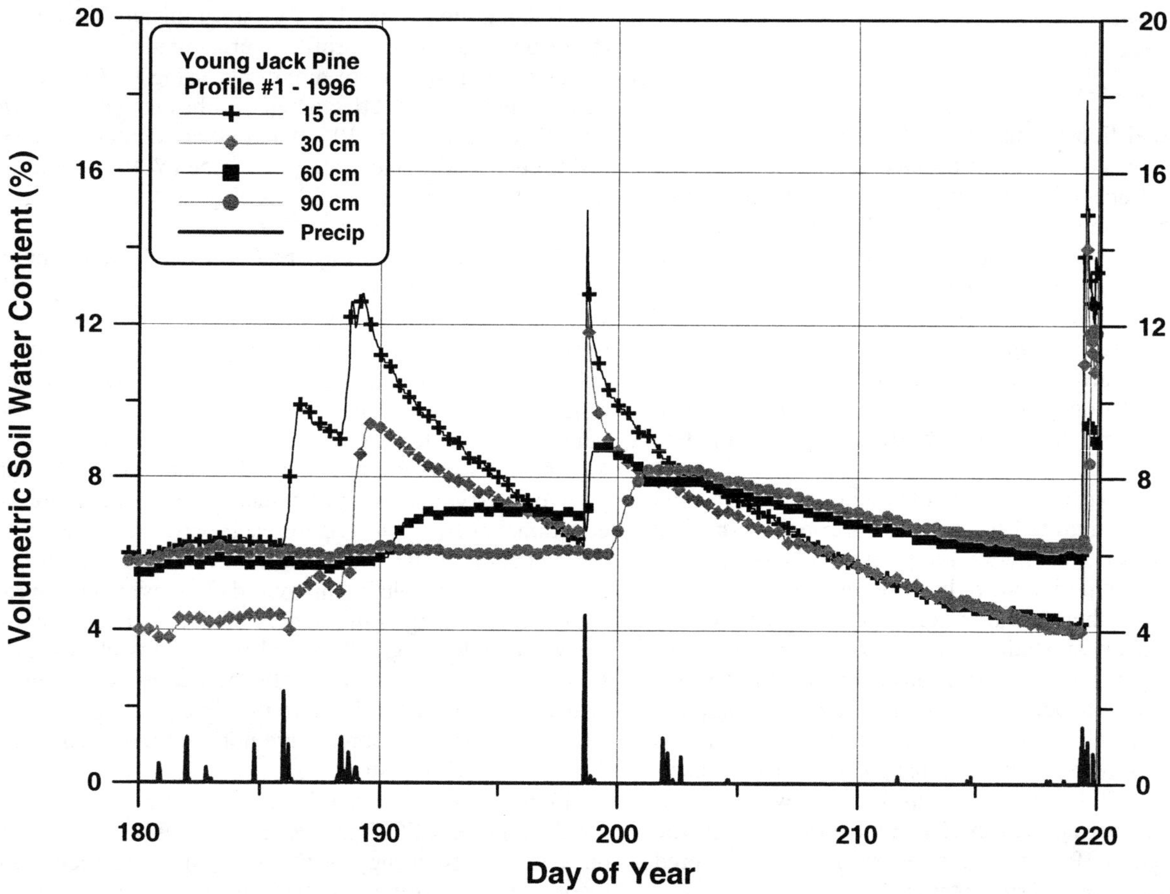

Figure 3a. Soil moisture content taken using CS615 probes for four depth layers and precipitation records from BOREAS, Northern Study Area, Young Jack Pine site in 1996.

ject to similar rainfall conditions, are seen to have quite different responses in various layers of the soil profile to similar rainfall events.

Soil Hydraulic Properties

As previously indicated, analysis for unsaturated hydraulic conductivity was carried out in HAPEX-Sahel using the tension infiltrometer. This work was continued in BOREAS but using a newly designed infiltrometer system that incorporated pressure transducers and a micrologger for automated data collection. Some 230 infiltration tests were run at a total of eight flux tower sites.

Numerous publications describe the operating principle of the tension infiltrometer. The most recent review article describing the theory and application of the tension infiltrometer is that by Angulo-Jaramillo et al. (2000) which should be referred to for details. Application of this method in BOREAS involved use of Wooding's equation (Wooding, 1968) which intrinsically assumes the validity of Gardner's (1958) relationship for hydraulic conductivity as an exponential function of soil water tension which is given as,

$$K(h) = K_{sat} \exp(\alpha h) \tag{13}$$

where K_{sat} = saturated hydraulic conductivity
α = fitting parameter

Under steady-state conditions for infiltration across the membrane of a single disk in non-deformable soils, Wooding's (1968) relationship is given as,

$$Q = \pi r_0^2 K(h)\left[1 + \frac{4}{\pi \alpha r_0}\right] \tag{14}$$

where r_0 = infiltration disk radius

Substituting the Gardner expression into Eq. (14) for two

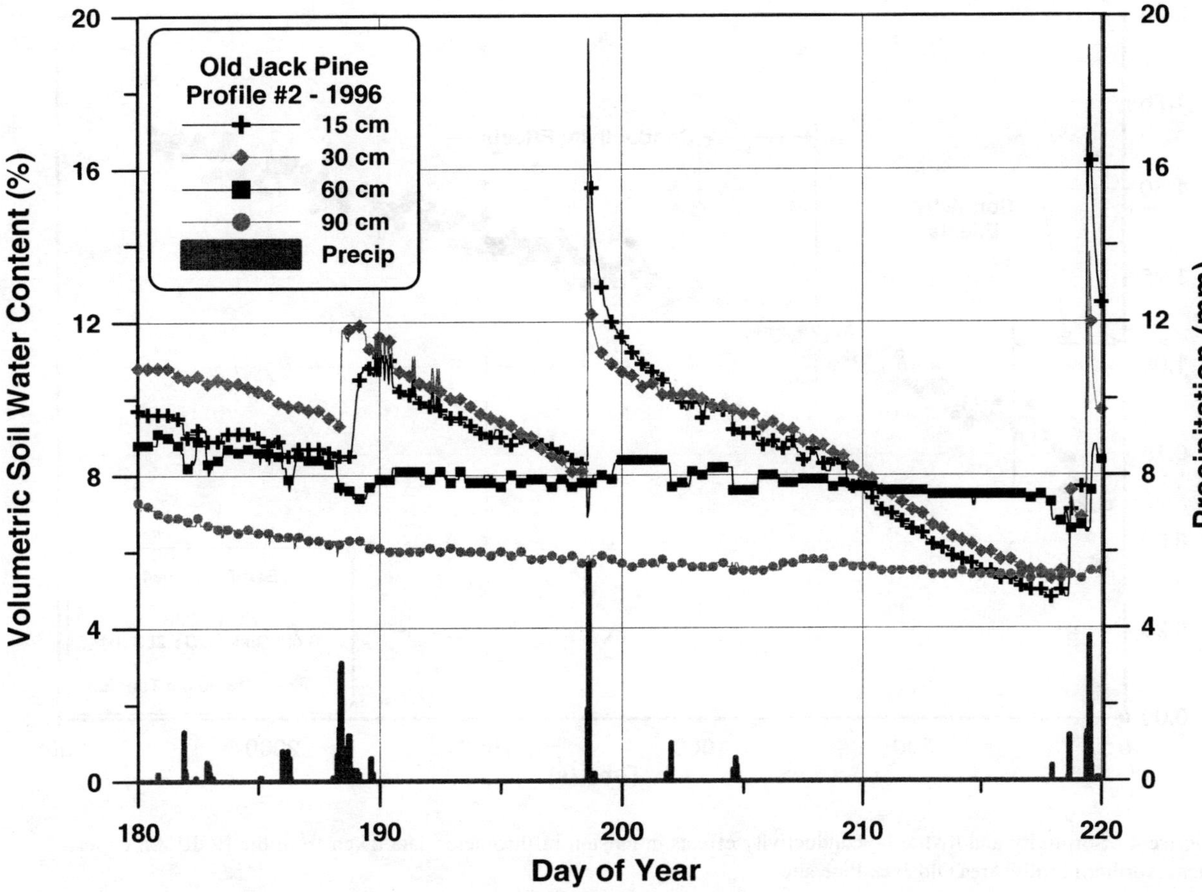

Figure 3b. Soil moisture content taken using CS615 probe for four depth layers and precipitation records from BOREAS, Northern Study Area, Old Jack Pine site in 1996.

steady-state flow rates, Q_1 and Q_2, at tensions h_1 and h_2, respectively, we have,

$$Q_1 = \pi\, r_0^2\, K_{sat} \exp\left(\alpha\, h_1\right)\left[1+\frac{4}{\pi\, \alpha\, r_0}\right] \tag{15}$$

$$Q_2 = \pi\, r_0^2\, K_{sat} \exp\left(\alpha\, h_2\right)\left[1+\frac{4}{\pi\, \alpha\, r_0}\right] \tag{16}$$

which yields two equations for the two unknowns K_{sat} and α. Rearranging Eqs. (15) and (16) and taking the natural logarithm of both sides yields the following expression for α,

$$\alpha = \frac{\ln\left(Q_2/Q_1\right)}{h_2 - h_1} \tag{17}$$

K_{sat} is then solved for using,

$$K_{sat} = \frac{Q}{\pi\, r_0^2}\, \frac{1}{\exp\left(\alpha\, h\right)}\, \frac{1}{\left[1+\frac{4}{\pi\, \alpha\, r_0}\right]} \tag{18}$$

Figure 4 indicates the data plot for a relatively high tension infiltrometer test in which the data scatter is basically due to the sensitivity of the pressure transducers employed. The initial curvilinear data demonstrate the sorptivity effects of the soil during which time the water is wetting the soil particles. The magnitude of sorptivity is a function of initial and boundary conditions, particularly the initial soil water content (Angulo_Jaramillo et la., 2000; Warrick and Broadbridge, 1992; Haverkamp et al., 1994). Later in the test the data demonstrate steady-state conditions required for application of the Wooding (1968) equation.

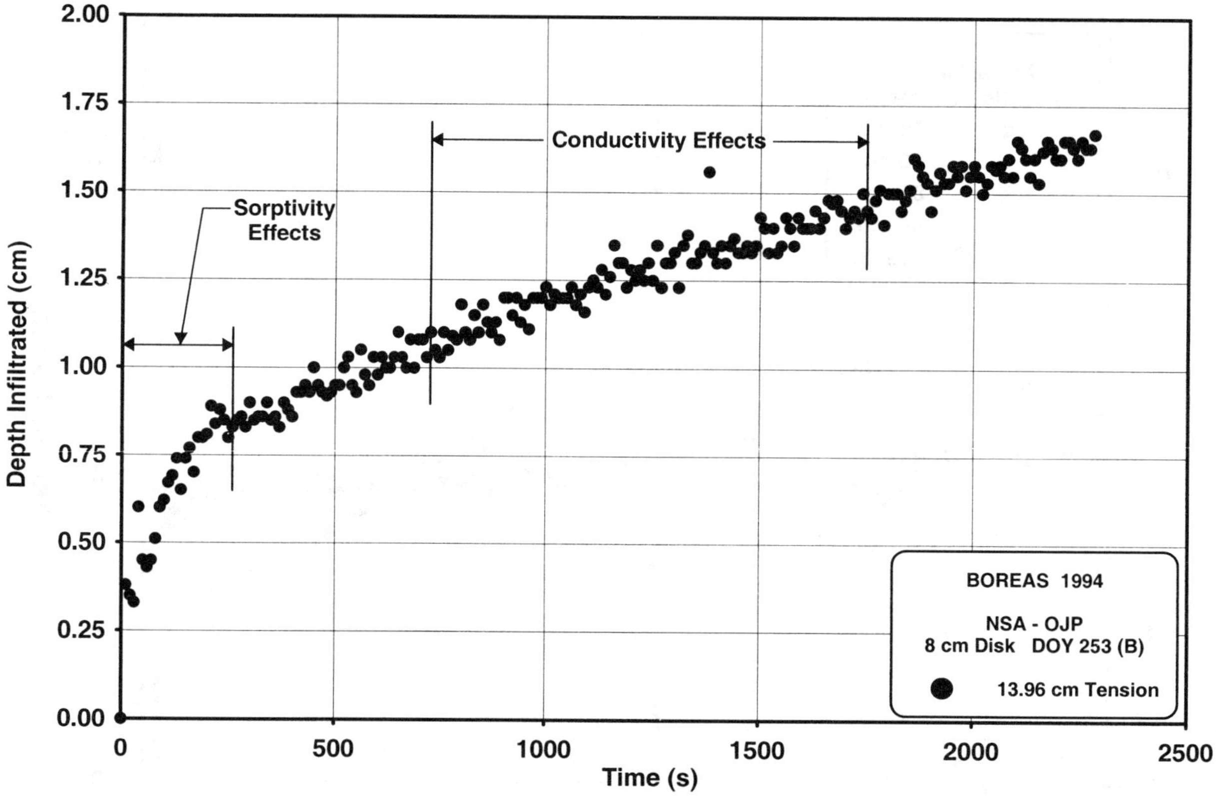

Figure 4. Sorptivity and hydraulic conductivity effects in tension infiltrometer data taken from the BOREAS experiment Northern Study Area Old Jack Pine site.

Multiple tests are required to solve this system of equations and Fig. 5 indicates the results of tests run at the same site at three different tensions. The 13.69 cm plot is of the same data indicated in Fig. 4 except every other data point has been removed for clarity, which was also done for plots of the two other tensions. Applying the above equations to the data plotted yields a saturated hydraulic conductivity for this coarse sand soil of 2.625 cm/h and an α of 0.285. Analyses such as these were conducted for all of the BOREAS sites and, combined with other data, were the basis for determination of the soil water retention and hydraulic conductivity functions published for the BOREAS flux tower sites (Cuenca et al., 1997b).

CASES

CASES (Cooperative Atmospheric Surface Exchange Study) is a multi-year, interdisciplinary effort to investigate linkages among the atmosphere, hydrosphere and terrestrial biosphere. The complex interactions among these components of the earth's ecosystem manifest themselves in the fields of meteorology, hydrology, climate, ecology and chemistry, and challenge current capabilities to understand, simulate and predict many aspects of our environment on time scales from minutes to years (LeMone et al., 2000). In support of this effort, a network of automated (TDR) sensors for soil water and thermistors for soil temperature was installed at the ABLE (Argonne Boundary Layer Experiment) Whitewater site in the Whitewater River sub-basin (approximately 30 by 35 km) of the Walnut River watershed in Kansas. The Whitewater site is the location of numerous other instruments operated by ABLE. ABLE is a research initiative devoted to atmospheric research and developed by the Atmospheric Section of Argonne National Laboratory (Wesely et al., 1997). Currently, ABLE is maintaining a 915 MHz radar wind profiler, a doppler minisodar, automatic weather station and an Energy Balance Bowen Ratio (EBBR) instrument system at this site as well as at several other sites located in the Walnut river watershed (LeMone et al., 2000).

Description of instrumentation and theory of operation.

Soil water content and soil temperature sensors record values in a profile to a total depth of 1.35 m comprised of three layers from the surface which are 15 cm each and three additional layers which are 30 cm each. (See LeMone et al., 2000 for a schematic of the instrument installation.)

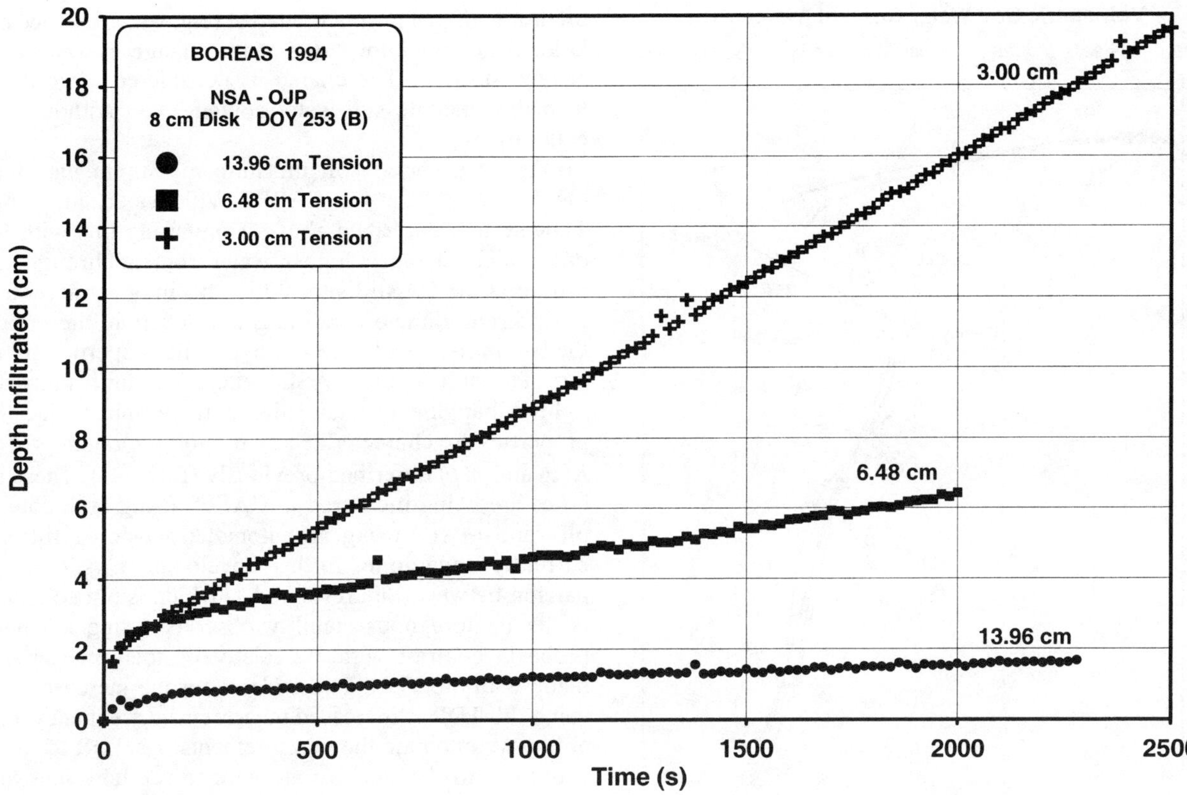

Figure 5. Tension infiltrometer data for three tensions taken from the BOREAS experiment Northern Study Area Old Jack Pine site.

This configuration was based on a compromise between instrumentation costs and the objective of more precise monitoring of relatively rapid changes near the soil surface. The monitoring network covers a field approximately 4 ha in size, large enough to be observed by remote sensing satellites, with 15 instrument pads distributed in the shape of a cross (LeMone et al., 2000). The automated TDR instrumentation is a composite system incorporating 5-segment, 1.2 meter long profiling probes and datalogging equipment from Campbell Scientific Inc. (Logan, UT). From past experience during the BOREAS experiment, we have found that four replicate readings of soil moisture every hour is an adequate sampling frequency to capture the time varying nature of field soil moisture conditions. Measurements can be made more intensely to capture rapid infiltration events if needed. Remote access to the automated system is provided through a fiber optic communication link provided by ABLE for routine data retrieval and near real time site monitoring. Surface soil moisture (0-15 cm depth) is measured using 30 cm long reflectometer probes (CS615) inserted horizontally into the soil at 7.5 cm depth. Temperature profiles from the soil surface to the ground corresponding to the mean soil moisture measurement depth are measured using thermistors attached to a fiberglass pole at 8 instrument pads and three infrared thermometers (Everest Interscience, Inc. Model 4000.4ZL) located at the central datalogger site (LeMone et al., 2000).

This system used a CR10 (Campbell Scientific, Inc.) datalogger to control external multiplexing switches and record the data. This custom built system is operated by sequentially measuring the fifteen, 5-segment probes switched individually to the MP-917 TDR meter under control of a program stored in the datalogger. The network of 15 probes is currently interrogated at 1-hour intervals. Each probe is measured four times sequentially and averaged to obtain the average volumetric soil moisture. The 5-segment probe effectively measures the moisture content within a cylinder of radius of approximately 4 cm from the center of the probe in the segment being measured. The measurement is weighted toward the regions closest to the sides of the probe.

Soil Moisture Measurement in CASES

The automated TDR instrumentation was challenging to design and install at the CASES site because of the size of the measurement area, the requirement to bury the sensors, and the clay loam soil texture. The instrument pad

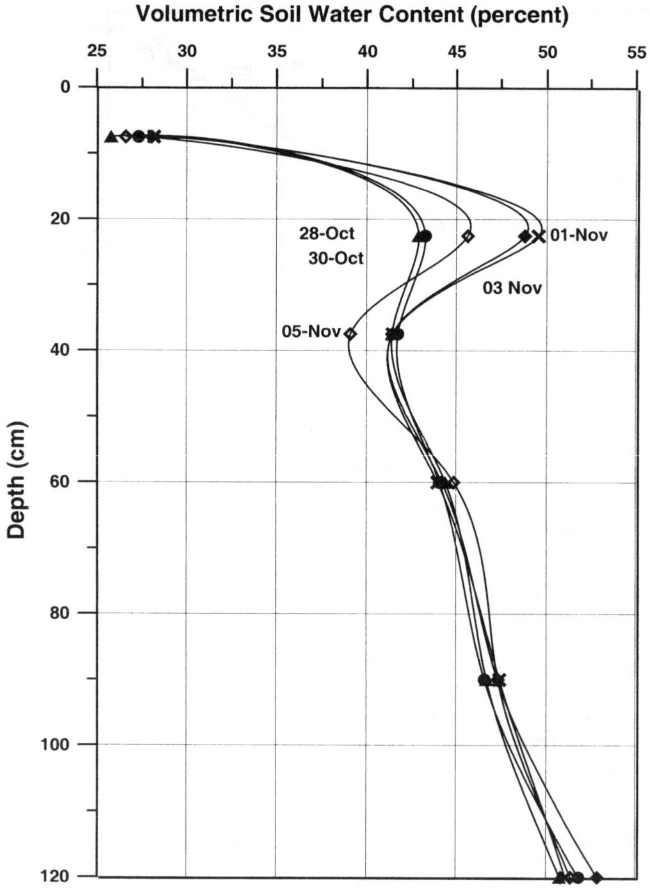

Figure 6. Volumetric soil water content as a function of depth at the CASES Whitewater site (Kansas, USA) for grass vegetative cover for five dates for a segmented TDR probe plotted as a daily average of measurements recorded at 30 minute intervals.

was designed so that all sensors and cables could be buried so that normal field operations (cutting and baling of hay) could be conducted, except at the central instrument box which houses the TDR meter and datalogger. In locating a suitable site for the automated soil moisture system we needed to take into account multiple criteria, such as power, the need for a long term site that was not going to be tilled, and suitable soil type without too many cobbles and rocks. The Whitewater site met all these criteria but was still a challenge for soil moisture observations because of the heavy soil texture.

In general, TDR soil moisture measurements are relatively independent of soil texture, but problems do occur in when making measurements in soils with an appreciable amount of either clay, organic matter or high salinity. From our experience it has been found that to confidently obtain an accurate absolute soil moisture better than about +/- 5%, it is necessary to carry out multiple calibrations by depth and/or probe location. While this may sound discouraging, often it is of just as much interest (e.g. water balance calculations) to have knowledge of the change in water content at one location. The change is considered more accurate than the absolute soil water content even without tedious calibration.

Figure 6 shows soil moisture measurements made in the fall of 1998 at the CASES Whitewater site using the TDR system described above. Comparing this with Fig. 1 one notices a much higher water content throughout the profile at the CASES site. While this may certainly be due to different climate conditions and location, the effects of the soil moisture and the soil hydraulic properties are striking. The soil at the CASES site holds onto more water tightly than the coarser soils at the Lubbon site. These properties are characterized by the soil hydraulic properties $K(\theta)$ and $h(\theta)$ described previously (Eqs. 1-7). These properties are being measured in CASES using both core samples and *in situ* using an automated tension infiltrometer currently undergoing further development. Another comparison between Figures 6 and 1, which is not so apparent, is the unambiguous stability observed using the neutron probe in contrast with the relatively noisy measurements made with the TDR. The tradeoff for the increase in noise using the TDR is increased measurement frequency and the ability to automate the measurements. As TDR technology continues to develop we are sure to see new innovations that will allow relatively undisturbed soil moisture profiles to be measured with higher accuracy and repeatability.

Figure 7 shows soil temperature measurements collocated with the soil moisture measurements. By combining the soil temperature and soil moisture data it is possible to obtain independent measurements of the soil heat flux and latent heat flux over the 4 ha measurement area. These

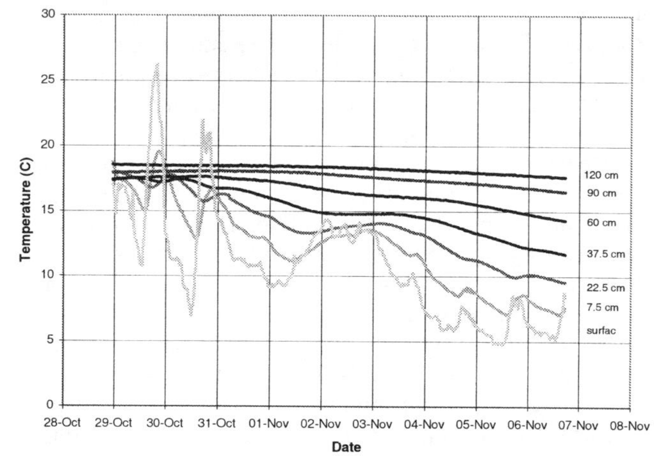

Figure 7. Soil temperatures as a corresponding to soil measurement depths at the CASES Whitewater site (Kansas, USA) for grass vegetative cover.

estimates will be compared to values obtained using other energy techniques such as the EBBR currently being operated by ABLE. CASES is different than the previous regional scale experiments in that it will provide a unique opportunity to observe and model inter-seasonal and inter-annual land surface hydrologic processes because it is a long-term, continuous experiment.

CONCLUSIONS AND LOOKING FORWARD

Standards for Automated, High Frequency, Long-Term Data Collection

We have seen the evolution of profile soil moisture measurements in large-scale experiments go from daily or weekly data collection to 15 min frequencies by application of current sensor technology and micro-electronics. These changes are due to developments in both sensor technology and data storage or transmission capabilities. The high frequency data systems can now be used to capture processes in large-scale experiments, such as movement of a wetting front through the soil profile, which could only be assumed or measured on extremely small-scale experiments a decade earlier. These new measurement systems come with additional challenges in calibration techniques and also open the question of optimum spatial sampling scales.

The HST instrumentation deployed in the CASES experiment is one of the better examples of the type of long-term, high frequency, spatially distributed systems that can be developed. The spatial distribution of the sensor network is adequate to be observed from satellite platforms while the almost continuous frequency of data collection is sufficient to capture the passage of wetting fronts due to infiltration of rainfall. These measurements coupled with monitoring of the diurnal energy balance and atmospheric profiling can finally bridge the continuum in time of the soil-water-atmosphere-plant environment.

Application of Various Soils Data Bases

The parameterization of soil hydraulic properties has also been shown to be of paramount importance in simulation of the diurnal energy balance and atmospheric boundary layer. The work of Fuentes et al. (1992) and others has indicated the importance of application of consistent and theoretically valid functions for soil water retention and unsaturated hydraulic conductivity that go beyond application of soil texture. A number of national and international data bases have evolved to fill this gap and characterize the soil hydraulic properties as a function of soil texture, soil bulk density and increasingly sophisticated properties as the data bases continue development. Among these are the UNSODA database maintained by the USDA Salinity Lab, the SORGO database maintained by the National Resource Conservation Service, and the GRIZZLY database maintained by LTHE. These databases can be expected to undergo significant development in upcoming years and should be looked upon as essential resources for modelers and others involved in decision making involving natural environments.

Acknowledgements. The authors would like to acknowledge the support of the NASA BOREAS Project (Project No. 2212-BOREAS-U112, Grant No. NAG 5 2377), the National Science Foundation for support of the soil moisture component of the CASES experiment (Award Number: 9614884), Argonne National Laboratory, ABLE research project and all the scientists actively involved in CASES.

REFERENCES

Angulo-Jaramillo, R., J-P. Vandervaere, S. Roulier, J-L. Thony, J-P. Gaudet and M. Vauclin. Field measurement of soil surface hydraulic properties by disc and ring infiltrometers – A review and recent developments, *Soil & Tillage Research*, Vol. 55, pp. 1 – 29, 2000.

André, J. C., J. P. Goutorbe, A. Perrier, F. Becker, P. Bessemoulin, P. Bougeault, Y. Brunet, W. Brutsaert, T. Carlson, R. Cuenca, J. Gash, J. Gelpe, P. Hilderbrand, J. P. Lagouarde, C. Lloyd, L. Mahrt, P. Mascart, C. Mazaudier, J. Noilhan, C. Ottlé, M. Payan, T. Phulpin, R. Stull, J. Shuttleworth, T. Schmugge, O. Taconet, C. Tarrieu, R. M. Thepenier, C. Valancogne, D. Vidal-Madjar, and A. Weill, Evaporation over land-surfaces, First results from HAPEX-MOBILHY Special Observing Period, *Annales Geophysicae*, Vol. 6, pp. 477-492, 1988.

Beljaars, A. C. M., P. Viterbo, M. J.Miller and A. K. Betts, The anomalous rainfall over the United States during July 1993; Sensitivity to land surface parameterization and soil moisture anomalies, *Monthly Weather Review*, Vol. 124, pp. 362-383, 1996.

Betts, A. K., S.-Y. Hong and H.-L. Pan, Comparison of NCEP/NCAR reanalysis with 1987 FIFE data, *Monthly Weather Review*, Vol. 124, pp. 1480-1498, 1996.

Betts, A. K., P. Viterbo and A. C. M. Beljaars, Comparison of the land-surface interaction in the ECMWF reanalysis model with the 1987 FIFE data, *Monthly Weather Review*, Vol. 126, pp. 186-198, 1998a.

Betts, A. K., P. Viterbo and A. C. M. Beljaars, H.-L. Pan, S.-Y. Hong, M. L. Goulden and S. C. Wofsy, Evaluation of the land-surface interaction in the ECMWF and NCEP/NCAR reanalysis over grassland (FIFE) and boreal forest (BOREAS), *Journal of Geophysical Research*, Vol. 103, pp. 23 709-23 085, 1998b.

Betts, A. K., M. L. Goulden and S. C. Wofsy, Controls on evaporation in a boreal spruce forest, *Journal of Climate*, Vol. 12, pp. 1601-1618, 1999.

Bolle, H. J., J.-C. Andre, J. L. Arrue, H. K. Barth, P. Bessemoulin, A. Brasa, H. A. R. de Bruin, J. Cruces, G. Dugdale, E. T. Engman, D. L. Evans, R. Fantechi, F. Fiedler, A. van de`Griend, A. C. Imeson, A. Jochum, P. Kabat, T. Kratzsch, J.-P. Lagouarde, I. Langer, R. Llamas, E. Lopez-Baeza, J.

Melia Miralles, L. S. Muniosguren, F. Nerry, J. Noilhan, H. R. Oliver, R. Roth, S. S. Saatchi, S. Sanchez Diaz, M. de Santa Olalla, W. J. Shuttleworth, H. Søgaard, H. Stricker, J. Thornes, M. Vauclin, and D. Wickland, EFEDA: European Field Experiment in a Desertification-threatened Area, *Annales Geophysicae*., 11 (2/3), 173-189, 1993.

Burman, R. D., R. H. Cuenca and A. Weiss, Techniques For Estimating Irrigation Water Requirements, in *Advances In Irrigation, Vol. 2*, D. Hillel (editor), Academic Press, New York, pp. 336-394, 1983.

Carrijo, O. A. and R. H. Cuenca, Precision of Evapotranspiration Estimates Using the Neutron Probe, *Journal of Irrigation and Drainage Engineering*, Vol. 28 (9), pp. 2437-2446, 1992.

Cuenca, R. H., Hydrologic Balance Model Using Neutron Probe Data, *Journal of Irrigation and Drainage Engineering*, Vol. 114, No. 4, pp. 644-663, 1988.

Cuenca, R. H. and J. Noilhan, Use of Soil Moisture Measurements in Hydrologic Balance Studies, in *Land Surface Evaporation - Measurement and Parameterization*, T. J. Schmugge and J. C. André (editors), Springer-Verlag, New York. pp. 287-299, 1991.

Cuenca, R. H., M. Ek and L. Mahrt, Impact of Soil Water Property Parameterization on Atmospheric Boundary-Layer Simulation, *Journal of Geophysical Research*, Vol. 101, pp. 7269 – 7277, 1996.

Cuenca, R. H., J. Brouwer, A. Chanzy, P. Droogers, S. Galle, S. R. Gaze, M. Sicot, H. Stricker , R. Angulo-Jaramillo, S. A. Boyle, J. Bromley, A. G. Chebhouni, J. D. Cooper, A. J. Dixon, J-C. Fies, M. Gandah, J-C. Gaudu, L. Laguerre, J. Lecocq, M. Soet, H. J. Steward, J-P. Vandervaere, M. Vauclin, Soil Measurements During HAPEX-Sahel Intensive Observation Period, *Journal of Hydrology - HAPEX-Sahel Special Issue*, Vols. 188-189, pp. 224-266, 1997a.

Cuenca, R. H., D. E. Stangel and S. F. Kelly, Soil Water Balance in a Boreal Forest, *Journal of Geophysical Research - BOREAS Special Issue*, Vol. 102, No. D24, pp. 29,355 - 29,365, 1997b.

Dean, T. J., J. P. Bell and A. J. B. Baty, Soil moisture measurement by an improved capacitance technique, Part 1. Sensor design and performance, *Journal of Hydrology*, Vol. 93, pp. 67-78, 1987.

Ek, M. And R. H. Cuenca, Variation in Soil Parameters: Implications for Modeling Surface Fluxes and Atmospheric Boundary-Layer Development, *Boundary-Layer Meteorology*, Vol. 70, pp. 369-383, 1994.

Fuentes, C., R. Haverkamp and J.-Y. Parlange, Parameter constraints on closed-form soil-water relationships, *Journal of Hydrology*, Vol. 134, pp. 117-142, 1992.

Gardner, W. R. Some steady-state solutions of the unsaturated moisture flow equation with application to evaporation from a water table, Soil Sci., Vol. 85, pp. 228-232, 1958.

Goutorbe, J-P, T. Lebel, A. Tinga, P. Bessemoulin, J. Brouwer, A. J. Dolman, E. T. Engman, J. H. C. Gash, M. Hoepffner, P. Kabat, Y. H. Kerr, B. Monteny, S. Prince, F. Said, P. Sellers, and J. S. Wallace, HAPEX-Sahel: A Large Scale Study of Land-Atmosphere Interactions in the Semi-Arid Tropics, *Annales Geophysicae*, Vol. 12, pp. 53-64, 1994.

Goutorbe, J. P., J. Noilhan, C. Valancogne and R. H. Cuenca, Soil Moisture Variations During HAPEX-MOBILHY. *Annales Geophysicae*, Vol. 7, No. 4, pp. 415-426, 1989.

Haverkamp, R., M. Vauclin and G. Vachaud, Error Analysis in Estimating Soil Water Content from Neutron Probe Measurements: 1. Local Standpoint, *Soil Science*, Vol. 137, No. 2, pp. 78-90, 1984.

Haverkamp, R., P. J. Ross, K. R. J. Smetten, J-Y. Parlange. Three-dimensional analysis of infiltration from the disc infiltrometer, Part 2. Physically-based infiltration equation, *Water Resources Research*, Vol. 30, pp. 2931-2935, 1994.

Jacquemin, B. and J. Noilhan, Sensitivity Study and Validation of a Land Surface Parameterization Using the HAPEX-MOBILHY Data Set, *Boundary-Layer Meteorology*, Vol. 52, pp. 93-134, 1990.

Lebel, T., J.-D. Taupin and N. D'Amato, Rainfall monitoring during HAPEX-Sahel. 1. General rainfall conditions and climatology, *Journal of Hydrology*, Vols. 188-189, pp. 74-96, 1997.

Lebel, T. and L. Le Barbé, Rainfall monitoring during HAPEX-Sahel. 2. Point and areal estimation at the event and seasonal scales, *Journal of Hydrology*, Vols. 188-189, pp. 97-122, 1997.

LeMone, M. A., R. L. Grossman, R. L. Coulter, M. L. Wesely, G. E. Klazura, G. S. Poulos, W. Blumen, J. K. Lundquist, R. H. Cuenca, S. F. Kelly, E. A. Brandes, S. P. Oncley, R. T. McMillen and B. B. Hicks, Land-Atmosphere Interaction Research and Opportunities In the Walnut River Watershed in Southeast Kansas: CASES and ABLE, *Bulletin of the Amer. Meteor. Soc.*, Vol. 81, No. 4, pp. 757-779, 2000.

Mahfouf, J. F., A Numerical Simulation of the Surface Water Budget During HAPEX-MOBILHY, *Boundary-Layer Meteorology*, Vol. 53, pp. 201-222, 1990.

Nichols, W. E., R. H. Cuenca, T. J. Schmugge and J. R. Wang. Pushbroom Microwave Radiometer Results from HAPEX-MOBILHY, *Remote Sensing of the Environment*, 1993.

Noilhan, J. and S. Planton, A Simple Parameterization of Land Surface Processes for Meteorological Models, *Monthly Weather Review*, Vol. 117, No. 3, pp. 536-549, 1989.

Parkes, M. E. and N. Siam, Error associated with measurement of soil moisture change by neutron probe, *Journal of Agricultural Engineering Research*, Vol. 24, No. 1, pp. 87-93, 1979.

Reynolds, W. D. and D. E. Elrick, *In situ* measurements of filed-saturated hydraulic conductivity, sorptivity and the α-parameter using the Guelph permeameter, *Soil Science*, Vol. 133, pp. 61-64, 1985.

Robinson, M. and T. J. Dean, Measurement of near surface soil water content using a capacitance probe, *Hydrological Processes*, Vol. 7, pp. 77-86, 1993.

Roth, K., R. Schulin, H. Fluhler and W. Attinger, Calibration of Time Domain Reflectometry for water content measurements using a composite dielectric approach, *Water Resources Research*, Vol. 26, pp. 2267-2273, 1990.

Rowntree, P. R. and J. A. Bolton, Simulation of the atmospheric response to soil moisture anomalies over Europe, *Quarterly Journal of the Royal Meteorological Society*, vol. 109, pp. 501-526, 1983.

Sellers, P. F. G. Hall, G. Asrar, D. E. Strebel and R. E. Murphy, An overview of the First International Satellite Land Surface Climatology Project (ISLSCP) Field Experiment (FIFE), *Journal of Geophysical Research*, 97 D17:18345-18373, 1992.

Sellers, P., F. G. Hall, H. Margolis, B. Kelly, D. Baldocchi, J. den

Hartog, J. Cihlar, M. Ryan, B. Goodison, P. Crill, J. Ranson, D. Lettenmaier and D. Wickland, The Boreal Ecosystem-Atmospheric Study (BOREAS): An overview and early results from the 1994 Field year, *Bulletin of the American Meteorological Society*, Vol. 76, No. 9, pp. 1549-1577, 1995,

Thony, J.-L., G. Vachaud, B. E. Clothier and R. Angulo-Jaramillo, Field measurement of the hydraulic properties of soil, *Soil Technology*, Vol. 4, pp. 111-123, 1991.

Topp, G. C., J. L. Davis and A. P. Annan, Electromagnetic determination of soil water content: measurements in coaxial transmission lines, *Water Resources Research*, Vol. 16, pp. 574-582, 1980.

Vachaud, G., A. Passerat de Silans, P. Balabanis and M. Vauclin, Temporal stability of spatially measured soil water probability density function, *Soil Science Society of America Journal*, Vol. 49, pp. 822-828, 1985.

Vauclin, M. R. Haverkamp and G. Vachaud, Error analysis in estimating soil water content from neutron probe measurements: 2. Spatial standpoint, *Soil Science*, Vol. 137, pp. 141-148, 1984.

Vandervaere, J.-P., M. Vauclin, R. Haverkamp and R. H. Cuenca, Error analysis in estimating soil water balance of irrigated fields during the EFEDA experiment: 1. Local standpoint, *Journal of Hydrology*, Vol. 156, pp. 351-370, 1994.

Vandervaere, J.-P., C. Peugeot, M. Vauclin, R. Angulo-Jaramillo and T. Lebel, Estimating hydraulic conductivity of crusted soils using disc infiltrometers and minitensiometers, *Journal of Hydrology*, Vol. 188-189, pp. 203-223, 1997.

Warrick, A. W., P. Broadbridge. Sorptivity and macroscopic capillary lengths relationships, *Water Resources Research*, Vol. 28, pp. 427-431, 1992.

Wesely, M. L., R. L. Coulter, G. E. Klazura, B. M. Lesht, D. L. Sisterson, and J. D. Shannon, A Planetary Boundary Layer Observational Capability in Kansas, Proc. 1st Symposium on Integrated Observing Systems. Amer. Meteor. Soc., 169-171, Long Beach, CA. 1997.

Wooding, R. A., Steady infiltration from a shallow circular pond, *Water Resources Research*, Vol. 4, p. 1259-1273, 1968.

Richard H. Cuenca, 116 Gilmore Hall, Bioresource Engineering Dept., Oregon State University, Corvallis, OR 97331.

Shaun F. Kelly, 208 Gilmore Hall, Bioresource Engineering Dept., Oregon State University, Corvallis, OR 97331.

Section 2

MODELING

Bounding the Parameters of Land-Surface Schemes Using Observational Data

Luis A. Bastidas, Hoshin V. Gupta, and Soroosh Sorooshian

Department of Hydrology and Water Resources, University of Arizona, Tucson, Arizona

The potential is explored for using observations of state variables (ground temperature and surface soil moisture) to bound (by optimization) the parameters of land surface models (LSMs), thereby improving the ability to simulate heat fluxes (sensible and latent heat) returned to the atmosphere. This problem is relevant because the heat fluxes cannot be measured directly at the GCM grid spatial scale at which land-surface models are used. However, satellite-based globally extensive remotely sensed estimates of ground surface temperature and soil moisture are expected to become available. Experiments were conducted with the Biosphere Atmosphere Transfer Scheme (BATS) using hydrometeorological data from the Oklahoma ARM-CART site and a multi-criteria approach. The model was found to be capable of replicating, with a high degree of accuracy, each of the fluxes or state variables separately. However, it was unable to simultaneously track the output fluxes and the state variables; improved tracking of the state variables resulted in deterioration in tracking of the output heat fluxes and vice versa. A likely explanation for this inconsistency is the lack of sufficient correspondence between what is observed and what is actually being modeled. Because the simulated state variables are used directly for estimating the output fluxes, this raises questions about the adequacy of the model structure.

1. INTRODUCTION

The last 30 years have witnessed the development of numerous models that attempt to represent the surface-atmosphere interactions. There is a wide variation in the complexity of the representation of the processes involved, from the simple bucket-type models *[Manabe, 1969]* to complex multilayered vertical representations such as BATS *[Dickinson et al., 1986]*, OSU-LSM *[Mahrt and Pan, 1984]*, SiB *[Sellers et al., 1986]*, VIC *[Wood et al., 1992]*, and many others. The models have been subjected repeatedly to improvements that include better representations of the vegetation physiology and attempts to represent surface heterogeneity at the GCM subgrid scale. Furthermore, new versions of some of the models have been developed, e.g., VIC-2L *[Nijssen et al., 1997]*, SiB2 *[Sellers et al., 1996]*, BATS2 *[Dickinson et al., 1998]*, NOAH-LSM *[Mitchell et al., 2000]*, and others. The increase in complexity of the process representation has resulted in large numbers of model parameters. However, the manner in which model parameter values are assigned has changed very little, namely, look-up tables based on literature review and ascribed to different vegetation and soil characteristics are still widely in use.

The large number of models currently in use, exceeding 30, led to the Project for Intercomparison of Land-Surface Parameterization Schemes (PILPS) [e.g., *Henderson-Sellers and Brown, 1992; Henderson-Sellers et al., 1995; Pitman and Henderson-Sellers, 1998]*. Originally, PILPS assumed that

Land Surface Hydrology, Meteorology, and Climate: Observations and Modeling
Water Science and Application Volume 3, pages 65-76

the parameters having the same physical interpretation should have the same value in all the models because some of them may be subject to measurement and/or estimation. It has been suggested, though, that this is not necessarily the case because LSMs are used on a GCM grid scale and, hence, effective values are required *[e.g., Bastidas, 1998; Beven, 1995; Brewer and Wheatcraft, 1994; Gupta et al., 1999b; Sorooshian et al., 1999]*. These effective values are, by their very nature, dependent on the particular parameterization. The fact that they might share the name and conceptual representation does not mean that they have the exact same meaning under different parameterizations. This difference in meaning and, thus, difference in the parameter value, should specifically be accounted for when carrying out a model intercomparison. Remote sensing has the potential to provide information about the space-time variations on the land-surface processes. This is of particular relevance because this kind of information can be used to parameterize LSMs and to derive estimates of the latent heat flux *[see, for example, Bastiaansen et al., 1994; Kustas and Humes, 1996; Pelgrum and Bastiaanssen, 1996; Wood and Lakshmi, 1993]*.

Available observational data can be used to constrain the models, i.e., to bound the parameter values so that the model outputs are consistent with the field observations. This consistency with observations provides the means not only to evaluate and test the model performance but also to help in the identification of proper parameter values. The assignment of values to the model parameters should guarantee consistency between the model outputs and the observational data. Only when this consistency is achieved can the models be properly compared to each other. To attain this consistency, different parameter sets should be obtained for different environmental conditions, hence the need for multiple observational data sets from different environments. The different levels of complexity and the different number of parameters influence the model sensitivity to the precise specification of the parameter values. The sensitivity level of the model outputs may vary from output to output and may depend also on the model forcings (model input), i.e., the models may show different sensitivity levels at different sites. Recently, a procedure named MOGSA (Multi-Objective Generalized Sensitivity Analysis) capable of performing a sensitivity analysis on all the considered outputs simultaneously and on an output by output basis, has been developed *[Bastidas et al., 1999]*. The MOGSA procedure is an extension of the Generalized Sensitivity Analysis (GSA) *[Spear and Hornberger, 1980]*. The procedure also allows for a pre-estimate of how much information is contained for each parameter in each observed flux.

Methodologies for a proper assignment of LSM parameter values by constraining the models with observational data, based on a multi-criteria calibration framework *[Gupta et al., 1998]*, are being developed *[e.g., Bastidas et al., 1999; Gupta et al., 1999b]*. The multi-criteria methods are specially suited for the calibration of the LSM because of their multiple output nature.

In this chapter, we illustrate and discuss how multi-criteria methods can be used in constraining the models and in evaluating the consistency between model outputs and observations. Criteria to discern between different model performances are also presented.

2. MODEL AND DATA

For this work we used the Biosphere Atmosphere Transfer Scheme (BATS) *[Dickinson et al., 1986; Dickinson et al., 1993]* because the number of parameters (22) is representative of the degree of complexity most commonly in use. The computer code used is the offline version available from the BATS home page (*www.atmo.arizona.edu/~bats*) and is named BATS 1e.

BATS consists of six interacting hydrometeorological phases (three soil layers, a canopy air phase, a canopy leaf-stem phase, and a snow-covered portion). At every time step, BATS computes 12 state variables namely, the temperature and water content for each of the six model components. However, two of these variables are not independent. The model assumes that the temperature of the lowest soil layer is constant and, when present, the snow cover has the same temperature as the upper soil layer. Therefore, 10 water-energy conservation equations are solved to compute the independent state variables. *Gao et al. [1996]* showed that errors in specification of initial values for these state variables tend to decay rapidly, after very few days, with the notable exception being the initial moisture contents of the soil layers, and in particular of the bottom layer which has an influence even after 20 years of spin up. In *Bastidas et al. [1999]*, the results of a multi-criteria parameter sensitivity analysis showed that the sensitivity of the model response to the initial soil-moisture contents is significant. To avoid problems caused by poor specification of these variables, the initial soil-moisture contents of the three soil layers must be considered as parameters to be estimated. A schematic of the inputs to and outputs from the model is presented in Figure 1. The multiple-input multiple-output characteristics of the model are emphasized.

The data set used corresponds to station E13 of the Atmospheric Radiation Measurement Cloud and Radiation Testbeds (ARM-CART) program in the Southern Great Plains site (SGP) in Oklahoma. The data set covers the period April-July 1995 with a time interval of 30 minutes and includes all the necessary atmospheric forcings for the model and observational information on sensible heat (H in W/m^2) and latent heat fluxes (λE W/m^2), soil temperature (T_g in K) as the

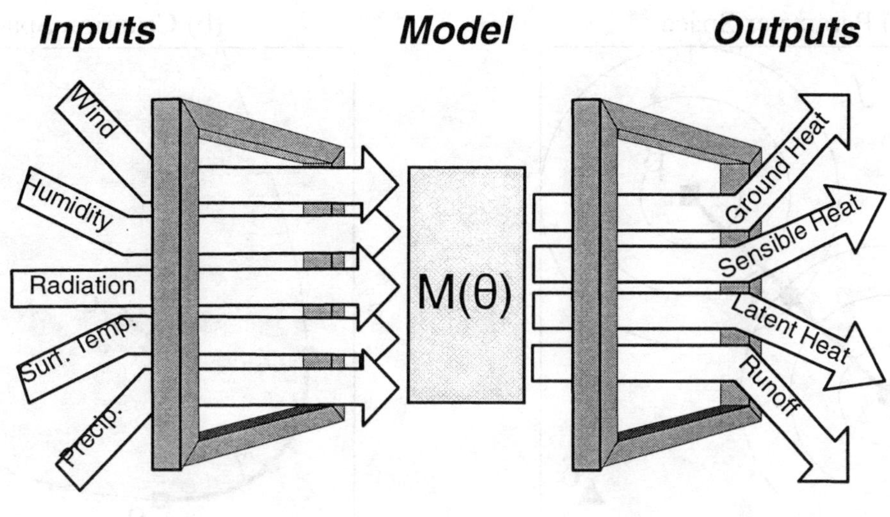

Figure 1. Schematic diagram of a multiple input-multiple output model with several state variables. θ represents the parameters.

average of five sensors that integrate the temperature over the top 5 cm, and the average of five soil-moisture content measurements (S_w in weight of water per weight of dry soil) at a depth of 2.5 cm. The data are representative of the local-(small) scale hydrometeorology and was collected over a flat cattle pasture plot with a Bowen ratio system.. All of this information is used to constrain the model parameters. The interested reader may also find a detailed description of the sensitivity of the BATS parameters at the ARM-CART site in *Bastidas et al. [1999]*.

3. ESTIMATING PARAMETERS USING MULTI-CRITERIA METHODS

3.1. Multi-Criteria Approach

Gupta et al. [1998] presented a framework for the application of the multi-criteria theory to the calibration of conceptual physically based models. In *Gupta et al. [1999b]*, the methodology is extended to LSMs. The method can be summarized as follows: Consider a model having the parameter vector $\theta = \{\theta_1, \ldots, \theta_p\}$ which is to be calibrated using time series observations collected on k different simulated response variables ($Z_j(\theta, t_j)$, $t_j = ta_j, \ldots, tb_j, j = 1, \ldots k$). The different responses represent the different model outputs, e.g., sensible heat flux, latent heat flux, ground heat flux, runoff, etc. To measure the distance between the model-simulated responses Z_j and the observations O_j, separate criteria $f_j(\theta)$ for each model response are defined. The criteria and their mathematical form depend on the goals of the users. It is common practice to use a measure of residual variance such as the root mean square error:

$$RMSE_j(\theta) = \sqrt{\frac{1}{n}\sum_{t=1}^{n}(Z_j(\theta,t) - O_j(t))^2}$$

For a discussion of this, see *Gupta et al. [1998]*. The multi-criteria model calibration problem can then be formally stated as the optimization problem:

$$\textit{Minimize } F(\theta) = \{f_1(\theta), \ldots, f_k(\theta)\} \textit{ subject to } \theta \subset \Theta$$

where the goal is to find the values for θ within the feasible set Θ that simultaneously minimize all of the k criteria.

The multi-objective minimization problem does not, in general, have a unique solution. Due to errors in the model structure (and other possible sources), it is not usually possible to find a single point θ at which all the criteria have their minima. Instead, it is common to have a set of solutions, with the property that moving from one solution to another results in the improvement of one criterion while causing deterioration in another. A case with two parameters (θ_1, θ_2) and two-criteria response functions $\{f_1, f_2\}$ is illustrated in Figure 2. In Figure 2a the feasible parameter space Θ is shown, and the corresponding projection of the parameter space into the function space (shaded area) is shown in Figure 2b. Criterion f_1 is minimized at point α, and criterion f_2 is

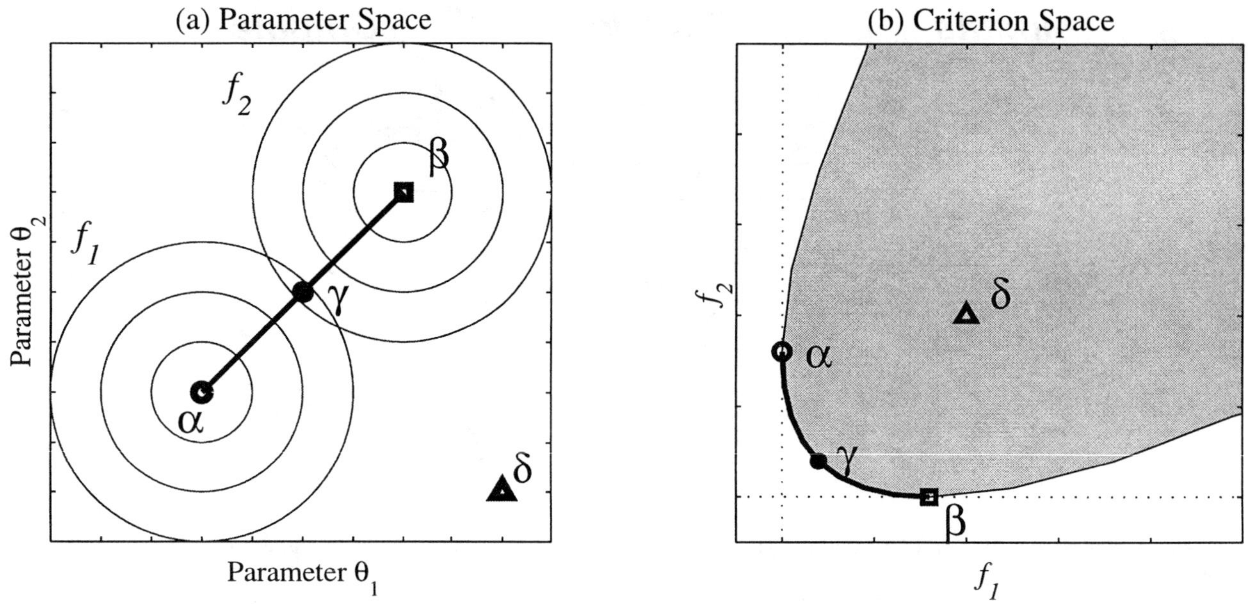

Figure 2. Example showing the Pareto solution set for two- parameter problem *(θ_1, θ_2)* and two criteria (f_1, f_2): (a) feasible parameter space and (b) criterion space. Point α minimizes f_1, and point β minimizes f_2. The thick line indicates the Pareto set P of multi-criteria minimizing points to the function $\{f_1, f_2\}$; $\gamma \in P$ is superior to any $\delta \notin P$.

minimized at point β. The thick line indicates the set P of multi-criteria minimizing points to the function $\{f_1, f_2\}$. If $\gamma \in P$ and $\delta \notin P$ are points selected arbitrarily, then every point γ is superior to every point δ in a multi-criteria sense because each point has the property that $f_j(\gamma) < f_j(\delta)$, for $j = 1,2$. However, it is not possible to find another point $\gamma^* \in P$ such that γ^* is superior to γ; instead γ^* will be superior to γ for one criterion but inferior for at least one other criterion. The set P of solutions is variously called the *trade-off* set, *non-inferior* set, *non-dominated* set, *efficient* set or *Pareto* set. Here, we call it the *Pareto* set.

3.2. The MOCOM-UA Algorithm

Because the Pareto set seldom consists of a finite number of solutions, most multi-objective techniques attempt to identify a countable number of distinct solutions distributed within the Pareto region. The MOCOM-UA (Multi-Objective COMplex evolution) is a general purpose global multi-objective optimization algorithm that provides an effective and efficient estimate of the Pareto solution space within a single optimization run and does not require the commonly used subjective weighting of the different objectives. MOCOM-UA is based on an extension of the SCE-UA population evolution method reported by *Duan et al. [1993]*. A detailed description and explanation of the method are given by *Yapo et al. [1997]*.

In brief, the MOCOM-UA method involves the initial selection of a "population" of p points randomly distributed within the n-dimensional feasible parameter space Θ. In the absence of prior information about the location of the Pareto optimum, a uniform sampling distribution is used. For each point, the multi-objective vector $\boldsymbol{F}(\boldsymbol{\theta})$ is computed, and the population is ranked and sorted using a Pareto-ranking procedure suggested by *Goldberg [1989]*. Simplexes of $n+1$ points are then selected from the population according to a robust rank-based selection method *[Whitley, 1989]*. A multi-objective extension of the downhill simplex method is used to evolve each simplex in a multi-objective improvement direction. Iterative application of the ranking and evolution procedures causes the entire population to converge towards the Pareto optimum. The procedure terminates automatically when all points in the population become non-dominated. The final population provides an approximation of the Pareto solution space $P(\Theta)$.

4. CASE STUDY

The above-described multi-criteria methodology is used to calibrate the offline version of BATS 1e by using data from the SGP ARM-CART site E13. Following *Bastidas [1998]*, the initial values of soil water content in the three different layers are considered as additional model parameters, i.e., subject to optimization. This reduces the influence of improper initialization of the model. To further limit the initialization effect on the model outputs, a one-month spin up is used. Therefore, the length of the series used for constraining the model is 75 days (May-July 1995). In an operational environment, it will be difficult to optimize for the initial conditions, and a good estimate with some period for spin up should be used. Throughout the present chapter, the minimized objective functions are based on the RMSE between the observed and simulated output series (λE, H, T_g, S_w).

In several previous studies, where the calibration approach was used, it has been a customary assumption to consider the latent heat time series as the most important [e.g., *Sellers et al., 1989; Rocha et al., 1996*] and in some others as the sole consideration for calibration [e.g., *Franks and Beven, 1997*]. For this reason, a single-criterion optimization on the latent heat flux is considered here as the benchmark for comparison. The notation of a closed pair of braces is used to denote a calibration run; for example, $\{\lambda E\}$ denotes a single-criterion calibration using λE as the calibration criterion, and $\{H, T_g, S_w\}$ denotes a multi-criteria calibration using H, T_g, S_w as the calibration criteria. The SCE-UA [*Duan et al., 1993*] and the MOCOM-UA algorithms were used for single- and multi-criteria calibrations, respectively.

Every optimization procedure requires the definition of the feasible parameter space, i.e., maximum and minimum allowable parameter values. In the present case, the values prescribed in the BATS model description [*Dickinson et al., 1993*] were used for the definition of the feasibility region. To preserve the physical soundness of the parameterization, additional constraints such as successively increasing thicknesses of the soil layers with depth were imposed. Along the same lines, the seasonal variations of the vegetation cover and leaf area index are forced to be smaller than the maximum value of the corresponding parameter.

Seven different calibration runs on all 22 parameters and the three soil water content initial values were carried out, namely, $\{\lambda E\}$ (the benchmark), $\{H\}$, $\{T_g\}$, $\{S_w\}$, $\{H, S_w\}$, $\{T_g, S_w\}$, and $\{H, T_g, S_w\}$. An additional calibration run was performed on the three initial soil water content values, while the remaining parameters were fixed to have the BATS prescribed (default) values for the region. This run is noted as *{Default}*. The results for the four time series available for comparison λE, H, T_g, and S_w and for each one of the calibration runs are presented in Figures 3-6. In a previous work, *Gupta et al. [1999b]* showed that the parameter estimates obtained by using the multi-criteria framework fall within "expert-defined" reasonable ranges for the area of interest. The same is true for the parameter sets obtained for the current exercise. *Gupta et al. [1999b]* also showed that the inclusion of λE does not result in an improvement and that λE and H provide similar information in terms of parameter identification. A summary of the optimized RMSE values obtained for the different calibrations is presented in Table 1, and the corresponding Nash-Sutcliffe efficiency (NSE) values are presented in Table 2. NSE is defined as follows:

$$NSE_j(\theta) = \sqrt{1 - \frac{\sum_{t=1}^{n} (Z_j(\theta,t) - O_j(t))^2}{\sum_{t=1}^{n} (O_j(t) - \bar{O}(t)_j)^2}}$$

a value of NSE = 1 is a perfect fit while negative values mean that the average of the observations performs better than the model.

Consider first the latent heat flux series. The time series obtained by running the model using the parameter sets obtained from the different calibration runs are presented in Figure 3. As expected, the best result is obtained by the $\{\lambda E\}$ calibration with an RMSE = 26.41 w/m². In Table 1b, this value is used for normalization of the RMSE so that the comparisons can be made more easily. The next result in quality of matching is the one obtained from the $\{H\}$ parameter set. The performance deterioration for the RMSE value is on the order of 25%. The deterioration in the model performance is worse if no information from the observed heat fluxes is used. For example, the use of $\{T_g\}$ results in a deterioration of 66% which, according to Table 1, is the fourth best. Some clipping effect is observed at the high values of the flux starting on day 163. The *{Default}* results in a deterioration of 110% with some underestimation occurring between days 162 and 172. Otherwise *{Default}* is performing relatively well; however, some difficulties with the energy partition are observed. The worst result is obtained for the $\{S_w\}$ with a deterioration of 240%. In Figure 3, it is clear that the simulated series is significantly underestimating the observed values. These eight time series plots suggest, however, that the model has a solid ability to track the latent heat flux, and that the inclusion of S_w into the calibration process results in a deterioration in the ability of the model to simulate the latent heat, particularly after days 164 and 185. These are periods where the soil-moisture content is at its lowest (see Figure 6).

Consider now the sensible heat time series which are plotted in Figure 4. As expected, the best result is produced by the $\{H\}$ calibration, but the improvement upon the performance yielded by $\{\lambda E\}$ is a marginal 4% (Table 1, column 2). Both this result and the previous one suggest that the sole use of latent heat provides enough information for

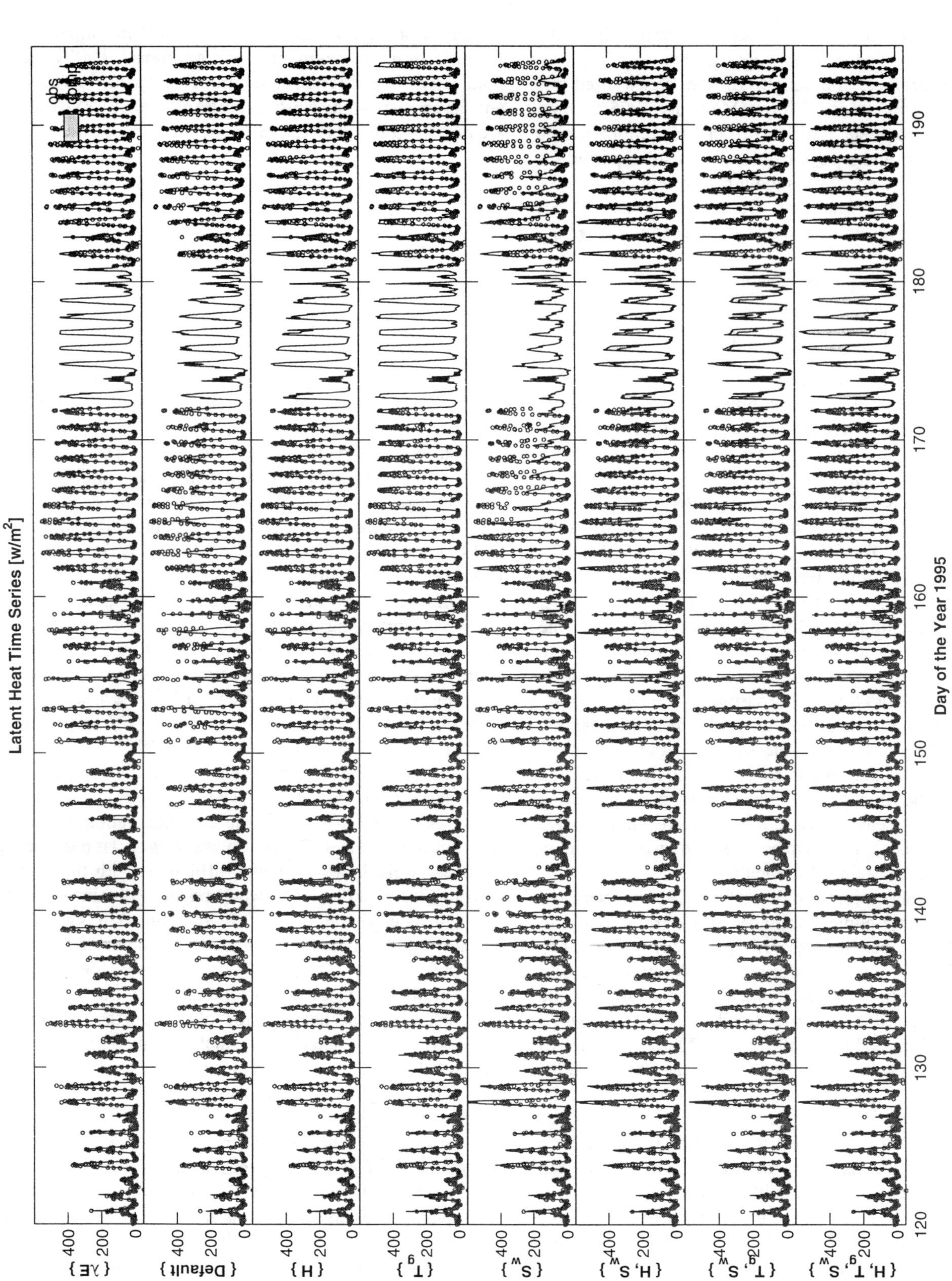

Figure 3. Observed (dots) and simulated (lines delimiting gray area) latent heat time series by using the parameter sets obtained from the *{λE}*, *{Default}*, *{H}*, *{T_g}*, *{S_w}*, *{H, S_w}*, *{T_g, S_w}*, and *{H, T_g, S_w}* calibrations.

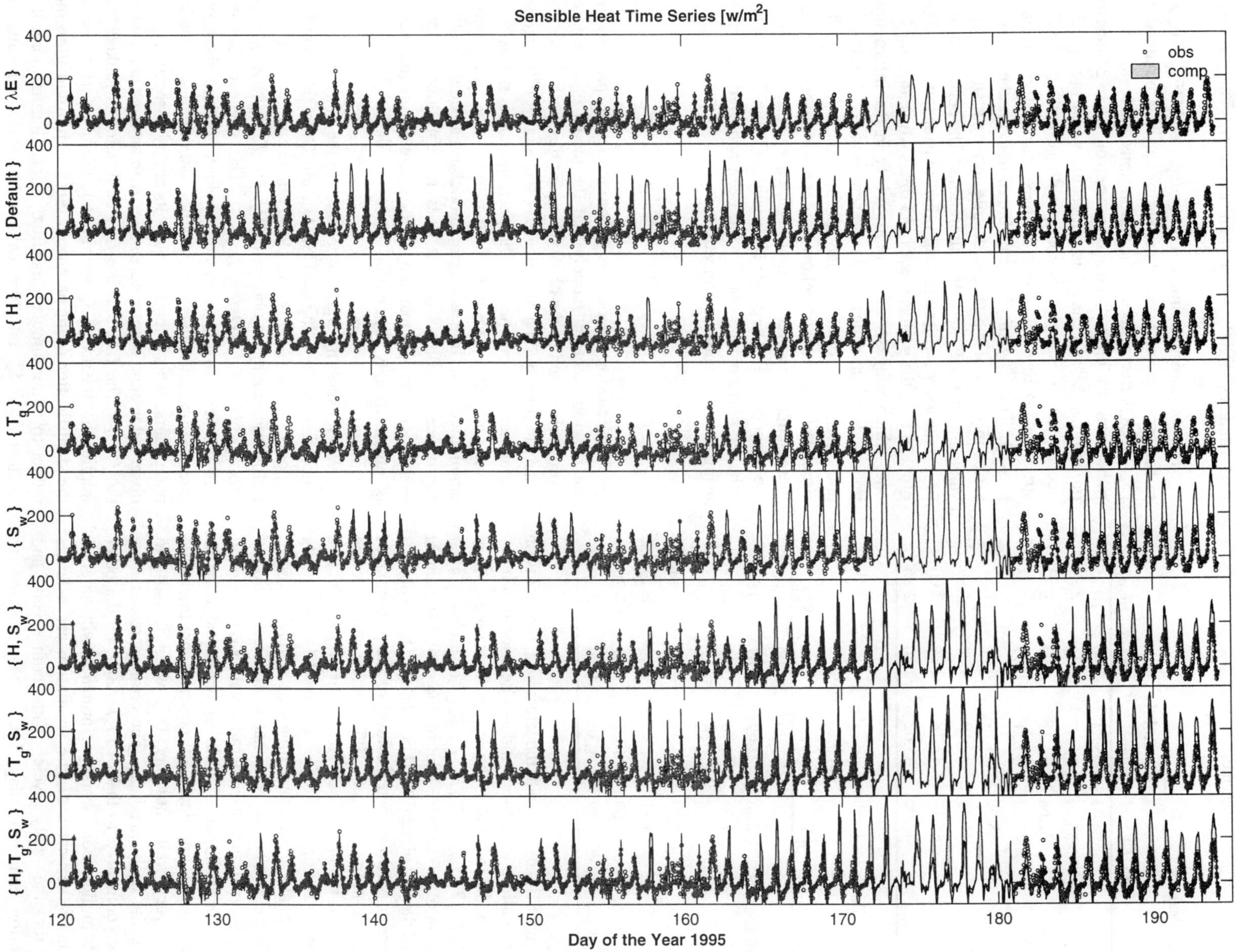

Figure 4. Observed (dots) and simulated (lines delimiting gray area) sensible heat time series by using the parameter sets obtained from the *{λE}*, *{Default}*, *{H}*, *{T_g}*, *{S_w}*, *{H, S_w}*, *{T_g, S_w}*, and *{H, T_g, S_w}* calibrations.

Table 1. Root Mean Square Error for different calibration schemes

(a) Actual Values

	RMSE			
	λE [W/m²]	H [W/m²]	T_g [K]	S_w [mm]
{ λE }	26.41	25.33	3.90	10.25
{ *Default* }	56.19	53.03	2.40	6.61
{ *H* }	32.66	24.31	1.93	8.97
{ T_g }	43.72	45.51	0.97	50.73
{ S_w }	88.90	74.14	4.03	1.41
{ H, S_w }	47.85	36.68	3.08	1.99
{ T_g, S_w }	55.17	47.28	2.24	2.00
{ H, T_g, S_w }	41.94	33.02	2.30	2.86

(b) Ratio to the { λE } calibration value

	Relative RMSE			
	λE	H	T_g	S_w
{ λE }	1.00	1.00	1.00	1.00
{ *Default* }	2.13	2.09	0.62	0.64
{ *H* }	1.24	0.96	0.49	0.88
{ T_g }	1.66	1.80	0.25	4.95
{ S_w }	3.37	2.93	1.03	0.14
{ H, S_w }	1.81	1.45	0.79	0.19
{ T_g, S_w }	2.09	1.87	0.57	0.20
{ H, T_g, S_w }	1.59	1.30	0.59	0.28

fitting both heat fluxes. As with the latent heat, the inclusion of S_w into the calibration process induces deterioration in the ability of the model to simulate the sensible heat series. All of {S_w}, {H, S_w}, {T_g,S_w}, and {H, T_g, S_w} result in a clear overestimation of the sensible heat for the period starting at day 162 until the end of the data. The {*Default*} has problems with the energy partition, and the level of the performance deterioration is on the order of 110%, a value similar to the deterioration for the latent heat. The use of {T_g} results in a deterioration on the order of 80%. The model is producing high negative peaks around dusk time starting on day 164.

From Figure 5, it can be seen that an underestimation of the low daily ground surface temperature values is the general characteristic for all the runs. This is barely noticeable for the {T_g} calibration. {*H*} produces a time series that tracks the observations very well until day 187, when it starts to overestimate the daily maxima. The error increases by a factor of two compared to that of {T_g}. {λE}, on the other hand, results in the second worst performance, with an error 4 times that of the best one. The simulation has a larger amplitude for the daily cycle, with both over- and underestimation until the day 160, when it starts to overestimate the daily maxima by a factor of two or more. The simulations from the calibrations where T_g was used yield performances slightly worse than that of *H*. The {*Default*} performance is of similar quality.

The soil-moisture time series are depicted in Figure 6. As in the previous cases, the best performance is obtained by {S_w}. The performances of the calibrations that include S_w are rather similar both in shape and in the error, 1.5 to 2 times the minimum. A "spikeness" is noticeable when precipitation occurs, which suggests some problems with the model structure, because only a saturation excess runoff mechanism is used. The {*Default*} is incapable of tracking the observed series in terms of the shape, despite having a smaller error value than the {λE} and {*H*} calibrations. The performance of {T_g} is the worst, with simulated values outside the scale of the plot. In fact, the performance is 35 times worse than the best one, which suggests an incompatibility in the information provided by T_g and the model representation of S_w. Both {λE} and {*H*} have errors of around six times the minimum. The simulated shapes are somehow similar to the observed, but the increments when precipitation occurs are two to three times the observed. The negative NSE values (Table 2) show that a better performance could have been achieved by simply using the observation mean value.

5. DISCUSSION

A conclusion can be drawn from the results presented in the previous section, namely that the model is capable of reproducing each of the observed quantities, both fluxes and state variables, with a high degree of accuracy, if the proper parameter sets are chosen. The performance is not as good when all the observed quantities are considered simultaneously. In general, the model has a better ability to reproduce the fluxes than it does to reproduce the state variables. A proper matching of the soil-moisture time series is obtained only when that series is the object of the optimization.

When ascribing equal importance to the matching of the four series and by using the USDA-Multi-Objective Decision Support System [*Imam et al., 1994*] ranking methodology, the following ranking is obtained: {*H*}, {H, T_g, S_w}, {λE}, {T_g}, {H, S_w}, {T_g, S_w}, {*Default*}, {S_w}. This suggests that the sensible heat flux is the quantity which contains the most information and allows for the better identification of model parameters. This is mainly due to its ability to reproduce the observed ground temperature series with an error half that of the {λE}. The fact that the {S_w} calibration is ranked last implies that the use of S_w does not add useful information for the identification of the parameter sets which would allow the model to better simulate the flux series. Furthermore, the inclusion of S_w into the calibration process results in deterioration of the model performance when compared with

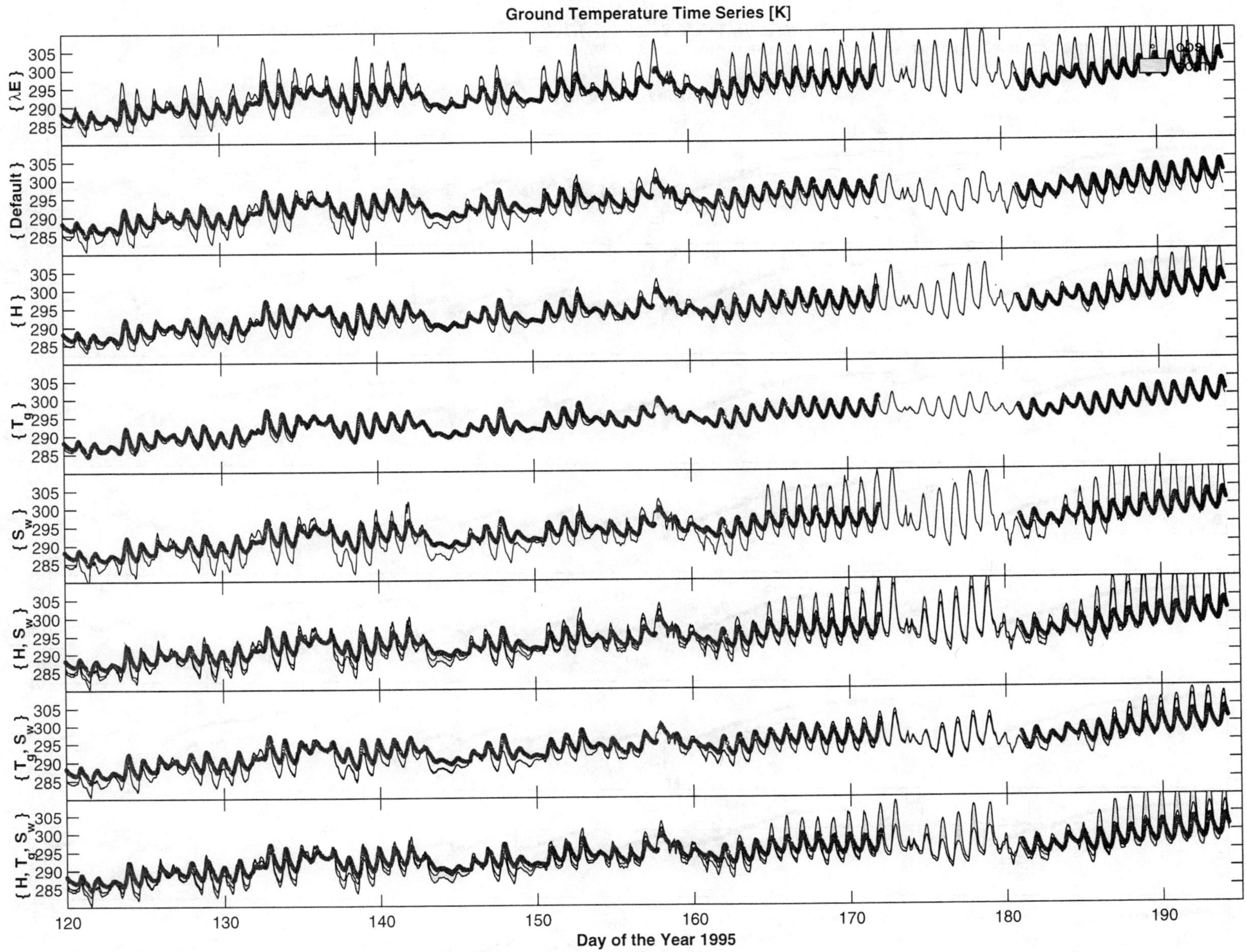

Figure 5. Observed (dots) and simulated (lines delimiting gray area) ground temperature time series by using the parameter sets obtained from the *{λE}*, *{Default}*, *{H}*, *{T_g}*, *{S_w}*, *{H, S_w}*, *{T_g, S_w}*, and *{H, T_g, S_w}* calibrations.

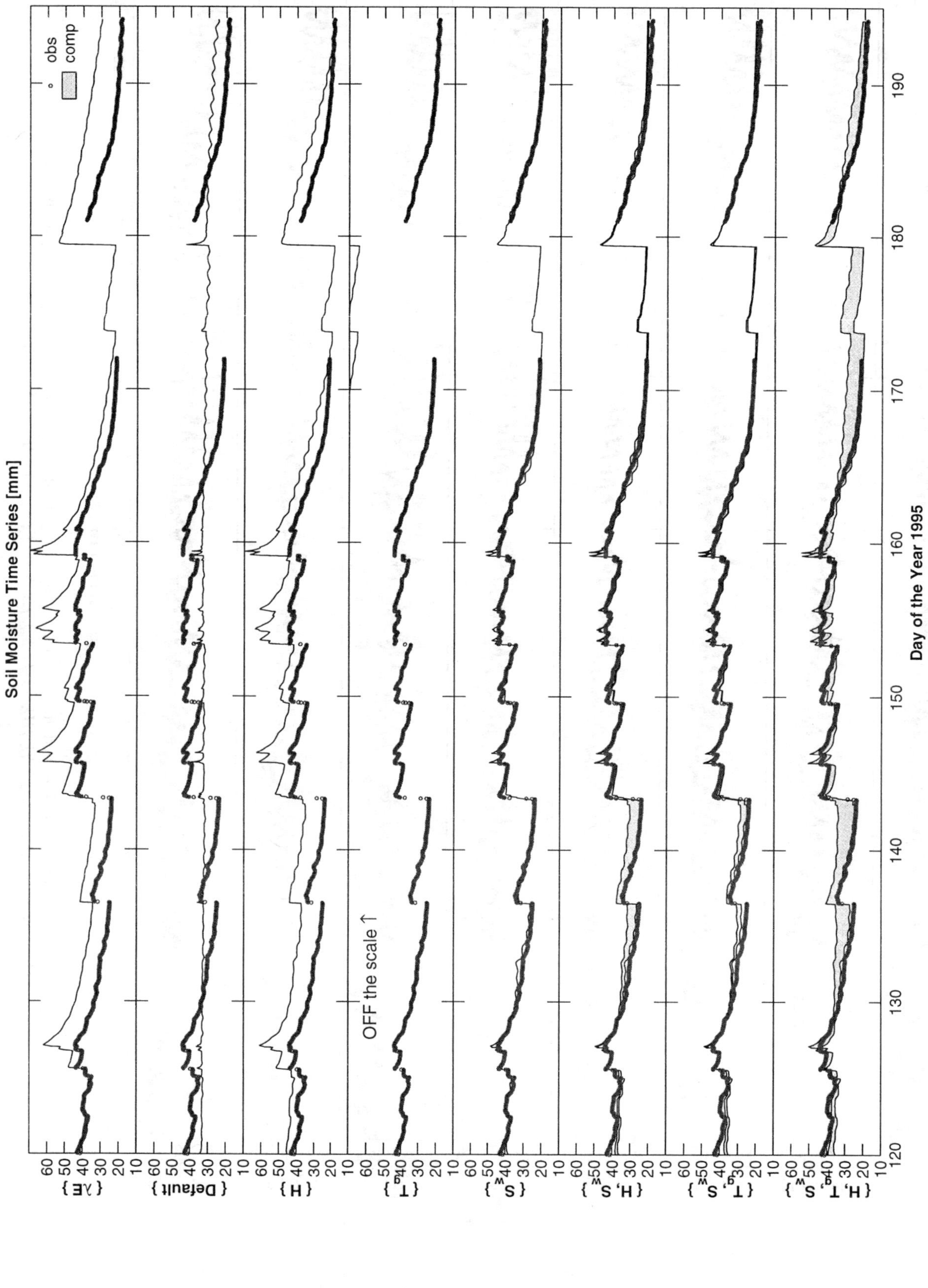

Figure 6. Observed (dots) and simulated (lines delimiting gray area) soil-moisture time series by using the parameter sets obtained from the $\{\lambda E\}$, $\{Default\}$, $\{H\}$, $\{T_g\}$, $\{S_w\}$, $\{H, S_w\}$, $\{T_g, S_w\}$, and $\{H, T_g, S_w\}$ calibrations.

Table 2. Nash-Sutcliffe Efficiency for different calibration schemes

	NSE			
	λE	H	T_g	S_w
$\{\lambda E\}$	0.82	0.53	-0.04	-0.31
$\{Default\}$	0.63	0.01	0.36	0.16
$\{H\}$	0.78	0.49	0.49	-0.14
$\{T_g\}$	0.71	0.15	0.74	-5.47
$\{S_w\}$	0.41	-0.38	-0.07	0.82
$\{H, S_w\}$	0.68	0.32	0.18	0.75
$\{T_g, S_w\}$	0.63	0.12	0.40	0.75
$\{H, T_g, S_w\}$	0.72	0.39	0.39	0.64

the sole use of the T_g. On the other hand, the use of the $\{T_g\}$ parameters induces a 60- 80% increase in the error of the simulations of both heat fluxes. However, the general pattern is very good, which suggests that spatially distributed remotely sensed information on ground temperature can be used to identify parameter sets that will simulate the flux series reasonably well. Some attempts in this direction have already been made [e.g. *Franks and Beven, 1999*]. This is particularly relevant in the context of four-dimensional data assimilation (4DDA) techniques because the identification of proper parameter values will significantly enhance the usefulness of such techniques.

The flux measurements represent an integrated response to surface characteristics over a relatively large area, unlike the measurements of soil moisture and soil temperature that can be considered point measurements, or representative of an area of a few square meters. The model was unable to simultaneously track the output fluxes and the state variables; improved tracking of the state variables resulted in deterioration in tracking of the output heat fluxes and vice versa. A likely explanation for this inconsistency is the lack of sufficient correspondence between what is observed and what is actually being modeled. Because the simulated state variables are used directly for estimating the output fluxes, this raises questions about the adequacy of the model structure.

The multi-criteria calibration approach was found to be effective at bounding the parameter values within physically meaningful ranges.

A paper using information from several data stations within the ARM-CART Southern Great Plains site where the issues of areal representation are specifically addressed is being prepared and will be soon submitted for publication.

Acknowledgements. Partial financial support for this research was provided by the National Aeronautics and Space Administration (NASA-EOS grant NAGW2425), the National Oceanic and Atmospheric Administration (NOAA-GCIP grant NA86GP0324), and the Hydrologic Research Laboratory of the National Weather Service (grants NA47WH0408 and NA57WH0575). Special thanks are due to Bisher Imam for the use of the USDA-DSS and to Corrie Thies for proofreading. We also wish to thank two anonymous reviewers whose comments improved this manuscript.

6. REFERENCES

Bastiaanssen, W. G. M., D. H. Hoekman, and R. A. Roebling, A methodology for the assessment of surface resistance and soil water storage variability at mesoscale based on remote sensing measurements, a case study with the HAPEX-EFEDA data, *IAHS Special Publications,* 2, 66 p., 1994.

Bastidas, L. A., Parameter estimation for hydrometeorological models using multi-criteria methods, Ph.D. Dissertation, Department of Hydrology and Water Resources, The University of Arizona, Tucson, Arizona, 1998.

Bastidas, L. A., H. V. Gupta, S. Sorooshian, W. J. Shuttleworth, and Z. L. Yang, Sensitivity analysis of a land surface scheme using multi-criteria methods, *Journal of Geophysical Research-Atmospheres,* 104(D16), p. 19,481-19,490, 1999.

Beven, K. J., Linking parameters across scales: subgrid parameterizations and scale dependent hydrological models, *Hydrological Processes*, 9, p. 507-525, 1995.

Brewer, K. E. and S. W. Wheatcraft, Including multi-scale information in the characterization of hydraulic conductivity distributions, in *Wavelets in Geophysics*, Edited by E. Foufoula-Georgiou and P. Kumar, pp. 213-248, Academic Press, San Diego, CA, 1994.

Dickinson, R.E., A. Henderson-Sellers, P.J. Kennedy, and M. Wilson, Biosphere Atmosphere Transfer Scheme (BATS) for the NCAR Community Climate Model. NCAR Technical Note, NCAR/TN-275+STR, 1986.

Dickinson, R.E., A. Henderson-Sellers, and P.J. Kennedy, Biosphere Atmosphere Transfer Scheme (BATS) Version 1e as Coupled to the NCAR Community Climate Model. NCAR Technical Note, NCAR/TN-387+STR, 72 p., 1993.

Dickinson, R.E., M. Shaick, R. Bryant, and L. Graumlich, Interactive canopies for a climate model, *Journal of Climate*, 11, pp. 2823-2836, 1998.

Duan, Q., V. K. Gupta, and S. Sorooshian, A shuffled complex evolution approach for effective and efficient global minimization, *Journal of Optimization Theory and Applications*, 76(3), pp. 501-521, 1993.

Franks, S. W., An evaluation of single and multiple objective SVAT model conditioning schemes: parametric, predictive and extrapolative uncertainty, Research Report No. 167.09.1998, The University of Newcastle, Newcastle, Australia, 1998.

Franks, S. W., and K. J. Beven, Bayesian estimation of uncertainty in land-surface-atmosphere flux predictions, *Journal of Geophysical Research - Atmospheres,* 102(D20), pp. 23,991-23,999, 1997.

Franks, S. W., and K. J. Beven, Conditioning a multiple-patch SVAT model using uncertain time-space estimates of latent heat fluxes as inferred from remotely sensed data, *Water Resources Research,* 35(9), pp. 2,751-2,761, 1999.

Gao, X., S. Sorooshian, and H. V. Gupta, Sensitivity analysis of the

Biosphere-Atmosphere Transfer Scheme, *Journal of Geophysical Research-Atmospheres*, 101(D3), pp. 7279-7289, 1996.

Goldberg, D. E., *Genetic Algorithms in Search, Optimization, and Machine Learning*, Addison-Wesley, Reading, Massachussetts, 1989.

Gupta, V.K. and S. Sorooshian, The relationship between data and the precision of parameter estimates of hydrologic models, *Journal of Hydrology*, 81, pp. 57-77, 1985.

Gupta, H.V., S. Sorooshian, and P.O. Yapo, Towards improved calibration of hydrologic models: multiple and non-commensurable measures of information, *Water Resources Research*, 34 (4), pp. 751-763, 1998.

Gupta, H.V., S. Sorooshian, and P. O. Yapo, Status of automatic calibration for hydrologic models: comparison with multilevel expert calibration, *J. of Hydrologic Engineering,* 4(2),pp. 135-143, 1999a.

Gupta, H.V., L. A. Bastidas, L., S. Sorooshian, W.J. Shuttleworth and Z.L. Yang, Parameter estimation of a land surface scheme using multi-criteria methods, *Journal of Geophysical Research*, Vol. 104, No. D16, p. 19,491-19,504, 1999b.

Henderson-Sellers, A., V. B. Brown, Project for intercomparison of landsurface parameterization schemes (PILPS): First science plan, GEWEX Technical Note, IGPO Publication Series No 5, 53 pp., 1992.

Henderson-Sellers, A., A. J. Pitman, P.K. Love, P. Irannejad, and T. H. Chen, The project for intercomparison of land surface parameterization schemes (PILPS): phases 2 and 3, *Bulletin of the American Meteorological Society,* 76, pp. 489-503, 1995.

Imam, B., D. S. Yakowitz, and L. J. Lane, Effects of Optional Averaging Schemes on the Ranking of Alternatives by the Multiple Objective Component of a U.S. Department of Agriculture Decision Support System, in *Multiple Objective Decision Making for Land, Water, and Environmental Management*, S. A. El-Sawify and D.S. Yakowitz. Editors, Lewis Publishers, pp. 217-232., 1996.

Kustas, W. P., and K. S. Humes, Sensible heat flux from remotely sensed data at different resolutions, in *Scaling up in Hydrology Using Remote Sensing*, J. B. Stewart, et al., Editors, pp. 255, John Wiley and Sons, New York, 1996.

Liang, X., D. P. Lettenmaier, E. F. Wood, and S. J. Burges, A simple hydrologically based model of land surface water and energy fluxes for general circulation models, *Journal of Geophysical Research*, 99(D7), pp. 14,415-14,428, 1994.

Manabe, S., Climate and the ocean circulation. I. the atmospheric circulation and the hydrology of the Earth's surface. *Monthly Weather Review*, 97, pp. 739-774, 1969.

Mahrt L. and H. Pan, A two-layer model of soil hydrology. *Boundary Layer Meteorology,* 23, pp.1-20, 1984.

Mitchell, K., Y. Lin, E. Rogers, C. Marshall, M. Ek, D. Lohmann, J. Schaake, D. Tarpley, P. Grunmann, G. Maninkin, Q. Duan, and V. Koren, Recent GCIP-sponsored advancements in coupled land-surface modeling and data assimilation in the NCEP ETA mesoscale model, AMS 15th Conference on Hydrology, 9-14 January, Long Beach, CA, pp 180-183, 2000.

Nijssen, B., D. P. Lettenmaier, X. Liang, S. W. Wetzel, and W. F. Wood, Streamflow simulation for continental-scale river basins, *Water Resources Research,* 33(4), pp. 711-724, 1997.

Pelgrum, H., and W. G. M. Bastiaansen, An intercomparison of techniques to determine the area-averaged latent heat flux from individual in situ observations: A remote sensing approach using the European Field Experiment in a desertification-threatened area data, *Water Resources Research*, 32(9), pp. 2775-2786, 1996.

Pitman, A. J., A. Henderson-Sellers, Recent progress and results from the project for the intercomparison of landsurface parameterization schemes, *Journal of Hydrology,* (212), pp. 128-135, 1998.

Rocha, H. R., P. J. Sellers, G. J. Collatz, I. R. Wright, and J. Grace, Calibration and use of the SiB2 model to estimate water vapour and carbon exchange at the ABRACOS forest, in *Amazonian Deforestation and Climate*, J. H. Gash, C. A. Nobre, J. M. Roberts, and R. L. Victoria, Editors, pp. 459-471, 1996.

Sellers, P. J., Y. Mintz, Y. C. Sud, and A. Dalcher, A Simple Biosphere Model (SIB) for use within general circulation models, *Journal of the Atmospheric Sciences,* 43(6), pp. 505-531, 1986.

Sellers, P. J., W. J. Shutleworth, J. L. Dorman, A. Dalcher, and J. M. Roberts, Calibrating the simple biosphere model (SiB) for Amazonian tropical forest using field and remote sensing data, part 1, average calibration with field data, *Journal of Applied Meteorology*, 28, pp. 728-756, 1989.

Sellers, P. J., D. A. Randall, G. J. Collatz, J. A. Berry, C. B. Field, D. A. Dazlich, C. Zhang, G. D. Collelo, and L. Bounoua, A revised land surface parameterization (SiB2) for atmospheric GCMs. part I: model formulation, *Journal of Climate*, 9, pp. 676-705, 1996.

Sorooshian, S., L. A. Bastidas, and H. V. Gupta, Application of multi-objective optimization algorithms for hydrologic model identification and parameterization, 2nd International Conference on Multiple Objective Decision Support Systems for Land, Water, and Environmental Management, Brisbane, Australia, August 1-5, 1999.

Spear, R.C. and G.M. Hornberger, Eutrophication in peel inlet - II. identification of critical uncertainties via generalized sensitivity analysis, *Water Research*, 14, pp. 43-49, 1980.

Whitley, D., The genitor algorithm and selection pressure: Why rank-based allocation of reproductive trials is best, Proceedings of the Third International Conference on Genetic Algorithms, pp. 116-121, 1989.

Wood, E. F., and V. Lakshmi, Scaling water and energy fluxes in climate systems: three land-atmosphere modeling experiments, *Journal of Climate,* 6, pp. 839-857, 1993.

Wood, E. F., D. P. Lettenmaier, and V. G. Zartarian, A land surface hydrology parameterization with subgrid variability for general circulation models, *Journal of Geophysical Research*, 97(D3), pp. 2717-2728, 1992.

Yapo, P.O., H.V. Gupta, and S. Sorooshian, Multi-objective global optimization for hydrologic models, *Journal of Hydrology*, 204, pp. 83-97, 1997.

A Priori Estimation of Land Surface Model Parameters

Qingyun Duan, John Schaake, and Victor Koren

NOAA/NWS, Hydrology Laboratory, Silver Spring, Maryland

Land surface models (LSM's) contain parameters (coefficients and exponents) that control model behavior. To apply a LSM successfully, model structure must be physically realistic and parameters must be properly estimated. This chapter addresses issues concerning *a priori* parameter estimation, with particular emphasis on estimation of runoff-related parameters. *A priori* relationships linking model parameters and land surface characteristics such as soil and vegetation classes are available for many LSM's. But these relationships have not been fully validated through rigorous testing using retrospective hydrometeorological data and corresponding land surface characteristics data. This chapter reviews existing *a priori* parameter estimation procedures used in current LSM's. A strategy for developing improved *a priori* procedures is outlined. Issues such as data requirements and parameter estimation techniques are also discussed. The proposed strategy is illustrated through a newly developed *a priori* parameter estimation procedure for the Sacramento model used in National Weather Service.

1. INTRODUCTION

Land surface models (LSM's) are essentially a set of equations that represent how land surface processes are believed to operate. These equations determine the model structure and they also contain coefficients and exponents called model parameters. Together, model structure and model parameters control the model behavior. To apply a model successfully, model structure must be physically realistic and parameters must be properly estimated. Some model parameters are unique to each grid point or computational element. Some may vary seasonally as well. Some depend on space/time scales of model application (Finnerty et al., 1997; Koren et al., 1999). This chapter is concerned with *a priori* parameter estimates, with particular emphasis on estimation of runoff-related parameters. Other chapters in this book address model calibration techniques and various model structural issues.

There are two general classes of LSM's. One is the class used widely to model streamflow for river forecasting. The other is the class used to represent surface water and energy fluxes in weather prediction and in climate models. Although these two classes of LSM's emerged from different scientific communities and serve different practical applications, the underlying physical processes are the same and many of the modeling issues are also similar. It is interesting to note that international programs such as Global Energy and Water Cycle Experiment (GEWEX) and Biospheric Aspects of Hydrological Cycle (BAHC) have brought scientists from both hydrologic and atmospheric communities closer together during last decade to improve our ability to model land surface processes.

Typical river forecasting models are conceptual models formulated using empirical relationships between hydrologic variables observed in nature or field experiments or derived based on abstract conceptualization of physical processes. Conceptual models attempt to account for both the physics and basin-scale spatial heterogeneity of surface and subsurface processes. Conceptual models are in contrast to so-called "physically-based" models (PBM's), that have origins in partial differential equations used to describe mass and energy processes. PBM's have parameters that are directly tied to local physical processes. But these parameters generally do not adequately account for basin-scale heterogeneities. Over the last two decades, a new generation of more physically based

Land Surface Hydrology, Meteorology, and Climate: Observations and Modeling
Water Science and Application Volume 3, pages 77-94

LSM's has emerged (Dickinson et al., 1986; Sellers et al., 1986; Ek and Mahrt, 1991; Xue et al., 1991; Chen et al. 1996). These LSM's typically are embedded in atmospheric models that run on grid cells regionally or globally. The resolution of grid cells ranges from about 10^2 to 10^5 square kilometers.

It is well known that hydrologists use model calibration to estimate parameters for the conceptual LSM's used in river forecasting because parameters of conceptual models generally are not directly observable and have less physical significance than usually attributed to those in PBM's. Even if parameters of conceptual models could be estimated *a priori*, river forecast applications would still require fine tuning of the *a priori* parameters where there are sufficient data to support model calibration. This additional fine tuning or calibration is needed because the physical information available to estimate *a priori* parameters is not adequate to define local physical properties of individual basins for accurate hydrologic forecasts. There exists extensive literature over the last three decades on the subject of model calibration (see Ibbitt, 1970; Johnston & Pilgrim, 1976; Sorooshian & Dracup, 1980; Brazil, 1988; Duan et al, 1992 & 1994; Kuczera, 1997, Yapo et al., 1997, Gutpa et al., 1999). Previous research has mostly focused on the following issues: (1) model structural identifiability, (2) quantity and quality of data needed for calibration, (3) techniques dealing with uncertainties associated with model structure, calibration data and model parameters, and (4) techniques to facilitate estimation of parameter values. Other chapters in this book offer in-depth discussion on automatic calibration in a multi-objective framework (Bastidas et al., in this book). To calibrate model parameters, many years of historical hydrometeorological data, including precipitation, streamflow discharge, and potential evaporation must be available. For ungaged basins where calibration is not possible, other procedures to define model parameters *a priori* must be employed.

Most LSM's used in atmospheric models are PBM's that have parameters related to observable physical properties and they can be estimated using assumed *a priori* relationships between model parameters and land surface characteristics. Some LSM's such as NOAH-LSM (Chen et al., 1996) and VIC-3L of Liang et al. (1994) have elements of both conceptual and physically based models. Some parameters in these LSM's are tied directly to soil/vegetation properties, while other parameters would be better determined using model calibration. Abdulla et al. (1996) developed an *a priori* parameter estimation technique for the VIC model. First, the automatic calibration method developed by Duan et al. (1992, 1994) was used to determine the parameters of VIC model for 34 basins in the Arkansas/Red river basin. Then regional regression equations were developed relating the calibrated parameters to soil and vegetation properties for the entire Arkansas/Red basin.

The estimation of LSM parameters in global atmospheric models presents a different challenge than that of conceptual models. First, there exists no hydrometeorological database to conduct model calibration globally. Second, even where data exist, there is not enough to use calibration techniques to estimate parameters for every grid cell. Therefore, alternative procedures to specify parameters *a priori* for each cell must be used.

Presently *a priori* relationships linking model parameters and land surface characteristics such as soil and vegetation classes are available for many LSM's, but these relationships have not been fully validated through rigorous testing using retrospective hydrometeorological data and corresponding land surface characteristics data. This is partly because of insufficient data for such testing. Moreover, generally available information about soils (e.g., texture) and vegetation (e.g., type or vegetation index) only indirectly relates to model parameters such as hydraulic properties of soils and rooting depths of vegetation. Also it is not clear how heterogeneity associated with spatial land surface characteristics data affects those characteristics at the scale of a model grid cell. Consequently, there is a considerable degree of uncertainty associated with the parameters given by existing *a priori* procedures. Recent studies have illustrated that these procedures do not necessarily produce proper parameter values and that improper model parameters result in poor model performance (see Liston et al, 1994; Duan et al., 1995; Gupta et al, 1999; Bastidas et al., this book). Because of the high sensitivity of model performance to model parameters, improved *a priori* procedures for parameter estimation are needed. Also, existing model calibration techniques tend to produce "noisy" parameter estimates because many combinations of model parameters produce very similar model response. Therefore, improved *a priori* estimation procedures might be used not only to provide initial parameter estimates, but also to provide uncertainty limits during model calibration.

The Project for Intercomparison of Land-surface Parameterization Schemes - Phase 2(c) (PILPS 2(c)) represented the first PILPS effort to allow participating LSM's to utilize streamflow data to estimate some model parameters (Wood et al., 1998; Liang, et al., 1998; & Lohmann et al., 1998). Their results indicated that the LSM's that used streamflow data to calibrate model parameters had better overall performance in both water balance as well as energy balance simulations (Wood et al., 1998; Liang et al.,1998; Lohmann et al, 1998). Figures 1 and 2 summarize some of the results of PILPS 2(c). Figure 1 shows the range of variability of components of the annual water balance among the models. Because each model conserved water, data points in Figure 1 lie along the same line. But there is considerable scatter along the line and much of this may be accounted for by uncertainty in parameter estimates. Figure 2 shows how the runoff

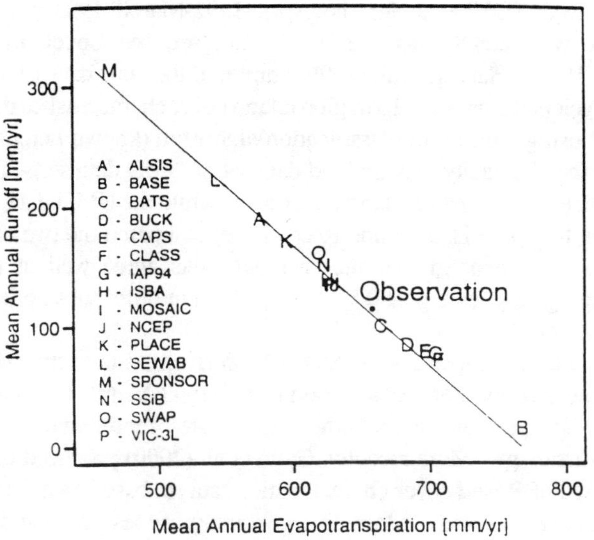

Figure 1. Runoff/Evapotranspiration partition of different models from PILPS 2c

produced by each model was partitioned between surface and subsurface runoff components. Clearly there is a wide range of results and much of this variability can be accounted for by uncertainty in the values of parameters that would be best to use in each model.

The purpose of this chapter is to examine existing *a priori* parameter estimation procedures for both conceptual models and PBM's. Also presented are strategies for developing improved *a priori* parameter estimation procedures. Section 2 reviews existing *a priori* parameter estimation procedures used by current LSM's. Section 3 examines the limitations of existing *a priori* procedures. Section 4 outlines a strategy for developing improved *a priori* procedures. Issues such as data requirements and parameter estimation techniques are discussed. Section 5 illustrates an implementation of the proposed strategy. Finally, Section 6 presents a summary and conclusions.

2. REVIEW OF *A PRIORI* PARAMETER ESTIMATION PROCEDURES

Early LSM's represented land surface hydrologic processes very simply and involved few parameters. Universal parameter values were often prescribed. For example, in Manabe's simple bucket scheme (Manabe, 1969), the land surface is treated as a 15cm bucket. The rate of evaporation is dependent on the water content in the bucket. Runoff occurs only when the bucket is full. The 15 cm bucket size is equivalent to assuming that the water holding capacity of the soil is 15 percent of its volume and that the root depth is one meter. Recent research suggests that the bucket scheme is too simplistic and leads to poor performance in simulating runoff and evapotranspiration (Chen et al., 1996; Schaake et al, 1996). Further, a value of 15 cm for the bucket size is much too small for many locations.

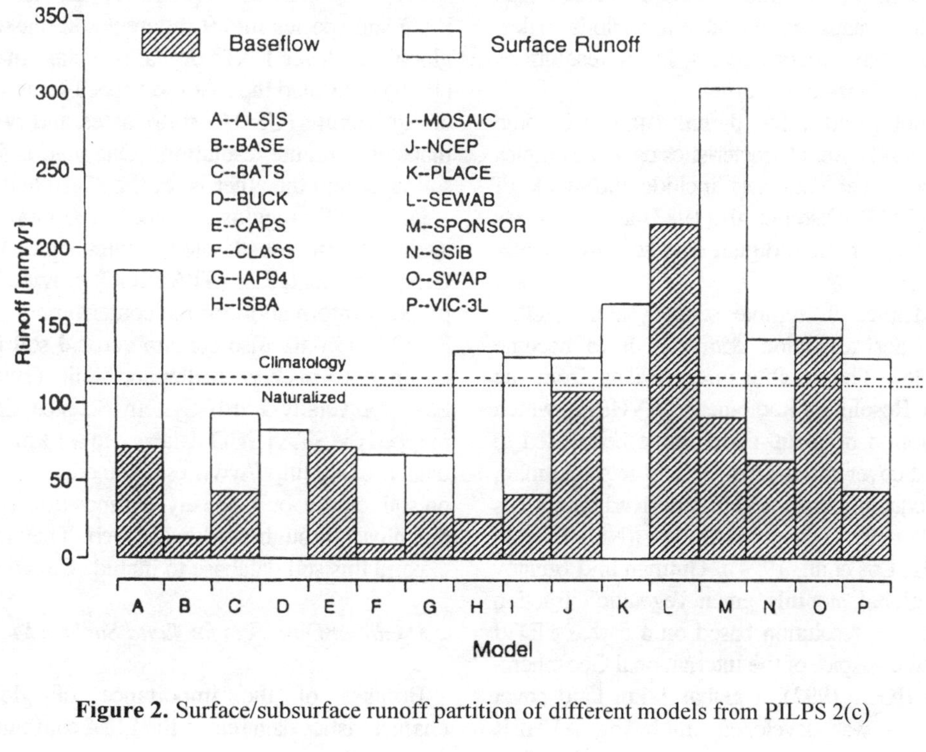

Figure 2. Surface/subsurface runoff partition of different models from PILPS 2(c)

As more sophisticated LSM's such as BATS (Dickinson et al., 1986), SiB (Sellers et al., 1986), and NOAH LSM (Chen et al., 1996) emerged, model structures became more complex and the number of parameters grew considerably. These LSM's have resulted in enhanced model performance, but the estimation of the parameters presents a huge challenge to model users (Cuenca, 1996). To run LSM's globally or regionally, model parameters are usually determined based on the knowledge of land surface characteristics including vegetation and soil quantities that are available globally or regionally (Dickinson et al., 1986; and Sellers et al.,1986). Global or regional land surface characteristics data sets are therefore essential in land surface modeling. Land surface characteristics data sets are generally divided into two categories: land cover characteristics (e.g., land use, vegetation type and phenology), and soil characteristics (e.g., soil texture and other physical properties). The land cover characteristics are believed to exert great influence on surface energy fluxes, while the soil characteristics are believed to impact how water moves within soil. Below these two categories of data sets are reviewed separately.

2.1 Land Cover Characteristics Data Sets

The early land cover characteristics data sets are in the form of printed maps for the entire world or for specific regions or countries. Examples of printed maps include the World Forestry Atlas (1961), World Atlas of Agriculture (1969), Oxford Economic Atlas (1972) and Goode's World Atlas (Kuchler, 1983). These maps are available at various scales. Generally global maps have coarser scales, lower resolutions than maps for specific regions or countries.

As computer technology advances, digital data sets become the preferred form of land cover characteristics data. Examples of early digital land cover data sets include the work of Hummel and Reck (1979), Olson et al. (1983) and Mathews (1983). The resolution for these digital data sets are usually low, at 1°x1° or lower.

Owing to the advance in remote sensing and satellite technology, much finer resolution data sets have become increasingly available. The satellite imagery data from the Advanced Very High Resolution Radiometer (AVHRR), which has a nominal resolution of about 1 km, have been used to produce various land cover characteristics data sets (Gutman, 1994). The most frequently used AVHRR derived data set is the normalized difference vegetation index (NDVI) data (Goward et al., 1991; Los et al., 1994). Gutman and Ignatov (1998) produced a global monthly green vegetation fraction data set at 0.15°x 0.15° resolution based on a 5-year NDVI climatology. Under the auspice of the International Geosphere-Biosphere Program (IGBP,1992), a global 1 km land cover characteristics data set was developed employing AVHRR derived NDVI data for the April 1992-March 1993 period (known hereafter as the IGBP data set, see Loveland et al.,1999). Hansen et al. (1999) employed the same data set and developed a separate 1 km global land cover characteristics data set using a different classification algorithm (known hereafter as the University of Maryland data set or UMD data set). The IGBP data set has 17 land cover types while the UMD data set has 14 types. Hansen and Reed (1999) compared the two data sets and reported that the two data sets agree well at the aggregate level (e.g. at 0.5°x0.5° scale), but differ substantially at pixel level.

The aforementioned AVHRR derived data sets were generated for a one-year period (April 1992-March 1993) and it is questionable whether they are representative of climatology. More recently, Gallo et al. (2000) produced on a new IGBP land cover characteristics data set based on a 5-year NDVI climatology. Monthly vegetation greenness fraction data sets for the first few dominant land cover types were created as well. This new data set should be more robust and reliable than previous AVHRR based data sets because it is based on a longer data period and is therefore more representative of land surface climatology.

2.2 Soil Characteristics Data Sets

Like land cover characteristics data sets, early soil data sets were in printed maps. The often cited soil data sets are the FAO soil maps which classify soils into more than 200 soil species (FAO, 1961, 1976, 1981). Zobler (1986) reduced the FAO soil species into 9 different soil classes and produced a global one-layer 1°x1° digital soil class map. Reynolds et al. (1999) translated the FAO soil species into the US Department of Agriculture (USDA) soil classes and produced 2 digitized maps at 5 minute resolution. One map is for the 0-30 cm top soil layer and the other is for the 30-100 cm deeper soil layer.

For specific regions and countries, more detailed maps are available. In the United States, the USDA State Soil Geographic Database (STATSGO) provide high resolution soil property information for the conterminous United States. The STATSGO data also contain vertical soil profile information for up to 2.5 m deep. Miller and White (1999) of Pennsylvania State University Earth System Science Center (PSU-ESSC) mapped the STATSGO database to a 1 km, 11-layer soil digital database (see http://www.essc.psu.edu/soil_info/). Information on soil texture, bulk density, composition, depth is available for download through worldwide web. They are also working to expand this soil database to include Canada and Mexico.

2.3 Global Data Sets for Land Surface Modeling

Because of the importance of global land surface characteristics data sets to the LSM community, nationally and internationally coordinated research efforts have been undertaken over the last 20 years to develop and assemble data sets needed for land surface modeling at global and regional scales. One key step in developing global data sets for LSM's is the interpretation of different classification schemes used to describe land cover and soils. Wilson and Henderson-Sellers (1985) were among the first in LSM community to compile global land surface characteristics data sets from various sources and to tailor them for running LSM's globally. They utilized FAO (1961, 1976, 1981) soil maps to derive a soil classification system that has 3 soil color classes, soil texture classes and soil drainage classes. They also employed various land cover maps to derive land cover data sets that can be easily used for land surface modeling. The Wilson and Henderson-Sellers data sets have a spatial resolution of 1°x1° or lower.

The International Satellite Land Surface Climatology Project (ISLSCP) Initiative I represented an unprecedented international collaborative effort to develop global land cover and soils data sets for land surface modeling. A 5-volume CD-ROM set was published as the result (Meeson et al., 1995; Sellers et al., 1996). Land cover (e.g., vegetation type, phenology, etc.) and soils data (e.g., texture, depth, porosity, slope, etc.) were conveniently put together and made available on a common 1°x1° grid structure. Many satellite (e.g., AVHRR and landsat) based data products, including NDVI, leaf area index (LAI), Fraction of Photosynthetically Active Radiation (FPAR), and vegetation greenness fraction, were included in the CD-ROMs. Parameters important for surface energy fluxes such as snow-free albedo and surface roughness are also available on the CD (Dorman and Sellers, 1989). Besides the land surface characteristics data, the ISLSCP I CD-ROM set also contains a 2-year atmospheric forcing data set needed to run GCM/LSM's. ISLSCP Initiative II is underway to develop comprehensive global data sets that will include more recent information and have higher spatial resolution.

2.4 Land Surface Characteristics and LSM Parameters

The diverse classification schemes used in the global data sets make them unsuitable for direct application in LSM's. Furthermore, land surface characteristics are related to model parameters only indirectly. To determine parameters, some procedures to estimate model parameters *a priori* based on land cover and soils information are used. There are many existing procedures on how to relate models parameters to land cover and soils information. The basic approach is to divide model parameters into two groups: one group related to land cover type and another group related to soil type.

Dickinson et al. (1986) synthesized the land cover data sets of Olson et al. (1983), Mathews (1983), and Wilson and Henderson-Sellers (1985) and mapped them into 18 land cover types. Parameters such as roughness length, rooting depth, minimum stomatal resistence, etc were prescribed for each land cover type. Dickinson et al. (1986) also used the soils data of Wilson and Henderson-Sellers (1985) to determine soil related parameters. The 3 texture classes were expanded to 12 classes to match the USDA soil classes. Soil parameters such as porosity, wilting point, field capacity, saturated hydraulic conductivity, and exponent "B" in soil moisture depletion curve, were assigned according to these texture classes. Soil albedo values were determined according to soil color. The 3 color classes were expanded to 8 colors and albedo values were assigned to each of the 8 colors.

Sellers et al. (1986) used the global data fields of Kuchler (1983) and Mathews (1983) to determine some vegetation parameters in SiB. Parameters such as surface roughness length and snow-free albedo were determined for 20 land cover types. They also used NDVI values to determine monthly FPAR and LAI. Soil parameters such as porosity, saturated soil matric potential, and saturated hydraulic conductivity, were associated with soil texture as defined by Zobler (1986).

To determine soil related parameters, the relationship between soil properties and soil classes must be established. Clapp and Hornberger (1978, known as the Clapp parameterization hereafter) developed a relationship between soil hydraulic properties and USDA soil texture using a soil pedon data set collected in the United States. Cosby et al. (1984, known as the Cosby parameterization hereafter) refined the Clapp parameterization by using an expanded soil pedon data set. Other researchers including Rawls et al. (1982, 1991) and Carsel and Parrish (1988) developed their own relationships between soil hydraulic properties and soil classes. In developing these relationships, various soil water retention models, including Brooks and Corey (1964), Campbell (1974) and van Genuchten (1978), were employed to fit the soil pedon data and to derive soil parameters. Cuenca et al. (1996) compared the soil parameters developed by different researchers and reported that diurnal energy balance of the land surface is highly sensitive to how soil parameters are defined.

Newly available soils and vegetation data are being used by the multi-agency/multi-institution Land Data Assimilation System (LDAS). The purpose of this project is to improve the accuracy of numerical weather/climate prediction models by providing improved soil and temperature initial conditions. This is done by using observed precipitation and solar radiation forcing data together with analyzed meteorological variables to drive offline versions of the LSMs in the coupled land/ocean/atmospheric models (Mitchell et al., 1999). The 1 km STATSGO data for the US, the 1 km UMD land cover data, and the 0.15° degree monthly greenness fraction data of Gutman and Ignatov (1998) were mapped to the LDAS grid. The data sets prepared for LDAS preserved the information on spatial distribution of land cover type and soil type. LDAS

Table 1. Vegetation Classifications Used in NOAH LSM

1	Broadleaf-evergreen trees (tropical forest)
2	Broadleaf-deciduous tress
3	Broadleaf and needleleaf tress (mixed forest)
4	Needleleaf-evergreen trees
5	Needleleaf-deciduous tress (larch)
6	Broadleaf tress with groundcover (savanna)
7	Groundcover only (perennial)
8	Broadleaf shrubs with perennial groundcover
9	Broadleaf shrubs with bare soil
10	Dwarf trees and shrubs with groundcover (tundra)
11	Bare soil
12	Cultivations (use Type 7 parameters)
13	Glacial

participants have the choice of several sets of soil parameters, including the Clapp, Cosby and Rawls parameterizations.

2.5 A Priori Parameter Estimation Procedure for NOAH LSM

The Eta model is a mesoscale numerical weather prediction model. It has been operational in National Center for Environmental Predictions (NCEP) since mid-1994. Because the Eta model is run on a grid that covers all of North America, it is necessary that *a priori* parameters be assigned to all parameters at each grid point. The NOAH (stands for NCEP-Oregon State University-Air Force-NWS Office of Hydrology) LSM is a component of the Eta model that simulates water and thermodynamic processes on land surfaces (Chen et al., 1996). NOAH-LSM is a modified version of OSU-CAPS (Mahrt and Pan, 1984). NOAH LSM parameters are determined by vegetation and soil classifications.

The vegetation classifications for NOAH are adapted from those of Simplified SiB (Xue et al., 1991). The soil classifications are from Zobler (1986). Tables 1 and 2 display the vegetation and soil classifications. Some of the parameters which are determined by vegetation and soil classifications are respectively listed in Tables 3 and 4. Note that some parameters such as *ALBEDO* and *SHDFAC* vary not only spatially but also seasonally.

The 1°x1° vegetation class map from Mathews (1983) and the soil class map from Zobler (1986) are used to make gridded fields of *a priori* values for vegetation and soil parameters in NOAH LSM. Table 5 lists the *a priori* values of soil parameters, which are determined based on the Cosby parameterization. Since the Cosby parameterization used the USDA soil classification which is different from the one used by Zobler, Table 2 shows how the two classifications are related in NOAH LSM. Table 6 lists the selected vegetation related parameters.

3. LIMITATIONS OF EXISTING *A PRIORI* PARAMETER ESTIMATION PROCEDURES

A limitation of *a priori* parameter estimates based on relationships such as given by Clapp and Hornberger (1978) and Cosby et al. (1984) is that these relationships were derived from point measurements, while typical LSM's are run on grid cells with areas between several hundred square kilometers to several hundred thousand square kilometers. A related problem is how to treat spatial heterogeneity within a grid cell. In using global soil and vegetation data sets, parameters are often determined from the most dominant soil or vegetation type. In many cases the dominant land cover type does not reflect the true land cover. For example in Zobler's global soil data base which has a resolution of 1°x1°, 64 percent of the continental US belongs to a single type: medium (which corresponds to USDA silty-clay loam in NOAH LSM). The STATSGO soil data base, which is available for the US only and has been digitized to a spatial resolution of 1x1 km^2, shows that only 5 percent of the continental US is silty-clay-loam. Similar problems exist for vegetation databases.

3.1 Comparison of A Priori Parameters and Partially Tuned Parameters

Liston et al. (1994) conducted a study of runoff produced by the VIC LSM in the Mississippi river basin. They compared runoff simulation using default (*a priori*) parameters as prescribed by VIC and using parameters that were partially tuned. Their conclusion is that partially tuned parameters

Table 2. Soil Classifications of Zobler and USDA

Class	Zobler (1986)	Cosby et al. (1984)
1	Coarse	Loamy Sand
2	Medium	Silty Clay Loam
3	Fine	Light Clay
4	Coarse-med	Sandy Loam
5	Coarse-fine	Sandy Clay
6	Medium-fine	Clay Loam
7	Coarse-med-fine	Sandy Clay Loam
8	Organic	Loam
9	Land Ice	

Table 3. Definition of Selected Vegetation Parameters in NOAH LSM

ALBEDO	Snow-free surface albedo
Z0	Roughness length
SHDFAC	Plant shade factor
NROOT	Rooting depth
RCMIN	Minimum stomatal resistance
RSMAX & RGL	Parameters in radiation stress function
HS	Parameter in vapor pres. deficit function
TOPT	Parameter in temp stress function

produced much more realistic runoff simulation when compared to observed runoff (see Figure 3). We conducted a similar study using the NOAH LSM on selected US basins and found similar results (Duan et al., 1995).

Three basins representing dry, moderate, and wet climate were chosen. These basins are Bird Creek in Oklahoma, Leaf River in Mississippi, and French Broad in North Carolina. All three basins have the same vegetation class (broadleaf trees with ground cover) and soil class (loamy sand). For more detailed information on these basins, see Schaake et al. (1996). In the first test, the NOAH LSM was run using *a priori* parameters as specified in previous section. In the second test, four of NOAH LSM parameters were selected for calibration. The choice of parameters for calibration in this particular test is arbitrary. To properly decide which parameters should be calibrated, it would be prudent to conduct sensitivity studies so that only the sensitive parameters are included in calibration. Performance measures of the test runs are recorded in Table 7. The results clearly indicated a remarkable difference in the performance of the two tests, no matter which measure is compared. In Table 8, the *a priori* parameters and the calibrated values are listed. The table indicates that the differences between *a priori* and calibrated parameter values are substantial.

3.2 *Effect of Heterogeneity on Interpretation of Model Parameters*

Most PBM's are based on an understanding of physics at a point scale or a plot scale (up to 10's of m^2). However, the spatial scales of interest in land surface modeling are several orders of magnitude larger (from 10^2 's to 10^5 's of km^2). Nevertheless, some hydrologic modelers simply assume that the same governing equations for point processes also govern large spatial processes.

But, the model physics and parameters describing behavior of aggregate variables are not the same as the point physics and parameters. For example, the original infiltration parameterization of the NOAH LSM was based on a point formulation of the diffusion equation (Ek and Mahrt, 1991). It did not consider spatial heterogeneity of soil moisture distribution, nor did it consider the sub-grid distribution of precipitation. That formulation led to excessive infiltration and consequently little surface runoff. The current NOAH LSM uses the infiltration formulation of the Simple Water Balance (SWB) model of Schaake et al. (1996) that was derived from an assumed point process by considering the spatial distributions of both infiltration capacity and precipitation. More runoff is generated by the SWB approach than by the original approach because some parts of the basin receive precipitation in excess of the local infiltration capacity even when the average precipitation rate is less than the average infiltration capacity. Table 9 shows that the SWB approach produced much superior runoff simulation statistics than the original approach.

Other LSM's use similar approaches to treat spatial heterogeneity of hydrologic variables. For example, the Xianganjiang (Zhao, 1992) and VIC (Liang et al., 1994) models use a nonlinear function to represent the effect of variable infiltration capacity on the partition of precipitation into runoff and infiltration. Koster and Suarez (1996) proposed the use of a tiled approach to handle spatial heterogeneity. In their approach, a grid cell is partitioned into several tiles, each representing a distinct land cover type. The model physics is applied to each tile. The total model response of the grid cell is the aggregated response of individual tiles.

3.3 *Scale Dependence of Model Parameters*

Some model parameters have been shown to be highly dependent on the space/time scale of model application. Such parameters estimated at one space/time scale can not be used at another scale without being re-calibrated. Finnerty et al. (1997) conducted a study to investigate the scale dependency of the Sacramento model. In the study, merged radar-gage precipitation observations were aggregated over different

Table 4. Definition of Selected Soil Parameters in NOAH LSM

SMCMAX	Porosity
SMCREF	Field capacity
SMCWLT	Wilting-point soil moisture content
DWSAT	Saturated soil hydraulic diffussivity
DKSAT	Saturated soil hydraulic conductivity
B	Clapp-Hornberger exponent
KDT	Infiltration parameter

Table 5. *A Priori* Values of Selected Soil Parameters in NOAH LSM

Model parameters	Soil classes							
	1	2	3	4	5	6	7	8
SMCMAX	0.421	0.464	0.468	0.434	0.406	0.465	0.404	0.439
SMCREF	0.283	0.387	0.412	0.312	0.338	0.382	0.315	0.329
SMCWLT	0.029	0.119	0.139	0.047	0.010	0.103	0.069	0.066
DKSAT	1.41E-5	0.20E-5	0.10E-5	0.52E-5	0.72E-5	0.25E-5	0.45E-5	0.34E-5
DWSAT	5.71E-6	2.33E-5	1.16E-5	7.96E-6	1.90E-5	1.14E-5	1.06E-5	1.46E-5
B	4.26	8.72	11.55	4.74	10.73	8.17	6.77	5.25
KDT	12.32	1.74	0.87	4.54	6.29	2.18	3.93	2.97

space/time scales. They found out that runoff timing and volumes of the Sacramento soil moisture accounting (SAC-SMA, see Burnash et al., 1973) model are highly dependent on how precipitation forcing is aggregated spatially and temporally. They showed that moving to finer spatial resolution to compute precipitation inputs would lead to an increase in both the fast runoff components (e.g., surface and interflow) and the slow runoff components (primary and supplemental baseflow) and a corresponding drop in total evapotranspiration over a 9-month period (Figure 4). They also reported that when the temporal resolution of precipitation inputs was reduced from 6 hours to 1 hour, a significant increase in fast runoff components and total runoff volume and a decrease in slow runoff components were observed (Figure 5). They further pointed out that adjusting certain model parameters would reduce the biases caused by the differing space/time scales. Schaake et al. (1998) confirmed the findings of Finnerty et al. (1997) regarding the sensitivity of runoff volume to temporal disaggregation. Figure 6 displays the difference in long term monthly runoff simulations of NOAH LSM for Arkansas/Red river basin. The solid line represents simulation using precipitation input that was disaggregated uniformly from daily data into hourly data, while the dash line represents simulation using hourly precipitation inputs. The maximum difference monthly runoff is 6.5mm. The annual runoff difference is 38mm (131mm to 169mm), representing 30% of the annual runoff total.

Koren et al. (1999) used the same data cited above and compared the degree of scale dependency of 4 different LSM's (Figure 7). They showed that the degree of scale dependency is related to model formulation. Model formulations based on infiltration excess equations derived for a single vertical column showed the most scale dependency. They reported that a reformulated version of the Sacramento model that considers spatial distribution of precipitation input performed much better than other LSM's. To deal with the scale dependency of LSM parameters, Shuttleworth (1998) developed some aggregation rules to compute effective parameters that would preserve aggregated surface energy balance from point scale to grid scale.

4. PARAMETER ESTIMATION STRATEGY

Because existing *a priori* estimation procedures may lead to unreliable parameter estimates, and because sufficient data do not exist to tune model parameters for many applications, improved *a priori* parameter estimation procedures are needed. A strategy to meet this need has been developed and is being used by an international Model Parameter Estimation Experiment (MOPEX) (Schaake et al., 1999).

The goal of the MOPEX parameter estimation strategy is to develop techniques for *a priori* estimation of parameters that can be applied globally. The first step of the strategy is to

Table 6. A Priori Values of Selected Vegetation Parameters in NOAH LSM

Type	ALBEDO	Z0	RCMIN	RSMAX	HS
1	0.11	2.653	150.0	30.0	41.69
2	0.19	0.826	100.0	30.0	54.53
3	0.16	0.563	125.0	30.0	51.93
4	0.13	1.089	150.0	30.0	47.35
5	0.19	0.854	100.0	30.0	47.35
6	0.19	0.856	70.0	65.0	54.53
7	0.19	0.075	40.0	100.0	36.35
8	0.29	0.238	300.0	100.0	42.00
9	0.29	0.065	400.0	100.0	42.00
10	0.14	0.076	150.0	100.0	42.00
11	0.15	0.011	999.0	999.0	999.0
12	0.19	0.075	40.0	100.0	36.35
13	0.15	0.011	999.0	999.0	999.0

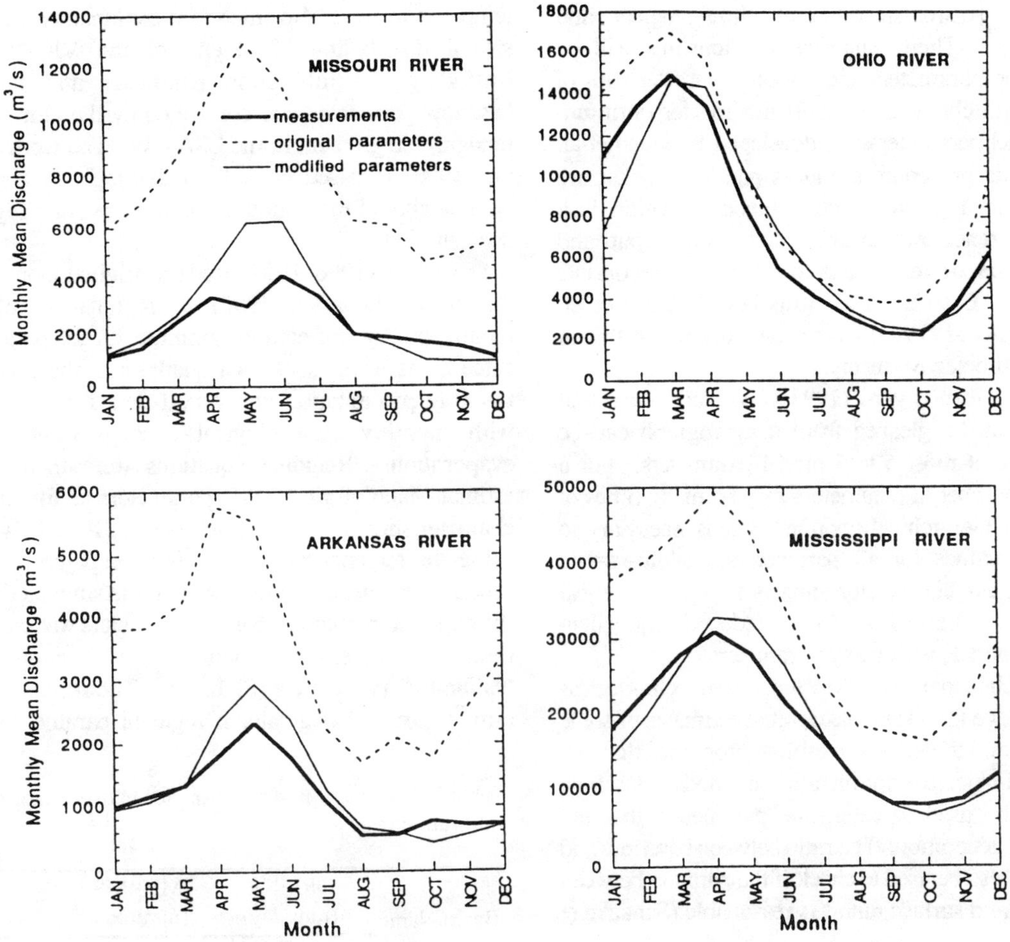

Figure 3. Comparison of simulated runoff and observed runoff (From Liston et al., 1993)

develop the necessary data sets. The strategy is then to use these data to study individual models using three parallel paths illustrated in Figure 8. The first path is to make control runs with model parameters estimated using existing *a priori* parameter estimation procedures. The second path is to make model runs using calibrated or tuned values of selected model parameters. Then, relationships would be developed between the calibrated parameters and basin climate, soils, vegetation and topographic characteristics. These relationships are used to define the new *a priori* parameters. The third path is to make new model runs using the new *a priori* parameter estimates. Achievement of the parameter estimation goal is then established in two steps. The first is to measure how much of the potential improvement in model performance when operated in calibration path is obtained when the model is operated using new *a priori* parameters. This step uses the same data sets as were used to develop the new *a priori* parameter estimates. The second step is to demonstrate that new *a priori* techniques produce better model results than existing *a priori* techniques for basins not used to develop the new *a priori* techniques.

4.1 Data Needed for Parameter Estimation

MOPEX is assembling the hydrometeorological data as well as land surface characteristics data that are needed to implement its parameter estimation strategy. MOPEX has been funded by GEWEX Continental-scale International Project (GCIP, see WMO, 1992). Data from at least 200 basins in the US and other parts of the world are being assembled. These basins cover a wide variety of climates. A minimum of 10 years of data are required for all basins.

4.2 Systematic Procedure for Automatic Calibration

A major effort in implementing the MOPEX strategy is to develop a systematic procedure for automatic calibration of selected model parameters and to apply this procedure to a large number of basins in different climatological and hydrologic settings. Then, empirical relationships will be sought between the parameters and various characteristics of soils, vegetation and climate. A robust method for optimum estimation of model parameters was developed by Duan et al. (1992, 1994). This procedure produces parameters that are optimum relative to a given objective function, which is a measure of the difference between simulated model outputs and observations. An important issue is to select the appropriate objective function. Bastidas et al. (this book), Yapo et al. (1997), and Gupta et al (1999) have been working on this in the context of multi-objective theory.

Jakeman and Hornberger (1993) pointed out that information that can be gleaned from a hydrograph can be useful to determine at most 5 to 6 model parameters. For a typical LSM, the number of parameters may be many times of that. Therefore, a hierarchical approach seems necessary to determine proper values for all parameters. Multi-variate techniques such as eigenvector analysis and canonical correlation analysis may lead to a reduced equivalent parameter set for models with many parameters.

Another approach is to use baseflow separation techniques as described by Rutledge (1993) to make initial estimates of some model parameters and to derive streamflow properties that can be used with multi-objective optimization methods. Baseflow hydrographs can be used to determine parameters that are related to baseflow generation. The ratio between baseflow and surface runoff can also be used to check if the partition between modeled baseflow and surface runoff is reasonable (Schaake et al., 1999).

5. TESTING THE PARAMETER ESTIMATION STRATEGY

Initial studies have been undertaken to test the feasibility of regional parameter estimation (i.e. a *priori* estimation). Abdulla et al. (1996) followed the above strategy to regionalize the parameters of the VIC-2L LSM for the GCIP Large Scale Area-Southwest (LSA-SW). Optimum models parameters for 34 catchments in the LSA-SW were used as dependent variables in regression equations. The independent variables were easily determinable basin characteristics. The approach was tested by application of the regionally estimated parameters to basins not in the data set used to develop the regional parameter regression equation. Of ten soil attributes and twelve climatological characteristics considered in the analysis, the saturated hydraulic conductivity, the average permeability and the hydrologic group B soil type were the most important soil attributes in estimating VIC-2L parameters. The inter-arrival time (i.e., the time between the end of the first storm and the beginning of the second storm) and mean annual temperature were also important explanatory variables. The spatial distribution of several of the VIC-2L parameters (including the infiltration parameter and the maximum baseflow parameter) are controlled by the climatological and topographic gradient in the LSA-SW. The work of Abdulla et al. (1996) is based on a limited sample of basins and the applicability of the regional equations to other regions has not been studied.

Duan et al. (1996, 1997) used the original monthly version of the SWB (Schaake, 1990) to regionalize all five SWB parameters by correlating optimum model parameters at 50 catchments in the southeast quadrant of the U.S. They used monthly precipitation and runoff data for 40 years together with monthly climatological average values of potential evaporation. Regional equations were then developed to estimate each of all 5 SWB parameters as functions of basin characteristics of the 50 catchments. Runoff was simulated using the regional parameters and compared to the observed runoff. The mean absolute error in the simulated runoff using the regional (regression-based) parameters was compared to the mean absolute error in simulated runoff using the optimum (calibrated) parameters (Figure 9). The results show that the error is somewhat greater if regional parameters rather than

Table 7. Comparison of performance between a priori and calibration tests

Basin Name	DRMS [1] (mm/day)		MVRMS [2] (mm/month)		E [3]	
	a priori	calib.	a priori	calib.	a priori	calib.
Bird Creek	1.52	1.11	13.30	12.95	0.690	0.832
Leaf River	1.48	0.88	13.15	12.41	0.421	0.793
French Broad	2.54	1.64	40.68	27.69	-0.130	0.548

Notes:
1. DRMS denotes daily root mean square errors
2. MVRMS denotes monthly volume root mean square errors
3. E is the Nash-Sutcliffe efficiency coefficient (≤ 1; a value of 1 indicates perfect fitting)

Table 8 Comparison of *a priori* and calibrated parameters

Names	Bird Creek		Leaf River		French Broad	
	a priori	calib.	a priori	calib.	a priori	calib.
Z2 (mm)	1.0	0.805	1.0	1.67	1.0	3.00
DKSAT(m/s)	.710e-5	.175e-4	.710e-5	.293e-3	.710e-5	.147e-5
DWSAT (m/s)	.827e-6	.509e-5	.827e-6	.239e-3	.827e-6	.850e-6
KDT(1/day)	12.32	2.59	12.32	4.66	12.32	1.75

Table 9. Comparison of performance statistics using original and current infiltration formulation in NOAH LSM

	DRMS (mm/day)		MVRMS (mm/mon)		E	
	original	current	original	current	original	current
Bird Creek	1.82	1.52	17.91	13.30	0.561	0.690
Leaf River	1.56	1.48	14.54	13.15	0.351	0.421
French Broad	3.02	2.54	43.21	40.68	-0.617	-0.130

optimal parameters are used. But there was a large difference in how well the SWB functioned from basin to basin. Figure 9 indicates that where SWB worked best with optimum parameters, it also worked best with regional parameters.

Below we present the recent work by Koren et al. (2000) on the *a priori* parameter estimates for the Sacrament model to illustrate the strategy outlined in Figure 8.

5.1 A Priori Parameter Estimates for the Sacramento Model

Sacramento Soil Moisture Accounting model (SAC-SMA) is an operational model used in National Weather Service for river and flood forecasts throughout the United States. SAC-SMA has a two-layer structure (see Figure 10). Each layer consists of tension and free water storages. The free water storage of lower layer is further divided into two sub-storages which control supplemental (fast) and primary (slow) groundwater flows. A detailed description of SAC-SMA is available in the literature (see Burnash et al., 1973). SAC-SMA has a total of 16 parameters (see Table 10). Even though there are strong physical arguments for them, SAC-SMA parameters can not be measured and are usually estimated using historical hydrometeorological data. The estimation of SAC-SMA parameters is very difficult and this has been demonstrated by numerous previous studies (Brazil, 1988; Duan et al., 1994). This section presents an *a priori* procedure to determine SAC-SMA parameters based on soil texture properties. The 1 km STATSGO soil database developed by Miller and White (1999) was used to determine the soil properties for 11 soil layers (from ground surface to 2.5m beneath).

To quantify relationships between SAC-SMA parameters and soil properties, assumptions were made that tension water storages are related to available soil water, and that free water storages are related to gravitational soil water. Available soil water and gravitational soil water can be estimated from soil properties such as porosity, θ_{max}, field capacity, θ_{fld}, and wilting point, θ_{wlt}. The combined depth of the upper and lower layers is assumed to be equal to the soil profile depth, Z_{max}. A concept of an initial rain abstraction (Linsley et al., 1958) is used to estimate the thickness of the upper layer. The initial rain abstraction is a function of land cover, interception, infiltration, depression storage, and antecedent soil moisture. The Soil Conservation Service (SCS) developed an approach to estimate the initial rain abstraction based on soil and vegetation type, as well as on soil moisture conditions (McCuen, 1982). Under the average soil moisture condition stipulated by SCS, one can assume that the upper tension water storage is full and the free water storage is empty. In this case, the initial rain abstraction should satisfy upper free water storage capacity. The upper layer thickness, Z_{up}, can then be calculated based on an SCS curve number, CN, for each soil profile:

$$Z_{up} = 5.08 \times \frac{1000/CN - 10}{\theta_{max} - \theta_{fld}},\ mm \tag{1}$$

The SAC-SMA storages can be estimated then based on the schematic in Figure 11:

$$UZTWM = (\theta_{fld} - \theta_{wlt}) Z_{up} \tag{2}$$

$$UZFWM = (\theta_{max} - \theta_{fld}) Z_{up} \tag{3}$$

$$LZTWM = (\theta_{fld} - \theta_{wlt})(Z_{max} - Z_{up}) \tag{4}$$

$$LZFWM = LZFSM + LZFPM = (\theta_{max} - \theta_{fld})(Z_{max} - Z_{up}) \tag{5}$$

To split the total free water storage of the lower zone into two components, one can assume that lighter soils (with a higher percentage of sand) have less supplemental storage/runoff than

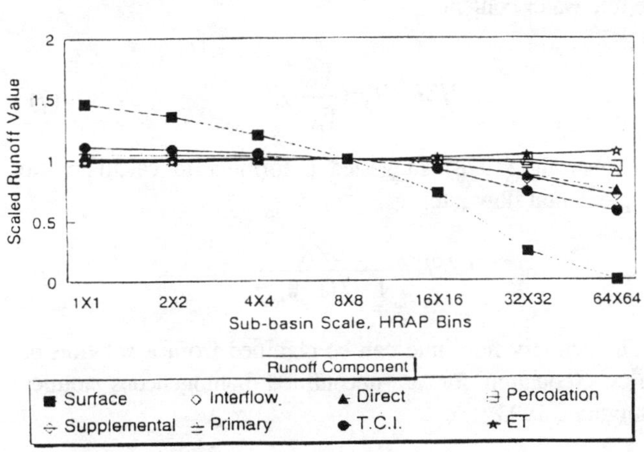

Figure 4. Scale dependence of SAC-SMA water balance components on spatial aggregation of precipitation input (From Finnerty et al., 1997)

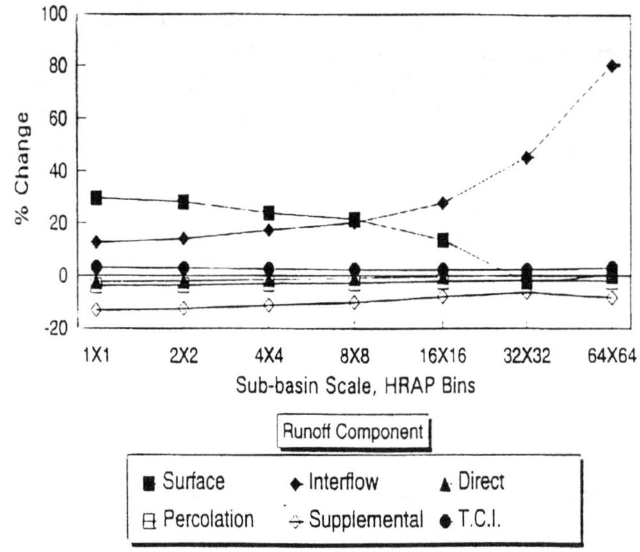

Figure 5. Scale dependence of SAC-SMA water balance components to temporal aggregation of precipitation input (From Finnerty et al., 1997)

heavier soils. The soil wilting point can be used as an index of how heavy a soil is:

$$LZFSM = LZFWM\left(\frac{\theta_{wlt}}{\theta_{max}}\right)^{n_1} \tag{6}$$

A value of 1.6 for n_1 was used in this analysis to keep an average ratio between supplemental and primary storage capacities close to 1/3.

Empirical relationships were derived to estimate runoff depletion coefficients. Interflow rate is calculated as a ratio of the free water content:

$$UZK = 1 - \left(\frac{\theta_{fld}}{\theta_{max}}\right)^{n_1} \tag{7}$$

Armstrong (1978) suggested a formula to calculate the supplemental flow rate:

$$LZSK = \frac{UZK}{1 + 2(1 - \theta_{wlt})} \tag{8}$$

The primary flow rate can be obtained from a solution of Darcy's equation for an unconfined homogeneous aquifer (Dingman, 1993):

$$LZPK = 1 - \exp\left[-\frac{\pi^2 K_s (Z_{max} - Z_{up}) D_s^2}{4\mu}\right] \tag{9}$$

where K_s is the saturated hydraulic conductivity, D_s is a stream channel density, and μ is the specific yield of soil. An empirical relationship was derived using data from Linsley et al. (1958) and Armstrong (1978):

$$\mu = 3.5(\theta_{max} - \theta_{fld})^{1.66} \tag{10}$$

The maximum percolation rate can be calculated from other parameters:

$$ZPERC = \frac{LZTWM + LZFSM\cdot(1 - LZSK)}{LZFSM\cdot LZSK + LZFPM\cdot LZPK} + \frac{LZFPM\cdot(1 - LZPK)}{LZFSM\cdot LZSK + LZFPM\cdot LZPK} \tag{11}$$

The ***REXP*** parameter, which defines the shape of the percolation curve, is associated with soil type. The minimum permissible value of 1.0 would indicate an almost constant decrease of percolation as the lower zone deficiency decreases and would be associated with sand. A large value would indicate a rapid decrease of percolation as the zones become saturated, as is expected in a clay. A reasonable approximation of recommended values for ***REXP*** is

$$REXP = \left(\frac{\theta_{wlt}}{\theta_{wlt,sand} - 0.001}\right)^{0.5} \tag{12}$$

The ***PFREE*** parameter relates to the water that follows paths through cracks, faults, etc. to escape the capillary demands of

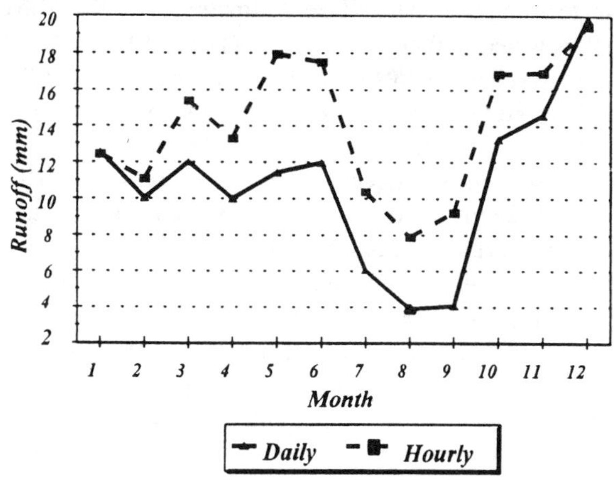

Figure 6. Long term monthly average runoff using daily and hourly precipitation

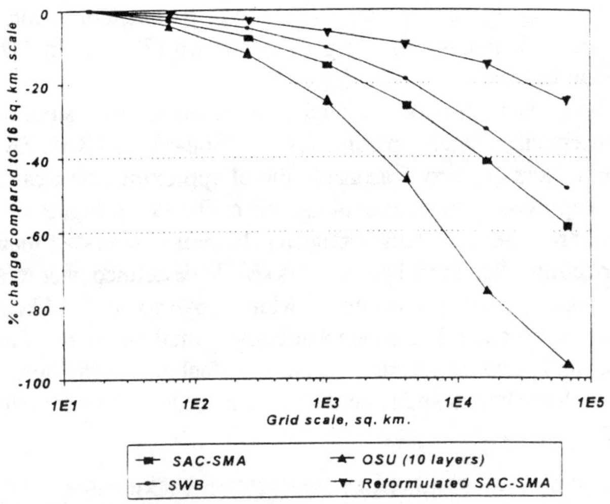

Figure 7. Scale dependence of difference LSM's on spatial aggregation of precipitation input (From Koren et al., 1999)

the soil. One can expect that clay type soils may have more cracks to recharge baseflow, and therefore ***PFREE*** should take on higher values. The ration of wilting point to porosity is a good index to represent this feature:

$$PFREE = (\frac{\theta_{wlt}}{\theta_{max}})^{n_1} \tag{13}$$

These derived relationships cannot account for some specific local conditions of river basins. Therefore, estimated parameters should be adjusted using calibration if there are observed rainfall/discharge data. The main objective of these relationships is to give reasonable initial values, and to reduce uncertainties in parameter ranges. Another benefit is that these relationships are based on available physical properties of soils and can be used on ungaged basins. They also allow a reduction in the number of calibrated parameters. Changing just a curve number *CN* can redefine all other model parameters. It is possible also to adjust/calibrate just four well defined characteristics of soil properties, θ_{max}, θ_{fld}, θ_{wlt}, Z_{max}, to get better estimates of 11 model parameters. This is very important for semi-distributed approach when measured discharge data are only available at the outlet of the basin.

5.2 Application of A Priori Estimates for the Sacramento Model to Test Basins

The approach delineated was used for a few river basins in different climatic regions (Arkansas-Red river basin, Des Moines basin, Conasauga river, Georgia). Carefully calibrated parameter sets were available for these basins. Soil characteristics for each soil class were calculated using the Cosby soil parameters (Cosby et al., 1984). SCS curve numbers were defined for four hydrologic soil groups (McCuen, 1982) under average soil moisture conditions: 80 mm for group A soil classes 1-3; 33 mm for group B classes 4-6; 20 mm for group C class 7; 13 mm for group D classes 8-12. SAC-SMA parameters calculated from Eqs. 1-13 were compared to calibrated parameters for selected river basins. Calibrated parameters and those estimated from soil texture are plotted in Figures 12a-b for the Illinois river, Oklahoma. Overall, most parameters derived from soil texture agreed reasonably well with calibrated parameters. The biggest difference was seen in ***ZPERC***. This difference is due to the fact that ***ZPERC*** was treated as an independent parameter in calibration and could take on any value over a wide range, subject to calibration procedures used. One the other hand, ***ZPERC*** was computed as a function of other SAC-SMA parameters in the soil-based approach (see Eq.11) and its values are determined by local soil properties. Noticeable differences were seen between calibrated and soil-based lower zone free water storages for basins with significant baseflow contribution from deep aquifers. This is because the soil-based approach is restricted to the top 2.5 m of the soil layer and can not account for deep groundwater storage. Future research may deal with this limitation by utilizing additional information such as outlet hydrographs.

Calibrated parameters were obtained using a manual calibration package which is based on visual fitting of simulated and observed hydrographs, and comparing different

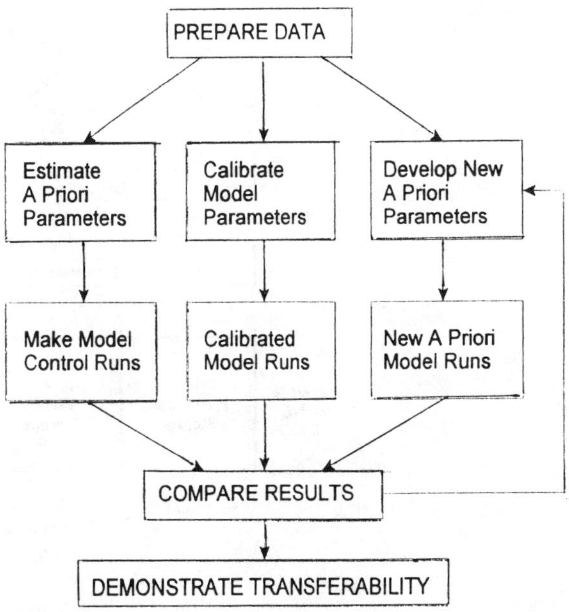

Figure 8. Strategy for a priori parameter estimation

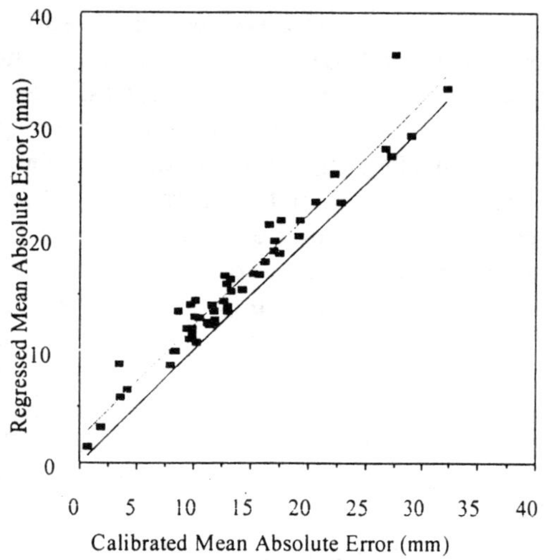

Figure 9. Comparison of objective functions using calibrated parameters and regionalized parameters

statistics. This is a subjective procedure and an 'optimal' parameter set can vary widely if calibrated by different hydrologists. Comparison of calibrated and soil derived parameters only can not provide conclusive results. Accuracy statistics of hydrographs simulated using calibrated and soil derived parameters were also calculated and are presented in Table 11. As shown in the table, calibrated parameters usually produce higher accuracy although the gain is not significant as compared to use of soil-based parameters. If parameters were not properly calibrated, soil-based parameters actually produced better statistics, e.g., Tilton basin (TLNG1 in Table 11) in Georgia.

Soil data provide valuable information in estimating conceptual model parameters. Soil-based SAC-SMA parameters are very reasonable initial approximation that can be improved by manual or automatic calibration if input/output data are available. This approach is helpful in semi-distributed modeling when model parameters should be defined over many ungaged small sub-basins. More analysis of soil-based parameters should be done in different climatic regions. Land use and land cover data contain valuable information on hydrologic basins and should be used in addition to soil texture data.

6. SUMMARY AND CONCLUSIONS

Land surface models contain parameters (coefficients and exponents) that control model behavior. *A priori* estimates of LSM parameters are needed to apply these models over large regions or globally where sufficient data do not exist to permit calibration of those model parameters that vary spatially. *A priori* parameter estimation techniques relate model parameters to land surface characteristics including climate, soils, vegetation and topography. Global land characteristics data sets are being developed with improved spatial resolution using satellite data. At present, relationships between model parameters and these characteristics are based on limited information from field experiments or from analyses of data

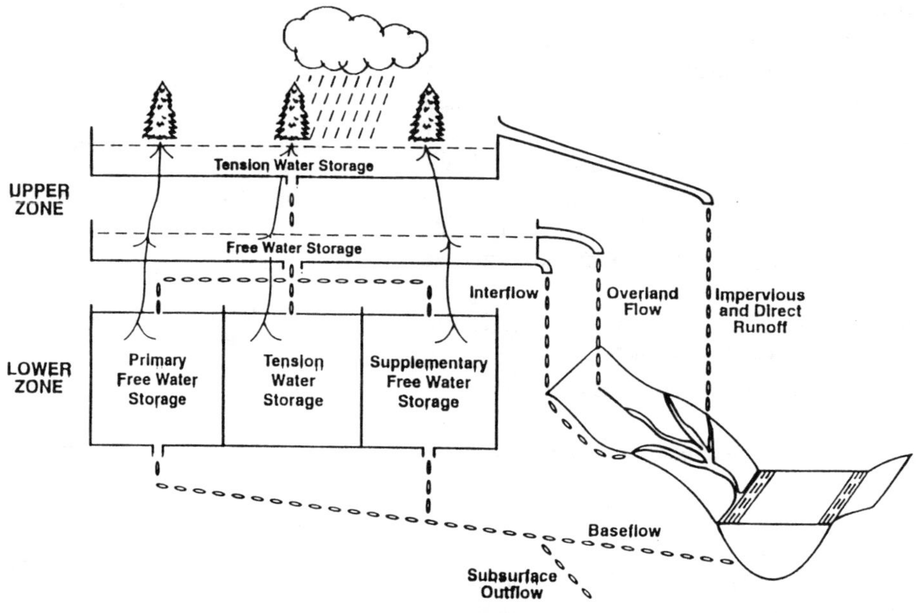

Figure 10. A schematic description of SAC-SMA

Table 10. Parameters of the NWSRFS-SMA Model

Parameters	Description
UZTWM	Maximum capacity of the upper zone tension water storage (mm)
UZFWM	Maximum capacity of the upper zone free water storage (mm)
LZTWM	Maximum capacity of the lower zone tension water storage (mm)
LZFPM	Maximum capacity of the lower zone free water primary storage (mm)
LZFSM	Maximum capacity of the lower zone free water supplemental storage (mm)
ADIMP	Additional impervious area (decimal fraction)
UZK	Upper zone free water lateral depletion rate (day^{-1})
LZPK	Lower zone primary free water depletion rate (day^{-1})
LZSK	Lower zone supplemental free water depletion rate (day^{-1})
ZPERC	Maximum percolation rate
REXP	Exponent of the percolation equation
PCTIM	Impervious fraction of the watershed area
RIVA	Riparian vegetation area
PFREE	Fraction of water percolating directly from upper zone to lower zone free water storage
SIDE	Ratio of deep recharge to channel baseflow
RSERV	Fraction of lower zone free water not transferrable to lower zone tension water

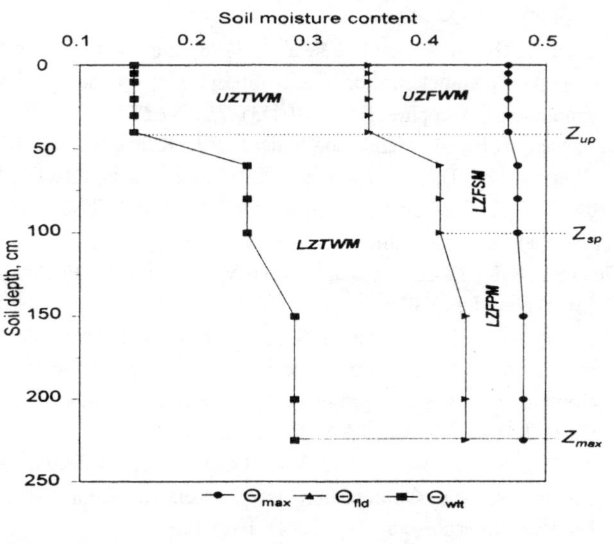

Figure 11. Schematic of SAC-SMA parameter definition based on soil profile properties

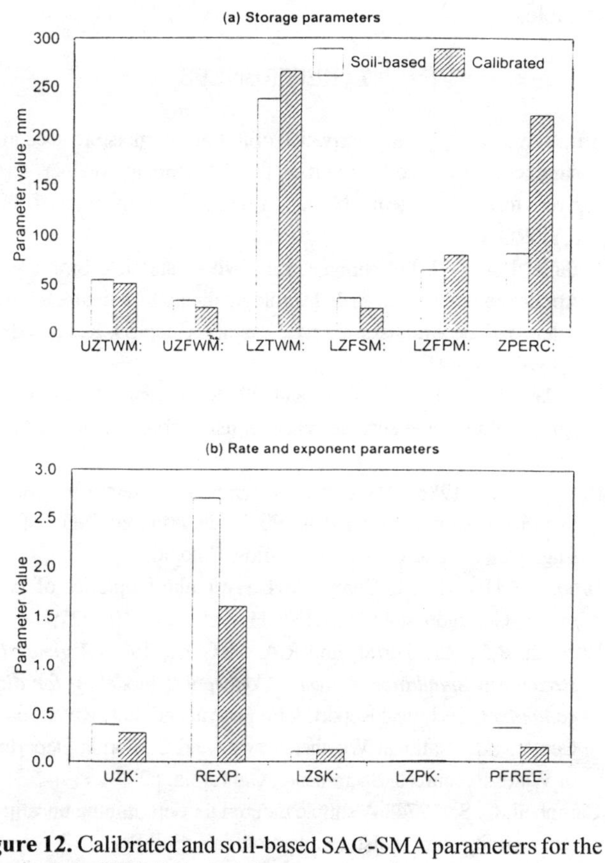

Figure 12. Calibrated and soil-based SAC-SMA parameters for the Illinois river (WTTO2), Oklahoma

reported in the literature. Available land characteristics data sets and existing *a priori* techniques for using them were summarized. Experiments conducted by the Project for Intercomparison of Land Surface Process Schemes (PILPS) show a wide range of model performance when existing *a priori* parameter estimates are used. Numerous studies have shown that existing LSM parameters can be tuned to observed data to achieve better performance. The fact that parameters

Table 11. Performance statistics of simulated hydrographs using calibrated and regional parameters

Basin ID/RFC	Calibrated/regional parameters				Soil derived parameters			
	DRMS	MVRMS	R^2	Bias, %	DRMS	MVRMS	R^2	Bias, %
WTTO2/ABRFC, calibrated	16.2	10.8	0.92	10.9	21.1	11.7	0.89	12.8
TIFM7/ABRFC, calibrated	22.3	8.7	0.91	-2.8	25.0	13.0	0.87	-21.8
BSSI4/NCRFC, calibrated	13.6	12.4	0.84	12.7	13.4	11.3	0.84	12.4
DLLI4/NCRFC, calibrated	13.3	11.3	0.84	3.1	13.1	13.2	0.81	-1.1
TLNG1/SERFC, calibrated	18.0	8.3	0.95	1.1	26.5	11.3	0.92	4.5
TLNG1/SERFC, regional	34.2	24.9	0.85	22.5	26.5	11.3	0.92	4.5

having the same apparent meaning in different models have different optimal values after being tuned suggests that model parameters are actually model dependent. An example was presented of how a priori estimates might be made for a conceptual model where relationships between model parameters and basin characteristics were inferred empirically. The international Model Parameter Estimation Experiment (MOPEX) is developing data sets and techniques for data analysis that should lead to improved a priori estimation of those LSM parameters that can be inferred by analyzing differences between observed and simulated streamflow variables.

7. REFERENCES

Armstrong, B. L., 1978. Derivation of initial soil moisture accounting parameters from soil properties for the national weather service river forecast system, NOAA Technical Memorandum NWS HYDRO 37.

Abdulla, F.A, D.P. Lettenmaier, E.F. Wood, and J.A. Smith, 1996. Application of a macroscale hydrologic model to estimate the water balance of the Arkansas-red river basin, J. Geophys. Res., 101(D3), 7,449-7459

Bastidas, L.A., H.V. Gupta, and S. Sorooshian, Bounding the parameters of land-surface schemes using observational data, this book

Brazil, L.E., 1988, Multilevel calibration strategy for complex hydrologic simulation models, Ph.D. dissertation, Dept. of Civil Eng., Color. State Univ., Fort Collins, Colorado

Brooks, R.H. and A.T. Corey, 1964, Hydraulic properties of porous media, Colorado State University Hydrol. Paper No.3 27p

Burnash, R.J., R.L. Ferral, and R.A. McGuire, 1973, *A generalized streamflow simulation system - Conceptual modeling for digital computers*, Technical Report, Joint Federal and State River Forecast Center, U.S. National Weather Service and California Department of Water Resources, Sacramento, California, p204

Campbell, G. S., 1974. A simple method for determining unsaturated conductivity from moisture retention data, Soil Sci., 117

Carsel, R.T., and R.S. Parrish, 1988, Developing joint probability distributions of soil water retention characteristics, Water Resourc. Res., 24(5), 755-769

Chen, F., K. Mitchell, J. Schaake, Y. Xue, H. Pan, V. Koren, Q.Y. Duan, M. Ek, and A. Betts, 1996, Modeling of land-surface evaporation by four schemes and comparison with FIFE observations. J. Geophys. Res., 101(D3), 7251-7268.

Clapp, R.B. and G.M. Hornberger, 1978, Empirical equations for some soil hydraulic properties, Water Resourc. Res., 14(4), 601-604

Cosby, B. J., G. M. Hornberger, R. B. Clapp, and T. R. Ginn, 1984. A statistical exploration of the relationships of soil moisture characteristics to the physical properties of soils, Water Resourc. Res., 20(6), 682-690

Cuenca, R.H., M. Ek and L. Mahrt, 1996, Impact of soil water property parameterization on atmospheric boundary layer simulation, J. Geophys. Res., 101(D3), 7,269-7,277

Dickinson, R.E., A. Henderson-Sellers, P.J. Kennedy, and M.F. Wilson, 1986, Biosphere-atmosphere transfer scheme (BATS) for the NCAR Community Climate Model, NCAR/TN-275+STR, National Center for Atmospheric Research, 69pp.

Dingman, S. L., 1993. Physical Hydrology, Prentice Hall, Englewood Cliffs, New Jersey 07632.

Dorman, J.L., and P.J. Sellers, 1989, A global climatology of albedo, roughness length and stomatal resistence for atmospheric general circulation models as represented by the Simple Biosphere model (SiB), J. Appl. Meteor., 28, 833-855

Duan, Q., S. Sorooshian, and V.K. Gupta, 1992, Effective and Efficient Global Optimization for Conceptual Rainfall-Runoff Models, Water Resour. Res., 28(4), 1015-1031

Duan, Q., S. Sorooshian, and V.K. Gupta, 1994, Optimal Use of the SCE-UA Global Optimization Method for Calibrating Watershed Models, J. of Hydro., 158, 265-284

Duan, Q., J. Schaake, V. Koren, F. Chen and K. Mitchell, 1995. Testing of a physically-based soil hydrology model: a priori parameters vs tuned parameters. EOS Trans. AGU, 76(46) Fall Meet. Suppl., 92.

Duan, Q., V. Koren, P. Koch and J. Schaake, 1996. Use of NDVI and soil characteristics data for regional parameter estimation of hydrologic models. EOS. AGU, 77(17), Spring Meet. Suppl., 138.

Duan, Q., V.I. Koren, J.C. Schaake, and S. Cong, 1997, Variability of

Hydrologic Model Parameters According to Space/Time Scales and Other Physical Characteristics, The 5th Scientific Assembly of the International Association of Hydrological Sciences, 23 April - 3 May, 1997, Morocco

Ek, M. and L. Mahrt, 1991, OSU1-D PBL Model User's Guide, College of Oceanic and Atmospheric Sciences, Oregon State Univ., Corvalis, Oregon

FAO, 1961, Production Year Book, FAO, Rome

FAO, 1976, Production Year Book, FAO, Rome

FAO, 1981, Production Year Book, FAO, Rome

Finnerty, B.D., M.B. Smith, D.-J. Seo, V. Koren, & G.E. Moglen, 1997, Space-time scale sensitivity of the Sacramento model to radar-gage precipitation inputs, J. of Hydrol., 203, 21-38

Gallo, K., D. Tarpley, K. Mitchell, I. Csiszar, and B. Reed, 2000, Monthly fractional green vegetation associated with land cover classes of the conterminous USA, paper submitted to Geophysical Research Letters

Goward, S.N., B. Markham, D.G. Dye, W. Dulaney, and J. Yang, 1991, Normalized Difference Vegetation Index measurements from the Advanced Very High Resolution Radiometer, Remote Sens. Env., 35, 121-136

Gupta, H.V., L.A. Bastidas, S. Sorooshian, W.J. Shuttleworth, and Z.L. Yang, 1999, Parameter estimation of a land surface scheme using multi-criteria methods, J. of Geophys. Res., 104(D16), 19,491-19504

Gutman, G.G., 1994, Global data on land surface parameters from NOAA AVHRR for use in numerical climate models, J. of Climate, 7, 669-680

Gutman, G., and A. Ignatov, 1998, The derivation of the green vegetation fraction from NOAA/AVHRR data for use in numerical weather prediction models, Int. J. Remote Sensing, 19(8), 1,533-1,543

Hansen, M., and B. Reed, 1999, A comparison of the IGBP DISCover and University of Maryland 1 km global land cover products, to appear in Int. J. Remote Sensing

Hansen, M., R.S. DeFries, J.R.G. Townshend, and R. Sohlberg, 1999, Global land cover classification at 1 km spatial resolution using a classification tree approach, To appear in Int. J. Remote Sensing

Hummel, J.R., and R.A. Reck, 1979, A global surface albedo model, J. Appl. Meteor., 18,239-253

IGBP, 1990, The initial core projects, IGBP Report 12, International Geosphere-Biosphere Program, Stockholm, Sweden

Ibbitt, R.P., 1970, Systematic parameter fitting for conceptual models of catchment hydrology, Ph.D. dissertation, Imp. Coll. Of Sci. and Technol., University of London, London, England

Jakeman, A. and G.M Hornberger, 1993, How much complexity is warranted in a rainfall-runoff model, Water Resourc. Res., 29(8), 2637-2649

Johnston, P.R. and D. Pilgrim, 1976, Parameter optimization for watershed models, Water Resourc. Res., 12(3), 477-86

Koren, V.I. , B.D. Finnerty, J.C. Schaake, M.B. Smith, D.-J. Seo, and Q.-Y. Duan, *1999, Scale dependencies of hydrologic models to spatial variability of rainfall*, J. of Hydrol., 217, 285-302

Koren, V.I., M. Smith, D. Wang, & Z. Zhang, 2000, Use of soil property data in the derivation of conceptual rainfall-runoff model parameters, Preprints, 15th Conference on Hydrology, Long Beach, CA, Amer. Meteor. Soc., 10-14 January 2000, Paper 2.16

Koster, R.D., and M.J. Suarez, 1996, Energy and Water Balance Calculations in the MOSAIC LSM, NASA Tech. Memo. 104606, Vol.9, NASA, Washington, DC, p58

Kuczera, G., 1997, Efficient subspace probabilistic parameter optimization for catchment models, Water Resourc. Res., 33(1), 177-185

Kuchler, D.S., 1983, World map of natural vegetation, Coode's World Atlas, 16th ed., Rand McNally, 16-17

Liang, X., D.P. Lettenmaier, E.F., Wood, and S.J. Burges, 1994, A simple hydrologically based model of land surface water and energy fluxes for GCM, J. of Geophys. Res., 99(D7), 14,415-14,428

Liang, X., E.F. Wood, D.P. Lettenmaier, D. Lohmann, A. Boone, S. Chang, F. Chen, Y. Dai, C. Desborough, R.E. Dickinson, Q. Duan, M. Ek, Y.M. Gusev, F. Habets, P. Irannejad, R. Koster, K.E. Mitchell, O.N. Nasonova, J. Noilhan, J. Schaake, A. Schlosser, Y. Shao, A.B. Shmakin, D. Verseghy, K. Warrach, P. Wetzel, Y. Xue, Z.-L. Yang, Q.-C. Zeng, 1998, The Project for Intercomparison of Land-surface Parameterization Schemes (PILPS) phase 2(c) Red-Arkansas River basin experiment: 2. Spatial and temporal analysis of energy fluxes, Global and Planetary Change, 19, 137-159

Linsley, R. K., M. A. Kohler, and J. L. H. Paulhus, 1958. Hydrology for Engineers, McGraw-Hill Book Co., Inc., New York, N.Y., p485

Liston, G.E., Y.C. Sud and E.F. Wood, 1994, Evaluating GCM land surface hydrology parameterizations by computing river discharge using a runoff routing model: Application to the Mississippi River basin. J. of Applied Meteorology, 33(3), 394-404.

Lohmann, D., D.P. Lettenmaier, X. Liang, E.F. Wood, A. Boone, S. Chang, F. Chen, Y. Dai, C. Desborough, R.E. Dickinson, Q. Duan, M. Ek, Y.M. Gusev, F. Habets, P. Irannejad, R. Koster, K.E. Mitchell, O.N. Nasonova, J. Noilhan, J. Schaake, A. Schlosser, Y. Shao, A.B. Shmakin, D. Verseghy, K. Warrach, P. Wetzel, Y. Xue, Z.-L. Yang, Q.-C. Zeng, 1998, The Project for Intercomparison of Land-surface Parameterization Schemes (PILPS) phase 2(c) Red-Arkansas River basin experiment: 3. Spatial and temporal analysis of water fluxes, Global and Planetary Change, 19, 161-179

Los, S.O., C.O. Justice, and C.J. Tucker, 1994, A 1°x1° global NDVI data set for climate studies derived from GIMMS continental NDVI data, Int. J. Remote Sensing, 15, 3493-3518

Loveland, T.R., B.C. Reed, J.F. Brown, D.O. Ohlen, Z. Zhu, L. Yang, and J.W. Merchant, 1999, Development of a global land cover characteristics database and IGBP DISCover from 1 km AVHRR data, To appear in Int. J. Remote Sensing

Mahrt, L. and H.L. Pan, 1984, A two-layer model of soil hydrology, Bound. Layer Meterol., 29, 1-20

Manabe, S., 1969, Climate and ocean circulation, I: The atmospheric circulation and the hydrology of the earth's surface, Mon. Wea. Rev., 97, 739-774

Mathews, E., 1983, Global vegetation and land use: new high resolution data bases for climate studies, J. Clim. Appl. Meteorol., 22, 474-487

McCuen, R. H., 1982. A guide to hydrologic analysis using SCS methods, Prentice-Hall, Inc., Englewood Cliffs, New Jersey 07632

Meeson, B.W., F.E. Corprew, J.M.P. McManus, D.M. Myers, J.W. Closs, K.-J. Sun, D.J. Sunday, and P.J. Sellers, 1995, ISLSCP Initiative I-Global data sets for land-atmosphere models, 1987-1988, Vol. 1-5, Published on CD-ROM by NASA (USA_NASA_GADDC_ISLSCP_001-005)

Miller,D.A. and R.A. White, 1999: A Conterminous United States multi-layer soil characteristics data set for regional climate and hydrology modeling. Earth Interactions, 2, (available at http://EarthInteractions.org).

Mitchell, K., P. Houser, E. Wood, J. Schaake, D. Tarpley, D. Lettenmaier, W. Higgins, C. Marshall, D. Lohmann, M. Ek, B. Cosgrove, J. Entin, Q. Duan, R. Pinker, A. Robock, F. Habets, and K. Vinnikov, 1999, The GCIP Land Data Assimilation System (LDAS) Project - Now underway, GEWEX News, 9(4), International GEWEX Project Office, Washington, DC

Olson, J.S., Watts, J.A., and Allison, L.J., 1983, Carbon in live vegetation of major world ecosystems, US Dept. of Energy, DOE/NBB-0037, No. TROO4, 152pp.

Oxford Economic Atlas of the World, 1972, Jones, D.B. (Ed.), Oxford University Press, 109-111

Rawls, W.J., D.L. Brakensiek, and K.E. Saxton, 1982, Estimation of soil properties, Trans. ASAE, 25, 1316-1320

Rawls, W.J., T.J. Gish, and D.L. Brakesiek, 1991, Estimating soil water retention from soil physical properties and characteristics, Advances in Soil Sci., 16, 213-234

Reynolds, C.A, T.J. Jackson, and W.J Rawls, 1999, Estimating available water content by linking the FAO soil map of the World with global soil profile database and pedotransfer functions, Proceedings of the AGU 1999 Spring Conference, Boston, MA, May 31-June4, 1999

Rutledge, A.T., 1993, Computer programs for describing the recession of groundwater discharge and for estimating mean ground water discharge from streamflow records, USGS Water Resources Investigations Report 93-4121, Richmond, VA, p45

Schaake, J.C., 1990, From climate to flow, In Climate Change and U.S. Water Resources, P.E. Wagonner (Ed.), J. Wiley & Sons, New York, 177-206

Schaake, J.C., V.I. Koren, Q.Y. Duan, K. Mitchell and F. Chen, 1996. *Simple water balance model for estimating runoff at different spatial and temporal scales.* J. Geophys. Res., 101(D3), 7461-7475.

Schaake, J., V. Koren, Q.Y. Duan, S. Cong, & A. Hall, 1998, Model Parameter Estimation Experiment (MOPEX): Data preparation and some experimental results, Presentation at GCIP Mississippi Climate Conference, St. Louis, MO, 8-12 June 1998

Schaake, J.C., Q. Duan and V. Koren, 1999, Opportunities for improved estimation of parameters of hydrological models, Presentation at the XXII General Assembly of IUGG, Birmingham, UK, 19-30 July 1999

Sellers, P.J, Y. Mintz, Y.C. Sud, and A. Dalcher, 1986, A simple biosphere model (SiB) for use within general circulation models, J. Atmos. Sci., 43 505-531.

Sellers, P.J., B.W. Meeson, J. Closs, J. Collatz, F. Corprew, D. Dazlich, F.G. Hall, Y. Kerr, R. Koster, S. Los, K. Mitchell, J. McManus, D. Myers, K.-J. Sun, and P. Try, 1996, The ISLSCP Initiative I global data sets: Surface boundary conditions and atmospheric studies, Bullet. of the American Meteor. Soc., 77(9), 1987-2005

Shuttleworth, W.J., 1998, Combining remotely sensed data using aggregation algorithms, Hydrol. and Earth Sys. Sci., 2(2-3), 149-158

Sorooshian, S., and J.A. Dracup, 1980, Stochastic parameter estimation procedures for hydrologic rainfall-runoff models: Correlated and heteroscedastic error cases, Water Resourc. Res., 16(2), 430-442

van Genuchten, M. Th., 1976, A closed-form equation for predicting the hydraulic conductivity of unsaturated soils, Soil Sci. Soc. J., 44(5), 892-898

Wilson, M.F., and A. Henderson-Sellers, 1985, A global archive of land cover and soil data for use in general circulation climate models, J. Climatol., 5, 119-143

WMO, 1992, Scientific plan for the GEWEX Continental-scale International Project (GCIP), WCRP-67/TD-No. 461, International GEWEX Project Office, Washington, DC

Wood, E.F., D.P. Lettenmaier, X. Liang, Lohmann, D., A. Boone, S. Chang, F. Chen, Y. Dai, C. Desborough, R.E. Dickinson, Q. Duan, M. Ek, Y.M. Gusev, F. Habets, P. Irannejad, R. Koster, K.E. Mitchell, O.N. Nasonova, J. Noilhan, J. Schaake, A. Schlosser, Y. Shao, A.B. Shmakin, D. Verseghy, K. Warrach, P. Wetzel, Y. Xue, Z.-L. Yang, Q.-C. Zeng, 1998, The Project for Intercomparison of Land-surface Parameterization Schemes (PILPS) phase 2(c) Red-Arkansas River basin experiment: 1. Experiment description and summary intercomparisons, Global and Planetary Change, 19, 115-135

World Atlas of Agriculture, 1969, Committee on the World Atlas of Agriculture, Inst. Geo. de Agos., Novara

World Forestry Atlas, 1961, Verlag Paul Parey, Hamburg

Xue, Y. P.J. Sellers, J.L. Kinter, and J. Shukla, 1991, A simplified biosphere model for global climate studies, J. Climate, 4, 345-364

Yapo, P., H.V. Gupta, and S. Sorooshian, 1997, Multi-objective global optimization for hydrologic models, J. of Hydrol., 204, 83-97

Zhao, R., 1992. The Xiangjiang model applied in China, J. of Hydrol., 135, 371-381

Zobler, L. 1986, A world soil file for global climate modeling, NASA Tech. Memo. 87802, NASA, Washington, D.C., 35 pp.

Qingyun Duan, Victor Koren, and John Schaake, Hydrology Laboratory, National Weather Service, 1325 East-West Highway, Silver Spring, MD 20910, USA.

Comparing GCM-Generated Land Surface Water Budgets Using a Simple Common Framework

Randal D. Koster,[1] Paul A. Dirmeyer,[2] P. C. D. Milly,[3] and Gary L. Russell[4]

Multi-decade climate simulations with four atmospheric general circulation models (GCMs) are analyzed to determine the extent to which the precipitation and net radiation forcing simulated by each model determines the simulated annual surface water budget in a given region. The success of two simple climatological relations in characterizing the mean and interannual variability of each GCM's surface water budget implies that the forcing does indeed impose a primary control. Intermodel differences in the forcing are found to be strongly related to intermodel differences in the normalized surface fluxes (i.e., the mean and variability of the annual evaporation and runoff normalized by annual precipitation), with the forcing differences explaining roughly half the variance in the normalized flux differences. The results imply that to understand the annual evaporation and runoff fluxes simulated by a GCM in a given region, a study of the forcing and its relationship to both the large-scale circulation and the land surface itself is at least as relevant as a focused analysis of the evaporation and runoff parameterizations used by the land surface scheme.

1. INTRODUCTION

Recent studies of the natural variability of global climate and its sensitivity to anthropogenic change have largely relied on numerical climate simulations with general circulation models (GCMs) of the atmosphere and ocean [*Houghton et al.*, 1996]. Given the importance of GCMs in climate research, differences in their behaviors can be a serious cause for concern. Why, for example, might one GCM suggest a small impact of Amazonian deforestation on local and far-field climate, while another suggests a large impact? Why might one suggest a larger increase in surface runoff for a given increase in precipitation? If GCM studies are to be truly useful for climate science, such discrepancies must be resolved. This requires a firm understanding of the underlying controls on the GCMs' simulated climates.

This paper addresses the controls on a major component of GCM-simulated climate, namely the global hydrological cycle. More specifically, it addresses the extent to which the simulated annual precipitation and net radiation forcing imposed on the surface determines the relative magnitudes of the simulated annual surface water fluxes (i.e., evaporation and runoff). Quantifying the impact of the simulated forcing on the surface fluxes has important implications for coupled model sensitiv-

[1]Hydrological Sciences Branch, Laboratory for Hydrospheric Processes, NASA/Goddard Space Flight Center, Greenbelt, Maryland

[2]Center for Ocean-Land-Atmosphere Studies, Calverton, Maryland

[3]U.S. Geological Survey and Geophysical Fluid Dynamics Laboratory/NOAA, Princeton, New Jersey

[4]Goddard Institute for Space Studies, NASA/Goddard Space Flight Center, New York, New York

Land Surface Hydrology, Meteorology, and Climate: Observations and Modeling
Water Science and Application Volume 3, pages 95-105

ity analyses focusing on land surface parameterization. If the impact is small (e.g., the relative partitioning of precipitation into evaporation and runoff is roughly the same regardless of how much rain falls), then the sensitivity of surface fluxes to imposed changes in surface parameterization can be easily predicted — an increase in surface evaporation conductance, for example, would allow a greater fraction of the precipitation to be converted to evaporation. If, on the other hand, the impact of the forcing on the fluxes is large, then the analysis of sensitivity becomes much more difficult. In this case, the impact of the surface parameterization change on the precipitation and net radiation forcing itself, through complex land-atmosphere feedback mechanisms, must be understood. Typically, a study of the runoff and evaporation formulations in isolation will not shed light on the nature and strength of this feedback.

In a similar vein, quantifying the impact of precipitation and net radiation forcing on the surface water balance is essential for understanding the contrasting hydrological behaviors of different GCMs. Consider, for example, two GCMs that behave differently over the Sahel: one GCM converts most of the rainfall there to runoff, whereas the other converts most of it to evaporation. This contrasting behavior might reflect, in a straightforward and easily identifiable way, the different evaporation and runoff parameterizations used in the Sahel by the two models. On the other hand, it might more directly reflect differences in the simulated Sahelian rainfall and net radiation forcing. This would complicate the analysis, since the controls on the forcing are usually difficult to isolate.

Thus, by quantifying the impact of the precipitation and net radiation forcing on the surface water budget, this paper effectively addresses whether intermodel differences in the simulated budgets can be easily explained in terms of details in the evaporation and runoff parameterizations alone. Simply put, a strong impact of the forcing on the water budget implies that intermodel differences cannot be related to the parameterization details in a simple, predictable way. Note that while the forcing has a clear, almost trivial impact on the absolute magnitudes of the surface fluxes — higher precipitation rates tend to produce higher runoff rates — the impact on the *relative* magnitudes of evaporation and runoff (the focus of this paper) is much less obvious. In this paper, these relative magnitudes will be characterized in terms of annual evaporation and runoff ratios, i.e., in terms of the annual evaporation and runoff fluxes normalized by the annual precipitation.

Section 2 of this paper presents a framework for characterizing the control of precipitation and net radiation forcing over (a) the mean evaporation and runoff ratios and (b) their interannual variability. In section 3, the surface water budgets generated by four GCMs are compared within the simple framework, allowing us to establish the extent to which intermodel differences in the forcing explain the intermodel differences in surface hydrological response. Section 4 offers a summary along with some further discussion.

2. MEAN AND VARIABILITY RELATIONS

To characterize the control of precipitation and radiation forcing on the partitioning of mean annual precipitation into evaporation and runoff, we use the climatological relation of *Budyko* [1958, 1974]. This relation has already proven useful in the separation of atmospheric and direct land effects in certain validation studies [*Koster et al.*, 1999]. To characterize atmospheric control over the interannual variability of evaporation and runoff fluxes, we use an extension of Budyko's relation derived by *Koster and Suarez* [1999].

Budyko [1958, 1974] asserted that the mean annual evaporation, $\overline{E}$, from a land surface is controlled mostly by the dual availability of water and energy for vaporization, represented respectively by the mean annual precipitation, $\overline{P}$, and the mean annual net radiation, $\overline{R}$. Simply put, $\overline{E}$ must be less than both $\overline{P}$ and $\overline{R}/L$, where L is the latent heat of vaporization. *Budyko* [1958, 1974] combined these ideas with some existing empirical equations to derive the following simple relationship:

$$\frac{\overline{E}}{\overline{P}} = \mathcal{F}(\phi) = \left[\phi\left(\tanh\frac{1}{\phi}\right)(1 - \cosh\phi + \sinh\phi)\right]^{\frac{1}{2}}, \qquad (1)$$

where ϕ is the "radiative dryness index":

$$\phi = \frac{\overline{R}}{\overline{P}L}. \qquad (2)$$

Of course, $\overline{E}$ is also influenced by other aspects of the system, notably the seasonal phasing of the precipitation and net radiation cycles and the complex physical mechanisms controlling evaporation and runoff at the land surface [*Milly*, 1994ab]. Still, Budyko's contention is that these other factors are secondary when considering the annual mean. He considered (1) to produce useful first-order estimates of the mean water balance

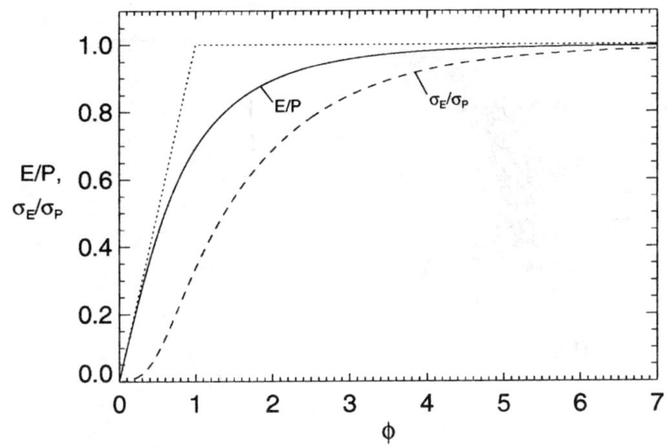

Figure 1. The solid curve represents *Budyko's* [1958, 1974] semi-empirical relationship between $\overline{E}/\overline{P}$ and the radiative index of dryness, ϕ. The dashed curve represents the derived relationship for σ_E/σ_P, as defined in Section 3. The dotted straight lines are the asymptotes implied by energy and water supply considerations.

at very long timescales, and he found support for his contention in available data.

Koster and Suarez [1999] made the assumption that interannual variations in moisture storage are small relative to the fluxes themselves, so that (1) can be used to approximate a given year's evaporation from that year's precipitation and net radiation. The further assumption that interannual changes in net radiation are small relative to those in precipitation allowed (1) to be transformed into:

$$\frac{\sigma_E}{\sigma_P} = \mathcal{F}(\phi) - \phi\mathcal{F}'(\phi), \qquad (3)$$

where σ_E and σ_P are the standard deviations of annual evaporation and annual precipitation, respectively. Eq. (3) was found to capture well the mean relationship between $\frac{\sigma_E}{\sigma_P}$ and ϕ in their GCM. *Koster and Suarez* [1999] further noted that, under this framework,

$$\frac{\sigma_E}{\sigma_P} = \frac{\Delta E_i}{\Delta P_i}, \qquad (4)$$

where ΔE_i and ΔP_i are the anomalies of annual evaporation and precipitation, respectively, in year i. In other words, (3) provides a means for estimating how a given annual precipitation anomaly is partitioned into annual evaporation and runoff anomalies.

The curves represented by (1) and (3) are shown in Figure 1. Note that the σ_E/σ_P curve always lies below the $\overline{E}/\overline{P}$ curve; this is consistent with the idea that the fraction of a given year's precipitation anomaly that gets converted into an evaporation anomaly is generally less than the ratio of the mean evaporation to the mean precipitation. The two equations are useful for the present study because they describe key aspects of the mean surface water budget and its interannual variability without any explicit reference to land surface physics. According to these equations, the only way that the land surface can have a first-order impact on $\overline{E}/\overline{P}$ and σ_E/σ_P is through its impact on the forcing itself.

Eqs. (1) and (3), by the way, can be easily transformed into corresponding equations for runoff, Q. The long term mean precipitation must be balanced by the long term means of evaporation and runoff, leading to

$$\frac{\overline{Q}}{\overline{P}} = 1 - \left[\phi\left(\tanh\frac{1}{\phi}\right)(1 - \cosh\phi + \sinh\phi)\right]^{\frac{1}{2}}. \qquad (5)$$

From the assumption that interannual variations in moisture storage are small relative to the fluxes themselves, (3) can be transformed into

$$\frac{\sigma_Q}{\sigma_P} = 1 - [\mathcal{F}(\phi) - \phi\mathcal{F}'(\phi)]. \qquad (6)$$

Since quantifying evaporation is, in fact, equivalent to quantifying runoff under the assumptions considered here, the analysis below focuses on the evaporation calculation alone.

3. GCM COMPARISON

Equations (1) and (3) are used as a general framework for comparing the behaviors of four GCMs, each of which was run for multiple decades (a minimum of 46 years) to produce reliable hydrological statistics. The designs and durations of the simulations vary because they were not produced specifically for the present analysis. Indeed, each simulation has already been discussed, in another context, in the literature.

The four GCMs examined here are: (1) the Aries GCM of the NASA/Goddard Space Flight Center; (2) the GCM of the Center for Ocean-Land-Atmosphere Studies (COLA); (3) the GCM of NOAA's Geophysical Fluid Dynamics Laboratory (GFDL); and (4) a global synchronously coupled atmosphere-ocean model developed at the NASA/Goddard Institute for Space Studies (GISS). A brief description of each GCM, the durations of the simulations performed with it, and relevant limitations (if any) of the resulting data is provided in Appendix A. Note that for each GCM, points consisting of

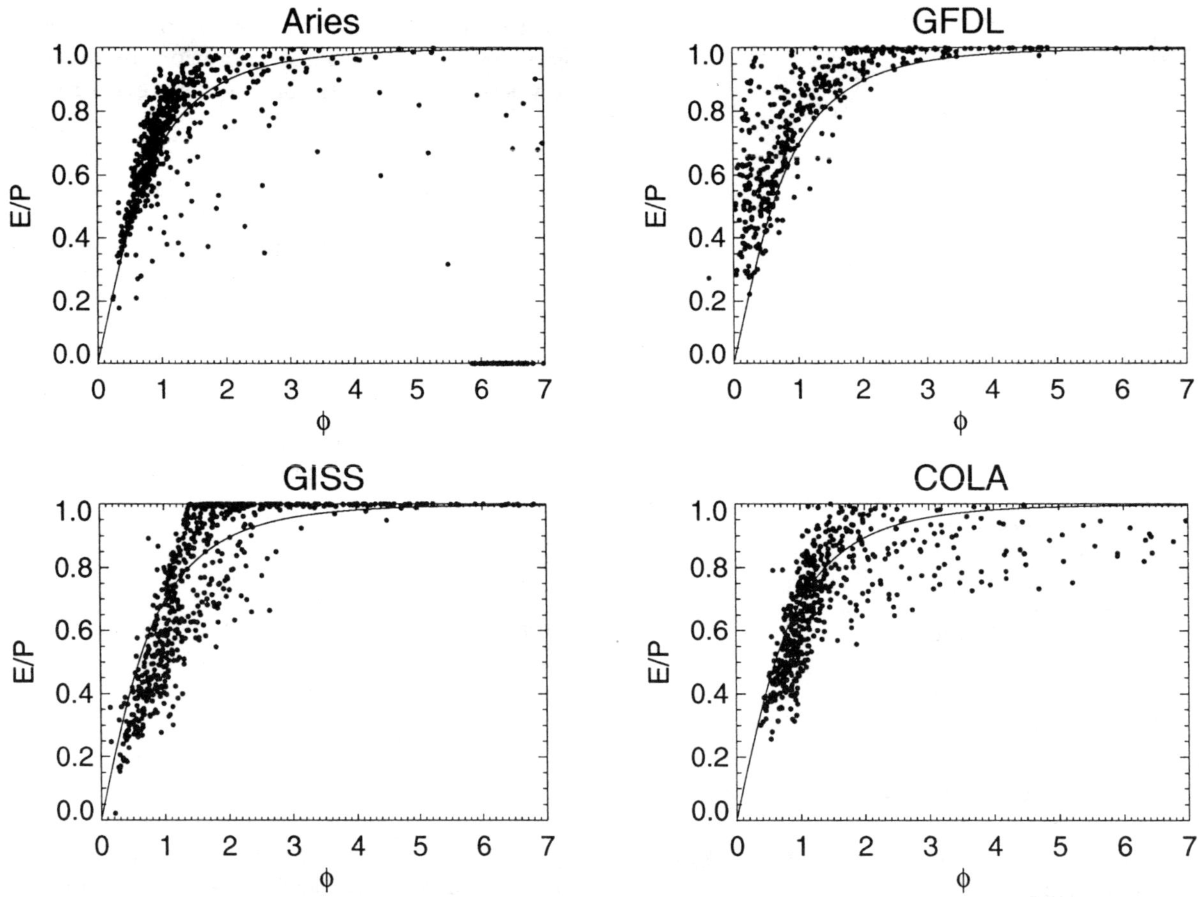

Figure 2. Variation of $\overline{E}/\overline{P}$ with the radiative index of dryness, ϕ, for each of the four GCMs. Each point in a plot represents conditions at a single ice-free land surface grid cell. The solid curve represents (1), *Budyko's* [1958, 1974] semi-empirical relationship.

ocean, ocean ice, or permanent land ice are not considered in the analysis.

The applicability of (1) and (3) to the annual data generated by each GCM is demonstrated in Figure 2 and Figure 3 — for each GCM, the plotted variations of $\overline{E}/\overline{P}$ and σ_E/σ_P with ϕ are described, to first order, by the two equations. As might be expected, the figures do show some intermodel differences in behavior that probably stem from differences in the land surface schemes. For example, at high ϕ, only the Aries and COLA GCMs tend to have $\overline{E}/\overline{P}$ values that are significantly lower than the Budyko estimates, presumably because these two GCMs include stronger mechanisms for producing runoff in very dry conditions. Reduced values of σ_E/σ_P at high ϕ are particularly pronounced for these two models. Also, at lower values of ϕ, the Aries and GFDL GCMs tend to produce $\overline{E}/\overline{P}$ values that exceed the Budyko estimates, whereas the $\overline{E}/\overline{P}$ values for GISS and COLA GCMs tend to fall below these estimates. Such variations, along with the scatter seen in each plot, serve as reminders that (1) and (3) are only meant to be approximate.

These variations, though significant, are nevertheless small enough to suggest that (1) and (3) characterize well the underlying relationship in each GCM between the precipitation and net radiation forcing and the annual surface water budget. Indeed, these equations can be used to predict intermodel differences in the normalized annual surface fluxes based on intermodel differences in ϕ alone. To demonstrate this, we compute, for each GCM, the average values of $\overline{E}$, $\overline{P}$, $\overline{R}/L$, σ_E, and σ_P in each of several regions representing a range of climatic regimes. These values are then combined to generate GCM-specific values of ϕ, $\overline{E}/\overline{P}$ and σ_E/σ_P for each region. (The regions, each consisting of several contiguous grid cells, are the same as those shown in

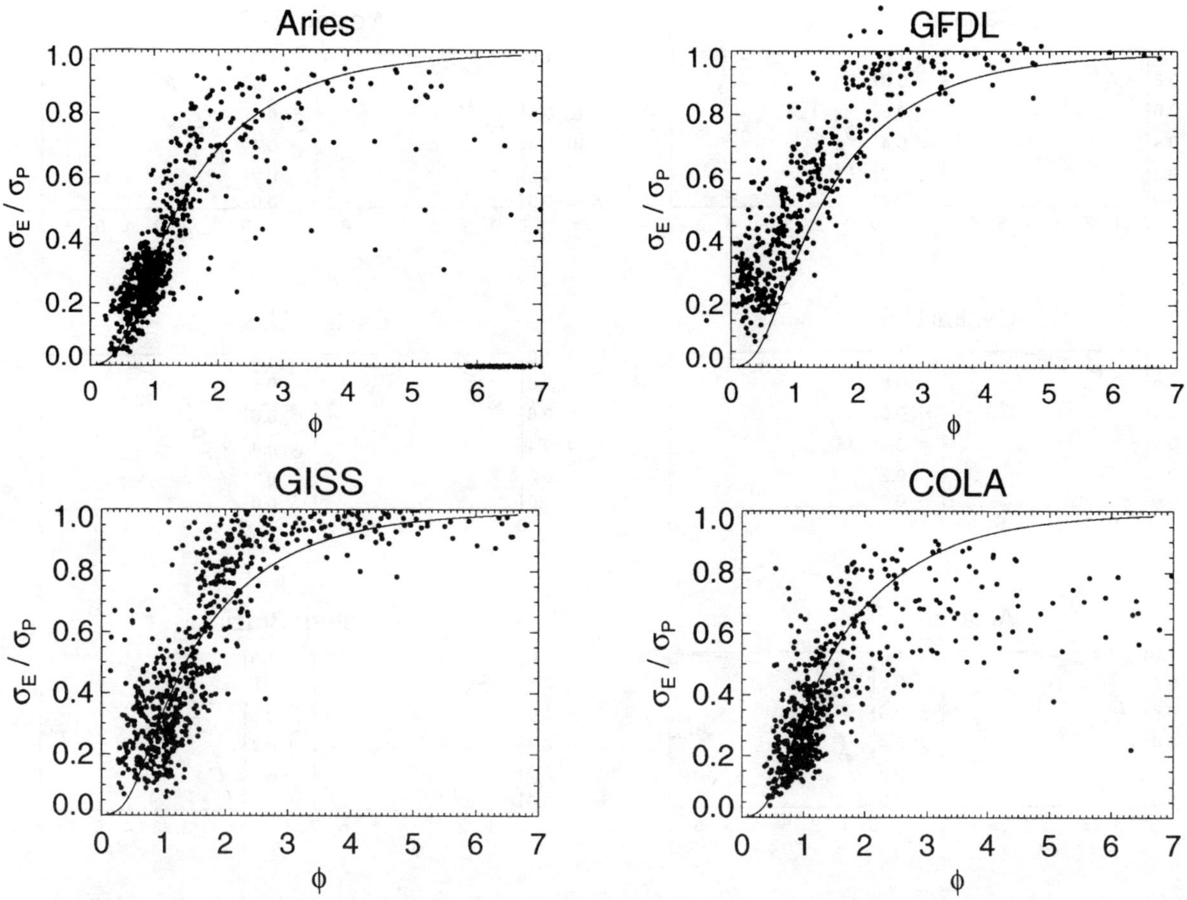

Figure 3. Variation of σ_E/σ_P with the radiative index of dryness, ϕ, for each of the four GCMs. Each point in a plot represents conditions at a single ice-free land surface grid cell. The solid curve represents (3), the relationship derived by *Koster and Suarez* [1999].

Figure 1 of *Koster and Suarez* [1996a].) Figure 4 shows how each GCM's regional $\overline{E}/\overline{P}$ and σ_E/σ_P values relate to its corresponding value of ϕ.

The salient result from figure 4 is that the points for the different GCMs generally fall quite close to the curves for (1) and (3), close enough, in fact, to suggest that the two equations can be used to predict first order differences in model behavior. In the central United States, for example, both the Aries and GFDL models have a ϕ of about 1, whereas both the COLA and GISS models have a ϕ of about 2. Eqs. (1) and (3) predict that the latter two models should have much higher values of both $\overline{E}/\overline{P}$ and σ_E/σ_P. The computed differences are qualitatively and (to first order) quantitatively correct. Intermodel differences in the character of the annual water balance in the central U.S. do appear to be determined primarily by differences in the precipitation and net radiation forcing there — predictions regarding the fundamental nature of the annual surface water budget can be made without knowledge of how land surface processes are parameterized in the models.

Can we more quantitatively describe the extent to which the two equations explain the intermodel differences? This is attempted in Figure 5. For each pair of GCMs (6 combinations in all), two scatter plots are shown. In the first, the actual intermodel difference in $\overline{E}/\overline{P}$ at each land surface grid cell is plotted against the difference predicted via (1). That is, we plot $\left(\overline{E}/\overline{P}\right)_j - \left(\overline{E}/\overline{P}\right)_i$ versus $\mathcal{F}(\phi_j) - \mathcal{F}(\phi_i)$, where j and i represent the two GCMs. The second plot for each combination shows the corresponding comparison of actual σ_E/σ_P difference versus that obtained with (3). Note that because the GFDL GCM has a coarser resolution than the others, the plots for the combinations that include it have fewer points.

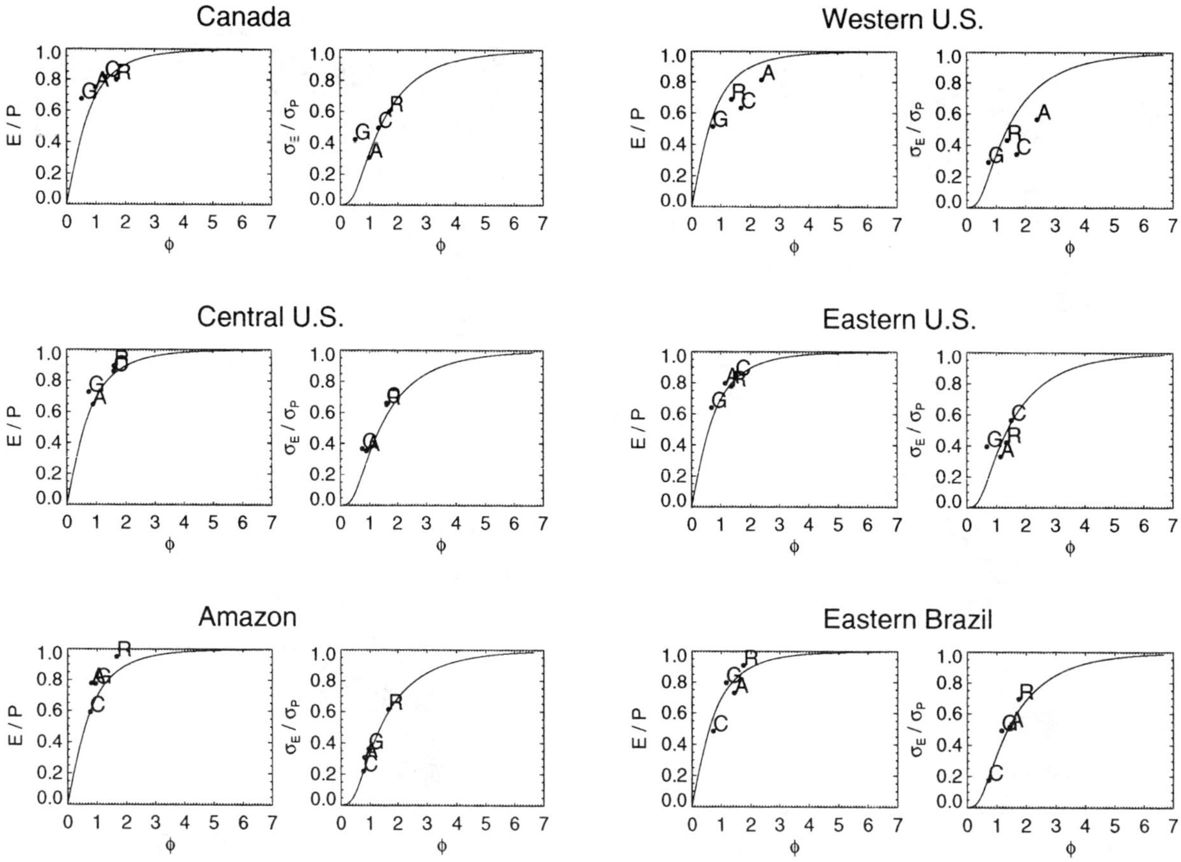

Figure 4. Plots of $\overline{E}/\overline{P}$ and σ_E/σ_P versus radiative index of dryness, ϕ, for each of the four GCMs (A=Aries, G=GFDL, R=GISS, and C=COLA) in various regions. (See *Koster and Suarez* [1996] for the precise boundaries of the regions.) The solid curves represent (1) and (3).

If (1) and (3) described the GCMs' behaviors perfectly, then the predicted differences would match the actual differences, and the points in each plot would lie along the indicated 1:1 line. This is not the case, of course. Still, the points generally cluster about the 1:1 line, and the r^2 values indicate that for each GCM combination, intermodel differences in precipitation and net radiation explain a large fraction (at least 35% and sometimes over 70%) of the intermodel differences in $\overline{E}/\overline{P}$ and σ_E/σ_P.

4. DISCUSSION

We naturally expect that if one GCM produces more precipitation at a given location than another GCM, then it will probably produce more evaporation there as well. The result discussed above, however, is less obvious and is, in that sense, more important. Intermodel differences in the *normalized* fluxes — i.e., in the *relative breakdown* of total precipitation into runoff and evaporation — can be predicted to a significant degree from the precipitation and net radiation alone. The contribution of the surface forcing to the surface water balance may even be underestimated by (1) and (3). The Budyko-based equations capture much of the forcing's impact on evaporation, but not all of it — not captured, for example, is the effect of intermodel differences in the phasing of the local precipitation and radiation seasonal cycles on the surface water balance. If all relevant atmospheric factors were included in a more comprehensive analysis, the ability to predict the surface water balance from the forcing alone might increase further.

This does not imply, however, that the impact of the land surface scheme on the annual water balance is negligible. Significant intermodel differences in annual evaporation and runoff under (essentially) identical precipitation and net radiation have been documented in

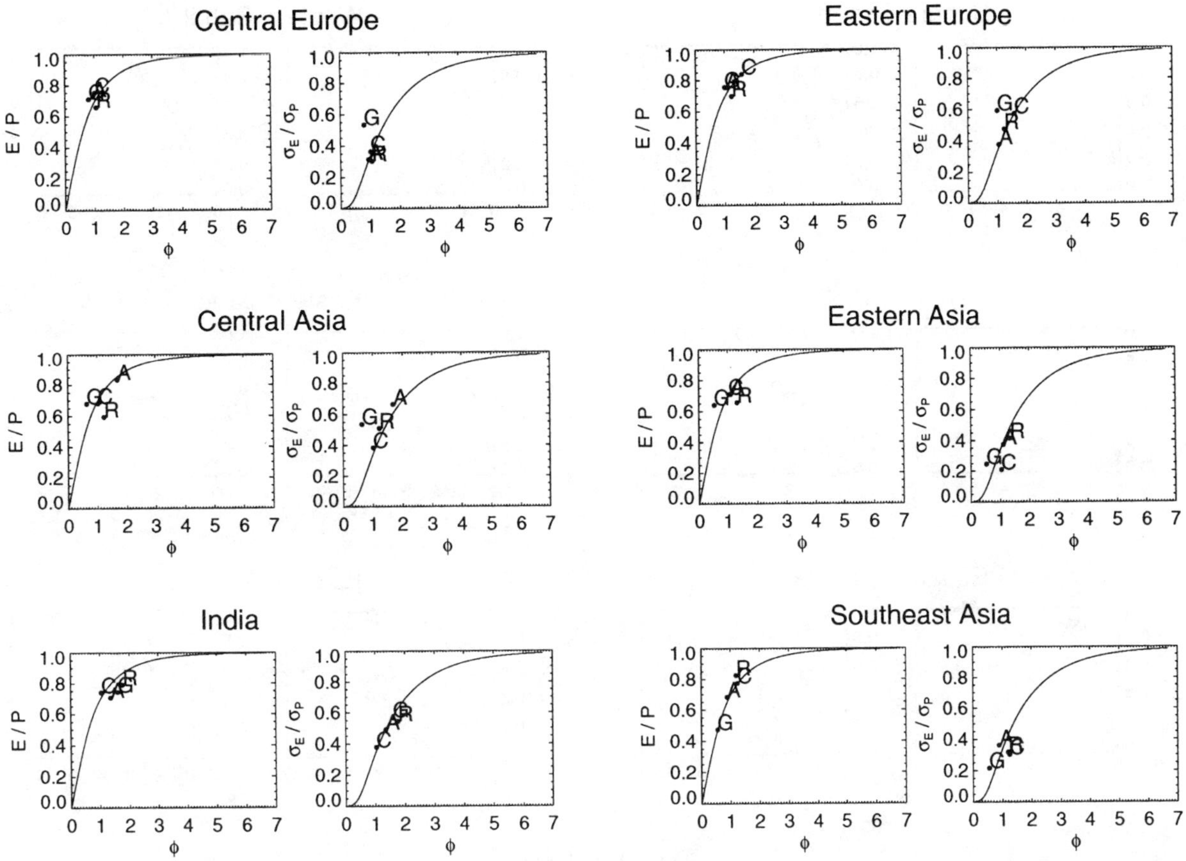

Figure 4. (cont.)

various intercomparison studies [*e.g., Chen et al.*, 1997; *Koster and Milly*, 1997; *Wood et al.*, 1998; *Oki et al.*, 1999], and a significant impact of land parameterization is indeed suggested in some of the plots in Figure 4. For example, in the western U.S. and the eastern Sahel, the points for the Aries and COLA models lie significantly below the curves, and in several other regions, the points for the GFDL model lie above them. The scatter in Figure 5 suggests that up to half of the intermodel differences may be caused by differences in land parameterization.

In addition, and perhaps more importantly, the formulation of land surface processes in a coupled system can significantly modify the forcing itself, in several ways. The surface albedo formulation, for example, has a clear and straightforward impact on the net radiation. Net radiation is also affected, in a much more complex way, by all aspects of the surface energy balance, since the energy balance determines surface temperature and thus the outgoing longwave radiation. The precipitation forcing can be strongly affected by the magnitudes and variability of the surface latent and sensible heat fluxes; these fluxes affect the character of the atmospheric boundary layer in complex ways, and precipitation in turn responds, in complex ways, to boundary layer conditions.

Thus, the lesson from Figures 4 and 5 is not that the land surface scheme has little control over the annual water balance. Rather, the lesson is that one cannot typically explain simulated differences in annual evaporation and runoff through a study of the evaporation and runoff formulations alone. If GCM A produces a higher evaporation ratio, and thus a lower runoff ratio, than GCM B in a given region, one might be tempted to explain this through a comparison of the transpiration functions used (perhaps GCM A uses a lower canopy resistance) and the runoff functions used (perhaps GCM A uses higher infiltration capacities). The present analysis suggests that this is not an appropriate approach. Figures 4 and 5 show that intermodel differences in wa-

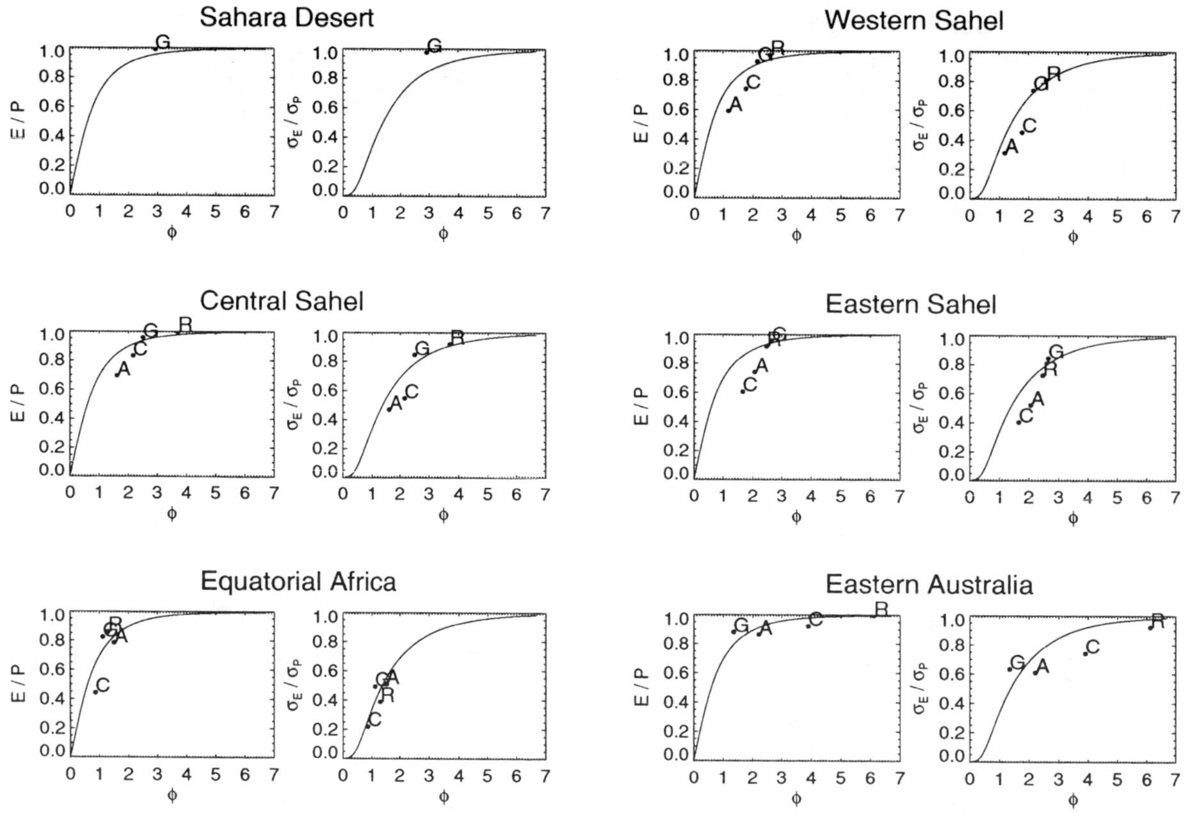

Figure 4. (cont.)

ter balance result in large part from differences in precipitation and net radiation forcing, and thus a focus on the forcing generated by GCMs A and B is at least as appropriate. Again, although the land may have an impact on this forcing, the nature of this impact is highly complex and not amenable to a study of the evaporation and runoff formulations in isolation.

An overarching lesson from the figures is that the climatological relation of *Budyko* [1958,1974], though several decades old, is still highly relevant to climate studies, at least for the consideration of the annual water balance. The curves in Figures 2 and 3 successfully characterize the simulated global variations in the annual surface water budget of each GCM, demonstrating that precipitation and net radiation forcing do impose a first order control on normalized evaporation and runoff. In closing, we note that our results pertain only to the annual timescale (or longer). Due to interseasonal moisture storage and other factors, Budyko's relation generally fails at shorter (diurnal, synoptic-scale, seasonal) timescales. Variations in evaporation and runoff at these shorter timescales may be tied to the imposed land surface formulations in a much more straightforward way.

APPENDIX A: DETAILS REGARDING GCM SIMULATIONS

A.1. Aries GCM

The Aries GCM (or GEOS-Climate GCM) is a grid point model that has been used extensively in various land-atmosphere feedback studies [*e.g., Koster and Suarez*, 1995, 1996a, 1999]. It computes penetrative convection with the relaxed Arakawa-Schubert scheme [*Moorthi and Suarez*, 1992], radiative transfer with detailed codes (*e.g.*, with the scheme of *Chou and Suarez* [1994] for the longwave component), and Richardson number-dependent fluxes in the surface layer. Vorticity and all scalars are advected with a fourth order scheme, and the atmospheric dynamics are coded as a dynamical core [*Suarez and Takacs*, 1995]. The land surface component is the Mosaic model [*Koster and Suarez*, 1996b], a standard soil-vegetation-atmosphere-transfer (SVAT) type land surface model that has performed well in re-from the native T30 grid to a slightly lower resolution 4°×5° global grid. This causes some grid boxes along the coasts to contain a mixture of land and ocean data. An amended land-sea mask was designed to try to remove as many of the contaminated coastal points as possible. However, oceanic information may still be present in some grid boxes, and may contribute to larger values of E/P for those points.

A.3. GFDL GCM

The GFDL climate model used for this analysis includes general circulation models of the atmosphere and ocean, coupled to a simple model of land water and energy balances. Details are provided by *Manabe and Stouffer* [1996]. The land model is the same as that described by *Manabe* [1969]. Land water is stored at each gridpoint in snow and soil-water stores, and the land has no heat capacity. (Solar forcing has a realistic seasonal cycle, but no diurnal cycle.) Land-atmosphere transfer is parameterized in terms of a bulk transfer coefficient, with no stability dependence, and stomatal resistance to vapor transport is neglected. Runoff is generated when soil-water storage capacity is exceeded. Land processes are resolved on a grid having a 7.5° longitude by approximately 4.5° latitude spacing.

The coupled atmosphere-ocean-land model was run (with "flux adjustment" to match the mean ocean-surface climate with observations) for 1000 years of simulated climate by *Stouffer et al.* [1994] in a study of natural variability. Data included here are from years 101-200 of that experiment.

A.4. GISS GCM

The global synchronously coupled atmosphere-ocean model developed by *Russell et al.* [1995] is designed for climate studies on decade to century time scales. There are nine vertical layers in the atmosphere and 13 in the ocean. The horizontal resolution for both the atmosphere and ocean is 4° in latitude by 5° in longitude. The effective resolution for heat, water vapor, and salt, however, is finer than this because the model maintains both the means and directional gradients of these quantities within each grid box. The six layer ground hydrology scheme of *Abramopoulos et al.* [1988] is applied at each 4°× 5° land surface grid cell. Surface and underground runoff that leaves the ground is fed into a river flow scheme [*Miller et al.*, 1994]. Based on comparisons of model and observed precipitation and river discharge, the ground hydrology scheme has too cent validation tests [*Chen et al.*, 1997; *Wood et al.*, 1998].

The simulations used for the present study are a subset (namely, "Ensemble ALO") of those recently analyzed by *Koster et al.* [2000]. Results from an ensemble of 16 simulations, each spanning 45 years and forced with realistic interannually-varying sea surface temperatures, are combined to produce 720 years of annual precipitation, evaporation, and net radiation data across the globe at a 4°×5° resolution.

Due to transient initialization issues, a few very dry desert points show evaporation rates that exceed precipitation rates. These are excluded from the analysis.

A.2. COLA GCM

The atmospheric GCM used is version 1.11 of the Center for Ocean-Land-Atmosphere Studies (COLA) GCM. It is a research version of the global spectral model described by *Sela* [1980] with modifications as described by *Kinter et al.* [1988] and *Schneider and Kinter* [1994]. This version of the model is very similar to that described by *Kinter et al.* [1997], the principal difference being that this version uses a relaxed Arakawa-Schubert convection parameterization [*DeWitt*, 1996]. The land surface scheme coupled to the GCM is a version of the simplified Simple Biosphere (SiB) model of *Sellers et al.* [1986], as described by *Xue et al.* [1991].

The model was integrated at a spectral resolution of T30 (3.75° latitude and longitude) with 18 vertical levels. A single 46-year integration was conducted, initialized at the beginning of November 1948, as part of the Climate of the Twentieth Century (C20C) experiment [*Shukla and Marx*, 1995]. The observed sea surface temperatures specified in the integration are from the GISST dataset [*Parker et al.*, 1995], version 2.2.

During post-processing the data were interpolated much evaporation, particularly at low latitudes.

Two 150 year control simulations with constant 1950 atmospheric composition, differing only in their initial conditions, are used for the present study. *Russell et al.* [2000] compared these control simulations and companion greenhouse gas experiments with the observed regional temperature record of the past 40 years and concluded that the model is faithfully representing the real world's climate changes due to greenhouse gas forcing in the northern hemisphere.

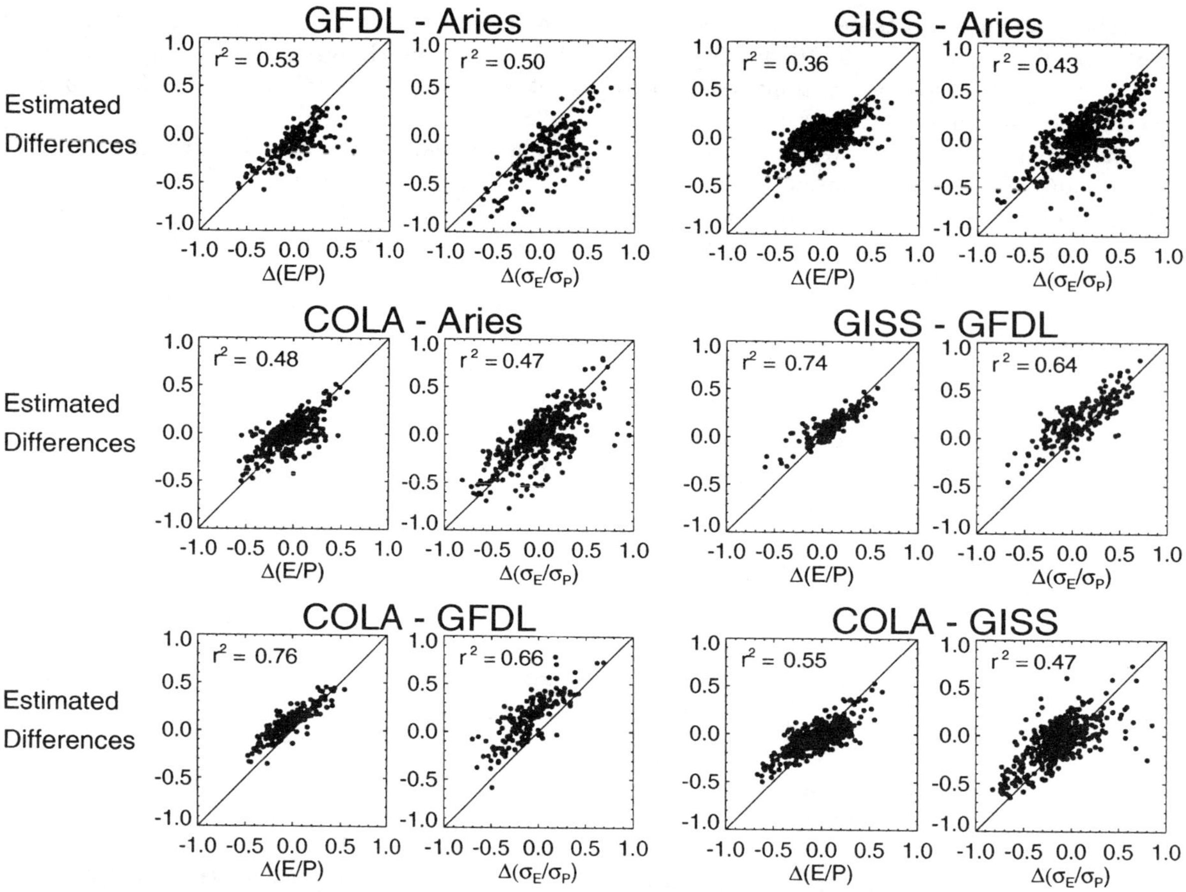

Figure 5. Scatter plots showing, for each combination of two GCMs, how the actual intermodel differences in $\overline{E}/\overline{P}$ and σ_E/σ_P compare to the estimated differences found by applying (1) and (3), respectively, to each GCM's dryness index. The abscissa represents the actual difference, and the ordinate represents the estimated difference. Each point in the plots represents a grid cell having approximately the same location in both GCMs.

Acknowledgments. NASA's EOS-Interdisciplinary Science Program and the NASA Seasonal-to-Interannual Prediction Project (NSIPP) provided support for the processing and analysis of the GCM data and for the production of the Aries simulations.

REFERENCES

Abramopoulos, F, C. Rosenzweig, and B. Choudhury, Improved ground hydrology calculations for global climate models (GCMs): Soil water movement and evapotranspiration, *J. Climate*, *1*, 921-941, 1988.

Budyko, M. I., *The Heat Balance of the Earth's Surface*, translated by Nina A. Stepanova, 259 pp., U. S. Dept. of Commerce, Washington, D. C., 1958.

Budyko, M. I., *Climate and Life*, 508 pp., Academic Press, New York, 1974.

Chen, T.H., et al., Cabauw experimental results from the Project for Intercomparison of Landsurface Parameterization Schemes, *J. Climate*, *10*, 1194-1215, 1997.

Chou, M.-D. and M. J. Suarez, An efficient thermal infrared radiation parameterization for use in general circulation models, *NASA Tech. Memo.* 104606, Vol. 3, 1994.

DeWitt, D. G., The effect of the cumulus convection on the climate of the COLA general circulation model, *COLA Report 27*, 58 pp., Center for Ocean-Land-Atmosphere Studies, Calverton, MD , 1996.

Houghton, J. T., L.G. Meira Filho, B.A. Callander, N. Harris, A. Kattenberg and K. Maskell, *Climate Change 1995:' The Science of Climate Change*, 572 pp., Cambridge University Press, 1996.

Kinter, J. L., J. Shukla, L. Marx and E. K. Schneider, A simulation of the winter and summer circulations with the NMC global spectral model, *J. Atmos. Sci.*, *45*, 2486-2522, 1988.

Kinter, J. L., D. DeWitt, P. A. Dirmeyer, M. J. Fennessy, B. P. Kirtman, L. Marx, E. K. Schneider, J. Shukla, and D. M. Straus, The COLA atmosphere-biosphere general circulation model, Volume 1: Formulation, *COLA Report 51*, 46 pp., Center for Ocean-Land-Atmosphere Studies, Calverton, MD , 1997.

Koster, R. D. and M. J. Suarez, Relative contributions of

land and ocean processes to precipitation variability, *J. Geophys. Res.*, *100*, 13775-13790, 1995.

Koster, R. D. and M. J. Suarez, The influence of land surface moisture retention on precipitation statistics, *J. Climate*, *9*, 2551-2567, 1996a.

Koster, R. D. and M. J. Suarez, Energy and Water Balance Calculations in the Mosaic LSM, *NASA Tech. Memo.* 104606, Vol. 9, 1996b.

Koster, R. D. and P. C. D. Milly, The interplay between transpiration and runoff formulations in land surface schemes used with atmospheric models, *J. Climate*, *10*, 1578-1591, 1997.

Koster, R. D., T. Oki, and M. J. Suarez, The offline validation of land surface models, Assessing success at the annual timescale, *J. Meteorol. Soc. Japan*, *77*, 257-263, 1999.

Koster, R. D. and M. J. Suarez, A simple framework for examining the interannual variability of land surface mosture fluxes, *J. Climate*, *12*, 1911-1917, 1999.

Koster, R. D., M. J. Suarez, and M. Heiser, Variance and predictability of precipitation at seasonal-to-interannual timescales, *J. Hydrometeorology*, *1*, 26-46, 2000.

Manabe, S., Climate and the ocean circulation, 1. The atmospheric circulation and the hydrology of the earth's surface. *Mon. Wea. Rev.*, *97*, 739-774, 1969.

Manabe, S., and R. J. Stouffer, Low-frequency variability of surface air temperature in a 1000-year integration of a coupled atmosphere-ocean-land surface model, *J. Climate*, *9*, 376-393 1996.

Miller, J. R., G. L. Russell, and G. Caliri, Continental-scale river flow in climate models, *J. Climate*, *7*, 914-928, 1994.

Milly, P. C. D., Climate, interseasonal storage of soil water, and the annual water balance, *Adv. Water Resour.*, *17*, 19-24, 1994a.

Milly, P. C. D., Climate, soil water storage, and the average annual water balance, *Wat. Resour. Res.*, *30*, 2143-2156, 1994b.

Moorthi, S., and M. J. Suarez, Relaxed Arakawa-Schubert, a parameterization of moist convection for general circulation models, *Mon. Wea. Rev.*, *120*, 978-1002, 1992.

Oki, T., T. Nishimura, and P. Dirmeyer, Assessment of annual runoff from land surface models using Total Runoff Integrating Pathways (TRIP), *J. Meteor. Soc. Japan*, *77*, 235-255, 1999.

Parker, D. E., M. Jackson, and E. B. Horton, The GISST sea surface temperature and sea ice climatology, *Hadley Centre for Climate Research Tech. Note 63*, 49 pp., available from the Hadley Centre Meteorological Office, London Road, Bracknell, Berkshire, RG12 2SY, United Kingdom, 1995.

Russell, G. L., J. R. Miller, and D. Rind, A coupled atmosphere-ocean model for transient climate change studies, *Atmos.-Ocean*, *33*, 683-730, 1995.

Russell, G. L., J. R. Miller, D. Rind, R. A. Ruedy, G. A. Schmidt, and S. Sheth, Comparison of model and observed regional temperature changes during the past 40 years, *J. Geophys. Res.* *105*, 14891-14898, 2000.

Schneider, E. K., and J. L Kinter III, An examination of internally generated variability in long climate simulations, *Climate Dyn.*, *10*, 181-204, 1994.

Sela, J. G., Spectral modeling at the National Meteorological Center, *Mon. Wea. Rev.*, *108*, 1279-1292, 1980.

Sellers, P. J., Y. Mintz, Y. C. Sud, and A. Dalcher, A simple biosphere model (SiB) for use within general circulation models, *J. Atmos. Sci.*, *43*, 505-531, 1986.

Shukla, J., and L. Marx, Simulation of atmospheric circulation with observed global SST and sea ice during 1949-1992, *Proc. Workshop on Simulations of the Climate of the 20th Century using GISST*, Hadley Centre, Bracknell, Berkshire, UK, 1995.

Stouffer, R. J., S. Manabe, and K. Y. Vinnikov, Model assessment of the role of natural variability in recent global warming, *Nature*, *367*, 634-636, 1994.

Suarez M. J., and L. L. Takacs, Documentation of the ARIES/GEOS Dynamical Core: Version 2, *NASA Tech. Memo.* 104606, Vol. 5., 1995.

Wood, E.F., et al., The Project for the Intercomparison of Land-Surface Parameterization Schemes (PILPS) phase 2(c) Red Arkansas River Basin experiment, 1, Experiment description and summary intercomparisons, *J. Global and Planet. Change*, *19*, 115-135, 1998.

Xue, Y., P. J. Sellers, J. L. Kinter and J. Shukla, A simplified biosphere model for global climate studies, *J. Climate*, *4*, 345-364, 1991

P. A. Dirmeyer, Center for Ocean-Land-Atmosphere Studies, 4041 Powder Mill Road, Suite 302, Calverton, MD 20705-3106 USA. (e-mail: dirmeyer@cola.iges.org)

R. D. Koster, Hydrological Sciences Branch, Laboratory for Hydrospheric Processes, Code 974, NASA/Goddard Space Flight Center, Greenbelt, MD 20771. (e-mail: randal.koster@gsfc.nasa.gov)

P. C. D. Milly, U.S. Geological Survey, Geophysical Fluid Dynamics Laboratory/NOAA, P.O. Box 308, Princeton, NJ 08542. (e-mail: pcm@gfdl.gov)

G. L. Russell, NASA/Goddard Institute for Space Studies, 2880 Broadway, New York, NY 10025. (e-mail: grussell@giss.nasa.gov)

Development and Application of Land-Surface Models for Mesoscale Atmospheric Models: Problems and Promises

Fei Chen

National Center for Atmospheric Research, Boulder, Colorado

Roger A. Pielke Sr.

Department of Atmospheric Sciences, Colorado State University, Fort Collins, Colorado

Kenneth Mitchell

Environmental Modeling Center, NCEP(NWS/NOAA), Camp Springs, Maryland

This paper reviews recent progress in the research area of coupling advanced land-surface/hydrology/ecology models with atmospheric mesoscale models. These coupled models have been utilized for real-time numerical weather predictions, air quality monitoring, and regional climate studies. The use of advanced land-surface models in numerical weather prediction models is still in its early stage, and it is not clear what level of physical complexity is needed in land-surface models coupled to weather forecast models. Nevertheless, preliminary results have shown promising success. The modern-era land-surface models, together with time varying soil moisture, improve the calculations of surface heat flux, and near-surface air temperature and moisture. They also improve the simulation of vertical profiles of temperature and moisture in the PBL. Long-term tests showed that they improve quantitative precipitation forecasts in the summer when convection is predominant. Some studies indicate that surface processes may enhance mesoscale predictability of summer convection. Correctly initializing the soil moisture fields and specifying the underlying surface vegetation and soil characteristics are critical in coupled modeling systems. Using coupled atmospheric-land surface models, it is shown that realistic simulations of climate require two-way, nonlinear interactions between the models. Such interactions limit our ability to predict future climate. These interactions occur at all time and space scales and include biophysical, biogeochemical and biogeographic two-way feedbacks with the atmosphere. The subgrid-scale variability in topography, soil moisture, snow cover, and vegetation characteristics should also be considered to some extent to reflect its control on area-averaged surface heat fluxes and on runoff processes. Using high-resolution remote sensing data will certainly help to specify these surface heterogeneities.

1. INTRODUCTION

Lorenz [1979] proposed the concept of forced and free variations of weather and climate. He refers to forced variations as those caused by external conditions, such as changes in solar irradiance. Volcanic aerosols also

Land Surface Hydrology, Meteorology, and Climate: Observations and Modeling
Water Science and Application Volume 3, pages 107-135

cause forced variations. He refers to free variations as those which "are generally assumed to take place independently of any changes in external conditions." Day-to-day weather variations are presented as an example of free variations. He also suggests that "free climatic variations in which the underlying surface plays an essential role may therefore be physically possible".

There have been no model experiments to assess regional climate prediction in which atmosphere-ocean-land surface processes were all included. Existing papers on this subject have been limited to coupled atmosphere-ocean global models [e.g., *Cubasch et al.*, 1994; *Larow and Krishnamurti*, 1998] or atmospheric models alone [e.g., *Bengtsson et al.* 1996]. In *Bengtsson et al.*, [1996], the ocean sea surface temperature is prescribed and vegetation effects, in their words, are "grossly simplified."

However, if the ocean surface and/or land surface changes over the same time period as the atmospheric changes, then the nonlinear feedbacks (i.e., two-way fluxes) between the atmosphere, land, and water prevent an interpretation of the ocean-atmosphere and land-atmosphere interfaces as boundaries. Rather than "boundaries", these interfaces become interactive mediums. The two-way fluxes that occur between the atmosphere and ocean, and the atmosphere and the land surface, must therefore necessarily be considered as part of the predictive system. On the time scale of what we typically call short-term weather prediction (days), important feedbacks include biophysical (e.g., vegetation controls on the Bowen ratio), snow cover, clouds (e.g., in their effect on the surface energy budget), and precipitation (e.g., which changes the soil moisture) processes. This time scale is already considered an initial value problem [*Sivillo et al.*, 1997]. Seasonal and interannual weather prediction include the following feedbacks: biogeochemical (e.g., vegetation growth and senescence); anthropogenic aerosols (e.g., through their effect on the long- and short-wave radiative fluxes); sea ice; and ocean sea surface temperature (e.g., changes in upwelling such as associated with an El Niño) effects. For even longer time periods (of years to decades and longer), the additional feedbacks include biogeographical processes (e.g., changes in vegetation species composition and distribution), human-caused land-use changes, and deep ocean circulation effects on the ocean surface temperature and salinity. In the context of Lorenz's [1979] terminology, each of these feedbacks are free variations.

Plate 1 schematically illustrates direct and feedback effects on global mean temperature, as an example of the complexity of climate prediction. Using Lorenz's terminology, the direct effects could be considered forced variations, while the feedbacks are part of the free variations.

Each of these time scales must be considered initial value problems because the predictions are dependent on the initial value for at least some aspects of the ocean-atmosphere-land surface coupled system. Recent work [*Claussen* 1994, 1998; *Claussen et al.*, 1998; *Foley*, 1994; *Texier et al.*, 1997] has shown that the initial specification of the land surface exerts a strong control on the subsequent atmospheric circulation in global climate prediction models. *Cubasch et al.*, [1994] suggest in a greenhouses gas warming experiment with a coupled ocean-atmosphere model that the time evolution of the modeled global mean warming is "strongly dependent on the initial state of the climate system."

In terms of its function in coupled atmospheric models, a land-surface model (LSM) serves to reproduce, as correctly as possible, the surface energy budget that forms the lower boundary conditions of the atmosphere. In other words, it partitions the available incoming energy at the land-surface into latent and sensible heat fluxes, into the soil, and into the atmosphere. The transport of heat and moisture from the land surface into the atmosphere modifies the thermodynamic structure in the planetary boundary layer (PBL), which often sets the stage for the development of warm-season convective systems. It is commonly believed that detailed land-surface boundary conditions in Numerical Weather Prediction (NWP) models improve the characterization of the PBL, and, thus, forecasts of precipitation. *Chen et al.* [1998] demonstrated that a new land-surface model, with a time-varying soil moisture field, increases the summertime-precipitation-forecast skill as much as nearly doubling the model resolution. However, the exact role of local and regional land-surface forcing, and the extent to which it affects precipitation processes, is not fully understood. For example, global model simulations by *Beljaars et al.* [1996] showed that increasing soil moisture can enhance precipitation via evaporation; but regional model results of *Paegle et al.* [1996] indicated that modification of surface evaporation is more important in changing the buoyancy than in providing additional water vapor.

For more than a decade, it has been widely accepted that land-surface processes and their modeling play an important role, not only in large-scale atmospheric models including general circulation models (GCMs)[e.g., *Mintz*, 1981; *Rowntree*, 1983], but also in regional and mesoscale atmospheric models [*Rowntree*

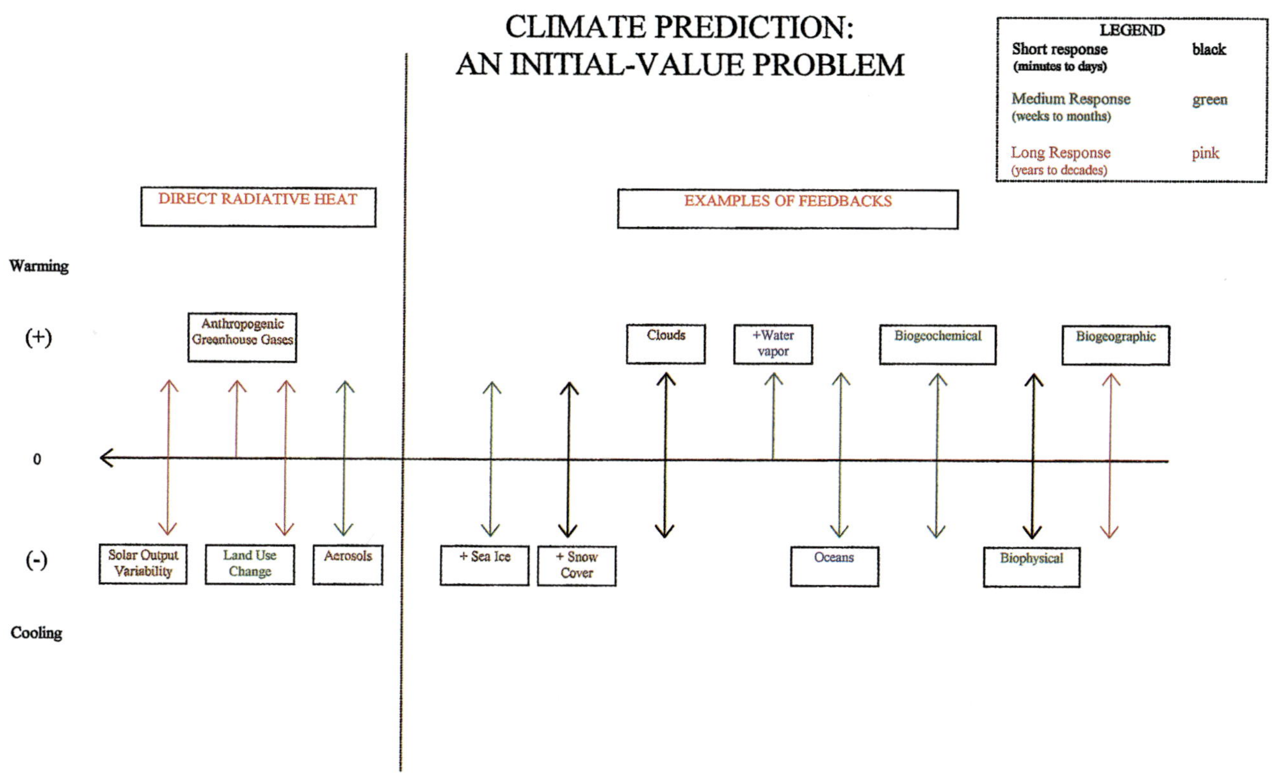

Plate 1. Schematic illustration of direct effects and feedback influences with respect to climate prediction (adapted from *Pielke et al.*, [1998a]). The length of the arrows has no meaning here, although the possible signs each effect could have are shown.

and Bolton, 1983; *Ookouchi et al.*, 1984; *Pielke*, 1984; *Mahfouf et al.*, 1987; *Avissar and Pielke*, 1989; *Chen and Avissar*, 1994a, 1994b]. Mesoscale models which resolve wavelengths of from 1-100 km (i.e., from meso-γ to meso-β scales) are often used for three applications: 1) regional climate simulations, 2) numerical weather prediction, and 3) air quality monitoring.

In early mesoscale models, the land-surface processes were either absent or given a prescribed diurnal cycle of surface heat fluxes and surface temperature. The work of *Physick* [1976] seems to be the first attempt to use an interactive soil model in a mesoscale model in which the sensible heat flux and the surface temperature were calculated from a surface-heat-balance equation. *McCumber and Pielke* [1981] coupled a mesoscale model with a soil model in which the surface heat fluxes depend on the moisture content and temperature within the soil column. These early attempts usually used very simple LSMs that did not account for the effect of vegetation, and the coupled mesoscale models ignored the spatial distribution of land-use, soil types, and seasonality of vegetation because of the lack of appropriate data. As mesoscale models continue to improve their physics schemes and model horizontal grid spacing, implementing advanced LSMs is critical in capturing the mesoscale structure at small scales forced by topography, soil moisture, snowpack, surface vegetation and soil characteristics.

During the last ten years, therefore, we have witnessed rapid progress in developing and testing land-surface models in mesoscale atmospheric models [e.g., *Bougeault et al.*, 1991; *Giorgi et al.*, 1993; *Bringfelt*, 1996; *Smirnova et al.*, 1997; *Chen et al.*, 1997a; *Pielke, et al.*, 1997]. Advances in LSM development have been largely supported by progress in two areas. First, field observations such as HAPEX-MOBILHY [*André et al.*, 1986], Cabauw [*Beljaars and Bosveld*, 1997], EFEDA [*Bolle et al.*, 1993], and First International Satellite Land Surface Climatology Project (ISLSCP) Field Experiment (FIFE) [*Sellers et al.*, 1992] have been conducted to investigate the interactions of the soil-vegetation-atmosphere continuum. These field campaigns often consisted of a dense network of surface stations, aircraft and satellite measurement platforms within a mesoscale domain, and provided rich measurements including not only traditional surface heat fluxes but also vegetation characteristics, soil properties, and soil moisture. The above data sets and some recently available data from BOREAS [*Sellers et al.*, 1995b] and Cooperative Atmosphere-Surface Exchange Study (CASES) [*LeMone et al.*, 2000] have been, and will be, critical for developing and validating LSMs.

Second, more numerical experiments performed in the context of both 1-D offline and 3-D coupled modes have been intended to verify/improve LSMs utilizing new data sets [e.g., *Braud et al.*, 1993; *Viterbo and Beljaars*, 1995; *Chen et al.*, 1996, *Noilhan et al.*, 1997]. One distinct example of these off-line numerical experiments is the Intercomparison of Land Surface Parameterization Schemes (PILPS) initiative [*Henderson-Sellers et al.*, 1993; *Chen et al.*, 1997b; *Wood et al.*, 1998]. For coupled model validation, for instance, *Betts et al.* [1997, 1998] used FIFE and BOREAS data to verify two versions of the NCEP Eta model and two global reanalyses.

Compared to the early attempts, these new developments in land-surface modeling dealt with much broader (and in much greater depth) topics, including advanced LSMs, the spatial distribution of vegetation and soil characteristics, the use of remote sensing data, and the initialization of soil moisture and snow depth. The purpose of this chapter is to briefly review the progress in coupling LSMs with mesoscale models and demonstrate their impacts on short-range numerical weather predictions and regional climate simulations. We also discuss problems and possible solutions for utilizing in-situ and remote-sensing data for specifying vegetation, soil, and soil-moisture conditions.

2. ADVANCES IN APPLYING LSMS FOR MESOSCALE WEATHER PREDICTION AND RESEARCH

2.1. Land-Surface Model Development

Undoubtedly, one major milestone in land-surface modeling was the introduction of a foliage layer (or "big leaf" model) in a soil model proposed by *Deardorff* [1978]. In later LSM developments, this concept of explicitly treating the plant canopy has been adopted with considerable variation. The explicit canopy treatment is modest in some models [e.g., *Deardorff*, 1978; *Pan and Mahrt*, 1987; *Noilhan and Planton*, 1989] but rather complex (including some biophysical processes) in other models [e.g., *Dickinson*, 1984; *Sellers et al.*, 1986]. Another type of LSM [e.g., *Entekhabi and Eagleson*, 1989; *Wood et al.*, 1992; *Schaake et al.*, 1996] is based on the understanding of the long-term hydrological cycle, and implicitly treats the effect of the vegetation canopy on transpiration. Some of these hydrological models are designed and calibrated over large-scale river basins and incorporate the effects of subgrid-scale variability in precipitation and soil moisture.

Some recent land-surface modeling studies, especially the Project for Intercomparison of Land Parameterization Schemes [PILPS, *Shao and Henderson-Sellers*, 1996; *Chen et al.*, 1997b; *Wood et al.*, 1998], focused on the evaluation of different LSMs using long-term observations. These uncoupled LSM intercomparison experiments ranged from single-site evaluations [e.g., Cabauw validation, *Chen et al.*, 1997b; HAPEX-MOBILHY, *Shao and Henderson-Sellers*, 1996] to sub-continental and global scale evaluations [e.g., Red-Arkansas River Basin, *Wood et al.*, 1998; Global Soil Wetness Project, *Dirmeyer et al.*, 1999]. One important fact learned from these experiments is that, given the same forcing conditions, the surface heat fluxes, soil moisture, and runoff simulated by different LSMs vary considerably. Partly motivated by these model intercomparison experiments, there are efforts to take the strengths of these LSMs originally designed for atmospheric applications and apply them to surface hydrologic models or vice versa [*Chen et al.*, 1996; *Liang et al.*, 1999]. These uncoupled LSM validations proved to be, and will continue to be, useful for understanding the physics of different LSM components and for identifying the techniques required to improve LSMs. In particular, the data obtained from a dense network of surface stations within a mesoscale domain such as FIFE, EFEDA, and CASES-97 are valuable for evaluating LSM's ability to provide area-averaged surface heat fluxes.

Given such a wide spectrum of land-surface models, it is a challenge for atmospheric modelers to select a land-surface scheme that is appropriately adapted to their needs. For example, the implementation of a land-surface scheme in operational numerical weather prediction (NWP) models is strongly dependent on the practical constraints of the computer environments of each NWP model [*Blondin*, 1991]. Even more challenging, given land-surface subgrid-scale variability, is the difficulty of operationally specifying a potentially vast set of physical parameters over continental domains on a daily, realtime, annual cycle basis, as required by most LSMs. This challenge makes the use of complex land-surface schemes quite unattractive. To investigate this issue, *Chen et al.* [1996] compared four LSMs simulations against a 5-month FIFE data set and concluded that a relatively simple LSM, with an explicit canopy resistance formulation had similar performance to more complex models in terms of providing "good" area-averaged surface heat fluxes. Again, as demonstrated by other PILPS studies [*Chen et al.*, 1997b; *Wood et al.*, 1998], this is partly due to the difficulty in accurately specifying a large set of physical parameters, especially at the local scales, due to the lack of proper measurements of such parameters.

Therefore, in most coupled mesoscale models [e.g., see *Chen et al.*, 1997a; *Smirnova et al.*, 1997; *Noilhan et al.*, 1997; *Chen and Dudhia*, 2000], the LSMs are relatively simple in physics but efficient in computation. They often have fewer parameters to be specified compared to LSMs used in some general circulation models intended for long-term climate simulations. Some preliminary evaluations of these coupled models indicate that the LSMs seem to perform adequately in most cases. Nevertheless, some physical processes are over simplified in these LSMs. For instance, the treatment of snow cover and its heterogeneous distribution is crude. Given the important effects of snow and frozen-ground processes on short-time weather forecasts, more realistic representations of these cold-season physics are being developed and tested [*Koren et al.*, 1999; *Smirnova et al.*, 1998; *Liston*, 1999]. For regional climate studies, however, the LSMs in coupled mesoscale models are quite sophisticated and include biological and ecological feedbacks [e.g., *Giorgi et al.*, 1993; *Lu et al.*, 2000; *Eastman*, 2000].

2.2. Specification of Vegetation and Soil Characteristics and Associated Parameters in Coupled Models

Modern-era land-surface models often require the specification of a number of parameters related to the underlying vegetation and soil characteristics. One basic concept in coupled models is to specify a spatial distribution of primary land-surface characteristics and relate the secondary parameters to the primary ones through look-up tables. In early coupled mesoscale models [e.g., *Bougeault et al.*, 1991], the primary parameters included dominant vegetation type and soil texture for each grid point. The secondary vegetation and soil parameters are numerous and variable, depending on the specific land-surface model. They often include the surface albedo, roughness length, vegetation fraction, leaf area index (LAI), canopy resistance, porosity, hydraulic conductivity, field capacity, wiling point, soil heat capacity, and thermal conductivity, etc.. More sophisticated land-surface models also need vegetation dependent photosynthesis and respiration parameters [e.g., *Bonan*, 1996].

Although some operational NWP models such as the NCEP Eta model and RUC model [e.g., *Betts et al.*, 1997; *Smirnova et al.*, 1997] still use the 1×1 degree resolution vegetation and soil maps (which is more coarse than the model grid spacing), higher resolution maps of surface characteristics are gradually integrated into coupled models owing to recently available data. *Loveland et al.* [1995] of the U.S. Geological Survey (USGS) derived a land-cover (vegetation) dataset from the 1-km satellite AVHRR data obtained from April 1992 through March 1993. Plate 2 shows the USGS 16-category (Sib land-cover legend) classification for North America. Several land-cover classifications were derived from the same AVHRR data set. Taking North America as an example, in addition to the relatively simple Sib 16-category classification, there is a very detailed 205-category (seasonal land-cover legend) classification. The USGS 16- and 24-category land-cover data have been used in RAMS and MM5 coupled models [*Pielke et al.*, 1997, *Chen and Dudhia*, 2000]. This 1-km resolution land-use dataset provides not only a detailed spatial distribution of vegetation, but also a delineation between water bodies and land surface for high-resolution applications.

Mapping of soil texture and soil properties is a more difficult problem, because they have to be based on ground surveys. *Miller and White* [1998] developed, based on the USDA State Soil Geographic Database (STATSGO), a 1-km resolution multi-layer 16-category soil characteristics dataset of the conterminous United States, that provides information on soil texture, bulk density, porosity, available water capacity, etc., at different soil depths. Plate 3 shows the spatial distribution of the surface soil texture over the conterminous United States. As we will discuss below, having detailed information about soil texture is critical in determining soil hydraulic properties and thus the determination of runoff and soil moisture evolution. At the global scale, besides the 1-degree 9-category soil texture data [*Zobler*, 1986], 5-minute soil maps with a classification similar to STATSGO are also available [*Reynolds et al.*, 1999]. These data are based on the original 5-minute Food and Agriculture Organization digital soil map of the world.

The list of primary parameters for use in a typical LSM seems to be expanding. For instance, the green vegetation fraction in the Eta and MM5 models is specified by 0.15-degree monthly data, which are derived from the aggregation of a 1-km, 5-year climatology AVHRR-NDVI [*Gutman and Ignatov*, 1998]. In the U.S., as shown in Plate 4, the area covered by more than 50% green vegetation starts in the spring in the west coast and south-eastern areas and spreads to the northern parts in the summer. Note that, in the central High Plains, the green vegetation fraction in some agricultural regions can be as high as 96% in July. The study of *Betts et al.* [1997] showed that model-derived summer evaporation in the FIFE field-study area of northeast Kansas is very sensitive to the specification of this parameter.

Other candidates for primary parameters can be seasonal variation of albedo and roughness length. Some of these data are based on the same original AVHRR images, but there are inconsistencies among these data due to different data processing techniques employed to remove atmospheric effects such as cloud contamination. Clearly, a coordinated effort of the land-surface modeling and remote sensing communities is necessary to develop and test these vegetation data sets with the same resolution, or at least from the same NDVI database. An excellent example of such effort is the recent work of Csiszar and *Gutman* [1999], in which they derived a monthly global albedo database consistent with and on the same output grid as the aforementioned greenness database of *Gutman and Ignatov* [1998], and in which they used the same underlying 5-year AVHRR database. For real-time NWP applications, using quasi-real-time vegetation data e.g., weekly or bi-weekly) in place of monthly climatology should improve LSM performance, because some vegetation parameters at a given time of year, such as greenness, can vary dramatically from year to year owing to anomalous stresses such as heat/cold stress and droughts or floods.

To some extent, specifying other secondary parameters such as minimal stomatal resistance is much more difficult. No experimental studies have been conducted to measure the large number of parameters needed for complex LSMs. These parameters, related to vegetation and soil characteristics, are usually based on limited laboratory experiments or field sampling. For instance, there are some measurements of canopy resistance, one fundamental factor controlling the vegetation transpiration process, for different kinds of vegetation at different stages of their development [e.g., *Monteith*, 1975, 1976]. Yet these data are not sufficiently comprehensive to cover even the modest set of Sib's 16-category vegetation. Hence, in some mesoscale models [e.g., *Mahfouf et al.*, 1995; *Chen and Dudhia*, 2000], the vegetation parameters of certain vegetation types for which there is no direct measurements have to be interpolated using the values from other vegetation categories. It also worth mentioning that vegetation parameters such as stomatal resistance, based on small-scale field observations, should be evaluated for use in the mesoscale modeling context.

The specification of soil parameters, like vegetation parameters, is similarly daunting and depends on the choice of formulations for computing soil physical properties. As reviewed by *Cuenca et al.* [1996], there are

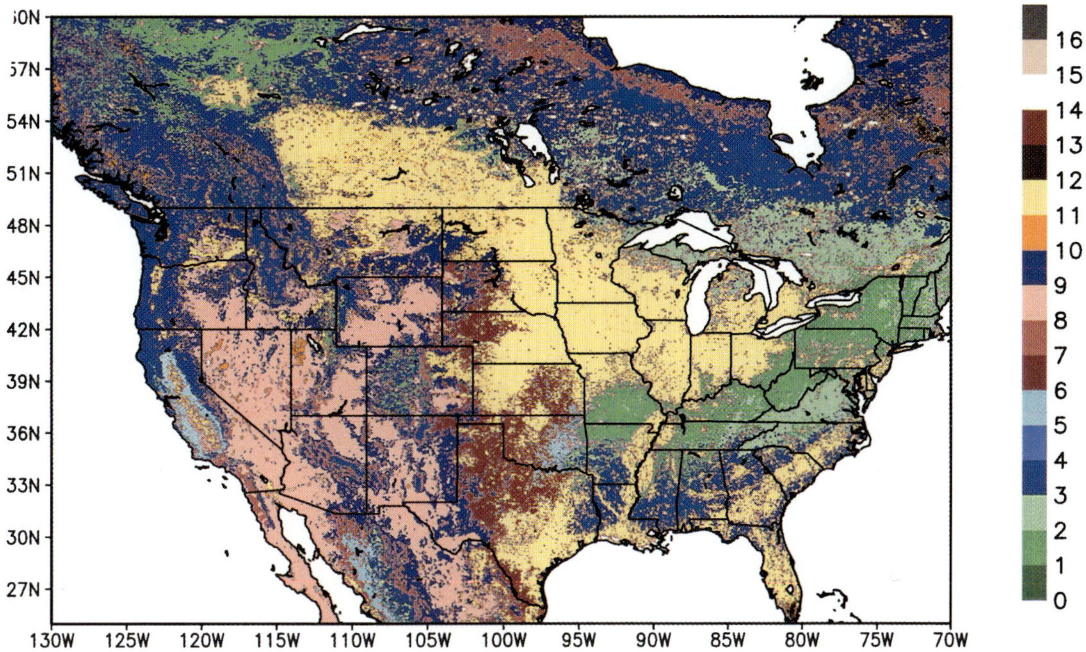

Plate 2. 1-km resolution USGS/EROS Sib 16-category land-use map. Category numbers are as follows: 1 - Broadleaf-evergreen tree, 2 - Broadleaf-deciduous tree, 3 - Broadleaf and needleleaf tree, 4 - Needleleaf-evergreen tree, 5 - Needleleaf-deciduous tree, 6 - Broadleaf tree and groundcover, 7 - Groundcover only, 8 - Broadleaf shrub and groundcover, 9 - Broadleaf shrub and bare soil, 10 - Tundra, 11 - Bare soil, 12 - Cultivation, 13 - Wetland, 14 - Coastal complex, 15 - Water, 16 - Land ice. Data are available at http://edcwww.cr.usgs.gov/landdaac.

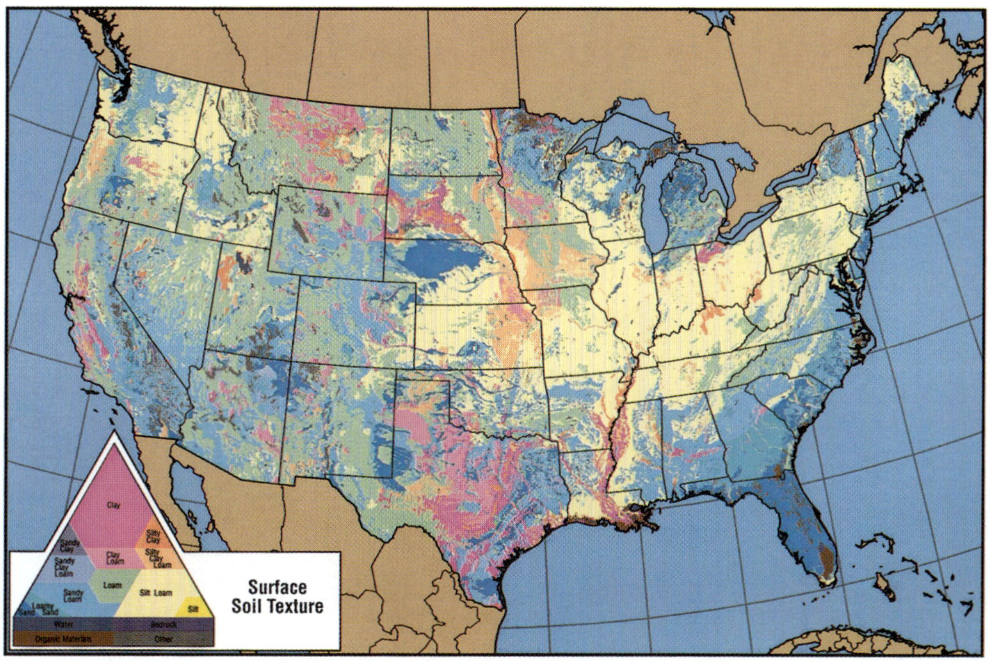

Plate 3. 1-km surface soil texture map available for the conterminous U.S. (by courtesy of D. Miller, Pennsylvania State University; data available at http://www.essc.psu.edu).

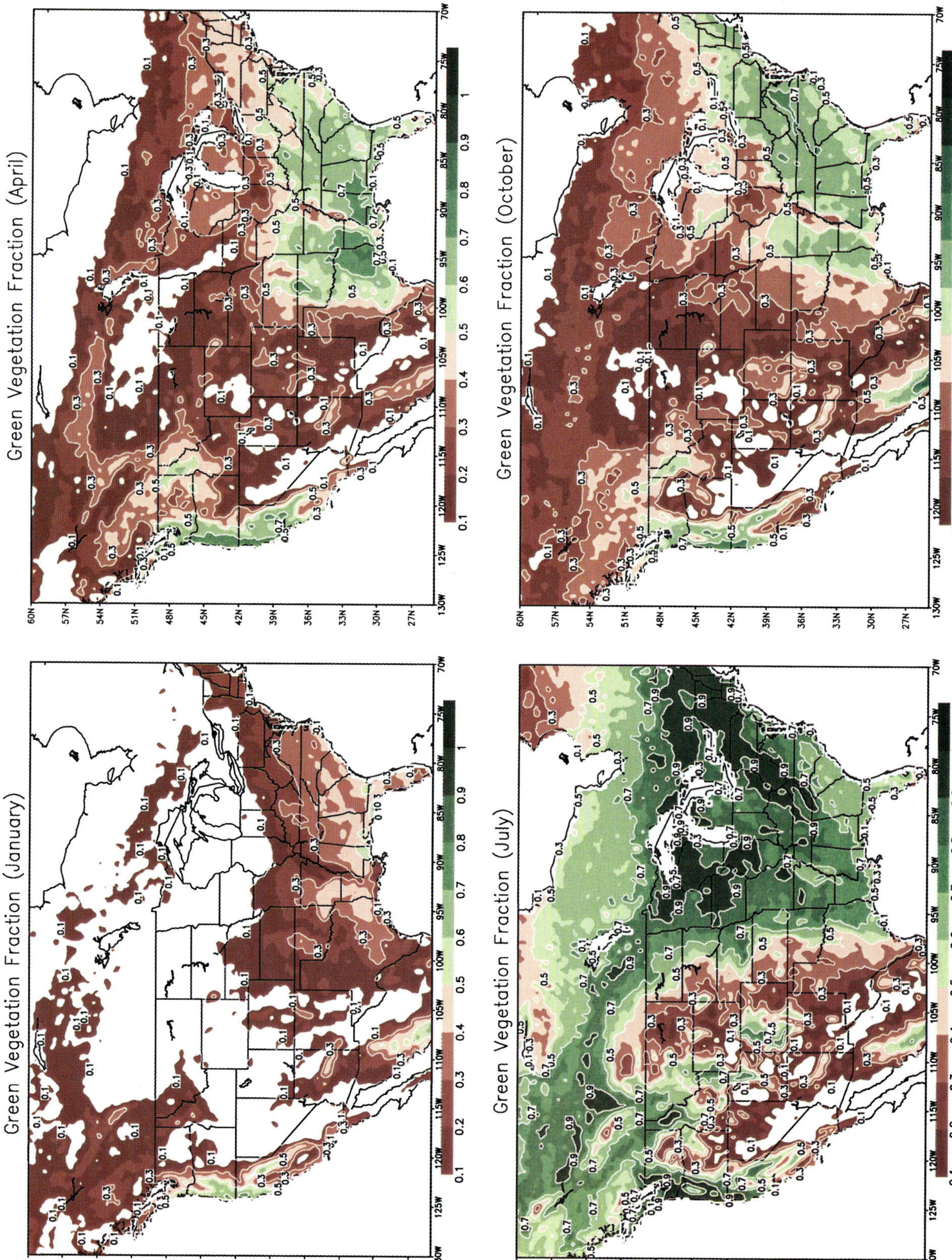

Plate 4. Seasonal variation of NOAA/NESDIS 0.15 degree climatology green vegetation-fraction for the month of January, April, July, and October.

two widely used parameterizations. One uses *Clapp and Hornberger* [1978] functions, which were developed from an analysis of 1446 soil samples taken from 34 localities in the U.S. and later expanded to include some additional data [*Cosby et al.*, 1984]. These functions and parameters are widely used in LSMs designed for atmospheric models [*McCumber and Pielke*, 1981; *Mahrt and Pan*, 1984; *Noilhan and Planton*, 1989]. The other uses *van Genuchten* [1980] functions, which have fewer parameters and are probably the most commonly used parameterization for the retention and conductivity functions in the soil physics community. *Cuenca et al.* [1996] showed that the magnitude of hydraulic conductivity, based on these two formulations and for the same soil texture and soil moisture content, is very different. They also demonstrated that for a 24-h simulation over a bare-soil site, the variations in the surface energy balance and boundary layer are sensitive to the different parameterizations of the soil water processes. In a recent study, *Walko et al.* [2000] describes an LSM in which either of the two different parameterizations discussed in *Cuenca et al.* [1996] can be used. Certainly, more studies must be pursued to include cases with vegetated areas and in a coupled model in order to evaluate and improve soil water parameterization.

2.3. Soil Moisture and Snow Initialization

There is a rich literature demonstrating the importance of soil moisture, particularly its spatial heterogeneity, in mesoscale processes [e.g., *Avissar and Pielke*, 1989; *Chen and Avissar*, 1994a,b; *Ziegler et al.*, 1994; *Avissar*, 1995]. Indeed, the soil moisture is a very important component of land-surface modeling, and it would not make much sense to implement a sophisticated LSM in mesoscale NWP models without a proper soil-moisture initialization procedure. However, the initialization of soil moisture in coupled regional models is jeopardized by the fact that there are no routine soil moisture observations over the large spatial domains typically spanned by mesoscale NWP models. In the context of modeling, soil moisture (wetness) refers to the amount of soil water computed from a LSM. As demonstrated by *Koster and Milly* [1997], different LSMs may have different soil moisture dynamic ranges due to various approaches in treating evaporation and runoff. Different LSMs can give different values of soil moisture while having similar estimates of water and energy exchange. It is, therefore, important to recognize that the soil moisture calculated for a given atmospheric forcing is model-dependent; the soil moisture in a coupled model is used primarily to calculate the water and energy fluxes between the land surface and the atmosphere and is meaningful, to a large extent, only within the context of a particular hydrological or ecological parameterization.

The soil moisture fields from the traditional 4-dimensional data assimilation (4DDA) systems of coupled land/atmosphere models often suffer substantial errors and drift owing to precipitation, temperature, and radiation biases in the coupled system. Such errors and drift often necessitate the application of soil moisture nudging techniques in coupled 4DDA systems. *Mahfouf* [1991] demonstrated that for some ground covers (e.g., crop area) and clear sky conditions, observations of screen-level temperature and humidity can be used to estimate a realistic soil moisture. Similar approaches have been adopted at various NWP centers [*Douville et al.*, 2000; *Giard and Bazile*, 2000]. These schemes seem to work well under certain atmospheric conditions but fail in situations where the soil moisture is insensitive to atmospheric conditions such as in cloudy and windy conditions.

An alternative strategy is to utilize precipitation observed by gage/radar/satellite, satellite-observed surface solar insolation, and meteorological analysis to drive an off-line simulation of a LSM, so that the evolution of soil moisture does not suffer from the model biases in surface forcing fields [e.g., *Smith et al.*, 1994; *Mitchell et al.*, 2000]. Studies show that if attention is closely paid to quality surface forcing, hydrological modeling aspects, and soil moisture dynamics, then uncoupled stand-alone LSMs seem to be able to capture the evolution of soil moisture. For example, *Chen et al.* [1997a] and *Chen and Mitchell* [1999] conducted two off-line numerical experiments to simulate the surface energy and soil moisture processes. In one numerical experiment, the LSM was driven by one-year-long [1986] atmospheric HAPEX-MOBILHY forcing data (with a 30-minute time interval] and was validated against the point measurement of soil moisture profiles from the surface down to 1.6 m. In the other experiment conducted under the context of the Global Soil Wetness Project, two-year-long (1987-1988) 1×1 degree ISLSCP atmospheric forcing and surface data (with a 60-minute time interval) were used to drive the LSM over the global land points. Also two years of soil moisture profiles from the surface down to 2 m, measured by the 17-station Illinois Soil Moisture Network, were used to evaluate the LSM simulations at regional scale. They found that the seasonal evolution of daily soil moisture in both the surface layer and the deep root-zone layer can be reasonably captured by the LSM. Similar results are also reported by *Calvet et al.* [1998].

Jointly supported by NOAA and NASA, a Land Data Assimilation System [LDAS, *Mitchell et al.*, 2000] is being developed. LDAS uses observed hourly precipitation and surface radiation to drive LSMs at NCEP, NASA/GSFC, Princeton University, and the University of Washington to simulate the soil moisture evolution. Hence, by forcing LDAS primarily with observations, the often severe biases (particularly precipitation and surface insolation) produced by atmospheric models can be avoided. This approach is appropriate for the regions where high-quality rainfall observations, both in terms of spatial and temporal resolution, are available.

On the other hand, increasingly improving remote sensing techniques hold the promise of monitoring the surface soil moisture at high resolution over a large domain up to global scale [*Jackson*, 1997]. The radiation emission in the microwave frequencies is, however, severely attenuated in the soil porous medium. As a result, the radiometric observations correspond to the dielectric constant profile only in the top few (less than 5) centimeters of the soil column. A number of studies have involved the subject of estimating soil moisture profiles down to depths beyond the penetration of remote sensing observations [*Jackson*, 1980; *Camillo and Schmugge*, 1983; *Milly and Kabala*, 1986]. But the effects of vegetation and soil were not incorporated in those assimilation procedures.

Utilizing remotely sensed surface characteristics to constrain unrealistic simulated soil moisture has been investigated (e.g. *Entekhabi et al.*, 1994; *Houser et al.*, 1998]. *Houser et al.* [1998] explored the feasibility of assimilating remotely sensed surface soil moisture by applying 4DDA in a hydrologic model, and tested several assimilation procedures. This data assimilation approach will be employed in the late stage of the LDAS development to validate the soil moisture fields simulated in LDAS. Therefore, given the pros and cons of various soil moisture data assimilation approaches, the ultimate solution may be a combination of off-line simulation by LSMs and assimilation of remotely sensed surface soil moisture.

As mentioned above, the current snow parameterization schemes in mesoscale NWP models need improvements to better describe the snow albedo and heterogeneity of snow cover. The use of realistic albedo over snow-covered forests has a strong influence on the atmosphere [*Douville and Royer*, 1996; *Viterbo and Betts*, 1999]. Accordingly, information on fractional snow cover and snow albedo derived from various observation platforms including satellite should be included in land-surface initialization procedures.

There are several high-resolution snow products available and some of them have been used in NWP models (e.g., NCEP Eta model). For instance, the 47-km daily U.S. Air Force global snow analysis provides information on the snow depth and sea ice, and the 23-km NESDIS/NOAA [*Ramsay*, 1998] daily analysis provides the spatial distribution of snow cover and sea-ice for the northern hemisphere. Near real-time quantitative assessments of snow cover and snow water equivalent throughout the United States and Canada have been developed at the National Operational Hydrologic Remote Sensing Center (NOHRSC/NOAA). This 10-km daily product is a result of composite use of multi-source satellite (i.e., Advanced Very High Resolution Radiometer (AVHRR) and Geostationary Operational Environmental Satellite (GOES)), point (i.e., manual and telemetered observations), and line (i.e., airborne gamma radiation surveys) data.

2.4. Impact of LSMs on Mesoscale Numerical Weather Prediction

As mesoscale models continue to improve their representation of the boundary layer physics, and as their spatial resolution increases, the inclusion of a more realistic LSM and initial soil moisture conditions should yield improved details of the lower boundary conditions. Relatively speaking, the application of advanced LSMs in real-time mesoscale systems is still in its early developmental stage, and few studies have been devoted to systematically evaluating the impact of LSMs on mesoscale weather predictions. Nevertheless, investigations to date seem to indicate an improvement in NWP through the use of advanced LSMs in coupled mesoscale models.

There is little doubt that a good LSM should improve the simulation of surface heat fluxes as demonstrated by *Lee et al.* [1995], *Betts et al.* [1997], *Noilhan et al.* [1997], and *Chen and Dudhia* [2000]. For instance, Figure 1 shows the comparison of 48-hour simulations of surface heat flux by two LSMs in MM5 for the FIFE June-4th case (clear sky situation). A LSM based on *Pan and Mahrt* [1987] and *Chen et al.* [1996] simulated the latent heat flux, which is more difficult to model than the sensible heat flux, very well, because of the complex interactions among plant physiology and transpiration processes. Although the new LSM's sensible heat flux is better than that from the simple slab model, both models overestimated the sensible heat flux. This discrepancy may be due to observational errors in the surface energy budget, even though the model produced

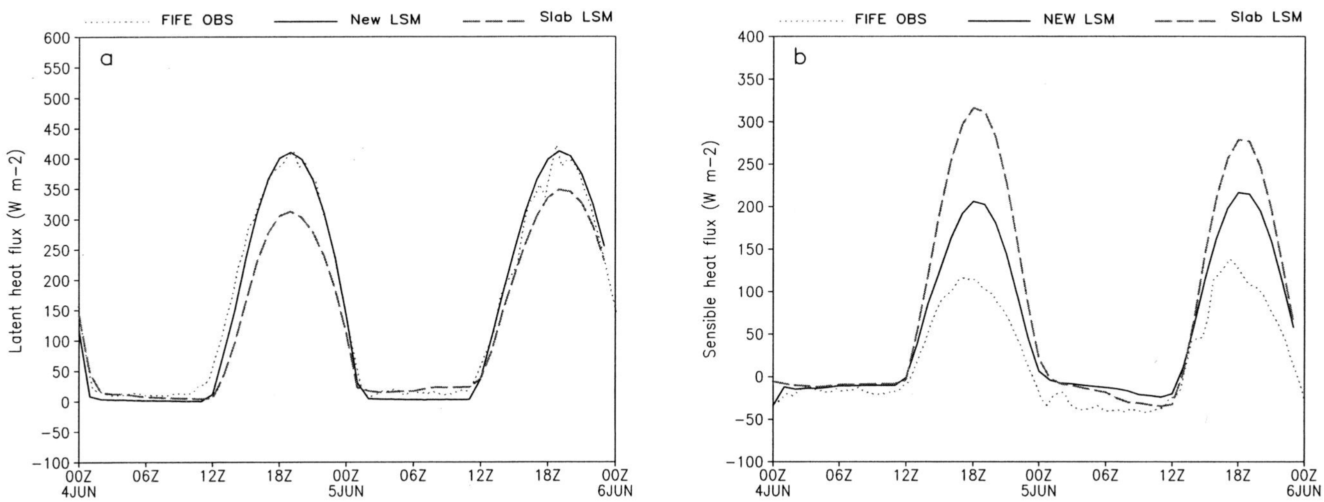

Figure 1. a: Comparison of MM5 simulated surface latent heat flux with FIFE observations, and b: Comparison of surface sensible heat flux. FIFE: FIFE observations; New LSM: MM5 simulation with the new LSM; and Slab LSM: MM5 simulation with the simple slab soil model.

nearly the same net radiation as observed. As pointed out by *Twine et al.* [2000], independent measurements of the major energy balance flux components are not often consistent with the principal of conservation of energy. It has been known that the sum of sensible and latent heat flux measured by eddy covariance to be less than the difference between net radiation and soil heat fluxes. Given such an uncertainty of observational errors in surface energy fluxes, we can not expect the LSMs to closely reproduce all observed surface heat flux terms simultaneously at all times. There is some evidence indicating that a good LSM can improve the simulation of the vertical structure of temperature and moisture in the PBL [e.g., *Betts et al.*, 1997]. However, as the transport of heat and moisture in the PBL is largely controlled by the particular PBL parameterization scheme used in the coupled model, a careful evaluation of the LSM in the coupled model must also involve the sensitivity of the PBL evolution to surface conditions.

The new land-surface model in the NCEP operational Eta model was validated against surface heat fluxes, PBL profiles, and near-surface air temperature and moisture [*Betts et al.*, 1997; *Chen et al.*, 1997a, *Yucel et al.*, 1998; *Marshall*, 1998; *Berbery et al.*, 1999; *Hinkelman et al.*, 1999]. It was also verified, alongside the former bucket model, with observed precipitation and surface fields over continental scales [*Chen et al.*, 1998]. Long-term parallel tests, as well as a number of case studies, with these two land-surface schemes indicate that the land-surface processes have a significant impact on short-range weather forecasts. Figure 2 shows a comparison, of an approximately one-month precipitation threat score for June 1991, between the bucket model (which is initialized by a fixed annual mean climatology soil moisture field) and the new land-surface model (which is initialized by the NCEP time-dependent Global Data Assimilation System (GDAS)). The horizontal grid increment used in the Eta model to make these forecasts was 80 km. The new and more advanced land-surface model significantly improved the quantitative precipitation forecast for the mid and heavy rainfall categories. In fact, this land-surface model increases the model precipitation forecast skill in summertime as much as nearly doubling the model resolution.

Chen et al. [1999] compared simulations of the 12 July 1996 flash flood event, using two land-surface models coupled to MM5, for an isolated thunderstorm moving across the Buffalo and Spring Creek watersheds in Colorado. During the four-hour storm period, precipitation accumulations exceeded 80 mm in places over both watersheds. Numerous factors were undoubtedly responsible for the mesoscale modulation of the large-scale, unstable, upslope flow such that the specific location and intensity of this storm were determined. They used MM5 in a two-way interacting quadruple nested computational domain: Grid 4 (the inner grid), grid 3, grid 2 and grid 1 (the outer grid) have mesh sizes of 58x52, 52x97, 76x76 and 84x73 points and grid increments of 1, 3, 9 and 27 km, respectively. Mesoscale models often have the option of using a series of nested

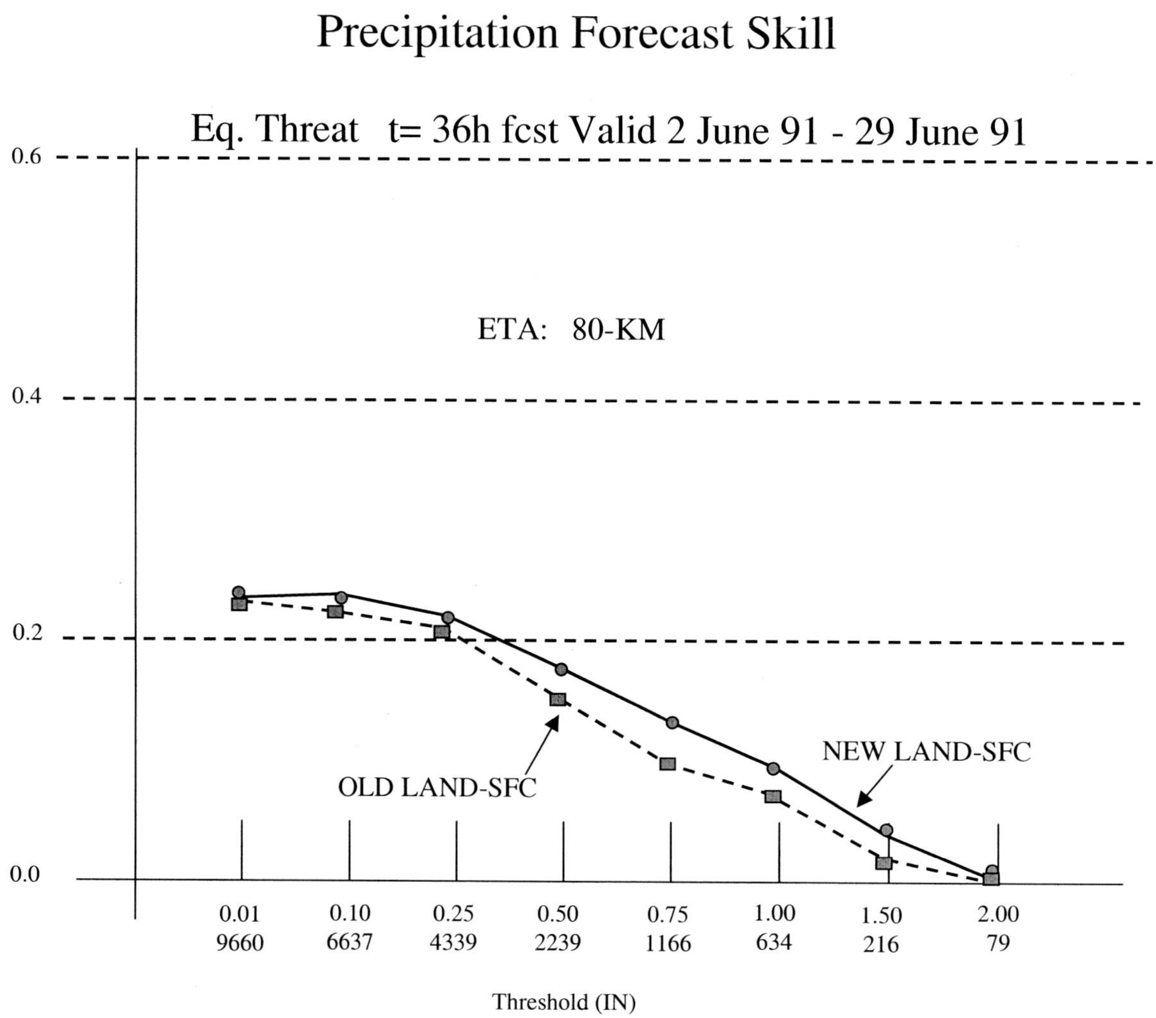

Figure 2. Comparison between the bucket model and the new LSM in the NCEP Eta model: precipitation equitable threat score for Eta model forecast at 36-hour from 2 to 29 June 1991. OLD LAND-SFC: bucket model results; New LAND-SFC: new LSM results. Adapted from *Chen et al.* [1998].

computational domains, where the horizontal resolution of the domain increases by a factor of 2 to 4 for each progressively smaller grid. This allows mesoscale models to better represent the complex terrain and to examine small-scale features of precipitation systems important for hydrological applications. In such nested grid systems, there is often an interactive interface or boundary conditions in which each grid can influence the next coarser one as well as the next finer one. One would expect that this truly interactive interface should allow the local forced system on the interior to interact with the longer domain-scale wavelengths. *Warner et al.* [1997] provided an excellent review of the effects of lateral boundary conditions on regional numerical weather prediction.

For grid 4, Plate 5 shows the S-Pol radar precipitation analysis with a rainfall maximum of about 80 mm centered over the northeast, lower-elevation, part of the Buffalo Creek watershed. As shown in Plate 6, the MM5 model with the LSM based on *Chen et al.* [1996] produced a slightly broader area of precipitation than observed, particularly in the southeastern part of the grid 4, than observed, but the maximum (90 mm) is in approximately the correct location and occurs at the right time, 10 hours into the simulation. By contrast, the MM5 model with the simple 'slab soil model' produced much lower precipitation (Plate 7), and the precipitation also seemed to cover a broader area. In particular, it incorrectly generated some weak convective cells in the southwestern area.

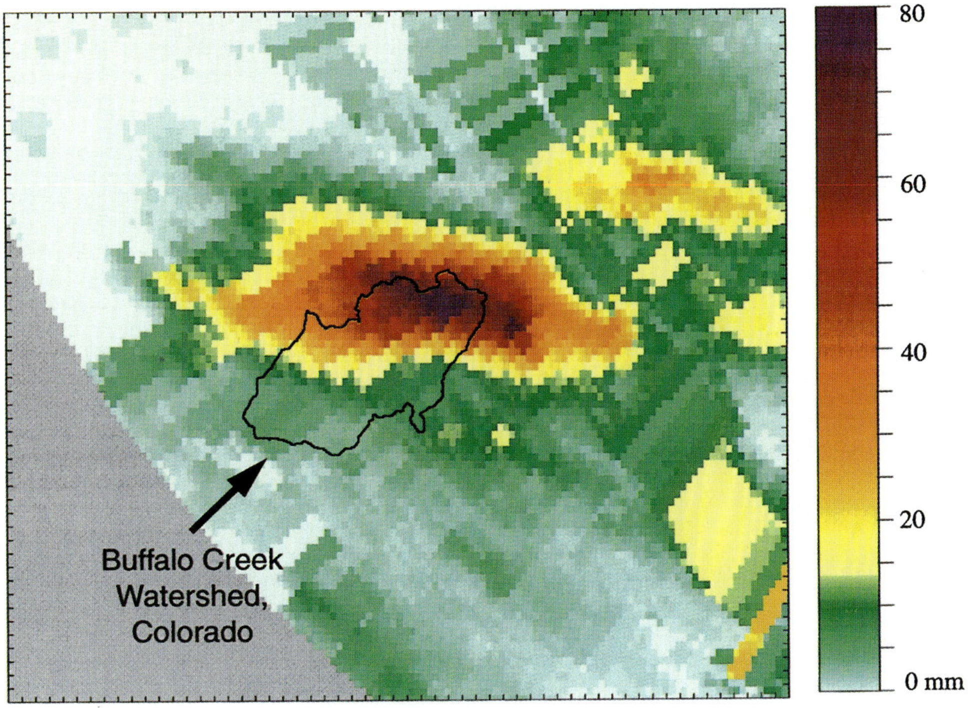

Plate 5. S-Pol rainfall analysis for the MM5 grid 4 for 00Z-0348Z 13 July 1996 (by courtesy of E. Brandes and D. Yates, National Center for Atmospheric Center, Colorado).

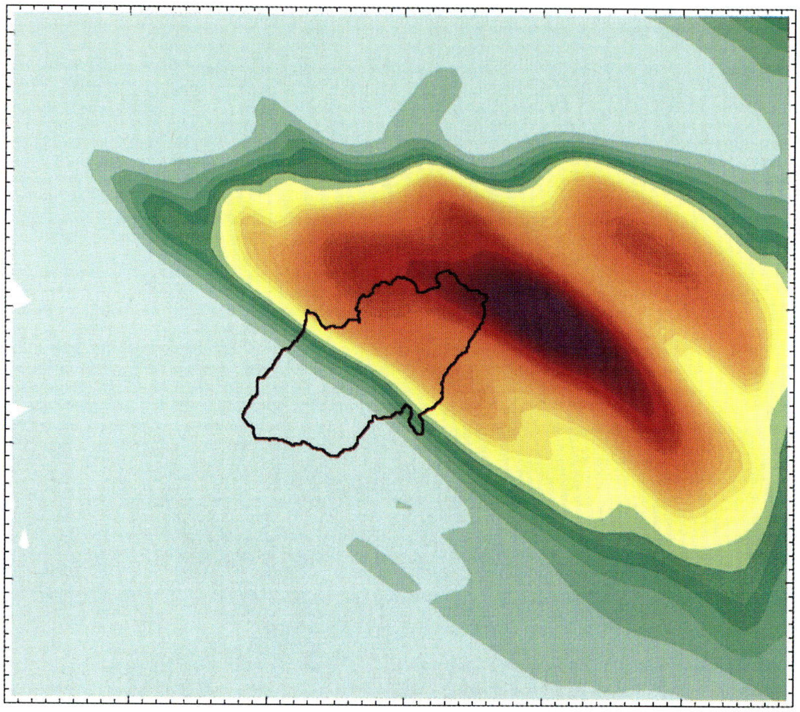

Plate 6. 4-h rainfall totals (00Z-04Z, 13 July 1996) for the gird 4 simulated by the MM5 model coupled with a new LSM.

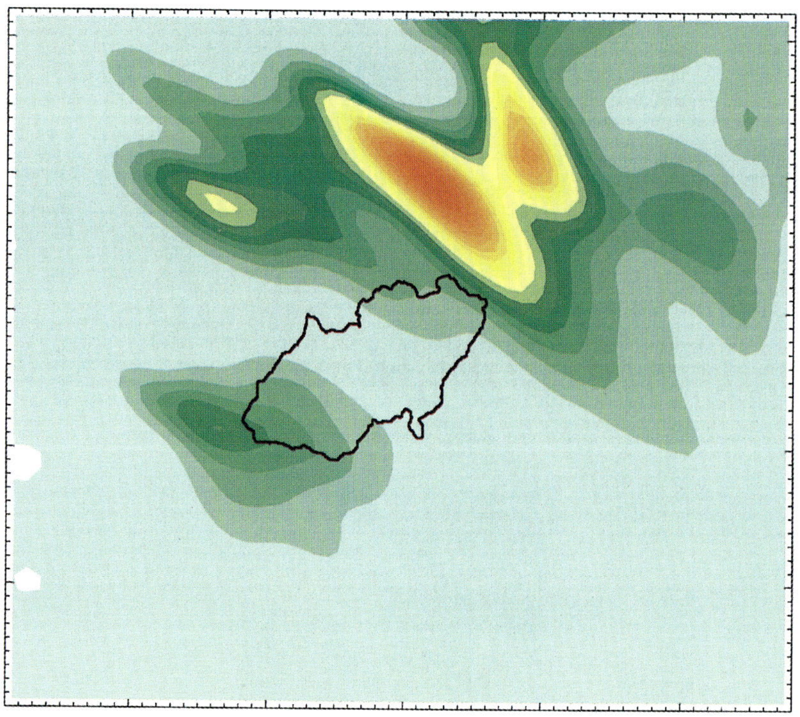

Plate 7. Same as Plate 7, but for the simulation with MM5 coupled to the traditional slab soil model.

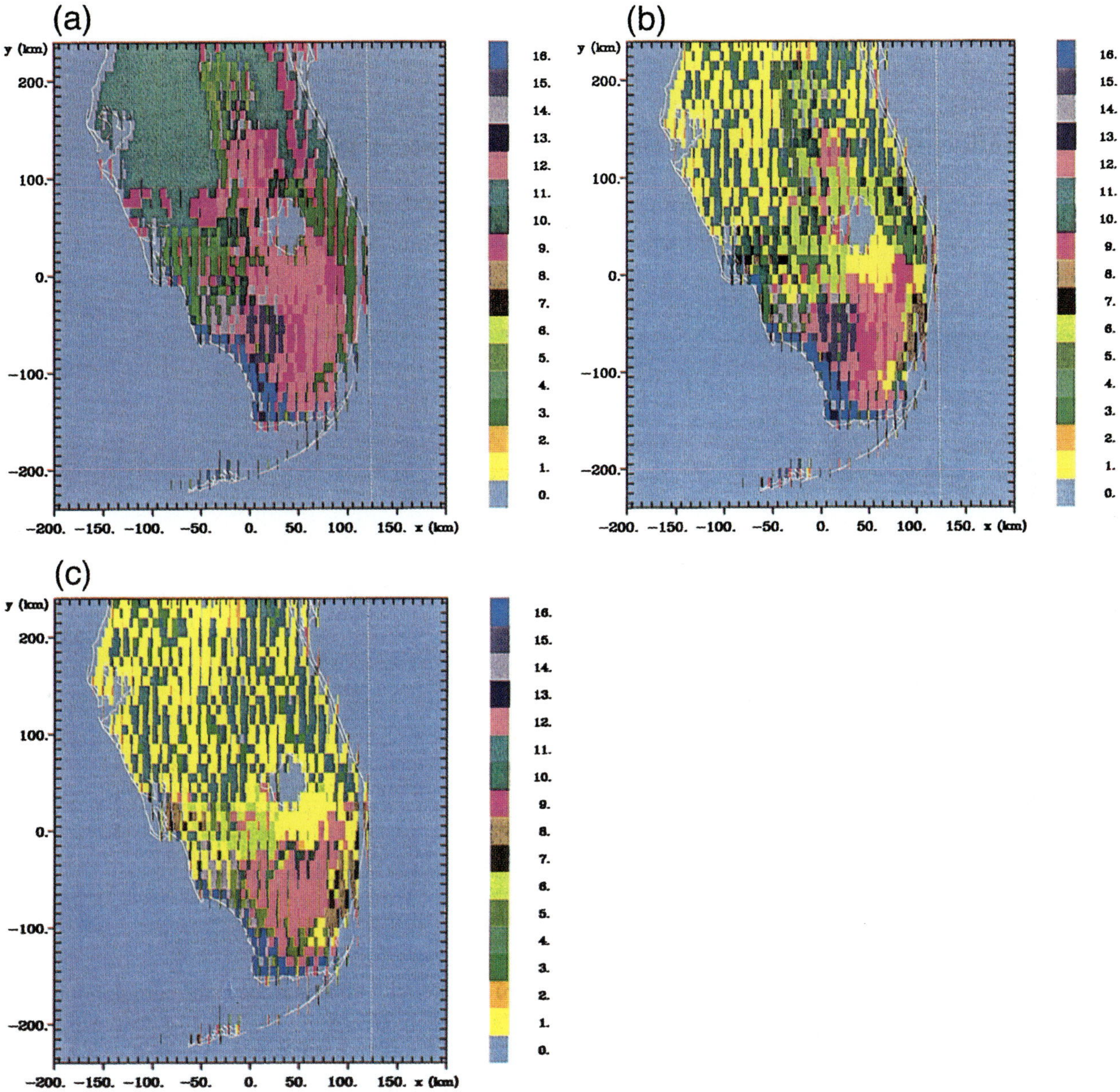

Plate 8. South Florida vegetation classes for (a) 1900, (b) 1973, and (c) 1993. Category numbers are as follows: 0 - lake, river, or ocean; 1 - agricultural crops/pasture-land; 2 - short grass; 3 - evergreen needle-leaf; 4 - deciduous broad-leaf tree; 5 - evergreen broadleaf; 6 - tall grass; 7 - desert; 8 - semi-desert; 9 - bog or marsh; 10 - evergreen shrub; 11 - mixed woodland; 12 -tall grass marsh; 13 - non-vegetated marsh; 14 - evergreen needle-leaf swampland; 15 - evergreen shrub with saturated soil; 16 - evergreen broadleaf with saturated soil. Within the meteorological grid (which is 10 km on a side), the individual vegetation classes are represented as the percentage of coverage. We have used rectangular shapes to illustrate this in the Figure, which is the reason for the linear north-south striping (from *Pielke et al.* [1999b]).

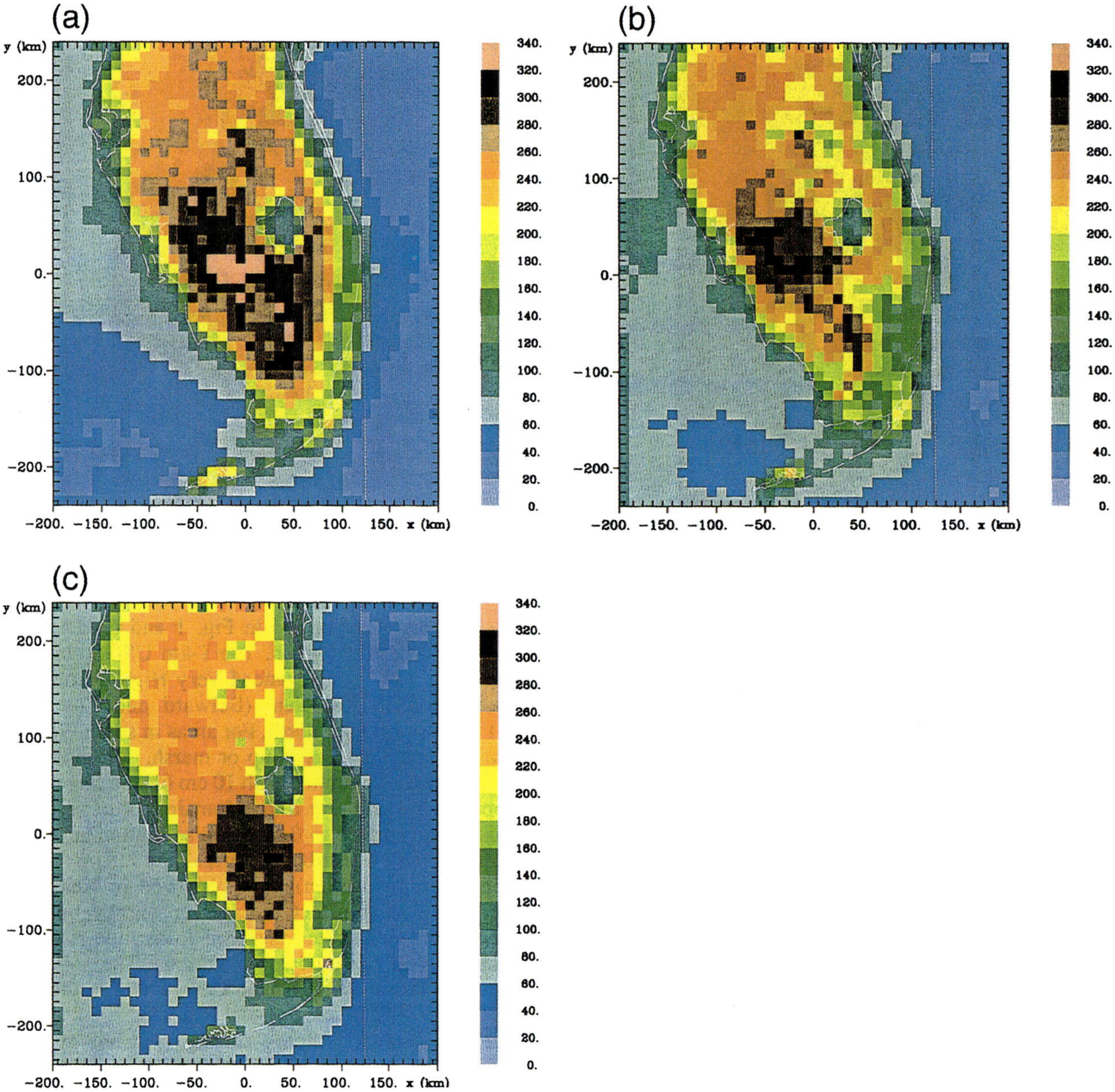

Plate 9. South Florida June-July total convective precipitation (mm) for (a) 1900, (b) 1973 and (c) 1993 landscapes. the spatially-averaged precipitation over land was 238 mm with the 1900 landscape, 216 mm with the 1973 landscape, and 213 mm with the 1993 landscape (from *Pielke et al.* [1999b]).

It is interesting to speculate on the predictability of the summer convection. If the perturbations that trigger convection within the large scale flow are not related to specific terrain or soil moisture features, the exact locations and timing of convective events will be relatively unpredictable. However, if the convective cells occur in response to the modulation of the large scale upslope flow by particular topographic or soil moisture features, it is reasonable to anticipate that this simulation will have deterministic skill in predicting the locations of specific events if the larger scales are simulated reasonably. In particular, a correct evolution of surface forcing conditions provided by surface sensible and latent heat fluxes may enhance the specification of surface thermal gradients, the PBL structure, and the predictability. That may be the case for the above Buffalo Creek flash-flood event simulation.

Another example on the possible enhancement of the coupled mesoscale model predictability by surface processes is given by *Davis et al.* [1999]. They used the MM5 model for real-time weather forecast for the U.S. Army Dugway Proving Ground (DPG), where the MM5 model had three nests inside the coarsest domain of 30-km, with 1.1 km grid spacing in the innermost grid. The DPG is surrounded by steep mountains whose peaks are 1-2 km above the desert floor. Also, there are salt flats, also known as playa, whose characteristics differ from those of the desert or nearby arid vegetation. These surface inhomogeneities play a major role in diurnally forced mesoscale circulations. From about 170 forecasts for the period 20 August 1997 to 31 July 1998, Figure 3 shows the statistics of those forecasts. The most notable feature of Figure is that model errors do not grow systematically with time. Model errors do grow for about the first 12 hours, but then remain nearly constant or actually decay. One probable reason for this performance is the dominance of local forcing and the decoupling of the boundary layer at night. While errors at upper levels may grow systematically with time, the behavior at the surface is, in effect, shielded from the behavior aloft. A correct specification of the local forcing induced by surface variability is crucial for enhancing this predictability.

2.5. Subgrid-Scale Variability

The extent to which the subgrid-scale surface heterogeneity impacts the surface heat flux calculation is currently debated. For example, *Avissar and Pielke* [1989], *Bonan et al.* [1993], *Chen and Avissar* [1994a,b], and *Grotzner et al.* [1996] have shown that subgrid-scale surface heterogeneity and its associated subgrid-

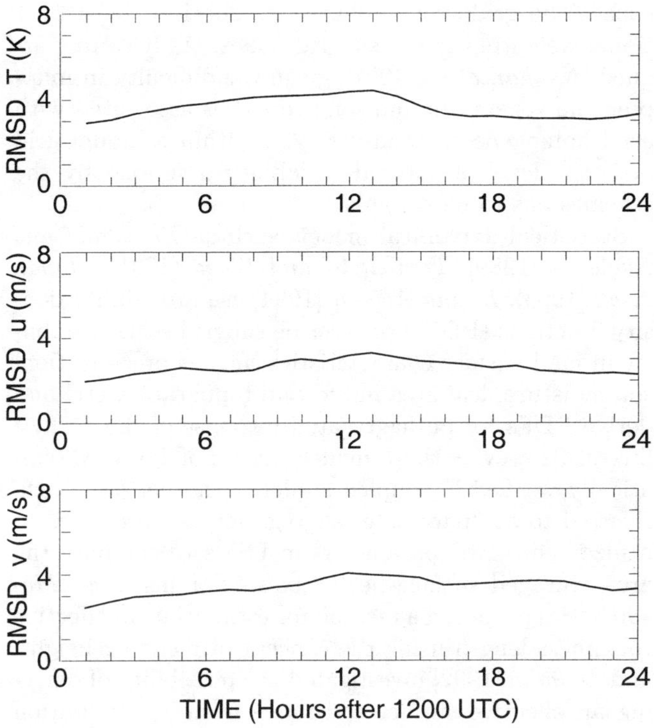

Figure 3. Root-mean-square difference (RMSD) derived from all MM5 forecasts for temperature (top), zonal wind (middle), and meridional wind (bottom). Adapted from *Davis et al.* [1999].

scale processes have significant influences on processes at global circulation model scales. As a result of nonlinearity in subgrid-scale processes, these effects can not be resolved by using only grid-averaged values and should be considered in large-scale models. On the other hand, *Garratt et al.*[1990] and *Wood and Lakshmi* [1993] suggested representing these processes at large scale through grid-averaged effective parameters. For example, the major result of *Wood and Lakshmi* [1993] was that fluxes and surface characteristics essentially scale linearly. In another study, *Sellers et al.* [1995a] concluded that using mean values of topography, vegetation conditions, and soil moisture to calculate the surface heat fluxes is sufficient for large-scale models.

Regardless of these debates, developing different approaches to represent the subgrid-scale variability effects for atmosphere-hydrology models has become a central focus of many recent studies. One approach employs so-called mosaic models [e.g., *Avissar and Pielke*, 1989; *Koster and Suarez*; 1992], which divides the model grid into finer subgrid elements and assumes they are homogeneous. The subgrid elements are then evaluated separately by applying surface energy equations to each. The grid-average fluxes are obtained by a fractional weighting of the subgrid fluxes. As it can be argued [*Noilhan et al.*, 1997], given the difficulty in mapping the vegetation and soil properties associated with each homogeneous land-use type within a model grid box, the level of subgrid variability described by the schemes has to be simplified.

Statistical-dynamical models such as *Entekhabi and Eagleson* [1989], *Famiglietti and Wood* (1991], *Bonan et al.* [1993], *Li and Avissar* [1994] use probability density functions (PDFs) to describe subgrid-scale variability in land-surface characteristics such as precipitation, soil moisture, leaf area index, and topography, etc, and derive PDFs for an aggregated response of the surface fluxes. However, these models have not been systematically verified in coupled models. Another approach, referred to as "parameter aggregation", is proposed to utilize "effective" parameters in LSMs to compute the area-averaged surface heat fluxes. For instance, *Mason* [1988] proposed a method for estimating an effective roughness length using the concept of blending height, and *Dolman* [1992] investigated the possibility of deriving an effective surface resistance from a distribution of stomatal resistance. *Noilhan et al.* [1997] examined the parameter aggregation approach for moderate land-surface variability of soil water and found that, despite the non-linear relations between surface fluxes and surface parameters, the effective surface fluxes match the area-averaged coupled model fluxes with relative little error.

It is certain that the development of new products such as time-varying maps of vegetation characteristics, albedo, and soil moisture, from high-resolution remote-sensing technology will improve the representation of subgrid-scale variability in mesoscale models. Also, some recent studies recognize the fundamental role of hydrologic processes such as runoff in the soil moisture simulation and hence in evaporation processes in LSMs. For instance, *Ducharne et al.* [1999] proposed to disaggregate the land-surface into a mosaic of hydrological catchments, with boundaries that are not dictated by the regular atmospheric grid but by topography. The soil moisture variability within each catchment is then related to topography. Such a trend of including topographic effects in the LSM will continue. To overcome the difficulty of specifying a potential large set of parameters in traditional LSMs, *Avissar* [1998] proposed an alternative approach. One distinct feature of this "high-order" LSM is that remote-sensed parameters can be directly used in this type of schemes, which do not need to be converted into pseudo vegetation and soil characteristics. Because of a lack of an appropriate data set, this new approach has not been tested.

3. ROLE OF LAND-SURFACE/ECOLOGY MODELS IN REGIONAL CLIMATE STUDIES

3.1. Climate Variability Associated With the Changes in Soil Moisture and Landscape

There have been important studies of the role of soil moisture and vegetation state in simulating weather and climate, as reported, for example, by *Dickinson* [1995], *Avissar* [1995], and more recently, in the following text by *Pielke et al.* [1999a]. *Eltahir* [1998] and *Zheng and Eltahir* [1998] have explored the role of soil moisture in precipitation processes, and demonstrated the crucial importance of radiative and dynamical feedbacks in regulating rainfall anomalies that result from soil moisture anomalies. *Zheng and Eltahir* [1997] have also concluded that in certain regions of the world, landscape type plays a crucial role in precipitation processes. They concluded that deforestation along the southern coast of west Africa could result in a complete collapse of the monsoon circulation in this region, with an associated significant decrease in rainfall. *Xue* [1997] has similarly found that a degradation of the land surface in tropical north Africa has a significant impact on the weather in this region. *Xue* [1996] also concluded that desertification of the Inner Mongolian grassland has had the effect of warming and drying the same area. For the Indian subcontinent, *Laval et al.* [1996] have shown that, if transpiration from vegetation is not included, their GCM model does not accurately simulate the interannual precipitation variation. *Liu and Avissar* [1999] have used a general circulation model (GCM) to determine the time period before which the initial soil moisture condition becomes unimportant. They concluded this memory lasts on the order of 200-300 days.

Fukutome et al. [1999] investigated the role of lateral boundaries in regional atmospheric models in simulating the synoptic evolution of individual low pressure systems. They found that the model simulations are strongly controlled by the lateral boundary conditions, with their influence becoming less as the model domain size is increased. *Warner et al.* [1997] and *Seth and Giorgi* [1998] have also discussed the major role of lateral boundary conditions in regional model simulations. The NCEP or ECMWF reanalysis (for past weather) or NCEP, ECMWF, or other global NWP models are used to provide lateral boundary conditions for current weather forecasts and regional climate simulations. With respect to seasonal and longer term climate predictions, however, the global models, with their coarse spatial resolution and incomplete modeling of the climate system, are unlikely to be able to provide lateral boundary conditions with the required spatial resolution. Without adequate resolution, the regional models cannot by themselves develop accurate predictive skill in the interior of the model.

These effects are superimposed on other seasonal and longer-term weather prediction influences such as El Niño. *Shabbar et al.* [1997] illustrate the major difference of winter weather patterns in the Great Plains and elsewhere, when the warm anomaly sea surface temperatures (SSTs) of an El Niño are replaced by low SSTs. *Chase et al.* [2000] found that landscape changes in the tropics, which have a major influence on the patterning and intensity of deep cumulus convection, actually may have an equivalent or even greater impact on Great Plains winter weather than do the SST anomalies. *Polcher* [1995] explored how different tropical landscape changes can influence deep tropical convection, while *Polcher and Laval* [1994] concluded that tropical deforestation weather effects are largely independent of El Niño influences in this region.

That soil moisture conditions and landscape changes influence weather and climate should not be surprising. *Lewis* [1998], from borehole temperatures, has shown, for instance, that deforested sites in western Canadian locations have an average ground surface temperature increase of about 1 deg C, as a result of an approximately 10% reduction in transpiration from the originally forested area. *Desborough* [1997] demonstrated the significant effect of roots on weather. He found that varying the near-surface (upper 10 cm) root fraction between 10% and 90% produces transpiration differences of up to 80 W m-2. He found that the entire seasonal cycle of transpiration was sensitive to the near-surface root fraction which will necessarily influence surface sensible heat fluxes as well.

Over the central United States, there have been several studies which have explored the importance of landscape on seasonal weather patterns. *Xue et al.* [1996], for instance, concluded that the effects of land-surface changes on atmospheric variables in the summer are pronounced and persistent, although largely limited to the area of land surface condition changes. *Copeland et al.* [1996] demonstrated, for a specific July, that temperature and precipitation patterns have been significantly altered as a result of changing much of the vegetative landscape in this region from its natural state to agriculture. *Giorgi et al.* [1996], however, concluded that local effects associated with surface evaporation played only a minor role in model simulations of the 1988 drought and 1993 flood conditions over the central United States.

On a smaller scale, *Pielke et al.* [1997] illustrated how land conversion in the Oklahoma-Texas Panhandle region, from its natural short-grass prairie condition to cropland, can provide sufficient energy input to cumulus clouds to produce tornadic thunderstorms. In that modeling study, only shallow cumulus clouds are simulated with the natural landscape despite identical meteorological initial and lateral boundary conditions. Using idealized experiments, *Emori* [1998] shows how the interaction between cumulus convection and soil moisture variations work to maintain a heterogeneous soil moisture distribution. *Pielke et al.* [1999a] also shows how soil moisture variations can feedback to cumulus cloud responses.

Pielke et al. [1999b] has explored how land-use change this century has influenced summer weather over south Florida. Using identical observed meteorology for lateral boundary conditions, an atmospheric model was integrated for July-August 1973 for south Florida. Three experiments were performed - one using the observed 1973 landscape, another the 1993 landscape, and the third the 1900 landscape when the region was close to its natural state (the 1900 and 1973 landscape are shown in Plate 8). Over the two-month period, there was a 9% decrease in rainfall averaged over south Florida with the 1973 landscape and an 11% decrease with the 1993 landscape, as compared with the model results when the 1900 landscape is used (Plate 9). Maximum temperature averaged over the two-month period over south Florida increased by about 0.5 deg C between the 1900 and 1993 simulations. The limited available observations of trends in summer rainfall and temperature over this region are consistent with these trends.

There is convincing modeling evidence, also, that mesoscale fluxes that result from landscape heterogeneity in flat terrain under some synoptic situations can often be as large as, and larger than, turbulent fluxes, as well as have a different vertical distribution than the turbulent fluxes averaged over the scale of a GCM or NWP grid increment [*Dalu et al.* 1991]. Examples of studies that document the modification of atmospheric boundary layer structure and/or the development of mesoscale flow due to land-surface inhomogeneity are reported in *Pielke et al.* [1991, 1993]. *Dalu et al.* [1991], *Cotton and Pielke* [1995], *Segal and Arritt* [1992], *Avissar and Pielke* [1989], *Manqian and Jinjun* [1993], *Raupach* [1991], *Guo and Schuepp* [1994], *Zhong and Doran* [1995], *Avissar and Chen* [1993], *Chen and Avissar*

[1994a,b], and *André et al.* [1989]. A number of these papers are summarized in *Cotton and Pielke* [1995]. These mesoscale fluxes have substantial coherent structure such that they are predictable features [i.e., *Zeng and Pielke* 1993, 1995]. *Zeng and Pielke* [1995] concluded that the important controls on mesoscale fluxes include planetary boundary layer depth, horizontal size of the surface heat patches, the potential temperature difference between different patches, the surface sensible, moisture, and momentum fluxes, and height above the surface.

Observations have also documented the importance of heterogeneous landscape in influencing boundary layer structures and mesoscale fluxes [*Mahrt et al.* 1994a,b; *Mahrt and Ek* 1993; *Doran et al.* 1992, 1995; *Smith et al.* 1992; *Beljaars and Holtslag* 1991; *Segal et al.* 1988, 1989; *Balling* 1988] and in affecting such related properties as soil water infiltration [*Wood et al.* 1992].

3.2. Including Hydrology and Ecology in Regional Climate Models

The dynamic coupling of mesoscale and regional atmospheric and land surface models opens a new research area [*Wang and Eltahir*, 2000a, b, c, d; and *Wang et al.*, 2000]. With this approach, models that were developed in the separate disciplines of atmospheric science, hydrology, and ecology are combined together into an integrated modeling system. *Pielke et al.* [1998a, b] and *Eugster et al.* [1999] provided recent reviews of the role of these interactions. Examples which show the importance of this coupled approach are reported in *Eastman et al.* [2000] and *Lu et al.* [2000].

Lu et al. [2000], for example, have used the RAMS-CENTURY coupled modeling system to investigate the relationship between weather and vegetation growth. That study has shown that the feedback between precipitation and above-ground vegetation growth results in wetter and cooler weather, than occurs if this feedback is excluded.

Eastman et al. [2000] have explored, using the RAMS-GEMTM modeling system, the influence of land-use change, and doubled CO2 on the weather over a season. The experiments performed were (i) changing the central Great Plains from the current condition to an estimate of the natural landscape; (ii) doubling CO2 in the radiation calculation in the RAMS model; and (iii) doubling CO2 in the GEMTM component of the modeling system. The model simulation was for 210 days during the growing season in 1989. The control experiment (with the current landscape and CO2 levels) was compared against observed weather and vegetation growth data. Figure 4 illustrates the spatial influence on the 210-day averaged maximum temperatures of each of the three effects shown above. Figure 5 and Figure 6 illustrate the domain-averaged effect of the three model changes on leaf area index and transpiration. Both the change to the natural landscape and the biological effect of doubled CO2 produced a cooling over the model domain.

4. CONCLUDING REMARKS

In the course of the last ten years, remarkable progress has been made in the development and validation of land-surface models in coupled mesoscale models owing to new field experiments and intensive numerical experiments. For instance, high-resolution land-use and soil data allow a better description of surface variability in coupled mesoscale models, and hence can take into account the fine mesoscale structures forced by those surface variabilities. In particular, the vegetation characteristics such as green vegetation fraction derived from satellite enable land-surface models to distinguish the vegetation growing season from senescence, which is important for an accurate calculation of vegetation transpiration. Still, considerable efforts must be undertaken to develop consistent data sets for seasonal variations in albedo, roughness length, etc. Also, for real-time weather forecasts it will be necessary to use quasi-real-time products, in place of today's climatology data, as some of these parameters (e.g., albedo) can rapidly change during a short period depending on the recent history of rainfall. It is expected that more and more high-resolution data obtained from remote sensing platforms will be integrated into the coupled mesoscale models to specify the surface characteristics including soil moisture and snow initialization.

Although, relatively speaking, the use of advanced land-surface models in mesoscale models is still in its early stage, and more assessment has to be done in the future to further improve the land-surface coupling, some results have shown promising success. The modern-era land-surface models, together with time varying soil moisture, improve the calculations of surface heat flux, and near-surface temperature and moisture. They also somewhat improve the simulation of vertical profiles of temperature and moisture in the PBL. Long-term tests showed that they add to the skill of quantitative precipitation forecasts in the summer when convection is predominant. Some studies suggest that surface processes may enhance mesoscale

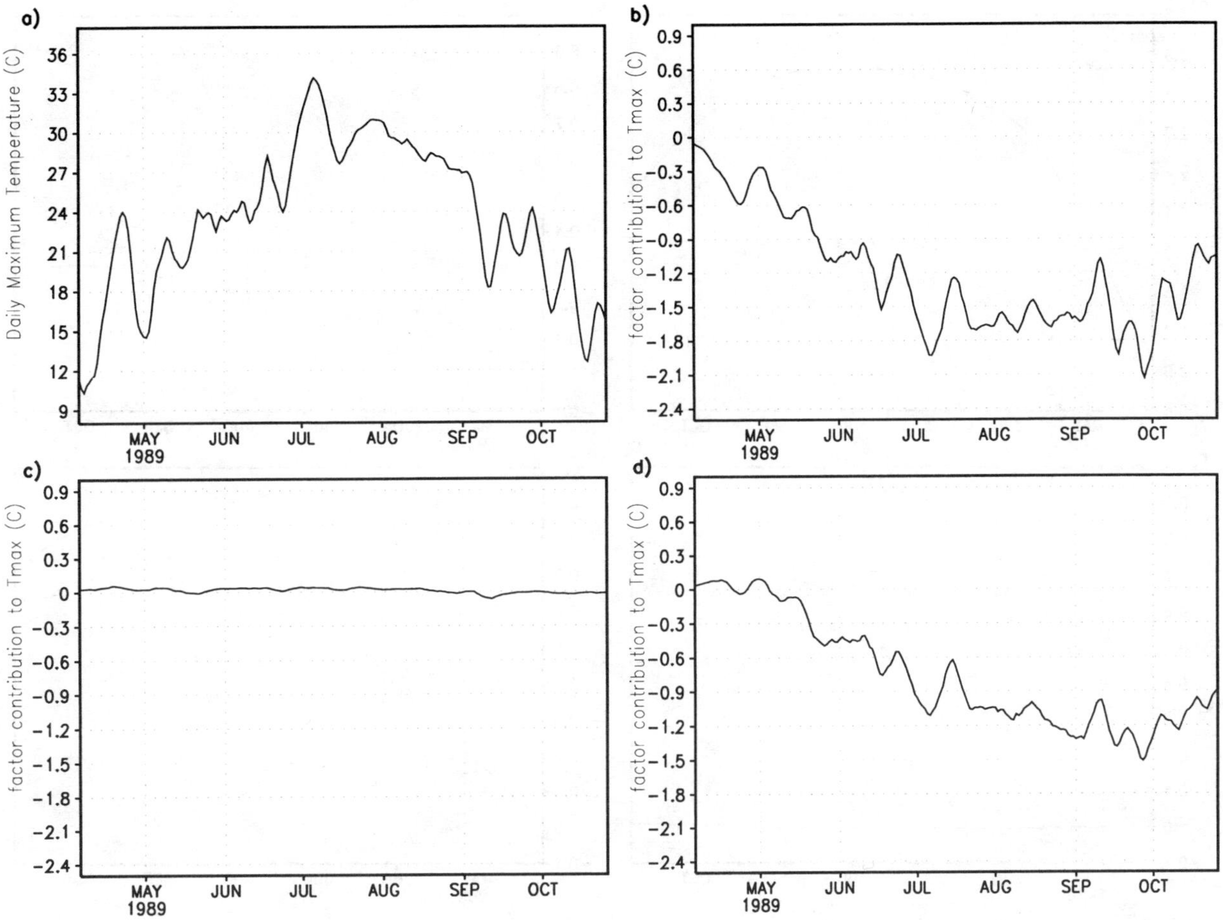

Figure 4. 210-day domain averaged maximum daily temperature where (a) is the control experiment; (b) is the change when the natural landscape is used; (c) is the change when doubled CO2 is specified in RAMS; and (c) is when doubled CO2 is used in GEMTM (from *Eastman* [2000]).

predictability of summer convection. In the situations where the convective cells occur in response to the modulation of large-scale upslope flow by particular topographic or surface-variability features, such as soil moisture and land-use, it may be reasonable to anticipate that the simulation of these storms will have deterministic skill in predicting their locations.

Regarding the future land-surface model development, the subgrid-scale horizontal inhomogeneity with various levels of complexity is most likely to be considered in the coupled models. This will be reflected, at least, in a few areas. One involves introducing the topographic control in the LSMs to better represent their effects on runoff and soil moisture evolutions. Approaches based on the disaggregation of LSMs into a mosaic of hydrological catchments are currently being tested. Another approach involves utilizing the "effective parameters" to take into account the variability of different landscapes to better simulate the area-averaged surface heat fluxes. Lastly, another involves considering the heterogeneity of snow cover and its impact on the albedo and snow melt calculation. Again, using remote-sensing products and developing more appropriate data sets for testing and improving these parameterization schemes is crucial. Initialization of soil moisture and snow in the coupled modeling systems needs further improvements, and the ultimate solution will most likely involve the combination of the off-line simulation of LSMs forced by observations and the assimilation of remotely sensed surface soil moisture.

Regional climate simulations have been shown to require two-way nonlinear interactions between the land surface and the atmosphere. The interactions involve clouds, precipitation, soil moisture, evaporation, and

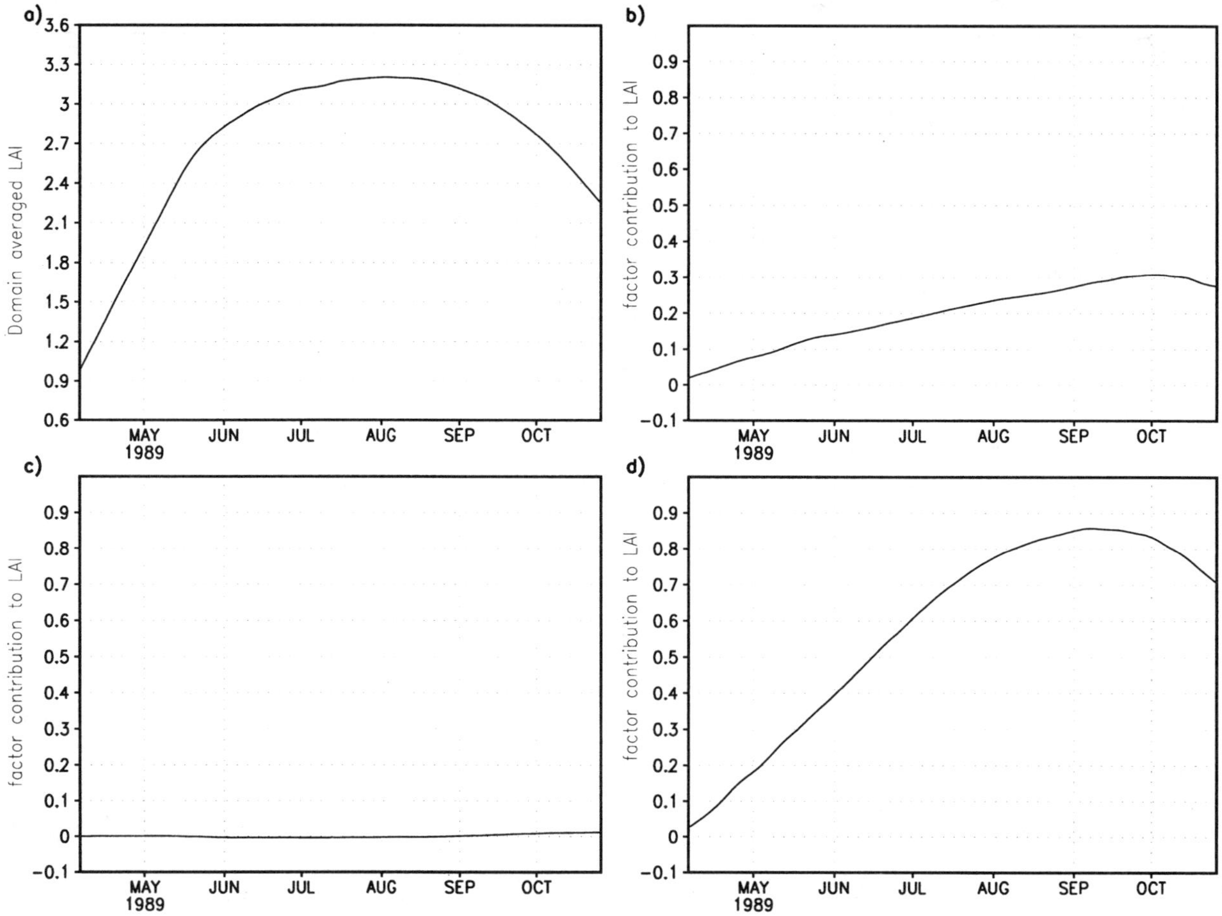

Figure 5. Same as Figure 4 but for LAI (from *Eastman* [2000]).

transpiration. Using coupled atmospheric-land surface models, it is shown, for example, that precipitation is influenced in a complex manner by soil and vegetation dynamics. Thunderstorms are harder to generate when evaporative and transpiration fluxes are larger, yet if thunderstorms are able to develop, they are more likely to be severe and to have heavy rain.

The role of carbon dioxide concentrations in controlling transpiration and vegetation growth were also explored. It was found for grasslands that, while increased carbon dioxide can make plants more water use efficient, there is an enhancement of plant cover which results over a growing season, in greater transpiration fluxes into the atmosphere.

Anthropogenic land cover change was also shown to exert a major influence in weather and climate. As demonstrated for south Florida and the central Great Plains of the United States, land use change likely has a more significant effect on regional climate than would the radiative effect of a doubling of carbon dioxide.

There are opportunities to improve regional climate simulations by recognizing that climate is an integration of atmospheric and land-surface processes. The predictability of regional seasonal and longer-term weather, for example, may be assisted by improved measurements of soil moisture at the beginning of the growing season. Higher levels of soil moisture can promote vegetation growth, which through transpiration, will increase regional rainfall during the summer. We must, however, develop procedures to monitor soil moisture within the root zone, in order to accomplish this goal.

The acceptance that regional climate involves an integration of atmospheric and land-surface processes necessarily brings into question the use of downscaling from general circulation models in order to attempt to predict future climate. The regional climate is not only a

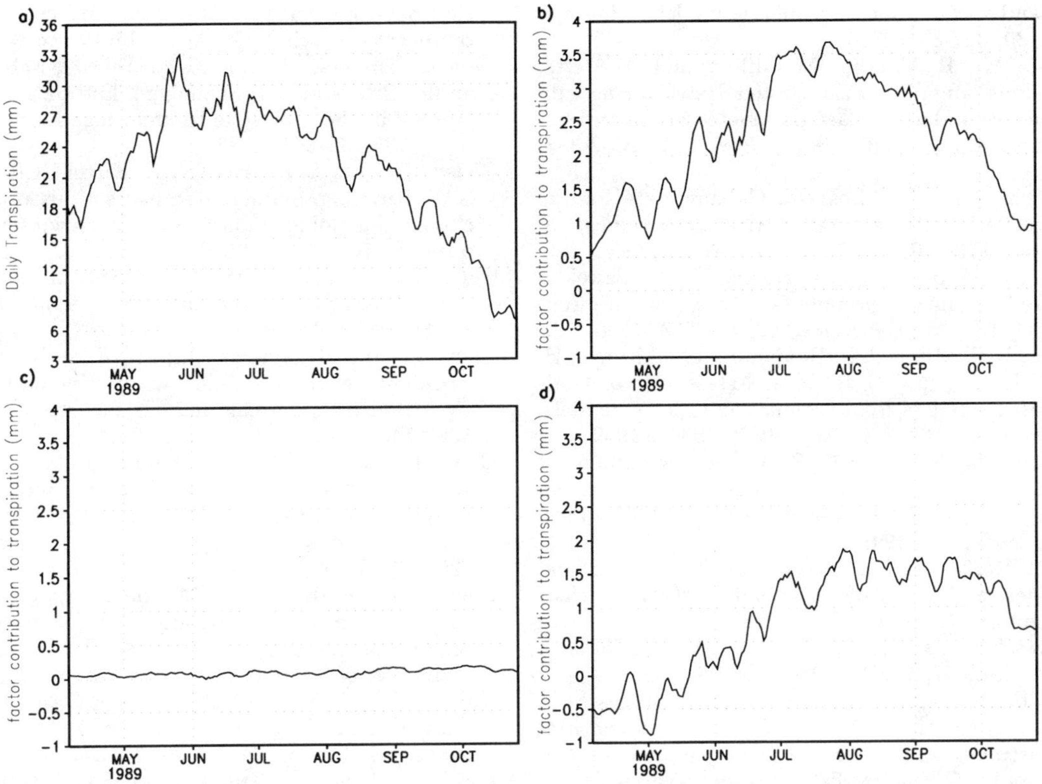

Figure 6. Same as Figure 4 but for transpiration (from *Eastman* [2000]).

response to larger-scale atmospheric inputs, but also involves complex nonlinear atmospheric, biophysical, biogeochemical, and hydrologic feedbacks within the regional climate system. The regional climate system in itself will feedback to influence the larger (global) climate system.

Acknowledgments. This project was supported by the Internal Research and Development funds from the Research Applications Program, NCAR, and by NASA under award No. NAG5-7593. Partial support for R. Pielke Sr. was provided by NASA Grant NAG8-1511 and Tulane University Contract TUL-062-98-99. We express our appreciation to Drs. David Yates, Thomas Warner, and an anonymous reviewer for reviewing this manuscript and their suggestions. We also appreciate the help of Kevin Manning and Wei Wang in preparing some figures shown in this paper. We thank Dr. Miller, Pennsylvania State University, for providing us with the 1-km surface soil texture map.

REFERENCES

André, J.C., J.P. Goutoube, and A. Perrier, HAPEX-MOBILHY: a Hydrologie Atmospheric Experiment for the study of water budget and evaporation flux at the climate scale, *Bull. Amer. Meteor. Soc.*, **67**, 138-144. 1986,

André, J.-C., P. Bougeault, J.-F. Mahfouf, P. Mascart, J. Noilhan, and J.-P. Pinty, Impact of forests on mesoscale meteorology. *Phil. Trans. Roy. Soc. London*, 324, 407-422, 1989.

Avissar, R., and R. Pielke, A parameterization of heterogeneous land surfaces for atmospheric numerical models and its impact on regional meteorology, *Mon. Wea. Rev.*, **117**, 2113-2136, 1989.

Avissar, R. and F. Chen, Development and analysis of prognostic equations for mesoscale kinetic energy and mesoscale (subgrid-scale) fluxes for large-scale atmospheric models, *J. Atmos. Sci.*, 50, 3751-3774, 1993.

Avissar, R., Recent advances in the representation of land-atmosphere interactions in general circulation models. *Rev. Geophys.*, Supplement, 1005-1010, U.S. National Report to International Union of Geodesy and Geophysics 1991-1994, 1995.

Avissar, R., Which type of soil-vegetation-atmosphere transfer scheme is needed for general circulation models: a proposal for a higher-order scheme, *J. Hydrol.*, 212-213, 136-154, 1998.

Balling, R.C., Jr., The climatic impact of a Sonoran vegetation discontinuity, *Climatic Change*, 13, 99-109, 1988.

Beljaars, A.C.M., and A.A.M. Holtslag, Flux parameterization over land surfaces for atmospheric models, *J. Appl. Meteorol.*, 30, 327-341, 1991.

Beljaars, A.C. M., P. Viterbo, M. Miller, and A. Betts, The anomalous rainfall over the United States during July 1993: sensitivity to land surface parameterization and soil moisture anomalies, *Mon. Wea. Rev.*, 124, 362-383, 1996.

Beljaars, A.C.M., and F.C. Bosveld, Cabauw data for the validation of land surface parameterization schemes, *J. Climate*, 10, 1172-1193, 1997.

Bengtsson, L., K. Arpe, E. Roeckner, and U. Schulzweida, Climate predictability experiments with a general circulation model, *Climate Dynamics*, 12, 261-278, 1996.

Berbery, E.H., K.E. Mitchell, S. Benjamin, T. Smirnova, H. Ritchie, R. Hogue, and E. Radeva, Assessment of land-surface energy budgets from regional and global models, *J. Geophys. Res.*, 104, No. D16, 19329-19348, 1999.

Betts, A., F. Chen, K. Mitchell, Z. Janjic, Assessment of land-surface and boundary-layer models in 2 operational versions of the Eta model using FIFE data, *Mon. Wea. Rev.*, **125**, 2896-2915, 1997.

Betts, A., P. Viterbo, A.C. Beljaars, H.L. Pan, S.Y. Hong, M.L. Goulden, and S.C. Wofsy, Evaluation of land-surface interaction in ECMWF and NCEP/NCAR reanalysis models over grassland (FIFE) and boreal forest (BOREAS), *J. Geophys. Res.*, 103, 23079-23085, 1998.

Blondin, C., Parameterization of land-surface processes in numerical weather prediction. in *Land-surface evaporation measurement and parameterization,* edited by T.J. Schmugge and J.C. André, Eds. Springer-Verlag Press, New York, 1991.

Bolle, H.J., J.C. Andre, J.L. Arrue, H.K. Barth, P. Bessemoulin, A. Brasa, H.A.R. de Bruin, J. Cruces, G. Dugdale, E.T. Engman, D.L. Evans, R. Fantechi, F. Fiedler, A. van de Griend, A.C. Imeson, A. Jochum, P. Kabat, T. Kratzsch, J.P Lagouarde, I. Langer, R. Llamas, E. Lopez-Baeza, J.M. Miralles, L.S. Muniosguren, F. Nerry, J. Noilhan, H.R. Oliver, H.R. Roth, S.S. Saatchi, J.S. Diaz, J. M. de Santa Olalla, W.J. Shuttleworth, H. Sogaard, H.Stricker, J. Thornes, M. Vauclin, and D. Wickland, European Field Experiment in a Desertification Threatened Area, *Annals Geophysicae*, 11, 173-189, 1993.

Bonan, G.B., D. Pollard, and S.L. Thompson, Influence of subgrid-scale heterogeneity in leaf area index, stomatal resistance, and soil moisture on grid-scale land-atmospheric interactions, *J. Climate*, 6, 1882-1897, 1993.

Bonan, G.B., A land-surface model (LSM version 1.0) for ecological, hydrological, and atmospheric studies: Technical description and user's guide, *NCAR Technical Note. TN-417+STR*, 150pp., NCAR, Boulder, CO., 1996.

Bougeault, P., J. Noilhan, P. Lacarrere, and P. Mascart, An experiment with an advanced surface parameterization in a mesobeta-scale model. Part I: Implementation, *Mon. Wea. Rev.*, 119, 2358-2373, 1991.

Braud, I., J. Noilhan, P. Bessemoulin, and P. Mascart, Bare-ground surface heat and water exchange under dry conditions: observations and parameterization, *Bound.-Layer Meteor.*, 66, 173-200, 1993.

Bringfelt, B., 1996: Tests of a new land surface treatment in HIRLAM. *HIRLAM Technical Report No. 23*, 72pp.,

Calvet, J.C., J. Noilhan, and P. Bessemoulin, Retrieving the root-zone soil moisture from surface soil moisture or temperature estimates: A feasibility study based in field measurements, *J. Appl. Meteorol.* 37, 371-386, 1998.

Camillo, P.J., and T.J. Schmugge, Estimating soil moisture storage in the root zone from surface measurements, *Soil Sci.*, 135 245-264, 1983.

Chase, T.N., R.A. Pielke, T.G.F. Kittel, R.R. Nemani, and S.W. Running, Simulated impacts of historical land cover changes on global climate in northern winter, *Climate Dynamics*, 16, 93-105, 2000

Chen, F., and R. Avissar, The impact of land-surface wetness on mesoscale heat fluxes, *J. Appl. Meteorol.*, **33**, 1324-1340, 1994a.

Chen, F., and R. Avissar, Impact of land-surface moisture variability on local shallow convective cumulus and precipitation in large-scale models, *J. Appl. Meteorol.*, **33**, 1382-1401, 1994B.

Chen, F., K. Mitchell, J. Schaake, Y. Xue, H.L. Pan, V. Koren, Q.Y. Duan, K. Ek, and A. Betts, Modeling of land-surface evaporation by four schemes and comparison with FIFE observations, J. Geophys. Res., **101**, 7251-7268, 1996.

Chen, F., Z. Janjic, and K. Mitchell, Impact of atmospheric surface layer parameterization in the new land-surface scheme of the NCEP mesoscale Eta numerical model, *Bound.-Layer Meteor.*, **185**, 391-421, 1997a.

Chen, T.H., A. Henderson-Sellers, P.C.D. Milly, A.J. Pitman, A.C.M. Beljaars, J. Polcher, F. Abramopoulos, A. Boone, S. Chang, F. Chen, Y. Dai, C.E. Desborough, R.E. Dickinson, L. Dumenil, M. Ek, J. Garratt, N. Gedney, Y.M. Gusev, J. Kim, R. Koster, E.A. Kowalczyk, K. Laval, J. Lean, D. Lettenmaier, X. Liang, J.F. Mahfouf, H.T. Mengelkamp, K. Mitchell, O.N. Nasonova, J. Noilhan, A. Robock, C. Rosenzweig, J. Schaake, C.A. Schlosser, J.P. Schulz, Y. Shao, A.B. Shmakin, D.L. Verseghy, P. Wetzel, E.F. Wood, Y. Xue, Z.L. Yang, Q. Zeng, Cabauw experimental results from the Project for Intercomparison of Land-surface Parameterization Schemes (PILPS), *J. Climate.*, **10**, 1194-1215, 1997b.

Chen, F., K. Mitchell, Z. Janjic, and M. Baldwin, Impact of Land-surface processes on the NCEP Eta model quantitative precipitation forecast, *Proc. Special Symposium on Hydrology. January 1998, Phoenix, Arizona, 1998.*

Chen, F., and K. Mitchell, Using GEWEX/ISLSCP forcing data to simulate global soil moisture fields and hydrological cycle for 1987-1988, *Journal of the Meteorological Society of Japan* Special Issue on Global Soil Wetness Project, 77, 167-182, 1999.

Chen, F. and J. Dudhia, Coupling an Advanced Land-Surface/Hydrology Model with the Penn State/NCAR MM5 Modeling System, Part I: Model implementation and Sensitivity. *Mon. Wea. Rev.*, in press, 2000.

Chen, F., T. Warner, and K. Manning, Simulation of the 1996 Buffalo Creek flash-flood event and its sensitivity to land-surface variability, *Proc. Workshop on land-surface modeling and applications to mesoscale models.* Boulder, CO, 24-25 June, 1999.

Csiszar and Gutman, Mapping global land surface albedo from NOAA AVHRR, *J. Geophys. Res.*, 104, No. D6, 6215-6228, 1999.

Clapp, R.B., and G.M. Hornberger, Empirical equations for

some soil hydraulic properties, *Water Resour. Res.*, **14**, 601-604, 1978.

Claussen, M., On coupling global biome models with climate models, *Climate Res.*, 4, 203-221, 1994.

Claussen, M., On multiple solutions of the atmosphere-vegetation system in present-day climate, *Global Change Biology*, 4, 549-559, 1998.

Claussen, M., V. Brovkin, A. Ganopolski, C. Kubatzki, and V. Petoukhov, Modeling global terrestrial vegetation – climate interaction, *Phil. Trans. Roy. Soc.*, 53-63, 1998.

Cosby, B.J., G.M. Hornberger, R.B. Clapp, and T.R. Ginn, A statistical exploration of the relationships of soil moisture characteristics to the physical properties of soils, *Water Resour. Res.*, **20**, 682-690, 1984.

Copeland, J.H., R.A. Pielke, and T.G.F. Kittel, Potential climatic impacts of vegetation change: A regional modeling study, *J. Geophys. Res.*, 101, 7409-7418, 1996.

Cotton, W.R. and R.A. Pielke, *Human Impacts on Weather and Climate*, Cambridge University Press, New York, 1995.

Cubasch, U., B.D. Santer, A. Hellbach, G. Hegerl, H. Hock, E. Maier-Reimer, U. Mikolajewicz, A. Stossel, and R. Voss, Monte Carlo climate change forecasts with a global coupled ocean-atmosphere model, *Climate Dynamics*, 10, 1-19, 1994.

Cuenca, R.H., M. Ek, and L. Mahrt, Impact of soil water property parameterization on atmospheric boundary-layer simulation, *J. Geophys. Res.*, 101, 7269-7277, 1996.

Dalu, G.A., R.A. Pielke, R. Avissar, G. Kallos, M. Baldi, and A. Guerrini, Linear impact of thermal inhomogeneities on mesoscale atmospheric flow with zero synoptic wind, *Ann. Geophys.*, 9, 641-647, 1991.

Davis, C., T. Warner, E. Astling, and J. Bowers, Development and application of an operational, relocatable, meso-gamma-scale weather analysis and forecasting system, *Tellus*, 51A, 710-727, 1999.

Deardorff, J.W., Efficient prediction of ground surface temperature and moisture, with inclusion of a layer of vegetation, *J. Geophys. Res.*, **83**, 1889-1903, 1978.

Dickinson, R.E., Modeling evapotranspiration for three-dimensional global climate models, *Climate Processes and Climate Sensitivity*, Geophysical Monogr. 29, Maurice Ewing Vol. 5, American Geophysical Union, 58-72, 1984.

Desborough, C.E., The impact of root weighting on the response of transpiration to moisture stress in land surface schemes, *Mon. Wea. Rev.*, 125, 1920-1930, 1997.

Dirmeyer, P.A., A.J. Dolman, and N. Sato, The pilot phase of the global soil wetness project, *Bull. Amer. Meteor. Soc.*, 80, 851-878, 1999.

Dolman A.J., A note on areally-averaged evaporation and the value of the effective surface conductance, *J. Hydrol.*, 138, 583-589, 1992.

Doran, J.C., F.J. Barnes, R.L. Coulter, T.L. Crawford, D.D. Baldocchi, L. Balick, D.R. Cook, D. Cooper, R.J. Dobosy, W.A. Dugas, L. Fritschen, R.L. Hart, L. Hipps, J.M. Hubbe, W. Gao, R. Hicks, R.R. Kirkham, K.E. Kunkel, T.J. Martin, T.P. Meyers, W. Porch, J.D. Shannon, W.J. Shaw, E. Swiatek, and C.D. Whiteman, The Boardman regional flux experiment, *Bull. Amer. Meteor. Soc.*, 73, 1785-1795, 1992.

Doran, J.C., W.J. Shaw, and J.M. Hubbe, Boundary layer characteristics over areas of inhomogeneous surface fluxes, *J. Appl. Meteorol.*, 34, 559-571, 1995.

Douville, H.P., and J.F. Royer, Influence of the temperature and boreal forests on the Northern Hemisphere climate within the Meteo-France GCM, *Climate Dyn.*, 13, 57-74, 1996.

Douville, H., P. Viterbo, J.F. Mahfouf, and A.C.M. Beljaars, Evaluation of optimal interpolation and nudging techniques for soil moisture analysis using FIFE data, *Mon. Wea. Rev.*, 128, 1733-1756, 2000.

Ducharne, A., R.D. Koster, M.J. Suarez, M. Stieglitz, and P. Kumar, Overview of an original catchment-based land-surface model for climate studies, *Proc. Workshop on land-surface modeling and applications to mesoscale models.* Boulder, CO, 24-25 June, 1999.

Eastman, J.L., M.B. Coughenour, and R.A. Pielke, The effects of CO2 and landscape change using a coupled plant and meteorological model, *Global Change Biology*, in press, 2000.

Eltahir, E.A.B., A soil moisture-rainfall feedback mechanism 1. Theory and observations, *Water Resources Res.*, 34, 765-776, 1998.

Emori, S., The interaction of cumulus convection with soil moisture distribution: An idealized simulation, *J. Geophys. Res.*, 103, 8873-8884, 1998.

Entekhabi, D., and P. S. Eagleson, Land surface hydrology parameterization for atmospheric general circulation models including subgrid scale variability, *J. Appl. Meteorol.*, **2**, 817-831, 1989.

Entekhabi, D., H. Nakamura, and E.G. Njoku, Solving the Inverse Problem for Soil Moisture and Temperature Profiles by Sequential Assimilation of Multifrequency Remotely Sensed Observations, *IEEE Trans. Geosci. Remote Sensing.* 32, 438-448, 1994.

Eugster, W., W.R. Rouse, R.A. Pielke, J.P. McFadden, D.D. Baldocchi, Y. Vaganov, T.G.F. Kittel, F.S. Chapin III, G.E. Liston, and P.L. Vidale, Energy balance feedbacks to climate: Integration and circumpolar extrapolations, *Global Change Biology*, in preparation, 1999.

Famiglietti, J.S., and E.F. Wood, Evapotranspiration and runoff from large land areas: land surface hydrology for atmospheric general circulation models, In: Wood E.F. (ed.). Land surface-atmosphere interactions for climate modelling. *Surveys in Geophysics*, 12, 179-204, 1991.

Foley, J.A., The sensitivity of the terrestrial biosphere to climate change: a simulation of the middle Holocene, *Global Biogeochemical Cycles*, 8, 505-525, 1994.

Fukutome, S., C. Frei, D. Luthi, and C. Schar, The interannual variability as a test ground for regional climate simulations over Japan, *J. Meteor. Soc. Japan*, 77, 649-672, 1999.

Garratt, J.R., R.A. Pielke, W.F. Miller, and T.J. Lee, Mesoscale model response to random, surface-based perturbation – A sea-breeze experiment, *Bound.-Layer Meteor.*, 52, 313-334, 1990.

Giard, D., and E. Bazile, Implementation of a new assimilation scheme for soil and surface variables in a global NWP model, *Mon. Wea. Rev.*, 128, 997-1015, 2000.

Giorgi, F, M.R. Marinucci, and G. Bates, Development of a second-generation regional climate model (RegCM2).

Part II: Convective processes and assimilation of lateral boundary conditions, *Mon. Wea. Rev.*, 121, 2814-2832, 1993.

Giorgi, F., L.O. Mearns, C Shields, and L. Mayer, A regional model study of the importance of local versus remote controls of the 1988 drought and the 1993 flood over the central United States, *J. Climate*, 9, 1150-1162, 1996.

Grotzner, A., R. Sausen, and M. Claussen, The impact of sub-grid scale sea-ice inhomogeneities on the performance of the atmospheric general circulation model ECHAM3, *Climate Dynamics*, 12, 477-496, 1996.

Guo, Y., and P.H. Schuepp, An analysis of the effect of local heat advection on evaporation over wet and dry surface strips, *J. Climate*, 7, 641-652, 1994.

Gutman, G., and A. Ignatov, The derivation of green vegetation fraction from NOAA/AVHRR data for use in numerical weather prediction models, *Int. J. Remote Sens.*, 19, 1533, 1998.

Henderson-Sellers, A., Z.L. Yang, and R.E. Dickinson, The project for intercomparison of land-surface parameterization schemes, *Bull. Amer. Meteor. Soc.*, Vol. 74, No. 7, 1335-1349, 1993.

Hinkelman, L.M., T.P. Ackerman, and R.T. Marchand, An evaluation of NCEP Eta model predictions of surface energy budget and cloud properties by comparison with measured ARM data, *J. Geophys. Res.*, 104, 19535-19549, 1999.

Houser, P. R., W.J. Shuttleworth, J.S. Famiglietti, H.V. Gupta K.H. Syed, and D.C. Goodrich, Integration of soil moisture remote sensing and hydrologic modeling using data assimilation, *Water Resources Research*, 34(12), 3405-3420, 1998.

Jackson, T.J., Profile soil moisture from surface measurements, *J. Irrigation Drainage Div., Proc. ASCE*, 106, 81-92, 1980.

Jackson, T.J., Soil moisture estimation using Special Sensor Microwave/Image satellite data over a grassland region, *Water Resources Research*, 18, 1475-1484, 1997.

Koren, V., J. Schaake, K. Mitchell, Q-Y. Duan, and F. Chen, A parameterization of snowpack and frozen ground intended for NCEP weather and climate models, *J. Geophys. Res.*, 104, 19569-19585, 1999.

Koster, R.D., and M.J. Suarez, Modeling the land-surface boundary in climate models as a composite of independent vegetation stands, *J. Geophys. Res.*, 97, 2697-2715, 1992.

Koster, R., and C.P. Milly, The interplay between transpiration and runoff formulations in land surface schemes used with atmospheric models, *J. Climate*, 10, 1578-1591, 1997.

Larow, T.E. and T.N. Krishnamurti, Initial conditions and ENSO prediction using a coupled ocean-atmosphere model, *Tellus*, 50A, 76, 1998.

Laval, K., R. Raghava, J. Polcher, R. Sadourny, and M. Forichon, Simulations of the 1987 and 1988 Indian monsoons using the LMD GCM, *J. Climate*, 9, 3357-3371, 1996.

Lee, T.J., R.A. Pielke, and P.W. Mielke, Jr., Modeling the clear-sky surface energy budget during FIFE87, *J. Geophys. Res.*, 100, 25585-25593, 1995.

LeMone, M. A., R. Grossman, R. Coulter, M. Wesely, G. Klazura, G. Poulos, W. Blumen, J. Lundquist, R. Cuenca, S. KElly, E. Brandes, S. Oncley, R. Mcmillen, B. Hicks, Land-atmosphere interaction research and opportunities in the Walnut River Watershed in Southeast Kansas: CASES and ABLE, *Bull. Amer. Meteor. Soc.*, 81, 757–780, 2000.

Lewis, T., The effect of deforestation on ground surface temperatures, *Global Planetary Change*, 18, 1-14, 1998.

Li, B. and E. Avissar, The impact of spatial variability of land-surface characteristics on land-surface heat fluxes, *J. Climate*, 7, 527-537, 1994.

Liang, X., E.F. Wood, and D.P. Lettenmaier, 1998: Modeling ground heat flux in land surface parameterization schemes, *J. Geophys. Res.*, 104, 9581-9602, 1999.

Liston, G.E., Interrelationships among snow distribution, snowmelt, and snow cover depletion: Implications for atmospheric, hydrologic, and ecological modeling, *J. Climate*, 38, 1474-1487, 1999.

Liu, Y., and R. Avissar, A study of persistence in the land-atmosphere system using a General Circulation Model and observations, *J. Climate*, 38, 2139-2153, 1999.

Lorenz, E.N., Forced and free variations of weather and climate, *J. Atmos. Sci.*, 36, 1367-1376, 1979.

Loveland, T.R., Merchant, J.W., Brown, J.F., Ohlen, D.O., Reed, B.C., Olson, P., and Hutchinson, J., Seasonal land-cover regions of the United States, *Annals of the Association of American Geographers*, 85(2): 339-355, 1995.

Lu, L., R.A. Pielke, G.E. Liston, W.J. Parton, D. Ojima, and M. Hartman, Implementation of a two-way interactive atmospheric and ecological model and its application to the central United States, *J. Climate*, in press, 2000.

Mahfouf, J.F., E. Richard, and P. Mascart, The influence of soil and vegetation on the development of mesoscale circulations, *J. Clim. Appl. Meteor.*, **26**, 1483-1495, 1987.

Mahfouf, J.F., Analysis of soil moisture from near-surface parameters: a feasibility study, *J. Appl. Meteorol.*, 30, 1534-1547, 1991.

Mahfouf, J.-F., A.O. Manzi, J. Noilhan, H. Giordani, and M. Deque, The Land Surface Scheme ISBA within the Meteo-France Climate Model ARPEGE. Part I: Implementation and Preliminary Results, *J. Climate*, 8, 2039-2057, 1995.

Mahrt, L., and H.L. Pan, A two-layer model of soil hydrology, *Bound.-Layer Meteor.*, **29**, 1-20, 1984.

Mahrt, L. and M. Ek, Spatial variability of turbulent fluxes and roughness lengths in HAPEX-MOBILHY, *Bound.-Layer Meteor.*, 65, 381-400, 1993.

Mahrt, L., J. Sun, D. Vickers, J.I. MacPherson, and J.R. Pederson, Observations of fluxes and inland breezes over a heterogeneous surface, *J. Atmos. Sci.*, 2484-2499, 1994a.

Mahrt, L., J.I. McPherson and R. Desjardins, Observations of fluxes over heterogeneous surfaces, *Bound.-Layer Meteor.*, 67, 345-367, 1994b.

Manqian, M., and J. Jinjun, A coupled model on land-atmosphere interactions – simulating the characteristics of the PBL over a heterogeneous surface, *Bound.-Layer Meteor.*, 66, 247-264, 1993.

Marshall, C.H., Evaluation of the new land-surface and planetary boundary layer parameterization schemes in the NCEP mesoscale Eta model using Oklahoma Mesonet observations, *Master's thesis*, School of Meteorology, University of Oklahoma, 175 pp, 1998.

Mason, P.J., The formation of areally-averaged roughness lengths, /QJRMS, **114**, 399-420, 1988.

McCumber, M.C., and R.A. Pielke, Simulation of the effects of surface fluxes of heat and moisture in a mesoscale numerical model soil layer, *J. Geophys. Res.*, 86, 9929-9938, 1981.

Miller, D.A. and R.A. White, A Conterminous United States Multi-Layer Soil Characteristics Data Set for Regional Climate and Hydrology Modeling, *Earth Interactions*, 2, 1998. [Available on-line at http://EarthInteractions.org]

Milly, P.C.D., and J. Kabala, Integrated modeling and remote sensing of soil moisture, *Hydrologic Application of Space Technology (Proc. Cocoa Beach Workshop, FL, Aug. 1985)*, IAHS Publ. 160, 331-339, 1986.

Mintz, Y., The sensitivity of numerically simulated climates to land surface boundary conditions, *Proc. JSC Study Conf. on Land Surface Processes in Atmospheric GCM*, Greenbelt, MD, 1981.

Mitchell, K., C. Marshall, D. Lohmann, M. Ek, Y. Lin, P. Grunmann, P. Houser, E. Wood, J. Schaake, D. Lettenmaier, D. Tarpley, W. Higgins, R. Pinker, A. Robock, B. Cosgrove, J. Entin, and Q. Duan, The collaborative GCIP Land Data Assimilation (LDAS) project and supportive NCEP uncoupled land-surface modeling initiatives, *Proc. 15th Conference on Hydrology, American Meteorological Society*, 09-14 January 2000, Long Beach, CA, 2000.

Monteith, J.L., Vegetation and the atmosphere. Vol. 1: Principals. Academy Press, 278 pp, 1975.

Monteith, J.L., Vegetation and the atmosphere. Vol 2: Case Studies. Academy Press, 439 pp, 1976.

Noilhan J., and S. Planton, A simple parameterization of land surface processes for meteorological models, *Mon. Wea. Rev.*, **117**, 536-549, 1989.

Noilhan, J., P. Lacarrere, A.J. Dolman, and E.M. Blyth, Defining area-average parameters in meteorological models for land surface with mesoscale heterogeneity, *J. Hydrology*, 190, 302-316, 1997.

Ookouchi, Y., M. Segal, R.C. Kesseler, and R. Pielke, Evaluation of soil moisture effects of the generation and modification of mesoscale circulations, *Mon. Wea. Rev.*, **112**, 2281-2292, 1984.

Paegle, J., K.C. Mo, and J. Nogues-Paegle, Dependence of simulated precipitation on surface evaporation during the 1993 United States summer floods, *Mon. Wea. Rev.*, 124, 345-361, 1996.

Pan, H-L., and L. Mahrt, Interaction between soil hydrology and boundary-layer development, *Bound.-Layer Meteor.*, **38**, 185-202, 1987.

Physick, W.L., A numerical model of the sea-breeze phenomenon over a lake or gulf, *J. Atmos. Sci.*, 33, 2107-2135, 1976.

Pielke, R.A. Mesoscale meteorological modeling, 612 pp., Academic Press, New York, N.Y., 1984.

Pielke, R.A., Climate prediction as an initial value problem, *Bull. Amer. Meteor. Soc.*, 79, 2743-2746, 1998.

Pielke, R.A., G. Dalu, J.S. Snook, T.J. Lee, and T.G.F. Kittel, Nonlinear influence of mesoscale land use on weather and climate, *J. Climate*, 4, 1053-1069, 1991.

Pielke, R.A., J.H. Rodriguez, J.L. Eastman, R.L. Walko, and R.A. Stocker, Influence of albedo variability in complex terrain on mesoscale systems, *J. Climate*, 6, 1798-1806, 1993.

Pielke, R.A., T.J. Lee, J.H. Copeland, J.L. Eastman, C.L. Ziegler, and C.A. Finley, Use of USGS-provided data to improve weather and climate simulations, *Ecological Applications*, 7, 3-21, 1997.

Pielke, R.A., R. Avissar, M. Raupach, H. Dolman, X. Zeng, and S. Denning, Interactions between the atmosphere and terrestrial ecosystems: Influence on weather and climate, *Global Change Biology*, 4, 461-475, 1998a.

Pielke, R.A. Sr., G. Dalu, J. Eastman, P.L. Vidale, and X. Zeng, Boundary layer processes and land surface interactions on the mesoscale, Chapter 7 in Clear and Cloudy Boundary Layers, A.A.M. Holtslag and P.G. Duynkerke, Eds., Royal Netherlands Academy of Arts and Sciences, PO Box 19121, 1000 GC Amsterdam, The Netherlands, 155-176, 1998b.

Pielke, R.A., G.E. Liston, J.L. Eastman, L. Lu, and M. Coughenour, Seasonal weather prediction as an initial value problem, *J. Geophys. Res.*, 104, 19463-19479, 1999a.

Pielke, R.A., R.L. Walko, L. Steyaert, P.L. Vidale, G.E. Liston, and W.A. Lyons, The influence of anthropogenic landscape changes on weather in south Florida, *Mon. Wea. Rev.*, 127, 1663-1673, 1999b.

Polcher, J., Sensitivity of tropical convection to land surface processes, *J. Atmos. Sci.*, 52, 3143-3161, 1995.

Polcher, J., and K. Laval, A statistical study of the regional impact of deforestation on climate in the LMD GCM, *Climate Dynamics*, 10, 205-219, 1994.

Ramsay, B.H., The interactive multisensor snow and ice mapping system, *Hydrological Processes*, 12, 1537-1546, 1998.

Raupach, M.R., Vegetation-atmosphere interaction in homogeneous and heterogeneous terrain: Some implications of mixed layer dynamics, In: Vegetation and climate interactions in semi-arid regions, Henderson-Sellers and Pitman, Eds., Dordrecht, The Netherlands, Kluwer Academic Publishers, 105-120, 1991.

Reynolds, C.A., T.J. Jackson, and W.J. Rawls, Estimating available water content by linking the FAO soil map of the world with global soil profile database and pedotransfer functions, *Proceedings of the AGU 1999 Spring Meeting*, Boston, MA. May 31-June 4, 1999.

Rowntree, P. R., Sensitivity of GCM to land surface processes, *Proc. Work. in Intercomparison of Large Scale Models for extended Range Forecasts*, ECMWF, Reading, England, 225-261, 1983.

Rowntree, P.R., and J.R. Bolton, Simulations of the atmospheric response to soil moisture anomalies over Europe, *Q. J. R. Meteorol. Soc.*, **109**, 501-526, 1983.

Schaake, J.C., V.I. Koren, Q.Y. Duan, K. Mitchell, and F. Chen, A simple water balance model (SWB) for estimating runoff at different spatial and temporal scales, *J. Geophys. Res.*, **101**, 7461-7475, 1996.

Segal M., R. Avissar, M.C. McCumber, and R.A. Pielke, Evaluation of vegetation effects on the generation and modification of mesoscale circulations, *J. Atmos. Sci.*, 45, 2268-2292, 1988.

Segal, M., W. Schreiber, G. Kallos, R.A. Pielke, J.R. Garratt, J. Weaver, A. Rodi, and J. Wilson, The impact of crop areas in northeast Colorado on midsummer mesoscale thermal circulations, *Mon. Wea. Rev.*, 117, 809-825, 1989.

Segal, M. and R.W. Arritt, Non-classical mesoscale circulations caused by surface sensible heat-flux gradients, *Bull. Amer. Meteor. Soc.*, 73, 1593-1604, 1992.

Sellers, P.J., Y. Mintz, Y.C. Sud, and A. Dalcher, A simple biosphere model (Sib) for use within general circulation models, *J. Atmos. Sci.***43**, 505-531, 1986.

Sellers, P.J., Hall, F.G., Asrar, G., Strebel, D.E., and F.F. Murphy, An overview of the First International Satellite Land Surface Climatology Project (ISLSCP) Field Experiment (FIFE), *J. Geophys. Res.*, **97**, 18345-18371, 1992.

Sellers, P.J., M.D. Heiser, F.G. Hall, S.J. Goetz, D.E. Strebel, S.B. Verma, R.L. Desjardins, P.M. Schuepp, and J.I. MacPherson, Effects of spatial variability in topography, vegetation cover and soil moisture on area-averaged surface fluxes: A case study using the FIFE 1989 data, *J. Geophys. Res.*, 100(D12), 25607-25629, 1995a.

Sellers, P.J., F. Hall, H. Margolis, B. Kelly, D. Baldocchi, G. den Hartog, J. Cihlar, M.G. Ryan, B. Goodison, P. Crill, K.J. Ranson, D. Lettenmaier, and D.E. Wickland, The Boreal Ecosystem-Atmosphere Study (BOREAS): An overview and early results from the 1994 field year, *Bull. Amer. Meteor. Soc.*, 76, 1549-1577, 1995b.

Seth, A., and F. Giorgi, The effects of domain choice on summer precipitation simulation and sensitivity in a regional climate model, *J. Climate*, 11, 2698-2712, 1998.

Shabbar, A., B. Bonsal, and M. Khandekar, Canadian precipitation patterns associated with the Southern Oscillation, *J. Climate*, 10, 3016-3027, 1997.

Shao, Y., and A. Henderson-Sellers, Modeling soil moisture: A project for for Intercomparison of Land Surface Parameterization Schemes Phase 2(b), *J. Geophys. Res.*, 101, 7227-7250, 1996.

Sivillo, J.K., J.E. Ahlquist, and Z. Toth, An ensemble forecasting primer, *Wea. Forecasting*, 12, 809-818, 1997.

Smirnova, T.G., J.M. Brown., and S.G. Benjamin, Performance of different soil model configurations in simulating ground surface temperature and surface fluxes, Mon. Wea. Rev., 125, 1870-1884, 1997.

Smirnova, T.G., J.M. Brown., and S.G. Benjamin, Impact of a snow physics parameterization on short-range forecasts of skin temperature in MAPS/RUC, 12th Conference on Numerical Weather Prediction, Phoenix, AZ, Amer. Meteor. Soc., 161-164, 1998.

Smith, E.A., A.Y. Hsu, W.L. Crosson, R.T. Field, L.J. Fritschen, R.J. Gurney, E.T. Kanemasu, W.P. Kustas, D. Nie, W.J. Shuttleworth, J.B. Stewart, S.B. Verma, H.L. Weaver, and M.L. Wesley, Area-averaged surface fluxes and their time-space variability over the FIFE experimental domain, *J. Geophys. Res.*, 97, 18599-18622, 1992.

Smith, C.B., M.N. Lakhtakia, W.J. Capehart, and T.N. Carlson, Initialization of soil-water content in regional-scale atmospheric prediction models, /BAMS, 75, 585-592, 1994.

Texier, D., N. de Noblet, S.P. Harrison, A Haxeltine, D. Jolly, S. Joussaume, F. Laarif, I.C. Prentice, and P. Tarasov, Quantifying the role of biosphere-atmosphere feedbacks in climate change: Coupled model simulations for 6000 years BP and comparison with paleodata for northern Eurasia and northern Africa, *Climate Dynamics*, 13, 865-882, 1997.

Twine, T.E., W.P. Kustas, J.M. Norman, D.R. Cook, P.R. Houser, T.P. Meyers, J.H. Prueger P.J¿ Starks, M.L. Wisely, Correcting eddy-covariance flux underestimates over a grassland, *Agri. For. Meteorol.*, 103, 279-300, 2000.

van Genuchten, M.Th., A closed-form equation for predicting the hydraulic conductivity of unsaturated soils, *Soil Science Society of American Journal*, **44**, 892-898, 1980.

Viterbo, P, and A. C. Beljaars, An improved land surface parameterization scheme in the ECMWF model and its validation, *J. Climate*, 8, 2716-2748, 1995.

Viterbo, P., and A. Betts, The impact on ECMWF forecasts of changes to the albedo of the boreal forests in the presence of snow, *J. Geophys. Res.*, 104, 7803-7825, 1999.

Walko, R.L., L.E. Band, J. Baron, T.G.F. Kittel, R. Lammers, T.J. Lee, D.S. Ojima, R.A. Pielke, C. Taylor, C. Tague, C.J. Tremback, and P.L. Vidale, Coupled atmosphere-biophysics-hydrology models for environmental modeling, *J. Appl. Meteorol.*, 39, 931-944, 2000.

Wang G, and E.A.B. Eltahir, Role of vegetation dynamics in enhancing the low-frequency variability of the Sahel rainfall, *Water Resources Research*, 36, 1013-1021, 2000a.

Wang G, and E.A.B. Eltahir, Ecosystem dynamics and the Sahel drought, *Geophysical Research Letters*, 27, 795-798, 2000b.

Wang G, and E.A.B. Eltahir, Biosphere-atmosphere interactions over West Africa. I: Development and validation of a coupled dynamic model, *Q. J. R. Meteorol. Soc.*, 126, 1239-1260, 2000c.

Wang G, and E.A.B. Eltahir, Biosphere-atmosphere interactions over West Africa, II: Multiple climate equilibria, *Q. J. R. Meteorol. Soc.*, 126, 1261-1280, 2000d.

Wang G, R.L., Bras, and E.A.B. Eltahir, The impact of observed deforestation on the mesoscale distribution of rainfall and clouds in Amazonia, *Journal of Hydrometeorology*, 1, 267-286, 2000.

Warner, T.T., R.A. Peterson, and R.E. Treadon, A tutorial on lateral boundary conditions as a basic and potentially serious limitation to regional numerical weather prediction, *Bull. Amer. Meteor. Soc.*, 78, 2599-2617, 1997.

Wood,E.F., D.P. Lettenmaier, and V.G. Zartarian, A land-surface parameterization with subgrid for general circulation models, *J. Geophys. Res.*, **97**, 2717-2728, 1992.

Wood, E.F., and E. Lakshmi, Scaling water and energy fluxes in climate systems: three land-atmospheric modeling experiments, *J. Climate*, 6, 839-857, 1993.

Wood, E.F., D.P. Lettenmaier, X. Liang, D. Lohmann, A. Boone, S. Chang, F. Chen, Y. Dai, C. Desborough, R.E. Dickinson, Q. Duan, M. Ek, Y.M. Gusev, F. Habets, P. Irannejad, R. Koster, K. Mitchell, O.N. Nasonova, J. Noilhan, J. Schaake, . Schlosser, Y. Shao, A.B. Shmakin, D. Verseghy, K. Warrach, P. Wetzel, Y.K. Xue, Z.L. Yang, and Q.C. Zeng, The project for Intercomparison of land-surface parameterization schemes (PILPS). Phase 2(c) Red-Arkansas River Basin Experiment: 1. Experiment Description and Summary Intercomparisons, *Global and Planetary Change* , 19, 115-136, 1998.

Xue, Y., The impact of desertification in the Mongolian and

the Inner Mongolian grassland on the regional climate, *J. Climate*, 9, 2173-2189, 1996.

Xue, Y., Biosphere feedback on regional climate in tropical north Africa, *Quart. J. Roy. Meteor. Soc.*, 123, 1483-1515, 1997.

Xue, Y., M.J. Fennessy, and P.J. Sellers, Impact of vegetation properties on U.S. summer weather prediction, *J. Geophys. Res.*, 101, 7419-7430, 1996.

Yucel, I., W.J. Shuttleworth, J. Washburne, and F. Chen, Evaluating NCEP Eta model derived data against observations, *Mon. Wea. Rev.*, 126, 1977-1991, 1998.

Zeng, X., and R.A. Pielke, Error-growth dynamics and predictability of surface thermally-induced atmospheric flow, *J. Atmos. Sci.*, 50, 2817-2844, 1993.

Zeng, X. and R.A. Pielke, Further study on the predictability of landscape-induced atmospheric flow, *J. Atmos. Sci.*, 52, 1680-1698, 1995.

Zheng, X., and E.A.B. Eltahir, The response to deforestation and desertification in a model of West African monsoons, *Geophys. Res. Lett.*, 24, 155-158, 1997.

Zheng, X. and E.A.B. Eltahir, A soil moisture-rainfall feedback mechanism 2. Numerical experiments, *Water Resources Research*, 34, 777-785, 1998.

Zhong, S., and J.C. Doran, A modeling study of the effects of inhomogeneous surface fluxes on boundary-layer properties, *J. Atmos. Sci.*, 52, 3129-3142, 1995.

Ziegler, C. L., W.J. Martin, R.A. Pielke, R. Walko, A modeling study of the dryline, *J. Atmos. Sci.*, 62, 263-285, 1994.

Zobler, L., A world soil file for global climate modeling, *NASA Tech. Memo.* 87802, NASA, 33pp, NASA Goddard Space Flight Center, Institute for Space Studies, 2800 Broadway, New York, NY, 10025, 1986.

F. Chen, National Center for Atmospheric Research, P.O. Box, 3000, Boulder, CO 80307. (e-mail: feichen@ncar.ucar.edu)

K. Mitchell, Environmental Modeling Center, 5200, Auth Road, Camp Springs, MD 20746-4304. (e-mail: kmitchell@ncep.noaa.gov)

R. Pielke, Sr., Department of Atmospheric Sciences, Colorado State University, Fort Collins, CO 80523. (e-mail: dallas@cobra.atmos.colostate.edu)

Evaluation of NCEP/NCAR Reanalysis Water and Energy Budgets Using Macroscale Hydrologic Model Simulations

Edwin P. Maurer[1], Greg M. O'Donnell[1], Dennis P. Lettenmaier[1], John O. Roads[2]

The land surface hydrologic and energy fluxes of the Mississippi River basin are simulated using the Variable Infiltration Capacity (VIC) macroscale hydrologic model for the 10-year period 1988-97 at 1/8 degree spatial resolution, and are compared with the same fluxes predicted by the NCEP/NCAR reanalysis. The VIC model, unlike the reanalysis, is driven with observed (and/or derived from observations) surface meteorology and radiative forcings and is validated with observed streamflow. It is therefore constrained to produce long-term mean evapotranspiration that closely balances observed precipitation and runoff. The observed precipitation, and VIC-derived evapotranspiration and surface energy fluxes, therefore provide a useful benchmark for evaluation of the reanalysis surface flux predictions. Comparison of reanalysis precipitation with observations indicates that the reanalysis precipitation has a high summertime bias relative to the observations, especially in the southeastern part of the basin, and winter snow accumulations are greatly underestimated along the western boundary of the basin, and in the Upper Mississippi basin. The precipitation bias is responsible for biases in many of the reanalysis surface fluxes, but another source of error in the reanalysis surface water budget is the use of nudging, which forces the reanalysis toward an assumed soil moisture climatology. Due both to errors in precipitation and nudging of the reanalysis soil moisture, evapotranspiration in the reanalysis is overpredicted in all seasons, by a factor of almost two on an annual basis averaged over the basin. However, residual evapotranspiration inferred from an atmospheric balance of the reanalysis, which arguably is more closely related to observed atmospheric variables, matches the hydrologic model prediction much more closely (within 20% on an annual basis averaged over the basin). However, the interannual persistence in evapotranspiration inferred from the atmospheric balance is much less than is predicted by the hydrologic model.

[1] Department of Civil and Environmental Engineering, University of Washington, Seattle, WA
[2] Scripps Institution of Oceanography, La Jolla, CA

Land Surface Hydrology, Meteorology, and Climate: Observations and Modeling
Water Science and Application Volume 3, pages 137-158

1. INTRODUCTION

The sensitivity of numerical climate models to land surface conditions is well established [e.g. *Mintz*, 1984; *Milly* and *Dunne*, 1994; *Beljaars*, 1996]. These studies and others like them have motivated climate modelers to use more sophisticated land surface parameterizations (LSPs), and to represent land-atmosphere interactions as coupled processes, rather than boundary conditions as was once the case. In this context, it is important to produce consistent, realistic estimates of those land surface properties (especially soil moisture and/or vegetation evaporative stress) that control the partitioning of net radiation into latent, sensible, and ground heat flux. This in turn requires accurate representation of the surface hydrologic cycle, specifically water balance processes such as the partitioning of precipitation into infiltration and direct runoff, which directly affects soil moisture. The coupled surface energy and water cycles are likewise closely linked to properties such as albedo and surface roughness, which influence evapotranspiration, surface temperature, and boundary layer properties in complex, non-linear relations.

Acceptance of the role and importance of land surface coupling has been somewhat slower in the numerical weather prediction community than in climate modeling. Although the two communities use atmospheric models with essentially similar formulations, climate models are run in "free-wheeling" mode for long periods of decades to centuries or more, which allows the land surface and atmosphere to reach a dynamic equilibrium. In such long simulations, it is especially important to properly incorporate land-surface interactions, which can have important implications for moisture recycling over the continents. For instance, *Koster and Suarez* [1995] and *Koster, et al.* [1999] have shown the importance of the land surface in controlling the variability and predictability of precipitation over the continents. They found land surface processes to be especially important, and in some cases of greater importance than land-ocean interactions, over the large land areas of the northern hemisphere in summer.

The need to represent land-atmosphere interactions in numerical weather prediction, which is an initial value problem, has been somewhat less obvious. Until the last decade or so, the traditional thinking was that land surface conditions could be prescribed, as they were unlikely to change much over the time horizon of weather forecasts (now typically four, to about ten, days). *Betts et al.* [1996a] however showed that the initial land surface conditions specified for numerical weather prediction models can have a profound influence on the simulated atmospheric dynamics and resulting computed fluxes, and on this basis the weather community has focused on improvement of land surface initial conditions. The quandary facing the community has been the absence of surface observational networks, e.g. of soil moisture, from which initial surface conditions could be extracted. If such observations were available, they might be used in the same manner as free atmosphere variables (typically soundings of temperature, humidity, and wind), to update the atmospheric state at the time of forecast.

The alternative approach has been to incorporate LSPs driven by model surface forcings to represent excursions of surface conditions from long-term climatologies, such as the global soil moisture fields of *Mintz and Serafini* [1992]. As we will show in this paper, this approach has problems as well, due in large part to the accumulation of errors in the land surface resulting from inaccurate specification of surface forcings, especially precipitation. An alternative approach now being pursued by NCEP is the Land Data Assimilation System (LDAS), which essentially makes a parallel off-line run of the same LSP that is coupled to the weather prediction model, using observed forcings up to the time of forecasting. The land surface states (soil moisture, snow extent and water equivalent or depth, and surface temperature) are then used as initial conditions for forecasting, in lieu of direct observations. An overview of the LDAS scheme being developed at NCEP is provided by *Mitchell et al.* [1999].

A comprehensive LSP captures the characteristics of the soil column and the vegetation, and simulates their interaction with the water and energy fluxes to and from the atmosphere. Soil moisture directly or indirectly controls several processes that affect this partitioning, specifically bare soil evaporative resistance, plant evaporative stress, and albedo. For instance, *Huang, et al.* [1996] summarize how soil moisture availability affects surface albedo and roughness, relative humidity, surface temperature, and upper-level atmospheric circulation, all of which affect simulated atmospheric dynamics. *Betts, et al.* [1996a] review several studies of the interaction of the land surface and atmosphere, and argue that soil moisture is analogous to, and therefore potentially as important as, sea surface temperature, which is the critical state variable defining the ocean boundary in global weather forecasts. Furthermore, *Dirmeyer* [1995] noted that soil moisture is, in general, more poorly specified than sea surface temperatures, due in part to the absence of global networks, and high spatial heterogeneity. *Robock, et al.* [1998] provide a detailed discussion of the many compelling motivations for studying the ability of models to realistically simulate soil moisture, not only to refine the specified conditions at the land atmosphere boundary, but also to produce physically meaningful hydrologic quantities at the land-surface.

It is important for both forecasting models and climate studies to establish the effect of soil moisture on numerical models, and improve parameterization of the land surface. *Beljaars, et al.* [1996] discuss the sensitivity of precipitation forecasts, specifically forecasts of the July 1993 Mississippi River flooding, to soil moisture initialization in the European Centre for Medium-Range Weather Forecasts (ECMWF) forecast system. They show an increase in the ability to correctly predict precipitation following incorporation of the improved land surface parameterization of *Viterbo and Beljaars* [1995] in the ECMWF model. *Viterbo and Betts* [1999] investigated forecast sensitivities with specific wet and dry soil moisture fields, using ECMWF reanalysis rather than the assumed climatology of *Beljaars, et al* [1996]. They show 40% changes in forecasts of precipitation when using the wet July 1993 ECMWF reanalysis soil conditions compared to using dry June 1988 values as initial conditions in their forecast model.

As noted by *Delworth and Manabe* [1988] soil moisture is a red noise process, due to the low pass filtering represented by moisture accumulation processes in the soil column applied to precipitation, which is nearly a white noise process. Therefore, soil moisture responds slowly to changes in hydrologic inputs, and provides a mechanism for persistence in medium and long range weather forecasts. For this reason, specification of initial conditions of soil moisture for a numerical model is important, since these conditions are "remembered" by the model. *Pielke, et al.* [1999] summarize studies indicating this memory may be capable of lasting on the order of 200-300 days. *Van den Dool et al.* [1986], *Huang and van den Dool* [1993], *Huang, et al.* [1996], *Durre, et al.* [2000], and others have shown that this long term memory can be exploited to improve long-range forecasts of air temperature over the central U.S. in summer, when soil moisture memory is the dominant process affecting persistence of weather.

In addition to establishing initial surface conditions for numerical weather forecast models, the LSP maintains a climatological balance that is now recognized to be important over the weather forecast time horizon. *Simmons and Bengtson* [1984] note that the climatological balance (between resolved and parameterized processes in general) of a forecast model becomes important after only several days of forecasts. This is quantified specifically for a LSP in a recent study by *Beljaars, et al.* [1996] which showed an increase in 2-3 day precipitation forecast accuracy through use of an improved LSP.

An important, and largely unresolved problem specific to the incorporation of LSPs in numerical weather prediction models is the effect of the LSP tendency to seek its own soil moisture equilibrium, which may not be consistent with the surface fluxes required by the boundary layer formulation to produce accurate forecasts. This problem has, in part, resulted from the past tendency to "tune" boundary layer formulations to prescribed soil moisture (such as the global fields of *Mintz and Serafini* [1981]), which are now replaced by the soil moisture produced by the coupled model. Arguably, production of global climatologies more consistent with the LSPs incorporated in the coupled models (see, e.g., *Nijssen et al.*, 2000) will resolve this problem in part. In any event, current practice is to counteract this tendency of soil moisture "drift" by injecting or extracting water from the soil column periodically as part of the forecast update (data assimilation) process. This procedure of soil moisture updating, or "nudging" is performed by both NCEP (at least in the case of its global simulations) and ECMWF [*Roads and Betts*, 2000]. In the case of the NCEP/NCAR reanalysis (hereinafter referred to as NCEP reanalysis) [*Kalnay, et al.*, 1996], we will show that soil moisture nudging results in significant non-closure of the surface water budget, and has implications for the ability of the coupled model to simulate inter-annual persistence, as well as the natural variability of the system.

In this paper, we examine the surface energy and moisture fluxes of the NCEP reanalysis in light of a concurrent retrospective hydrologic model simulation similar in concept to what will be produced in real-time by LDAS. We show that the off-line LDAS-like model simulations accurately represent runoff, a key surface water balance component, over large sub-basins of the Mississippi River. In addition, where soil moisture data exist, the off-line model is shown to reproduce

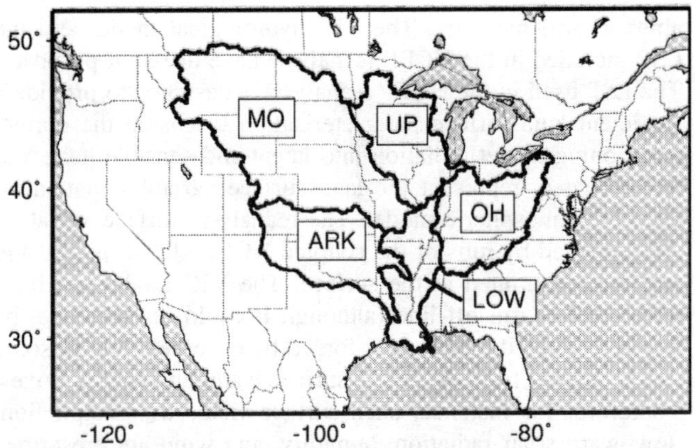

Figure 1. Location map of the major sub-basins in the Mississippi River basin. Designations are Lower Mississippi (LOW), Arkansas-Red system (ARK), Missouri River (MO), Upper Mississippi (UP), and Ohio River (OH).

the observed monthly fluxes. Furthermore, given that precipitation fields are derived from observations and hence prescribed, we argue that the hydrologic model simulations of evapotranspiration, which are produced as space-time fields, should be reasonably accurate, as they are forced by construct to close the water balance, at least in the long-term. We therefore argue that the resulting simulated fields of moisture and energy fluxes can be viewed as a baseline of pseudo-observations for purposes of evaluating the surface fluxes predicted by the coupled model runs archived in the reanalysis. This provides a basis for diagnosing deficiencies in the coupled model, and identifying future potential improvements in the model physics. Furthermore, examination of the resulting soil moisture fields and comparison with those produced by the reanalysis offers insights into the potential improvements that can be realized by utilizing LDAS soil moisture to initialize the forecast model.

2. MODELING APPROACH

We compare land surface fluxes and state variables as represented by the LSP used in the NCEP reanalysis with predictions of the same variables using an off-line simulation of a hydrologically-based LSP, the Variable Infiltration Capacity (VIC) model [*Liang et al.*, 1994; 1996]. The VIC model was run to solve the energy and water balance at the model time step of 3 hours. The domain for the analysis is the Mississippi River Basin, which is evaluated in its entirety, as well as for five separate sub-basins (Figure 1), with a grid cell resolution of 1/8 degree (approximately 140 km^2). A 10-year period of simulation (1988-1997) is used, which is sufficient to identify major differences between the two sets of model-derived fields. Nonetheless, we intend to extend the period of analysis eventually to cover the period 1950 on, which will allow evaluation of multi-decadal excursions of such surface variables as soil moisture. The VIC hydrological model and the LSP included in the NCEP reanalysis have different purposes. The LSP used in the NCEP reanalysis is designed to provide a sufficient land-surface parameterization to ensure the correct partitioning of net radiation into latent and sensible heat, and to adequately represent the land surface variables that influence the atmospheric model. The reanalysis surface variables are produced by runs of the coupled NCEP global model, and are not constrained at the surface. The VIC model has been developed to run off-line (although it could, in principle, be run in a coupled mode). It is forced by observed (and in some cases, derived from observed) surface water and energy fluxes. Therefore, for instance, such surface fields as precipitation, downward solar radiation, humidity, and wind are prescribed from observations in the VIC runs, but are simulated by the reanalysis. In this respect, the hydrological model derived surface variables offer an opportunity to verify the reanalysis fields.

2.1 Meteorological Forcing Data for Hydrologic Model

The VIC model is forced with observed meteorological data, which ideally would include temperature, precipitation, wind, vapor pressure, and incoming longwave and shortwave radiation. However, only temperature and precipitation are measured routinely at a reasonably large number of locations within the Mississippi River basin. Therefore, the forcing fields are all derived from observations of daily minimum and maximum temperature, precipitation, and wind speed. Dew point temperature (and consequently vapor pressure) is calculated using the method of *Kimball et al.* [1997], which relates the dew point to the daily minimum temperature, with an adjustment for arid climates. Downward shortwave radiation is calculated based on the daily temperature range and the dew point temperature using a method described by *Thornton and Running* [1999], which is a modification of the algorithm by *Bristow and Campbell* [1984]. The *Kimball et al.* [1997] method for dew point requires an estimate of solar radiation and the Thornton and Running method for estimating solar radiation requires an estimate of dew point temperature. Therefore, an iterative scheme was used to solve for both variables, as suggested by *Thornton and Running* [1999]. These algorithms produce daily average atmospheric transmittance, which allows computation of the shortwave radiation at the model time step. These have been shown to produce reliable estimates of shortwave radiation at various points throughout the continental U.S. [*Thornton and Running*, 1999] as well as at sites in other locations [Nijssen, et al., 2000]. Net longwave radiation at the surface is calculated following *Bras* [1990].

The precipitation data consist of daily precipitation totals from the NOAA Cooperative Observer (Co-op) Stations. The selection of precipitation stations to include in the gridded data set is based on the conditions that each station must have at least 20 years of record and 50 percent record completeness over the period of record. This is to avoid the inclusion of stations with unreliable or short records. In some data poor subbasins the minimum period of record requirement is relaxed to 10 years. Over the entire Mississippi River basin the station density varies considerably, with an approximate average of one precipitation station (meeting the above filtering criteria) for approximately every five 1/8 degree grid cells. The raw precipitation data is gridded to the 1/8 degree VIC grid size using the SYMAP algorithm of *Shepard* [1984]. This method grids the data using an inverse-square-distance rule, and includes subroutines for reducing the weight of clusters of points.

The second step in developing the gridded daily precipitation data for use in the VIC model is to scale the long term average of the gridded data to the long term average of the Parameter-elevation Regressions on Independent Slopes Model (PRISM) precipitation data set [*Daly, et al.* 1994; *Daly, et al.*, 1997]. The PRISM product is a gridded (at 1/24 degree) data set of monthly mean precipitation for the U.S. The precipitation means are based on the period 1961-90, and incorporate data from over 8000 Co-op sites, SNOTEL sites, and selected state network stations. Data-sparse areas were supplemented by a total of about 500 shorter-term stations. A station was included if it had at least 20 years of valid data, regardless of its period of record. The PRISM means are statistically adjusted to capture local variations due to rain shadows, high mountains, coastal regions, and other complex terrain. By adjusting the monthly means of the gridded VIC precipitation to the PRISM monthly mean precipitation (by the ratio of the mean precipitation for each month for the coincidental period 1961-90) we adjust for any bias introduced due to gridding with precipitation stations more densely populated at lower elevations. A detailed discussion of the gridding and rescaling procedures is included in *Widmann and Bretherton* [2000].

For the minimum and maximum daily temperature data for the Mississippi River Basin, also obtained from the Co-op stations, there is approximately one station with temperature data (meeting the same filtering criteria as for precipitation stations) for every seven 1/8 degree grid cells. Before beginning the gridding of the temperature data, we aggregated the Global 30 Arc Second Elevation Data Set (GTOPO30) produced by the U.S Geological Survey's EROS Data Center to the model resolution to obtain a mean elevation for each cell. In the gridding process, the temperatures were lapsed from each of the neighboring stations to the elevation of the grid cell using a lapse rate of 6.5°C per 1000 meters.

The final forcing data specified as input to the VIC model is wind speed. Surface observations of wind speed are very sparse and are biased toward certain geographical settings (e.g., airports), so the process used for obtaining the gridded data sets of precipitation and temperature will not provide a reliable representation of wind speed variation in the basin. Daily wind data were obtained from the NCEP reanalysis [*Kalnay, et al.*, 1996]. Daily gridded 10 meter wind fields are available on a T62 Gaussian grid (approximately 1.9 degrees square), in separate u-wind and v-wind vectors. These were linearly interpolated to the centers of the VIC model cells to create a full daily time series of wind speed.

2.2 Off-line Hydrologic Model Implementation

The VIC model is described in detail by *Liang et al.* [1994] and *Liang et al.* [1996]. It is a macroscale hydrologic model that balances both energy and water over a grid mesh, typically of resolution from a fraction of a degree to several degrees latitude by longitude (1/8 degree resolution was used in this study). It has been successfully applied to a number of large-scale river basins [e.g., *Abdulla et al.*, 1996; *Lohmann et al.*, 1998; *Nijssen et al.*, 1997; *Wood et al.*, 1997]. The VIC model computes the vertical energy and moisture fluxes in a grid cell based on a specification at each grid cell of soil properties and vegetation coverage, including a description of the sub-grid variability in each. Distinguishing characteristics of the VIC model include the representation of subgrid variability in soil infiltration capacity, specification of a mosaic of vegetation classes in any grid cell, and spatially varying subgrid precipitation. These subgrid characterizations are performed statistically, without assigning specific locations within the grid cell. It is this aspect of the subgrid characterization that distinguishes this macroscale model from a fully distributed hydrology model (see e.g. *Wigmosta et al.*, 1994, for an example of the latter). At the 1/8 degree resolution, the model represents about 23,000 computational grid cells within the Mississippi River Basin. The model was run for the 10-year simulation with full water and energy balance solutions at a three hour time step. Temperatures at each time step were interpolated by fitting an asymmetric spline through the daily maxima and minima. The daily precipitation total is distributed evenly over each time step.

Drainage between soil layers is modeled as gravity driven, with the unsaturated hydraulic conductivity a function of the degree of saturation of the soil [*Campbell*, 1974]. Baseflow is produced from the lowest soil layer using the nonlinear ARNO formulation [*Todini*, 1996]. To account for subgrid variability in infiltration at the land surface, the VIC uses a variable infiltration capacity scheme based on the Xinanjiang model [*Zhao, et al.*, 1980]. This scheme uses a spatial probability distribution to characterize available infiltration capacity as a function of the relative saturated area of the grid cell. Precipitation in excess of the available infiltration capacity forms surface runoff.

Land cover characterization over the Mississippi River Basin was based on the data set described in *Hansen, et al.* [2000], developed at the University of Maryland. This data set classifies the land surface at a resolution of 1 km, and has a total of 14 different vegetation classes. From this global data set, we identify the vegetation types present in each 1/8th degree grid cell in the model domain, and the proportion of the grid cell occupied by each. The primary characteristic of the vegetation that affects the hydrologic fluxes simulated by the VIC model is leaf area index (LAI). LAI is derived based on a gridded (¼ degree) monthly global LAI database based on *Myneni, et al.* [1997]. By overlaying this LAI database on the vegetation classification database, using a moving 2-degree window and determining the LAI assigned to cells with one predominant class, we derived the monthly LAI corresponding to each vegetation classification for each grid cell. This allows not only a spatially varying vegetation classification, but a variation in the LAI of each vegetation type with location, based on observations. Rooting depth is specified for each plant type, typically with shorter crops and grasses drawing their water from the upper soil layers, with tree roots extending into the deeper layer. Infiltration, the flux of moisture between the soil layers and runoff vary with each vegetation cover type within a grid cell, so that surface runoff and baseflow are computed for each vegetation type and then summed over the grid cell each time step.

The VIC model as applied in this study uses a three layer soil column, with a fixed upper layer of 0.10 m and depths of the deeper layers specified for each grid cell. The soil characteristics used in the VIC model for the Mississippi River Basin are derived from the 1 km resolution Pennsylvania State University data set [*Miller and White,* 1998], which is based on the USDA-NRCS State Soil Geographic Database (STATSGO). This classifies the soil texture into 16 classes for each of 11 layers. As part of the LDAS project, a gridded 1/8 degree data set has been developed for selected soil characteristics using the *Miller and White* [1998] database, and interpreting specific soil characteristics (e.g. field capacity, wilting point, saturated hydraulic conductivity) based on the work of *Cosby, et al.* [1984] and *Rawls et al.* [1998]. For these specific soil characteristics, the gridded 1/8 degree LDAS data were used directly in the soil description files of the VIC model. For the remaining soil characteristics (e.g., soil quartz content), values are specified using the soil textures in the 1 km database, indexing the values for different textures to published values, primarily from *Rawls, et al.* [1993], and aggregating the values to the 1/8 degree resolution of the model.

2.3 NCEP Reanalysis

With the increasing capabilities of computers and a wider array of observed and remotely sensed data sources, numerical weather prediction and climate models have become finer in resolution, have increased their capabilities for assimilating observations, and have improved their ability to capture local variability in climate and weather. Unfortunately, the progressive improvements in model and data result in long-term changes in the so-called analysis fields (essentially forecasts for relatively short lead times, over which the weather prediction model integrates the effects of atmospheric observations that are assimilated into the model at the forecast times). Otherwise, the analysis fields have the potential to address climate-related questions, as they produce continuous space-time fields representing a best estimate of atmospheric (and surface) conditions over the long term. The intent of the reanalysis project of NCEP and NCAR [*Kistler, et al.*, 1995], as well as other global weather forecast centers, was to produce long-term analysis fields using a "frozen" state-of-the-art version of a data assimilation and operational forecast model to produce continuous, consistent data sets from 1958 through 1997. The NCEP reanalysis has subsequently been extended back to 1948 and continues appending recent analyses with ongoing updates every month. Reanalysis model output is archived every 6 hours, with surface flux fields saved on a T62 Gaussian grid (about 1.9 degrees). Reanalysis data show great promise for use in climate studies, not only due the use of a consistent data assimilation and forecast model, but also because the data assimilation includes more observations than the forecast model can, as a result of inclusion of data sources not available in real time.

The NCEP reanalysis archive includes surface fluxes of both water and energy. These variables are all denoted as Class "C", which indicates that they are derived entirely from the data assimilation model, and have no direct relationship to observations. The variables included are precipitation, soil moisture content, runoff, downward and upward shortwave and longwave fluxes, latent and sensible heat transfers. Class "A" variables are those strongly linked to observed data, and class "B" variables are influenced by observations, but are also strongly influenced by the model. As reported by *Kalnay, et al.* [1996] variables in this class "C" should be used with caution due to the high influence of the model on the predicted values. Nonetheless, reanalysis data, including the surface variables noted above, have been widely used in lieu of (or perhaps more accurately in the absence of) observations by studies such as the Atmospheric Model Intercomparison Project [*Glecker*, 1996].

The LSP in the NCEP reanalysis model is based on the model described by *Mahrt and Pan* [1984] and *Pan and Mahrt* [1987], with later modification by *Pan* [1990]. The soil column has two layers, a thin top layer of 10 cm thickness and a lower layer 190 cm in depth. In addition to the globally constant soil depth, additional parameters are fixed globally, including wilting point (0.12), critical point (0.25), and porosity (0.47), and the hydraulic conductivity is a function only of the moisture content of the soil column. The percent of vegetation canopy coverage is also fixed at 70% for all grid cells. The NCEP LSP includes a representation of free drainage from the bottom of the soil column, which is a controlled by the hydraulic conductivity of the lower soil layer, which is a function of its water content. The water exiting the soil column as free drainage is not archived separately, and is included in the reported runoff.

2.4 *Hydrologic Routing to Sub-Basin Outlet*

We use the method of *Lohmann, et al.* [1996] to route runoff generated by both the VIC model and the NCEP LSP at each grid point or cell to the basin outlet. The NCEP runoff data are interpolated to the 1/8 degree VIC resolution prior to routing. The resulting predicted outlet hydrographs can then be compared with observed streamflows.

2.5 *Soil Moisture Adjustment*

Because the NCEP and VIC LSPs use different numbers of soil layers (two for NCEP and three for VIC), and have different soil depths and moisture storage capacities, direct comparisons between the two would be misleading. In order to facilitate comparisons of soil moisture from the two models, the reported soil moistures for each grid cell were adjusted by subtracting the hydrologically inactive column soil moisture, which is analogous to the dead pool storage in a reservoir:

$$SM_i = \sum_{j=1}^{N_L} d_{ij} f_{ij} - min \left\{ \sum_{j=1}^{N_L} d_{ij} f_{ij} \right\} \qquad (1)$$

where SM_i is the adjusted soil moisture for grid cell i, *min* denotes the minimum daily soil moisture value in the 10-year period of simulation for the grid cell, d_{ij} is the depth of layer j in cell i, f_{ij} is the fractional volumetric soil moisture in layer j in cell i, and N_L is number of layers in the soil column. While SM_i is averaged over only one month or a season, the minimum is still fixed as the minimum daily volumetric soil moisture over all days in the simulation. This adjustment applies equally when discussing total surface water (soil water plus snow water), since the minimum snow water is zero for all grid cells. All figures and data presented below use these adjusted soil moistures or surface waters, except where noted.

3. METHODS OF COMPARISON

We compare the concurrent results from a 10-year VIC model run and the NCEP reanalysis fields for the same period. To make the model domains comparable, the NCEP reanalysis data were overlaid onto the same 1/8 degree grid used in the VIC simulation using a simple inverse-distance relation with the four nearest neighbors. The results are aggregated to monthly, seasonal, and annual totals for each of the surface water budget components.

3.1 *Budgets*

The surface water budget for the land surface can be expressed as [*Roads, et al.*, 1999]:

$$\frac{dW}{dt} = P - ET - N + U \qquad (2)$$

which represents the balance of precipitation, P, evapotranspiration, ET, runoff, N, and the nudging term, U, with the change in total moisture storage in the grid cell, dW/dt, where W includes both soil moisture and snow water content. As shown in a time series analysis by *Roads, et al.* [1999], the

NCEP global spectral model used in the reanalysis has a tendency to drift to its own climatology, which seemingly produced too dry soil water content in some regions. By assuming a climatology, which for the NCEP reanalysis is the average monthly soil moisture climatology of *Mintz and Serafini* [1981], as described in detail in *Mintz and Serafini* [1992], the nudging term, *U*, was incorporated into the reanalysis model to maintain soil moistures close to the assumed climatology. Certainly, this procedure must be considered a "patch", and it tends to defeat the water balance performed by the LSP. In any event, the amount of the nudging term, *U*, must be accounted for to assure closure of the water balance. As noted by *Betts, et al.* [1998], incorporation of nudging prevents any long-term drift in the soil moisture conditions. However, it also reduces the inter-annual variability in the soil moisture content with subsequent negative impact on the ability of the model to represent persistent wet or dry periods. A reduction in variability resulting from the use of soil moisture nudging is also reported by *Viterbo and Betts* [1999] who evaluated the ECMWF reanalysis. We will show that the nudging term also seriously affects the reliability of other surface fields in the reanalysis.

Because the VIC model balances the surface water budget by construct, there is no non-closure term (*U*) in its budget. Unlike the reanalysis surface budget, the VIC model is calibrated by comparison of streamflow at the outlet of the major sub-basins with observations, (or, in the case of highly regulated sub-basins like the Missouri, through comparison with naturalized flows, which have had the major anthropogenic effects removed). The ability of the model to reproduce runoff hydrographs, taken together with the use of observed precipitation, and the physical representations of soil moisture and runoff generation processes within the model, suggests that the model simulations of other surface flux and state variables (e.g., *ET*, total soil moisture storage, and snow) are probably reasonable representations of the true system. We argue, therefore, that the space time fields of these water budget components can be used as a benchmark for evaluation of coupled model predictions, e.g., those of the NCEP reanalysis.

In order to compare the basin-wide and sub-basin average monthly water budget components, we present monthly average values for each variable. To examine the effects of the NCEP LSP on inter-annual variations in the surface water budget we evaluate time series of monthly values for the 10-year period of this study for each sub-basin as well as for the basin as a whole. We also use spatial plots of the water budget components, as well as the resulting latent and sensible heat fluxes from VIC and NCEP reanalysis for qualitative analysis of model differences by season.

3.2 Modeled and Derived Evaporation Comparison

To estimate the degree to which the biases in the reanalysis ET are caused by biases in the reanalysis precipitation fields, we follow a method described in *Trenberth and Guillemot* [1998]. The method is based on atmospheric water budget, which in its simplest form can be expressed as:

$$\frac{dP_W}{dt} = MC - (P - ET) + U_q \tag{3}$$

where *MC* is the horizontal convergence of vertically integrated atmospheric water vapor, U_q is an atmospheric nudging term similar to the surface water nudging term, and P_w is the precipitable water in the atmosphere:

$$P_W = \frac{1}{g} \int_0^{p_{sfc}} q dp \tag{4}$$

where q is specific humidity, p is pressure, g is gravitational acceleration, and p_{sfc} designates the pressure at the ground surface. Moisture convergence is defined by:

$$MC = -\nabla \bullet \frac{1}{g} \int_0^{P_{sfc}} q\mathbf{v} dp \tag{5}$$

where $\mathbf{v}$ is the horizontal wind velocity. All of the variables are included in, or are readily derived from the NCEP reanalysis fields. In particular, similar to the actual model computations in the NCEP reanalysis model, horizontal and vertical moisture advection was first computed spectrally for each atmospheric model level (sigma). This spectral advection was then converted to physical space on the associated Gaussian transform grid (192 columns x 94 rows globally). The horizontal and vertical advection was then summed vertically and multiplied by the surface pressure at each grid point. To reduce spatial noise, the resulting integrated moisture divergence was spectrally transformed, filtered with a 4th order Laplacian, and then once again transformed back to physical space.

Trenberth and Guillemot [1998] use two methods of estimating the residual ET-P in the NCEP reanalysis, in order to determine the ability of the NCEP model to close its atmospheric water budget. One method is to use model-generated precipitation, with ET derived from the archived surface latent heat fluxes (as mentioned above, both are class "C" variables in the NCEP reanalysis). The second method computes the residual of the atmospheric water balance, as the rate of change in precipitable water minus the atmospheric moisture convergence (both of which are derived from class "B" variables). We use a similar method to isolate the atmospheric water balance, which by NCEP reanalysis classification should be more reliable, from the water balance produced by the NCEP LSP.

As used in the NCEP reanalysis, Equation (3) includes an atmospheric nudging term, similar to the surface water balance nudging term. Using NCEP reanalysis fields, all other components of Equation (3) are defined, and the nudging term can be computed as the amount by which the atmospheric water balance fails to close. The atmospheric nudging term is not used in the computation of residual ET, however. Following *Roads, et al.* [1994], we apply Equation (3) to derive the atmospheric moisture convergence and rate of change in precipitable water from the NCEP reanalysis, and combine this with the gridded observed precipitation to compute values for ET. This "residual" ET estimate is not independent of the NCEP LSP and the NCEP precipitation fields, since the effects of these are reflected in atmospheric conditions through model feedback. However, it does provide a convenient method of largely separating the NCEP LSP and the NCEP precipitation fields from the ET estimate, and can be used to compare the VIC model and NCEP reanalysis ET in order to assess the possible sources of any biases in ET predictions by the respective LSPs.

4.0 RESULTS AND DISCUSSION

By comparing surface water budgets, we assess the spatial and temporal differences between the VIC and the NCEP reanalysis over the Mississippi River basin. The mean annual statistics, with the variation of annual values for the 10-year study period, are shown in Table 1.

4.1 Characterization of the Mississippi River Basin

To evaluate the ability of the VIC model to reproduce the essential, hydrologically important characteristics of the Mississippi River basin, the simulated daily runoff from each grid cell was routed to points near the outlets of four of the sub-basins. The comparisons between the simulated flow and the observed flows (naturalized flows, adjusted for anthropogenic effects, in the case of the Missouri) are shown in Figure 2. Also included in this figure is the daily runoff from the NCEP reanalysis, routed to the same point using the same routing algorithm. It should be noted that the Arkansas River has significant withdrawals, and naturalized flows were not available for the period of study. Therefore, the model high flows are expected to be higher than the observations. Elsewhere though, the VIC model is quite successful in capturing the peak flows, the autumn low flows, and the inter-annual variation throughout the Mississippi River Basin. This gives us some confidence in using the simulated soil moistures and surface fluxes as benchmarks against which to compare the reanalysis products. It is also notable in Figure 2 that, especially in the Missouri and Upper Mississippi Basins, the NCEP reanalysis has unrealistically early peaks with extremely high magnitudes, which is suggestive of possible problems with the NCEP LSP, as well as the precipitation forcings. Also, the routed NCEP runoff produces almost no flow at all for the Arkansas River.

As a further verification of the VIC model output, we compare the soil moisture predicted by the model to observed data. In the Mississippi River basin, there are few systematic soil moisture records of a length sufficient for comparison to the 10-year VIC model simulation. The soil moisture climatology *of Hollinger and Isard* [1994] is unique in the length and detail of collected soil moisture measurements, including regular observations of soil moisture at 19 sites in Illinois. Figure 3 compares the monthly average data from *Hollinger and Isard* [1994], extended through August 1996 as described by Robock, et al. [2000], with the VIC model simulation for 1988-1997. The VIC simulation includes an average for 800 grid cells, bounded by latitudes 42.5 and 37.5, and longitudes –88 and –90.5 degrees, which is compared to an averaging of 19 point measurements throughout the state for the observations. Despite an apparent negative bias in column soil moisture, the VIC model accurately captures the annual cycle in observed soil moisture variation, and shows relatively small error in the magnitude of fluxes, the actual component of the surface water balance.

4.2 Water Balance Comparison

Figure 4a shows the average monthly variation of the contributions of the components of the surface water budget for the entire Mississippi basin. The figure shows an overall tendency of the NCEP reanalysis to overestimate precipitation in the summer months, to overpredict ET in all months, to simulate earlier runoff, and to exhibit greater annual fluctuation in soil moisture as compared with the VIC simulations. The VIC model produces an ET (3.1 mm/d) that exceeds precipitation (2.8 mm/d) in the summer months (JJA), whereas in the NCEP reanalysis the summer precipitation (4.7 mm/d) is exaggerated so it exceeds even the model's overpredicted ET (4.0 mm/d). The magnitude of the nudging term, U, is also large, with an average annual (absolute) magnitude of 1.6 mm/d, which is comparable to the basin-wide average precipitation of 2.2 mm/d. This confirms that the nudging, by injecting water into the system in the winter and removing water in the summer, has an impact on the water budget nearly as large as the principal forcing mechanism for the land surface. Finally, the soil moisture has a larger annual fluctuation in the NCEP than in the VIC simulations.

Figures 4b through 4f show the water budget components for the major sub-basins. The NCEP budgets show substantial regional biases in some sub-basins, most notably a 150% overestimation in summer (JJA) precipitation over the Ohio basin, and a 125% overestimation for the Lower Mississippi basin. High regional bias over the southeastern U.S. in precipitation in the NCEP reanalysis has been recognized in several global studies [*Mo and Higgins*, 1996; *Janowiak, et al.*, 1998; *Trenberth and Guillemot*, 1998], as well as in studies focused over the central U.S. [*Higgins, et al.*, 1996; *Betts, et al,* 1996b]. ET is consistently overestimated in all sub-basins. Runoff is underestimated in all sub-basins, with the exception of the months January through April in the Upper Mississippi and January through March in the Missouri basin. The exaggerated reanalysis annual fluctuation in the soil moisture cycle

Table 1. Summary of annual average water balance components, with standard deviation (S.D.) and Coefficient of Variation (C.V.), from the VIC model and NCEP reanalysis. Included are precipitation (P), evapotranspiration (ET), runoff (N), soil moisture (SM), and snow water equivalent (SWE).

Basin	Statistic	P, mm VIC	P, mm NCEP	ET, mm VIC	ET, mm NCEP	N, mm VIC	N, mm NCEP	SM, mm VIC	SM, mm NCEP	SWE, mm VIC	SWE, mm NCEP
Entire Basin	Mean	800	1021	535	977	259	180	128	188	6.4	2.7
	S.D.	73.4	88.9	20.8	17.6	40.2	22.8	20.6	10.2	2.2	0.1
	C.V.	0.09	0.09	0.04	0.02	0.16	0.13	0.16	0.05	0.34	0.05
Lower Basin	Mean	1384	1593	710	1410	673	218	121	176	0.5	0.0
	S.D.	115.4	152.4	21.5	45.9	118.2	51.7	5.3	22.0	0.4	0.0
	C.V.	0.08	0.10	0.03	0.03	0.18	0.24	0.04	0.12	0.87	0.70
Arkansas-Red	Mean	821	951	532	927	281	50	113	131	2.0	0.4
	S.D.	73.4	101.9	14.6	28.1	44.2	12.4	12.3	10.2	0.8	0.1
	C.V.	0.09	0.11	0.03	0.03	0.16	0.25	0.11	0.08	0.41	0.22
Missouri	Mean	539	677	460	806	72	118	123	216	8.8	4.8
	S.D.	84.9	92.5	31.2	23.3	33.4	13.8	36.9	11.4	2.8	0.4
	C.V.	0.16	0.14	0.07	0.03	0.46	0.12	0.30	0.05	0.32	0.08
Upper Miss.	Mean	827	1093	547	1016	279	336	156	204	13.1	3.0
	S.D.	146.8	142.4	27.9	26.2	106.2	41.5	30.4	15.1	5.9	0.2
	C.V.	0.18	0.13	0.05	0.03	0.38	0.12	0.20	0.07	0.45	0.08
Ohio Basin	Mean	1125	1618	634	1201	484	325	128	165	2.0	0.5
	S.D.	135.5	119.9	46.4	20.1	92.9	60.3	12.1	11.1	1.3	0.2
	C.V.	0.12	0.07	0.07	0.02	0.19	0.19	0.09	0.07	0.65	0.38

is also shown to be most predominant in the Upper Mississippi and Missouri Basins, with overestimation in the winter and spring, and underestimation in the late summer and fall. As in the entire basin, the nudging term has a substantial influence on the water budget in each sub-basin. Also of note is that, though the runoff values are generally smaller in comparison with precipitation and ET, modest basin-wide differences in runoff production will produce large differences in routed streamflows, as shown in Figure 2.

4.3 Time Series Analysis of Water Budget Components

To assess the inter-annual variability of the NCEP reanalysis and VIC data the time series of monthly values for the 10-year simulation is shown for the entire Mississippi River Basin in Figure 5a. As was seen in the monthly average plots, the tendency for the NCEP reanalysis to overestimate the summer precipitation in the Mississippi basin is apparent, as is the overestimation of ET. The soil moisture (or the terrestrial water content; in Figures 5a, 5b, 5c, and 5f the values include a small contribution from snow water equivalent) has a much larger intra-annual fluctuation in the NCEP reanalysis as compared to VIC. This indicates that the climatology to which the LSP is being nudged may overestimate the range of soil moisture variation for this basin. Also, the inter-annual variability in the NCEP reanalysis is lower than that simulated by the VIC model, a characteristic that probably is a manifestation of the nudging, which pushes the NCEP soil moisture to the prescribed climatology, which itself has no interannual variability. The other notable difference in the annual time series for the entire basin is that the NCEP simulated runoff is close to zero for the late summer and autumn of nearly every year. This effect was also seen in Figure 2, where the NCEP routed hydrographs could not reproduce the late season, baseflow-dominated portion of the hydrographs. This is probably a characteristic of the NCEP LSP (rather than the surface forcings), resulting from the LSP runoff formulation having difficulty retaining column moisture for release as baseflow much later in the year.

The time-series plots for each sub-basin are included in Figures 5b through 5f. In this way the origin of the characteristics of the entire basin that were observed in Figure 5a can be tracked upstream. The precipitation overestimation in the NCEP reanalysis data is seen most dramatically in the Lower Mississippi and Ohio Basins (Figures 5b and 5f). As was seen in the monthly average time series, little seasonal pattern is evident in the gridded observed data used in the VIC model, however the NCEP reanalysis data include a regular annual cycle. Although the magnitudes of the annual precipitation for the entire Mississippi basin vary between VIC and NCEP, the monthly correlation is strong (r=0.84), revealing the general pattern of annual fluctuation is well represented on a basin-wide level. By contrast, the Lower basin VIC and NCEP monthly precipitation data are poorly correlated (r=0.09), indi-

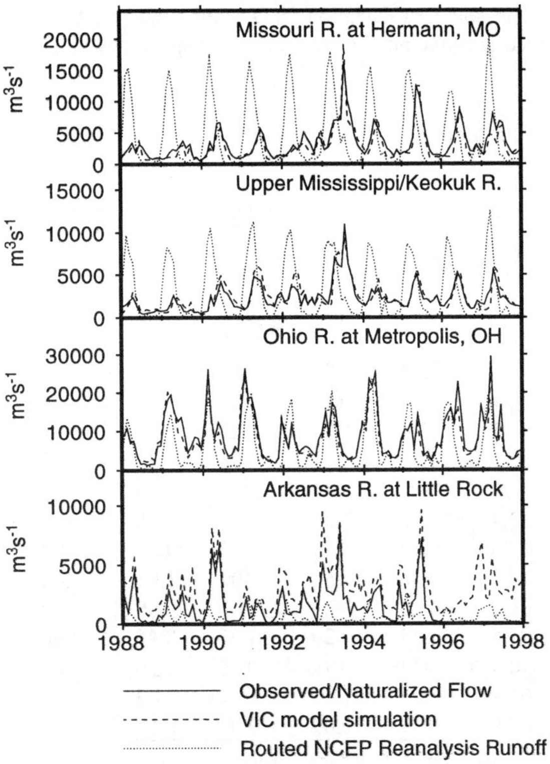

Figure 2. Hydrographs of monthly routed flows for the 10-year study period at outlet points in the Mississippi River basin for VIC model, NCEP reanalysis, and observed or naturalized flows.

cating that for this sub-basin the annual occurrence of precipitation in the NCEP reanalysis is not well represented. Likewise, for the Ohio basin the monthly correlation is low (r=0.31). This illustrates the general success of the NCEP reanalysis to capture large-scale patterns, but with considerable regional errors.

In Figures 5d and 5e, the snow and soil water are represented separately in the bottom frame, because the snow contribution to the average annual water budget is significant only in these basins. These figures show that the NCEP reanalysis does not accurately model the accumulation and melting of snow as represented by VIC. A portion of this difference can be attributed to a known error, in which the snow cover extent updating scheme in the NCEP reanalysis used 1973 data for the period 1974-1994. However, the underprediction of snow water equivalent by the NCEP reanalysis is consistent both with this error and from 1995-1998, which included the corrected snow initialization.

The annual average snow coverage (expressed as a water equivalent over the entire basin) for the VIC model over the Missouri basin is 8.8 mm, while the NCEP reanalysis estimates it as only 4.8 mm. For the Missouri basin the underestimation of snow in the NCEP reanalysis, relative to the VIC model, results primarily from the inability of the reanalysis to capture the intense precipitation and very deep winter snow pack over spatially limited areas in the Rocky mountains at the eastern edge of the basin. Since the NCEP reanalysis is at a much coarser scale than the VIC simulation, it is not surprising that the reanalysis has difficulty capturing localized extremes. However, for the Upper Mississippi basin the difference in snow water equivalent is larger, with an average annual snow accumulation of 13.1 mm and 3.0 mm for VIC and NCEP reanalysis data, respectively, and the bias is distributed relatively evenly throughout the basin. The snow as modeled in the VIC simulation in these two basins does not typically begin to melt significantly until April, as opposed to February for the NCEP data. This has implications on soil moisture accumulation, with the VIC soil moisture continuing to rise through the snow melt period while the NCEP soil moisture rapidly decreases during these same months. It is this delayed snow melt, along with the subsequent infiltration into the soil column and later emergence as baseflow that helps allow the VIC model to accurately simulate the late season baseflow typical of the region. In addition to the hydrologic impact of the differences in snow modeled by VIC and NCEP, snow has a profound effect on the surface energy balance as well, through increased albedo, changed surface roughness, insulation of the ground surface, and ultimately the transfer of latent and sensible heat to the atmosphere.

The runoff in the NCEP reanalysis mirrors the high levels of soil moisture in the winter. This is especially evident in the Upper Mississippi and Missouri basins, where the NCEP reanalysis soil moistures are much higher than the VIC values. Since gravity drainage and surface runoff are not archived

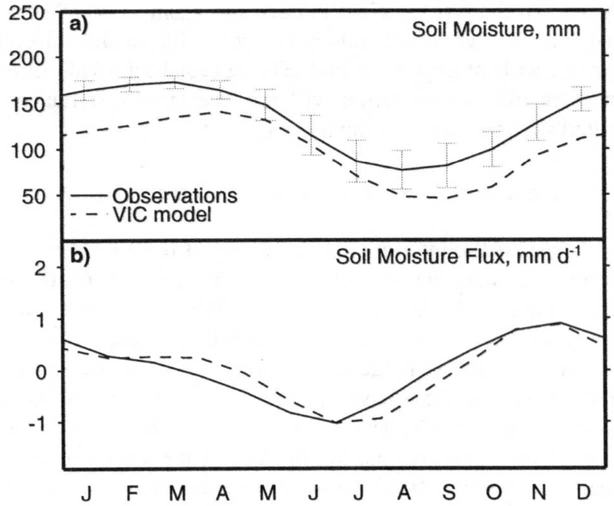

Figure 3. Comparison of monthly average soil climatology between Illinois data of *Hollinger and Isard* [1994] and the VIC simulation for a) Volumetric soil moisture, adjusted as described in the text (95% confidence limits for the observations are shown for each month) and b) soil moisture flux.

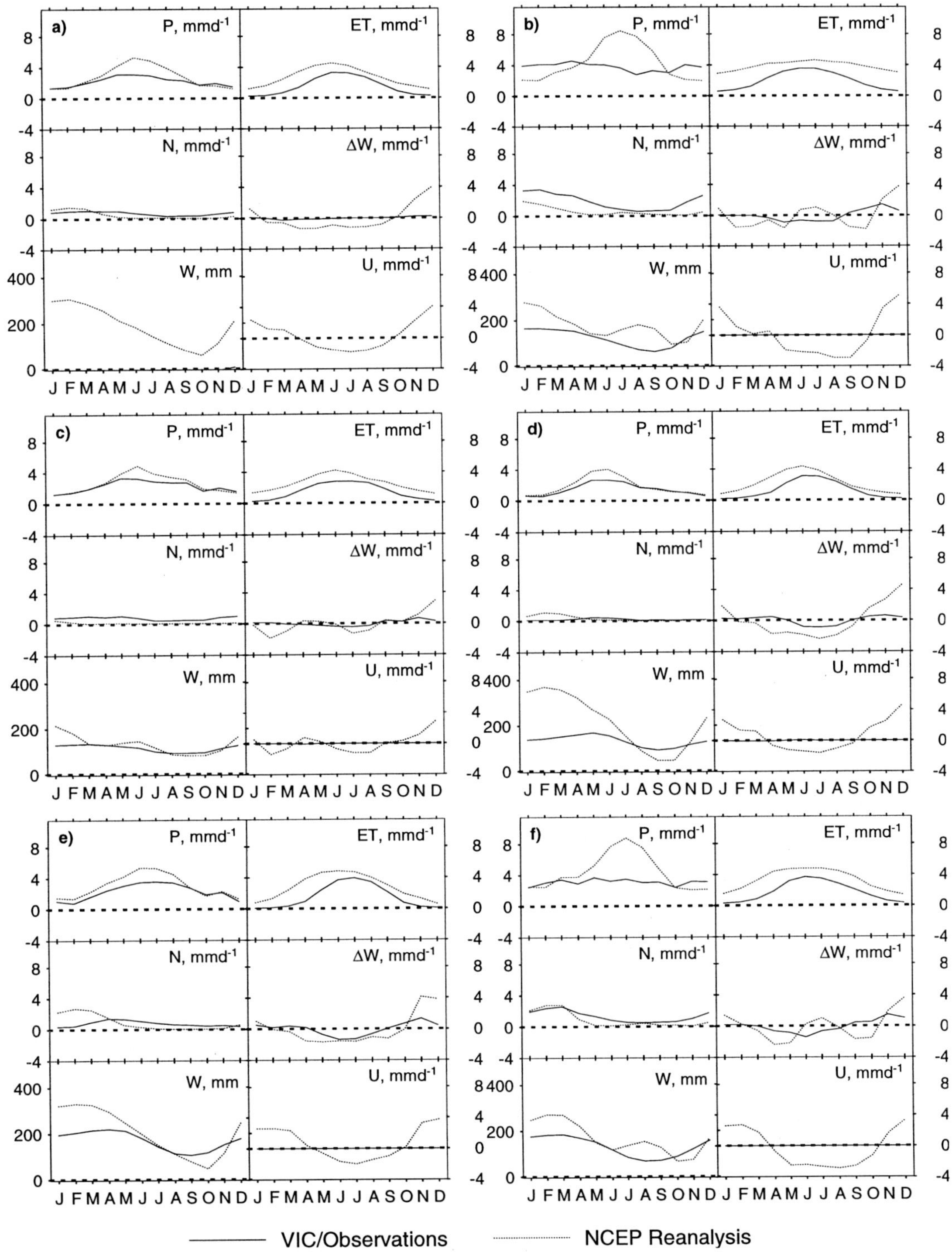

Figure 4. Average monthly surface water balance components for VIC simulation and NCEP reanalysis at a) Entire Mississippi River basin, b) Lower basin, c) Arkansas-Red system, d) Missouri River basin, e) Upper Mississippi basin, and f) Ohio River basin. Variables are precipitation (P), evapotranspiration (ET), runoff (N), change in water storage (ΔW), volumetric surface water (W), and nudging/non-closure term (U).

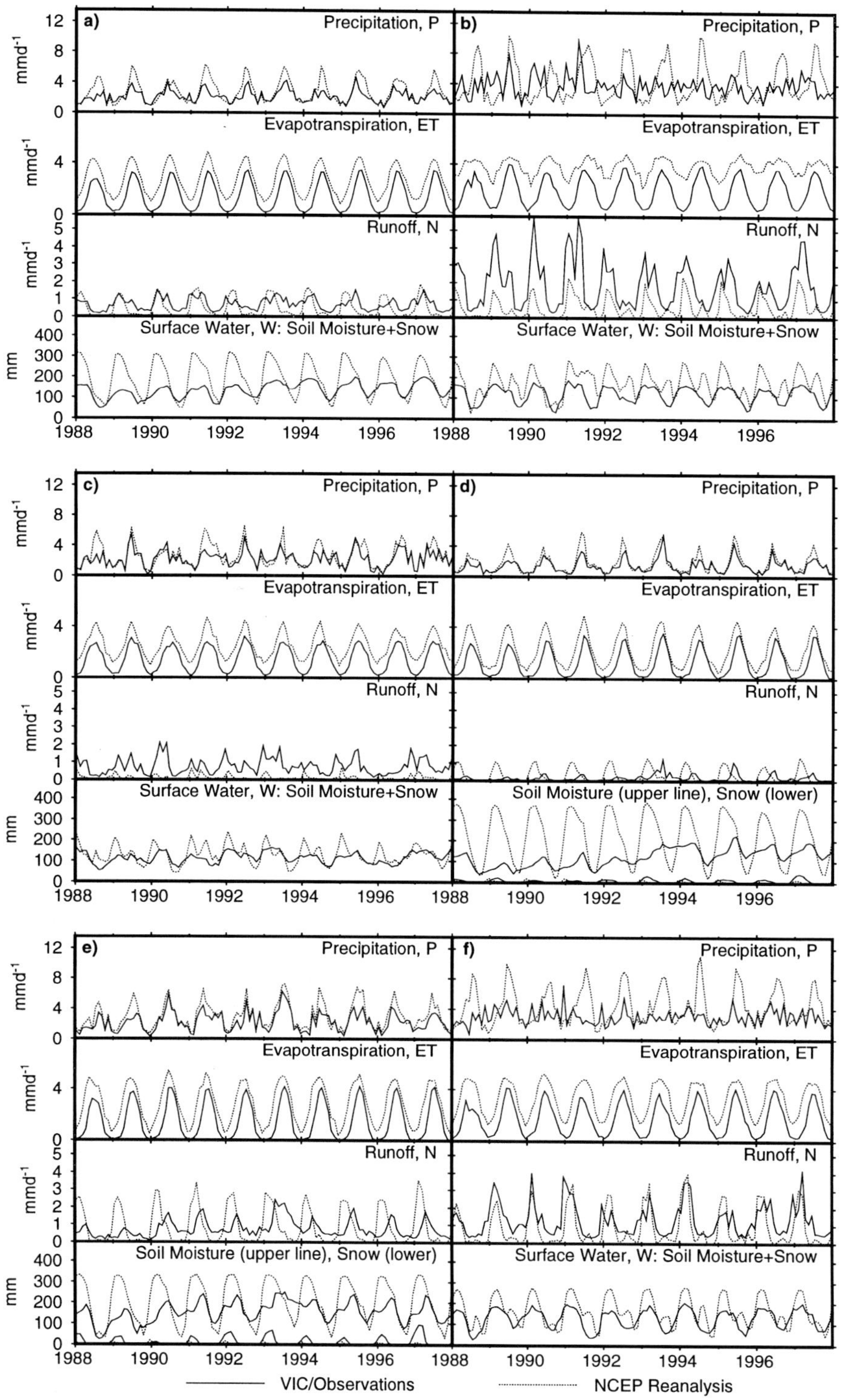

Figure 5. Timeseries of monthly average quantities for the water balance components for the VIC model and NCEP reanalysis for the 10-year study period for a) Entire Mississippi River basin, b) Lower basin, c) Arkansas-Red system, d) Missouri River basin, e) Upper Mississippi basin, and f) Ohio River basin.

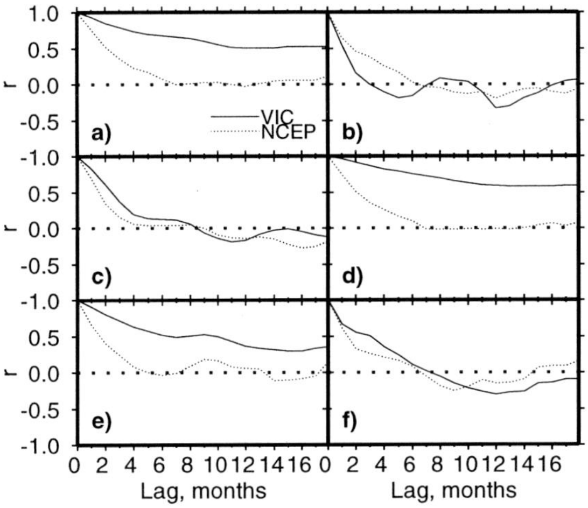

Figure 6. Autocorrelation of monthly soil moisture anomalies for VIC model results and NCEP reanalysis for the Mississippi River basin and major sub-basins: a) Entire Mississippi River basin, b) Lower basin, c) Arkansas-Red system, d) Missouri River basin, e) Upper Mississippi basin, and f) Ohio River basin.

separately in the NCEP reanalysis, it is difficult to determine whether the high soil moisture levels enhance saturation excess runoff or rapid gravity drainage through the bottom of the soil column. In the more southern sub-basins, especially the Lower and Arkansas-Red basins, the NCEP reanalysis runoff is very low throughout the year, even with soil moistures that are close to the VIC model results.

4.4 Soil Moisture Variability and Persistence

As indicated above, the inter-annual basin-wide soil moisture variability is small in the NCEP reanalysis, although the intra-annual fluctuations are much larger than those inferred by the VIC model. The VIC mean annual soil moisture is 128 mm, with an average annual fluctuation amplitude of 69 mm, while the NCEP reanalysis mean is higher at 188 mm, with an average amplitude of 242 mm. The inter-annual variation is described by the standard deviation of annual average soil moisture, which is higher for VIC (20.6 mm), than for NCEP (10.2 mm), resulting in coefficients of variation for the annual averages of 0.16 for the VIC model and 0.05 for NCEP data. The soil moisture time series in Figures 4d and 4e show that the large intra-annual soil moisture fluctuations in the NCEP reanalysis for the entire Mississippi Basin are derived predominantly contributed by Missouri and Upper Mississippi sub-basins. For the Missouri Basin, the coefficients of variation for annual average soil moistures are 0.3 and 0.05 for VIC and NCEP, respectively. For the Upper Mississippi Basin, the coefficients of variation for annual average soil moistures are 0.2 and 0.07 for VIC and NCEP. This illustrates the effect on inter-annual variability of imposing a climatology on the soil moisture conditions in the NCEP LSP, especially with the relatively short relaxation time constant of 60 days [*Roads, et al.,* 1999].

The low inter-annual variability in the NCEP reanalysis reflects the inability of the NCEP LSP to simulate low frequency variations in the hydrologic system. This is seen in the contrasting levels of persistence provided by the memory of soil moisture conditions in the NCEP reanalysis and the VIC model. The difference in these two models is shown in Figure 5a, where the high soil moisture conditions in 1993 in the VIC simulation are carried into the following year, while in the NCEP reanalysis the annual cycle is forced to its assumed climatology. This is quantified in Figure 6, which plots the autocorrelation function for soil moisture anomalies for the Mississippi River basin and each sub-basin. A useful variable for describing the persistence in the modeled soil moisture is the "decay time scale" concept presented by *Delworth and Manabe* [1988], which is derived using a stochastic model of soil moisture as a system forced by precipitation and damped by ET. The decay time scale is the e-folding time, or the lag at which the autocorrelation function reduces to 1/e (0.37). The decay time scales associated with the monthly soil moisture anomalies for each model are shown in Table 2. The sub-basins showing the greatest difference between the two models are the Missouri, which is also the driest basin, and the Upper Mississippi. The decay time scale in the NCEP reanalysis varies little between sub-basins, showing almost identical values in the driest and wettest sub-basins. The VIC model, by contrast, shows considerable variation, with the northernmost and drier Missouri and Upper Mississippi Basins having the longest decay time scales. This is consistent with the global study of *Delworth and Manabe* [1988], who identify a general trend of increasing decay time scale with latitude, and with *Huang, et al.* [1996] who conclude that areas with lower temperatures (hence lower potential evapotranspiration) and lower precipitation will experience higher soil moisture persistence.

4.5 Spatial Analysis

Plates 1 through 5 show the spatial distribution of seasonal averages of the precipitation, soil moisture, snow water equivalent, ET (latent heat flux), and sensible heat flux, re-

Table 2. Decay time scale for the monthly soil moisture anomalies in NCEP reanalysis and VIC model.

	Decay Time Scale, months	
Basin	VIC Model	NCEP Reanalysis
Entire Mississippi	> 18	3.0
Lower	1.5	3.0
Arkansas-Red	3.1	1.9
Missouri	> 18	2.9
Upper Mississippi	11.5	1.9
Ohio	3.9	1.8

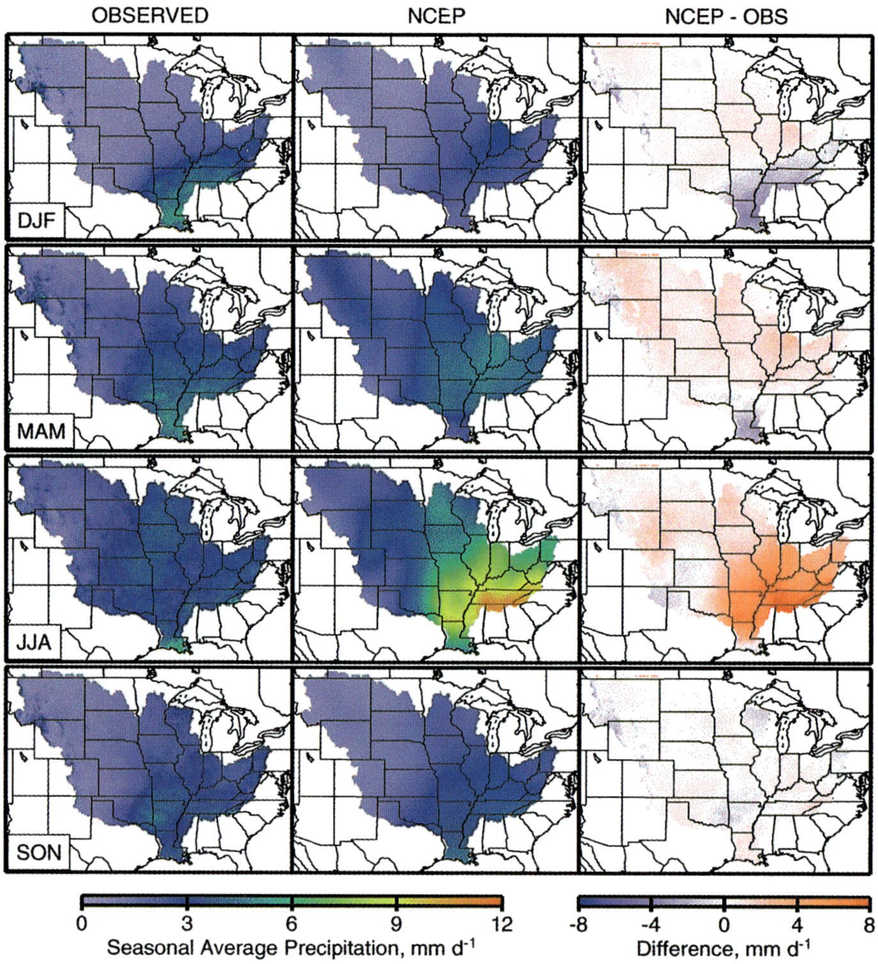

Plate 1. Comparison of seasonal average precipitation for the gridded observations and the NCEP reanalysis.

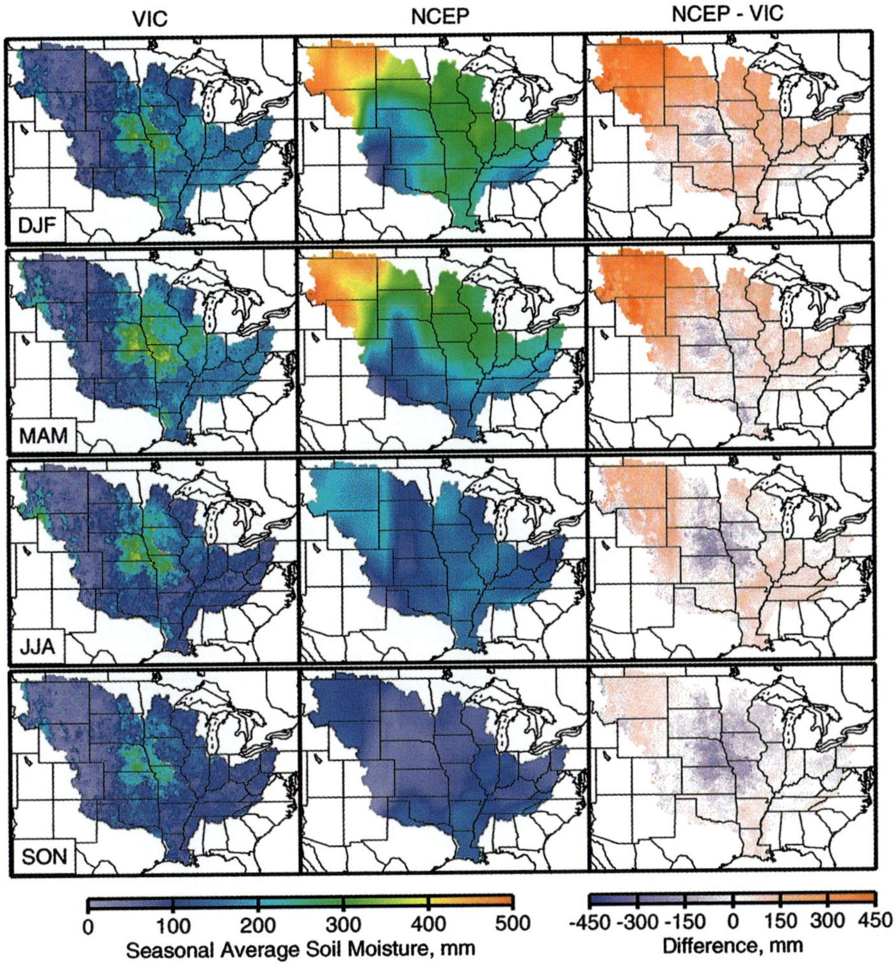

Plate 2. Comparison of seasonal average soil moisture, adjusted to represent the hydrologically active column soil moisture (see text), for the VIC simulation and the NCEP reanalysis.

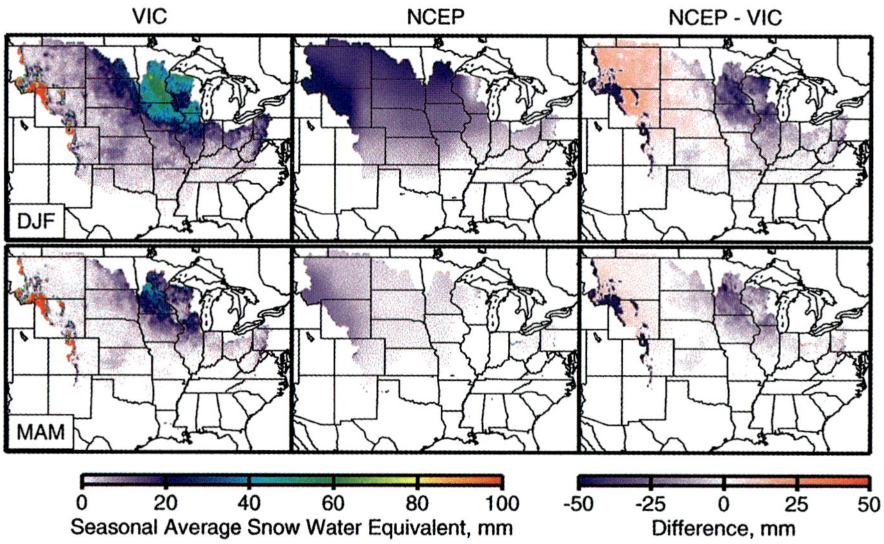

Plate 3. Comparison of seasonal average snow water equivalent for the VIC simulation and the NCEP reanalysis. Values above 100 mm are shown in red (center and left panels), and differences less than -50 mm are blue (right panels).

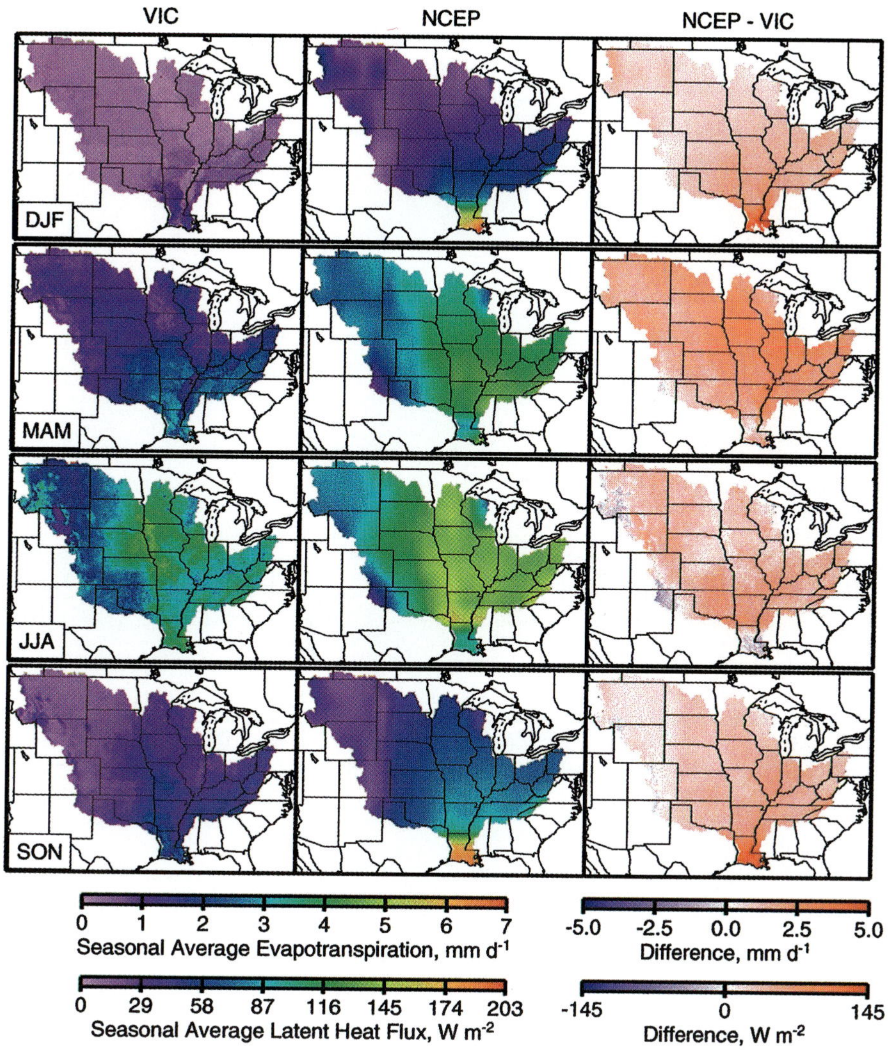

Plate 4. Comparison of seasonal average evapotranspiration and latent heat flux for the VIC simulation and the NCEP reanalysis.

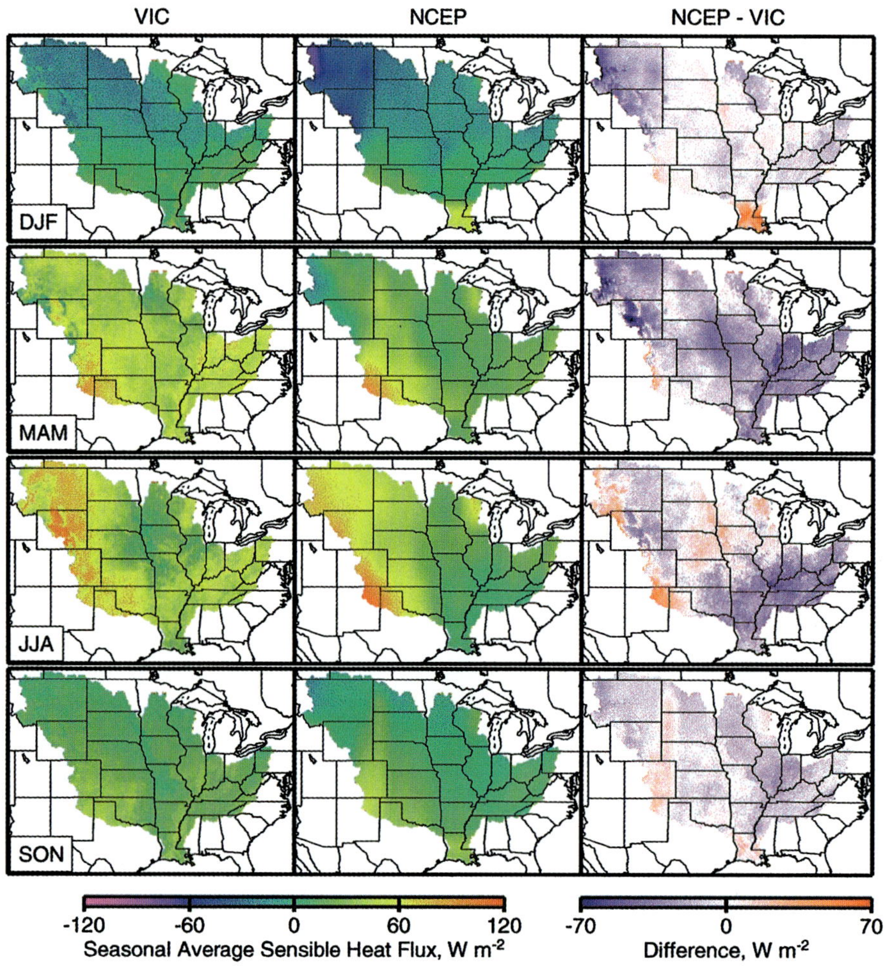

Plate 5. Comparison of seasonal average sensible heat flux for the VIC simulation and the NCEP reanalysis.

spectively. For each plot, we show the results of the VIC simulation, the NCEP reanalysis values, and the difference between the two for winter (December, January, February – DJF), spring (March, April, May – MAM), summer (June, July, August – JJA), and autumn (September, October, November – SON), with the exception of snow water equivalent, which is only shown for winter and spring. These plots allow the interpretation of the variation of the different components of the water balance within each sub-basin, and they reveal patterns that cross sub-basin boundaries. Plate 1 clearly shows the bias of the NCEP precipitation toward overpredicting the summer precipitation over the Ohio and Lower basins, and underestimating the winter precipitation over the southernmost areas of the Mississippi basin. It is also evident that the reanalysis model does not capture the smaller scale features such as the high precipitation in the mountainous Northeastern portion of the Missouri basin, due in large part to the coarser model resolution.

Plate 2 shows the soil water content in the Mississippi basin. This shows the dramatic overestimation of soil moisture for the upper Missouri basin, and in portions of the Upper Mississippi in the NCEP reanalysis as compared to the VIC data. This provides a possible explanation for the rapid response of runoff in the winter in the NCEP model for these sub-basins. It can also be seen that in the lower Missouri Basin the NCEP model underpredicts the soil moisture, especially in the summer and fall. Plate 3 highlights the differences in snow simulation between the VIC model and the NCEP reanalysis, with slight overprediction of snow by the NCEP reanalysis over broad areas of the Missouri basin, and the extreme snow accumulation shown by VIC in the western Missouri basin and the substantial winter snow in the Upper Mississippi basin being largely absent in the NCEP reanalysis.

Because ET is the final product of the NCEP LSP, and controls the partitioning of atmospheric net radiation at the surface into latent and sensible heat, any bias is of great concern for forecasting or climate studies. Plate 4 shows the spatial distribution of the ET (and latent heat flux) by season, and illustrates the basin-wide overprediction of ET by NCEP reanalysis in all seasons, relative to the VIC simulated ET. This results in the overestimation of latent heat flux and underestimation of sensible heat flux (Plate 5) throughout the majority of the basin and for all seasons.

4.6 *Evapotranspiration from Atmospheric Water Balance*

In order to address the question of whether a better estimation of ET could be obtained by assimilation of observed precipitation, and bypassing the NCEP LSP, we compute the ET as an atmospheric residual. Since NCEP reanalysis ET is overestimated throughout the year, throughout the entire basin, and since precipitation is overestimated predominantly in the summer months, the difference between the ET derived from the NCEP reanalysis assimilation model and the ET derived using this method are due largely to the NCEP LSP. From the moisture convergence and change in precipitable water from

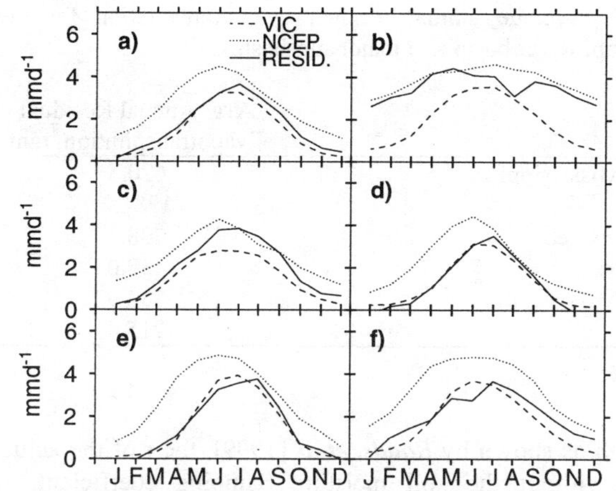

Figure 7. Average monthly residual evapotranspiration, compared with NCEP reanalysis and VIC model evapotranspiration for: a) Entire Mississippi River basin, b) Lower basin, c) Arkansas-Red system, d) Missouri River basin, e) Upper Mississippi basin, and f) Ohio River basin.

the NCEP reanalysis and the observed precipitation a residual ET is calculated, which is plotted along with ET from the VIC model and the NCEP reanalysis. Figure 7 shows the residual ET computed from the atmospheric water balance for the entire Mississippi River basin and major sub-basins. Table 3 summarizes the total residual ET computed for each sub-basin.

The significant change in computed ET is evident, with values derived from the NCEP reanalysis atmospheric variables and observed precipitation closely following the VIC simulated evaporation. The average basin-wide VIC ET is 1.47 mm/d, NCEP reanalysis is 2.68 mm/d, whereas the residual ET is 1.70 mm/d. This decrease in bias relative to the VIC model (82% to 16%) is interesting on several levels. First, residual ET is computed excluding the least reliable NCEP class "C" variables, and is arguably more accurate, so it is comforting that the agreement becomes closer, given that the VIC surface budget is constrained to closure. Second, the greatest improvement in the residual ET estimate relative to the VIC values is in the Missouri and Upper Mississippi basins, whereas the greatest precipitation bias is in the Ohio basin. This is evidence that it is the NCEP LSP and the use of nudging, and not simply the precipitation bias, that is responsible for errors in the reanalysis ET.

This raises the question of whether bypassing the NCEP LSP altogether, by assimilating precipitation observations into the weather prediction model, would yield improved estimates of latent heat flux, and therefore more accurate partitioning of the net radiation at the surface. The problem with this approach is that the LSP produces not only the soil moisture fields that control latent and sensible heat partitioning, but also the persistence in the system. The monthly ET anomalies in the VIC model at a lag of one month have an autocorrelation, r of 0.30, while the NCEP reanalysis ET has an r of 0.15

Table 3. Average annual residual evapotranspiration for the Mississippi River basin and major sub-basins.

Basin	Avg. Annual Residual Evapotranspiration, mm
Entire Mississippi	620.5
Lower	1281.2
Arkansas-Red	708.1
Missouri	438.0
Upper Mississippi	532.9
Ohio	715.4

(although as shown by *Roads, et al* [1999], the r of the actual model, without the soil moisture damping coefficient, is somewhat higher), and the computed residual ET has an r of 0.06. This shows that while the magnitude of the mean ET is improved with the assimilation of precipitation values, the persistence of the system is lost in the absence of an LSP.

5.0 CONCLUSIONS

The NCEP/NCAR reanalysis land surface water and energy fluxes have been shown to have important regional and temporal biases, as compared with observations and to the same fluxes predicted by a macroscale hydrology model that is closely constrained to preserving the long-term river basin scale water balance. On a basin-wide scale, summer precipitation is overpredicted by the reanalysis, especially in the southeast. Evapotranspiration exceeds the off-line hydrologic model predictions in all months. Though snow water equivalent is slightly overestimated in wide areas, both snow extent and duration on a basin-wide level are generally underestimated by the NCEP reanalysis, especially in the upper Missouri headwaters in the upper Mississippi River basin where the majority of the annual snowfall occurs. Intra-annual variations in soil moisture are too large in the reanalysis, and inter-annual variation and persistence of soil moisture are low as compared with the hydrologic model simulations. Interpretation of sources for these biases in the reanalysis are complicated by the incorporation of a soil moisture nudging term, which is used to maintain an assumed land surface climatology at the expense of a realistic water budget. The nudging is intended to improve the prediction of prognostic variables, such as surface air temperature, but grossly distorts the surface flux fields. The large magnitude of the nudging term relative to the other water balance produces a nearly negligible inter-annual variability in soil moisture, and greatly reduces the interannual persistence in the predicted soil moisture conditions. Furthermore, the assumed climatology to which the model is nudged does not appear to be appropriate for much of the basin, especially the Missouri and Upper Mississippi sub-basins.

The consistent overestimation of ET in the NCEP reanalysis results in underestimation of sensible heat. Estimation of evapotranspiration from the reanalysis atmospheric moisture budget, using observed precipitation, significantly improves the estimated ET. This is encouraging as the atmospheric moisture budget is arguably more closely linked to observations than is the surface budget. However, while this approach produces ET values closer to the hydrologic predictions, the predicted interannual persistence of the atmospheric budget estimates is much less than of those produced by the hydrologic model. Furthermore, the atmospheric budget method, although producing apparently better results than the surface budget of the reanalysis, is not independent of the reanalysis surface ET predictions, due to the interaction between the LSP and the atmospheric model. More study may elucidate the source of differences in ET persistence, and could evaluate the potential benefits of assimilating precipitation observations into schemes to update surface flux predictions derived from coupled land-atmosphere models.

Acknowledgments. This is Contribution No. 736 from the Joint Institute for the Study of the Atmosphere and Ocean, funded under NOAA Cooperative Agreement Number NA67RJ0155. This study was also supported by NOAA Grant NA77RJ0453 and NASA Grant NAG81516 to the fourth author. The authors wish to thank Keith Cherkauer and Andrew Wood at the University of Washington for their contributions to the hydrologic modeling effort for the Upper Mississippi and Ohio River basins, respectively, and to Bart Nijssen for his assistance in the soil moisture interpretation and analysis. NCEP reanalysis data was provided by the NOAA-CIRES Climate Diagnostics Center in Boulder, CO, from their web site at http://www.cdc.noaa.gov/.

REFERENCES

Abdulla, F.A., D.P. Lettenmaier, E.F. Wood, and J.A. Smith, Application of a macroscale hydrologic model to estimate the water balance of the Arkansas-Red River Basin, *J. Geophys. Res.,* 101(D3), 7449-7459, 1996

Beljaars, A.C., P. Viterbo, M.J. Miller and A.K. Betts, The anomalous rainfall over the United States during July 1993: sensitivity to land surface parameterization and soil moisture anomalies. *Mon. Weather Rev.,* 124, 362-383, 1996

Betts, A.K., J.H. Ball, A.C.M. Beljaars, M.J. Miller and P.A. Viterbo, The land surface-atmosphere interaction: a review based on observational and global modeling perspectives. *J. Geophys. Res.* 101(D3), 7209-7225, 1996a

Betts, A.K., S.-Y. Hong, and H.-L. Pan, Comparison of NCEP-NCAR reanalysis with 1987 FIFE data, *Mon. Weather Rev.,* 124, 1480-1498, 1996b

Betts, A.K., P. Viterbo, A. Beljaars, H.-L. Pan, S.-Y. Hong, M. Goulden and S. Wofsy, Evaluation of land-surface interaction in ECMWF and NCEP/NCAR reanalysis models over grassland (FIFE) and boreal forest (BOREAS). *J. Geophys. Res.* 103(D18), 23,079-23,085, 1998

Bras, R. L., Hydrology, an introduction to hydrologic science, Addison Wesley, Inc., Reading, North Carolina, 1990

Bristow, K. L. and G. S. Campbell, On the relationship between incoming solar radiation and daily maximum and minimum temperature. *Agric. For. Meteorol.,* 31, 159-166, 1984

Campbell, G. S., A simple method for determining unsaturated conductivity from moisture retention data, *Soil Sci.,* 117, 311-314, 1974

Cosby, B.J., G.M. Hornberger, R.B. Clapp, and T.R. Ginn, A Statistical Exploration of the Relationships of Soil Moisture Characteristics to the Physical Properties of Soils, *Water Resour. Res.*, 20(6), 682-690, 1984

Daly, C., R.P. Neilson, and D.L. Phillips, A statistical-topographic model for mapping climatological precipitation over mountainous terrain. *J. Appl. Meteorol.* 33, 140-158, 1994

Daly, C., G.H. Taylor, and W.P. Gibson, The PRISM approach to mapping precipitation and temperature, In *Reprints: 10th Conf. on Applied Climatology*, Reno, NV, American Meteorological Society, 10-12, 1997

Delworth, T.D. and S. Manabe, The influence of potential evaporation on the variabilities of simulated soil wetness and climate, *J. Clim.* 1, 523-547, 1988

Dirmeyer, P. A., Problems in initializing soil wetness, *Bull. Am. Meteorol. Soc.,* 76(11), 2234-2240, 1995

Durre, I., J.M. Wallace, and D.P. Lettenmaier, Dependence of extreme daily maximum temperatures on antecedent soil moisture in the contiguous United States during summer. *J. Clim.* (in press), 2000

Glecker, P. (ed.), AMIP II Guidelines, *AMIP Newsletter No. 8.* Lawrence Livermore National Laboratory, Livermore, CA, 1996

Hansen, M.C., DeFries, R.S., Townshend, J.R.G., and Sohlberg, R., Global land cover classification at 1km spatial resolution using a classification tree approach, *International J. Remote Sensing* (in press), 2000

Higgins, R.W., K.C. Mo and S.D. Schubert, The moisture budget of the Central United States as evaluated in the NCEP/NCAR and the NASA/DAO reanalyses, *Mon. Weather Rev.*, 124, 939-963, 1996

Hollinger, S.E. and S.A. Isard, A soil moisture climatology of Illinois. *J. Clim.*, 7, 822-833, 1994

Huang, J. and H.M. van den Dool, Monthly precipitation-temperature relations and temperature prediction over the United States. *J. Clim.* 6, 1111-1132, 1993

Huang, J., H.M. van den Dool and K.P. Georgakakos, Analysis of model-calculated soil moisture over the United States (1931-1993) and applications to long-range temperature forecasts. *J. Clim* 9, 1350-1362, 1996

Janowiak, J.E., A. Gruber, C.R. Kondragunta, R.E. Livezy and G.J. Livezy, A comparison of the NCEP NCAR reanalysis precipitation and the GPCP rain gauge satellite combined dataset with observational error considerations. *J. Clim.*, 11, 2960-2979, 1998

Kalnay, E., M. Kanamitsu, R. Kistler, W. Collins, D. Deaven, L. Gandin, M. Iredell, S. Saha, G. White, J. Woolen, Y. Zhu, A. Leetma, R. Reynolds, M. Chelliah, W. Ebisuzaki, W. Higgins, J. Janowiak, K.C. Mo, R. Jenne, and D. Joseph, The NCEP/NCAR 40-Year Reanalysis Project, *Bull. Am. Meteorol. Soc.*, 77, 437-471, 1996

Kimball, J.S., S.W. Running, and R. Nemani, An improved method for estimating surface humidity from daily minimum temperature. *Agric. For. Meteorol.*, 85(1997), 87-98, 1997.

Kistler, R., M. Kanamitsu, J. Woolen, S. Saha and E. Kalnay, Atmospheric reanalysis project at the National Meteorological Center. *Cray Channels* 17(2), 2-7, 1995

Koster, R.D. and M.J. Suarez, Relative contributions of land and ocean processes to precipitation variability. *J. Geophys. Res.*, 100(D7), 13,775-13,790, 1995

Koster, R.D., M.J. Suarez and M. Heiser, Variance and predictability of precipitation at seasonal-to-interannual timescales. *J. Hydrometeorology* (in press), 1999

Liang, X., D. P. Lettenmaier, and E. F. Wood, One-dimensional statistical dynamic representation of subgrid spatial variability of precipitation in the two-layer variable infiltration capacity model, *J.Geophys. Res.,* 101(D16), 21,403-21,422, 1996

Liang, X., D.P. Lettenmaier, E. Wood, and S.J. Burges, A simple hydrologically based model of land surface water and energy fluxes for general circulation models. *J. Geophys. Res.*, 99(D7), 14,415-14,428, 1994

Lohmann, D. R. Nolte-Holube, and E. Raschke, A large-scale horizontal routing model to be coupled to land surface parameterization schemes, *Tellus* 48A(5), 708-721, 1996

Lohmann, D., E. Raschke, B. Nijssen, and D.P. Lettenmaier, Regional Scale Hydrology: II. Application of the VIC-2L model to the Weser River, Germany. *Hydrol. Sci. J.*, 43(1), 131-141, 1998.

Mahrt, L. and H. Pan, A two-layer model of soil hydrology. *Boundary-Layer Meteorol.*, 29(1984), 1-20, 1984

Miller, D.A. and R.A. White, A conterminous United States multi-layer soil characteristics data set for regional climate and hydrology modeling. *Earth Interactions, 2*, 1-15, 1998

Milly, P.C.D. and K.A. Dunne, Sensitivity of the global water cycle to the water-holding capacity of land, *J. Clim.* 7:506-526, 1994

Mintz, Y., The sensitivity of numerically simulated climates to land-surface boundary conditions. In *The global climate*, J.T. Houghton, ed., Cambridge Univ. Press, 1984

Mintz, Y. and Y.V. Serafini, Global fields of soil moisture and land surface evapotranspiration. *NASA Goddard Flight Center Tech. Memo 83907.* Research Review 1980/81, 178-180, 1981

Mintz, Y. and Y. V. Serafini, A global monthly climatology of soil moisture and water balance, *Clim. Dynamics*, 8, 13-27, 1992

Mitchell, K., P. Houser, E. Wood, J. Shaake, D. Lettenmaier, D. Tarpley, W. Higgins, C. Marshall, D. Lohmann, L. Min, M. Elk, B. Cosgrove, J. Entin, Q. Duan, R. Pinker, A. Robock, H. van den Dool, and F. Habets, The GCIP Land Data Assimilation (LDAS) Project – Now Underway. *GEWEX News* 9(4), International GEWEX Project Office, Silver Spring, MD, 1999

Mo, K.C. and R.W. Higgins, Large-scale atmospheric moisture transport as evaluated in the NCEP/NCAR and the NASA/DAO reanalyses. *J. Clim.*, 9:1531-1545, 1996

Myneni, R.B., R.R. Nemani. and S.W. Running, Estimation of Global Leaf Area Index and Absorbed Par Using Radiative Transfer Models, *IEEE Transaction on Geoscience and Remote Sensing*, 35(6), 1380-1393, 1997

Nijssen, B., D.P. Lettenmaier, X. Liang, S.W. Wetzel, and E. Wood, Streamflow simulation for continental-scale basins. *Water Resour. Res.*, 33(4), 711-724, 1997

Nijssen, B., R. Schnur, and D.P. Lettenmaier, Global retrospective estimation of soil moisture using the VIC land surface model, 1980-1993, *J. Clim.* (in press), 2000

Pan, H.-L. A simple parameterization scheme of evapotranspiration over land for the NMC medium-range forecast model, *Mon. Weather Rev.* 118, 2500-2512, 1990

Pan, H.-L. and L. Mahrt, Interaction between soil hydrology and boundary-layer development. *Boundary-Layer Meteorol.*, 38(1987), 185-202, 1987

Pielke, R.A., G.E. Liston, J.L. Eastman, and L. Lu., Seasonal weather prediction as an initial value problem. *J. Geophys. Res.* 104(D16), 19,463-19,479, 1999

Rawls, W.J., L.R. Ahuja, D.L. Brakensiek, A. Shirmohammadi, Infiltration and Soil Water Movement, *In Handbook of Hydrology*, D. Maidment, ed., McGraw-Hill. 1993

Rawls, W.J., D. Gimenez, R. Grossman, Use of Soil Texture, Bulk Density, and Slope of the Water Retention Curve to Predict Saturated Hydraulic Conductivity, *Trans. ASAE*, 41(4), 983-988, 1998

Roads, J. and A. Betts, NCEP/NCAR and ECMWF Reanalysis surface water and energy budgets for the Mississippi River basin, *J. Hydrometeorol.* 1(1), 88-94, 2000

Roads, J.O., S.-C. Chen, A.K. Guelter, and K.P. Georgakakos, Large-scale aspects of the United States hydrologic cycle. *Bull Am. Meteorol. Soc.*, 75(9), 1589-1610, 1994

Roads, J.O., S.-C. Chen, M. Kanamitsu and H. Juang, Surface water characteristics in NCEP global spectral model reanalysis. *J. Geophys. Res.* 104(D16), 19,307-19,327, 1999

Robock, A., K.Y. Vinnikov, G. Srinivasan, J.K. Entkin, S.E. Hollinger, N.A. Speranskaya, S.Liu, and A. Namkhai, The global soil moisture data bank, *Bull Am. Meteorol. Soc.*, 81(6), 1281-1299, 2000

Robock, A., C.A. Schlosser, K.Y. Vinnikov, N.A. Speranskaya, J.K. Entin, S. Qiu, Evaluation of the AMIP soil moisture simulations, *Global and Planetary Change* 19(1998), 181-208, 1998

Shepard, D.S., Computer Mapping: The SYMAP Interpolation Algorithm, In *Spatial Statistics and Models*, G.L. Gaile and C.J. Willmott, eds., D. Reidel Publishing Co., 1984

Simmons, A.J. and L. Bengtsson, Atmospheric general circulation models their design and use for climate studies. In *The global climate*, J.T. Houghton, ed., Cambridge Univ. Press, 1984

Thornton, P.E. and S.W. Running, An improved algorithm for estimating incident daily solar radiation from measurements of temperature, humidity, and precipitation. *Agric. For. Meteorol.*, 93(1999), 211-228, 1999

Todini, E., The ARNO rainfall-runoff model, *J. Hydrol.*, 175, 339-382, 1996

Trenberth, K.E. and C.J. Guillemot, Evaluation of the atmospheric moisture and hydrological cycle in the NCEP/NCAR reanalysis. *Clim. Dynamics* 14(1998), 213-231, 1998

Van den Dool, H.M., W.H. Klein, and J.E. Walsh, The geographical distribution and seasonality of persistence in monthly mean air temperatures over the United States, *Mon. Weather Rev.*, 114:546-560, 1986

Viterbo P. and A.C.M. Beljaars, An improved land surface parameterization scheme in the ECMWF model and its validation, *J. Clim.* 8, 2716-2748, 1995

Viterbo, P. and A.K. Betts, Impact of the ECMWF reanalysis soil water on forecasts of the July 1993 Mississippi Flood, *J. Geophys. Res.*, 104(D16), 19,361-19,366, 1999

Widmann, M. and C.S. Bretherton, Validation of Mesoscale Precipitation in the NCEP reanalysis using a new grid-cell data set for the northwestern United States, *J. Clim.* (in press), 2000

Wigmosta, M.S., L.W. Vail and D.P. Lettenmaier, A distributed hydrology-vegetation model for complex terrain. *Water Resour. Res.* 30(6):1665-1679, 1994

Wood, E.F., D. Lettenmaier, X. Liang, B. Nijssen, and S.W. Wetzel, Hydrological modeling of continental-scale basins, *Ann. Rev. Earth Planet. Sci.*, 25, 279-300, 1997

Zhao, R.-J., L.-R. Fang, X.-R. Liu, and Q.-S. Zhang, The Xinanjiang model. In *Hydrological Forecasting* Proceedings, Oxford Symposium, IASH 129:351-356, 1980

Edwin P. Maurer, Dennis P. Lettenmaier, Greg M. O'Donnell, Department of Civil & Environmental Engineeing, Univ. of Washington, Box 352700, Seattle, WA 98195-2700

John O. Roads, Scripps Institution of Oceanography, Nierenberg Hall, 8605 La Jolla Shores Drive, La Jolla, CA 92037

Section 3

INTEGRATION OF OBSERVATION AND MODELING

The Effect of Sub-Grid Variability of Soil Moisture on the Simulation of Mesoscale Watershed Hydrology: A Case Study From the Southern Great Plains 1997 Hydrology Experiment

Karen I. Mohr[1] and James S. Famiglietti

Department of Geological Sciences, University of Texas at Austin

Aaron Boone

Météo-France/Centre National de Recherche Météorologique, Toulouse Cedex, France

The Parameterization for Land-Atmosphere-Cloud Exchange (PLACE), a general purpose land surface parameterization, was used in a case study of a 2500 km^2 area in southwestern Oklahoma July 9-16, 1997. The research objective was to assess how much incorporating variability of soil moisture at sub-grid scales (< 1 km) improves simulation results. Sub-grid variability was expressed as the coefficient of variation, the ratio of the standard deviation of soil moisture to the mean. Simulations with and without sub-grid variability were conducted. The effect of the coefficient of variation on the simulation of the spatial distribution of soil moisture was greatest for wet soils, diminishing over time during a dry down. For simulations with sub-grid variability, the non-linear relationship between soil moisture and evapotranspiration built into PLACE resulted in a significant improvement in the simulated spatial distribution of surface fluxes for both wetting and drying cycles. Using a coefficient of variation had little effect on the area-average soil moisture and surface fluxes, which may be an artifact of the maintenance of energy balance in a closed system.

1. INTRODUCTION

Through evaporation and transpiration, soil and plants, respectively, exchange heat and moisture with the atmosphere thereby influencing the growth of the turbulent atmospheric boundary layer. Soil moisture is directly involved in determining the magnitude of surface heat transfer and its partitioning into latent and sensible heating [*Anthes*, 1984; *Pan and Mahrt*, 1987; *Entekhabi et al.*, 1996]. Because of the many interactive ecological, atmospheric, hydrological, and anthropogenic processes influencing the spatial variability of soil moisture, the relationship between soil moisture and evapotranspiration is strongly non-linear, making scaling a significant issue among land surface modelers.

In numerical experiments that examined the effect of sub-grid variability in vegetation, precipitation, soil type, and meteorology, the value of area-average evapotranspiration varied greatly from more homogeneous to more heterogeneous simulations for *Wetzel and Chang* [1988], *Avissar* [1992], *Famiglietti and Wood* [1995], *Noilhan and Lacarrere* [1995],

[1]Now at the Department of Earth and Atmospheric Sciences, University at Albany, State University of New York, Albany, New York.

Land Surface Hydrology, Meteorology, and Climate: Observations and Modeling
Water Science and Application Volume 3, pages 161-176

and *Ghan et al.* [1997]. As heterogeneity increased, evapotranspiration increased for *Famiglietti and Wood* [1995], but it decreased for *Ghan et al.* [1997]. Although different models and different initial conditions may change the direction of the trend, the implication common to these sensitivity studies remains clear: increasing landscape complexity noticeably modifies model-derived heat transfer.

The objective of this validation study is to assess whether incorporating sub-grid variability of a dominant process control produces a meaningful, significant improvement in simulation results for a mesoscale study area. Since soil moisture variability integrates the effect of other process controls such as vegetation density and soil texture, a surface-vegetation-atmosphere (SVAT) model is used to run simulations with and without sub-grid variability of soil moisture during an 8-day period in the Little Washita watershed, Oklahoma during the Southern Great Plains 1997 Hydrology Experiment (SGP97). The SGP97 field campaign is the longest and largest remote sensing soil moisture mapping mission to date [*Jackson et al.*, 1999], taking place June 18, 1997 to July 16, 1997 in central Oklahoma (map in Figure 1). The analysis takes advantage of the rich data set available from SGP97 to evaluate the simulated evolution of the spatial and temporal variability of soil moisture and heat fluxes in the study area. Details on the SGP97 experiment and soil moisture sampling activities can be found in *Famiglietti et al.* [1999] and *Jackson et al.* [1999].

2. BACKGROUND

There are at least two common approaches for expressing spatial heterogeneity. Spatial heterogeneity can be treated explicitly by a mosaic or spatially distributed method in which grid cells are divided into a number of sub-regions of different surface characteristics [*Avissar and Pielke*, 1989; *Famiglietti and Wood*, 1994; *Wetzel and Boone*, 1995]. Alternatively, spatial heterogeneity can be treated implicitly by statistical methods. The statistical method applies an analytical or numerical solution to assumed distributions of variables in the process equations. To use this method, the user supplies statistical moments relevant to specified distribution (e.g., mean and standard deviation of the normal distribution) for each variable treated stochastically [*Wetzel and Chang*, 1988; *Entekhabi and Eagleson*, 1989; *Famiglietti and Wood*, 1991; *Famiglietti and Wood*, 1994; *Wetzel and Boone*, 1995; *Gusev et al.*, 1998]. The statistical method is preferable to the spatially distributed method for two reasons. It requires only information about the first two or three statistical moments rather than specific values for variables within the study area. It is computationally less intensive, requiring only a few extra calculations per time step rather than added iterations of prognostic equations. The SGP97 data set provided field observations of statistical moments for input and testing.

A number of authors [*Hills and Reynolds*, 1969; *Bell et al.*, 1980; *Hawley et al.*, 1983; *Francis et al.*, 1986; *Nyberg*, 1996; *Famiglietti et al.*, 1998] who undertook field studies of the distribution of soil moisture concluded that the distribution within their study areas was normal. *Loague* [1992] noted that his sampling strategy affected the resulting distribution. A linear transect resulted in a normal distribution whereas a grid did not. *Charpentier and Groffman* [1992] found that the distributions within 66×66 m remote sensing footprints were often positively skewed. During SGP97, volumetric soil moisture observations were taken nearly every day by impedance probes in 0.8×0.8 km grids and in 0.8 km linear transects at six different locations in the SGP97 region. From this data set, *Famiglietti et al.* [1999] showed that the distribution of soil moisture is non-normal under very wet (negatively skewed) and very dry (positively skewed) conditions, but normal in the mid-range of soil conditions. Hence, under a wide range of conditions, the normal distribution is a reasonable approximation for the distribution of soil moisture within an area.

Some field studies have also considered the relationship between the mean soil moisture and its standard deviation. *Hills and Reynolds* [1969], *Bell et al.* [1980], and *Famiglietti et al.* [1998] observed that the standard deviation generally decreases as the mean decreases. In contrast, *Hawley et al.* [1983] and *Charpentier and Groffman* [1992] detected no systematic variation of the standard deviation with the mean. In *Famiglietti et al.* [1999], the standard deviation increased with decreasing soil water content. The coefficient of variation (CV) is defined as the ratio of the standard deviation to the mean and can be considered the variability relative to the amount of soil water present. *Bell et al.* [1980], *Owe et al.* [1982], *Charpentier and Groffman* [1992], and *Famiglietti et al.* [1999] found an inverse relationship between the CV and the mean. For *Famiglietti et al.* [1999] this relationship was stronger than the relationship between the standard deviation and the mean. Equation 1 is the equation *Famiglietti et al.* [1999] derived between the CV and mean soil moisture from field data, representing a variety of soil types, land covers, and rainfall amounts, at all three ground sampling locations in SGP97 (Figure 1). Here, Θ is the mean volumetric soil moisture ($L^3\,L^{-3}$).

$$CV = 1.5 \exp(-7.43\,\Theta) \tag{1}$$

The Pearson correlation coefficient (R) for this expression is 0.83. The CV decreases sharply for volumetric soil moisture from 0.0 to 0.15 $cm^3\,cm^{-3}$ and then decreases slowly above 0.15 $cm^3\,cm^{-3}$. Because the range of mean soil moisture content is nearly six times greater than the range of the standard deviation, *Famiglietti et al.* [1999] attributed the decrease in CV more to increasing mean moisture content than to decreasing standard deviation.

3. DATA AND METHODS

3.1. Numerical Model

The numerical model used is *Wetzel and Boone*'s [1995] Parameterization for Land-Atmosphere-Cloud Exchange (PLACE) model because it is a general-purpose land surface parameterization designed to be coupled to atmospheric models. PLACE is broadly representative of the many land surface schemes available in the community today [*Henderson-Sellers et al.*, 1995]. PLACE consists of individual, linked modules parameterizing key components of surface and sub-surface water and energy exchange. Water and energy are transmitted through a vertical column consisting of an interception/dew reservoir, a plant storage reservoir, five soil moisture reservoirs, seven soil heat reservoirs and are exchanged with the atmosphere through turbulent sensible and latent heat fluxes. Details on the solution of PLACE's energy and water balance equations can be found in *Wetzel and Boone* [1995].

Following the parameterizations of *Wetzel and Chang* [1988], PLACE grid cells may be either vegetated or non-vegetated (bare soil or open water), and evapotranspiration may occur at either demand-limited (potential) or supply-limited (stressed) rates depending on soil water status. Latent and sensible heat fluxes are calculated by bulk aerodynamic formulae. Although PLACE has simple parameterizations for runoff and base flow, it is not intended for applications requiring detailed surface and sub-surface cell-to-cell horizontal water transport. PLACE emphasizes the interaction of the heterogeneous land surface with the overlying turbulent atmosphere through the vertical column, an appropriate strength for this study.

PLACE is capable of separate or simultaneous application of the statistical and spatially distributed methods of representing sub-grid variability, although the statistical method is used for simulations. For this research soil moisture is described by the normal distribution. In simulations with sub-grid variability (CV > 0), for each grid cell, PLACE multiplies the CV by the previous time step's mean soil moisture to derive the standard deviation. The standard deviation determines the difference between the lowest and highest values of the volumetric soil moisture within the grid cell. PLACE uses a "point model" in calculating evapotranspiration. It calculates the evapotranspiration at discrete values of soil moisture and then executes a weighted average to find the grid cell total evapotranspiration. Over time, this process may produce significant differences between simulations with and without sub-grid variability.

3.2. Study Environment

The study area, depicted in Figure 1, is a 50×50 km area surrounding the 610 km^2 Little Washita watershed in southwestern Oklahoma. The study time period is 192 h from July 9 to July 16, 1997. The gray lines within the SGP97 experiment area in Figure 1 are the lines of flight of the NASA P-3B aircraft fitted with the Electronically Scanned Thinned Array Radiometer (ESTAR), an L-Band (1.413 GHz) passive microwave sensor. Most days during SGP97, the P-3B aircraft over-flew the Little Washita at approximately 1600 Universal Time Coordinated (UTC) or 1000 Central Standard Time. The footprint of the raw brightness temperature data is 400 m, but the raw data were re-sampled to 800 m to derive soil moisture maps. Further details on the ESTAR instrument and the inversion of ESTAR brightness temperatures to volumetric soil moisture can be found in *Jackson et al.* [1995], *Jackson and LeVine* [1996], and *Jackson et al.* [1999]. During SGP97, 23 fields in the Little Washita were gravimetric soil moisture sampling sites providing ground truth for the ESTAR soil moisture estimates. ESTAR-derived 0–5 cm soil moisture estimates were within 3% of estimates of volumetric soil moisture (0–5 cm) from SGP97 ground samples [*Jackson et al.*, 1999]. Simulation results will be compared to both the ground observations and the ESTAR soil moisture maps.

As Figure 1 shows, the Little Washita is heavily instrumented, containing 4 meteorological towers of the Oklahoma Mesonet and a 5–10 km-resolution micronet (42 towers) of the U.S. Department of Agriculture Agricultural Research Service (USDA-ARS). Atmospheric boundary conditions to force simulations were obtained from these meteorological towers. Since rain gauges did not cover the entire study area, the radar-derived Stage III product (accumulated rainfall) available from the National Weather Service Arkansas-Red River Forecast Center was substituted. All simulations used a 5-minute time step, and meteorological data were provided at each time step for each grid cell. Continuous fields of radiation, pressure, air temperature (2 m), wind speed, and humidity observations were linearly interpolated (Cressman weighting) from the Mesonet and micronet observations. Because the Stage III product is 4-km and 1-hour resolution, rain gauges nearest to each Stage III cell were used to interpolate the temporal distribution of the Stage III rain amounts for each hour.

In a related study of the Little Washita using the same meteorological data set, *Mohr et al.* [2000] provide further detail on the processing. During the study time period, three rainfall events occurred on July 9, July 10–11, and July 15. The

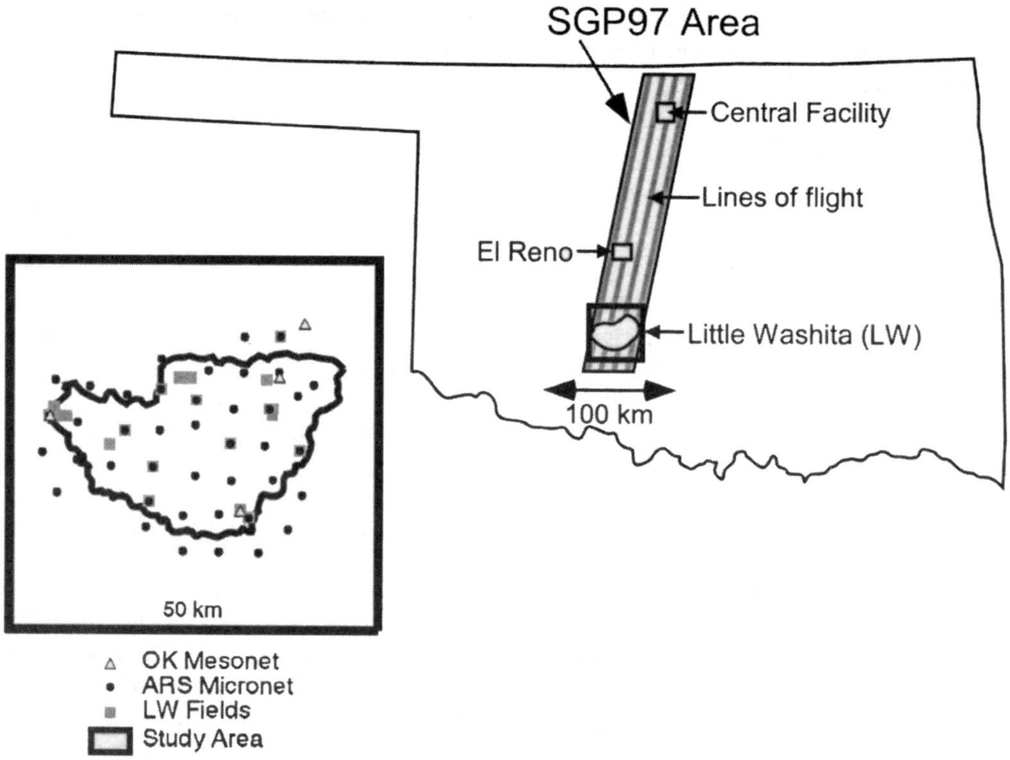

Figure 1. SGP97 ground sampling activities took place at the Central Facility, El Reno, and Little Washita watershed. In the study area (insert, left) surrounding the Little Washita, are 4 Oklahoma Mesonet sites (open triangle), 42 USDA-ARS micronet sites (black dot), and 23 gravimetric sampling fields in the Little Washita watershed (gray squares).

events on July 9 and July 15 were brief and confined to the western and southwestern sections of the study area. The July 10–11 event covered the entire watershed but, as Figure 2 shows, had a significant southwest to northeast gradient in accumulated rainfall.

3.3. Experimental Design

There are three sets of simulations in this study, and each set is comprised of three scenarios. Table 1 lists the three simulation sets and their scenarios. The first set is called the "FULL" set because it uses a 2500-cell (1-km resolution) grid of the study area and is run for the entire 8-day period. The FULL set isolates the effect of using a CV by comparing scenarios with and without a CV through several wetting and drying cycles. For the FULL set of scenarios, the initial profile of soil moisture was assigned by soil type, and based on (but not necessarily equal to) an average profile of fields with the same soil type on July 9. There was only one representative initial soil temperature profile assigned to all soil types. Because the rain event of July 10–11 thoroughly wets the upper to mid soil layers, the signal from the initial surface (i.e., 0–5 cm) soil moisture profile is eliminated early in the simulation period. Since soil moisture data are limited outside of field experiments such as SGP97, the simple, generalized initialization scheme of the FULL set of scenarios represents the common problem of limited soil moisture data. The U scenario has no sub-grid variability (CV = 0). The HC scenario uses a constant CV to represent a situation in which the relationship between the CV and the mean is not well known. The HP scenario represents a situation for which there has been previous field work by using the empirically derived Equation 1 to update the CV at each time step with the previous time step's mean soil moisture.

The second set of scenarios, the ESTAR set, uses the same 2500-cell grid and initializes and/or updates simulations with ESTAR-derived soil moisture to assess the effect of assimilating remote sensing data. Because there are no ESTAR maps for July 9 or July 10, these simulations start on July 11 at 1600 UTC, after the rainfall on July 10–11. The procedure used to initialize the soil temperature and sub-surface (> 5 cm) soil moisture profiles is the same as in the FULL set of

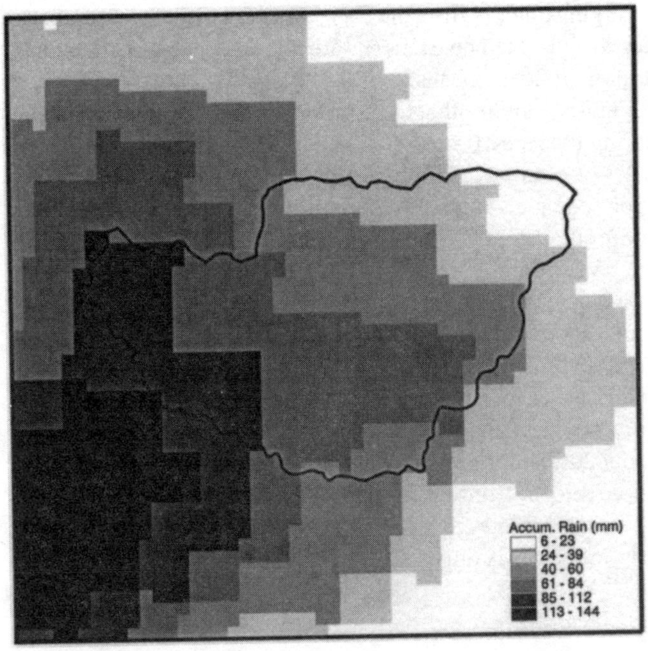

Figure 2. Accumulated rainfall (mm) for July 10–11 rain event.

scenarios, although more moisture was added. The third set of scenarios, the S set, use a single grid cell (50×50 km) to explore the issue of scaling by comparing S results to the averaged results of simulations with a 2500-cell (high-resolution) grid. Fifty kilometers is just under a half-degree, and the S set of scenarios can be compared to a grid cell within the mesh of a larger-scale weather prediction model. To a large-scale model, the Little Washita might appear to be homogenous grassland. The use of the CV in the ESTAR and S sets is similar to the FULL set.

In the high-resolution simulations, each grid cell was assigned a soil texture and a land cover type. Soil textures are based on the U. S. Geological Survey's State Soil Geographic Database (STATSGO). In Figure 3, five soil types occur within the study area with loam and sandy loam making up over half of the study area. Each soil type was assigned fixed percentages of sand/silt/clay from which PLACE calculated soil hydraulic properties using the empirical relations of *Cosby et al.* [1984]. Not shown is the land cover grid based on satellite data compiled by *Doraiswamy et al.* [1998] for the SGP97 time period. Rangeland is approximately 75% of the study area with small areas of crops, primarily in the western and northeastern sections of the study area, and surface water. Tables 2a-b list important parameters used in the simulations. For the HC scenario, the CVs in Table 2a are representative of each soil type but are not actual field observations. In the S (single cell) set of scenarios, loam soil and 70% grass cover were specified. Since there are no calibration curves relating the values in the Normalized Difference Vegetation Index (NDVI) imagery for SGP97 to specific values of percent cover by green vegetation, the assignments in Table 2b are only relative estimates of vegetation density.

4. RESULTS

4.1. Surface Soil Moisture

Figures 4a-b compare the ESTAR soil moisture maps and the model results for the U, HC, and HP scenarios for July 11, just after soaking rain, and July 14, after three days of drying. Compared to the ESTAR map for July 11, all of the model scenarios reproduce the general pattern of soil moisture, consistent with the rainfall gradient in Figure 2. However, by July 14, all of the model scenarios are noticeably wetter than the ESTAR map, particularly north and east of the watershed. The HC scenario more closely resembles the HP scenario on July 14 than on July 11. Comparing both the model and ESTAR soil moisture maps in Figure 4 to the soil texture map in Figure 3 reveals a strong control of soil texture over the soil moisture pattern. *Jackson et al.* [1995], *Hollenbeck et al.* [1996], and *Mattikalli et al.* [1998] have previously observed a similar control of soil texture over soil moisture in remotely sensed soil moisture images. Although this control is apparent in the ESTAR map, it is more obvious in the model-derived maps, particularly in the HC and HP maps.

The histograms in Figure 5 complement the maps in Figure 4. In general, the model scenario histograms are narrower, with peaks right-shifted from the peak of the ESTAR histogram. The model histograms further emphasize that the model scenarios have lower spatial variability than the ESTAR histogram. The driest (wettest) areas of the ESTAR map in Fig-

Table 1. Summary of Sets and Scenarios

Set	Scenario	Sub-grid Variability	Use of ESTAR
FULL	U	None	
2500 cell grid	HC	Constant value	
192 h	HP	Updated value	
ESTAR	EI	None	Initialized
2500 cell grid	EC	Constant value	Updated
128 h	EP	Updated value	Updated
S (Single Cell)	SU	None	
1 cell grid	SC	Constant value	
192 h	SP	Updated value	

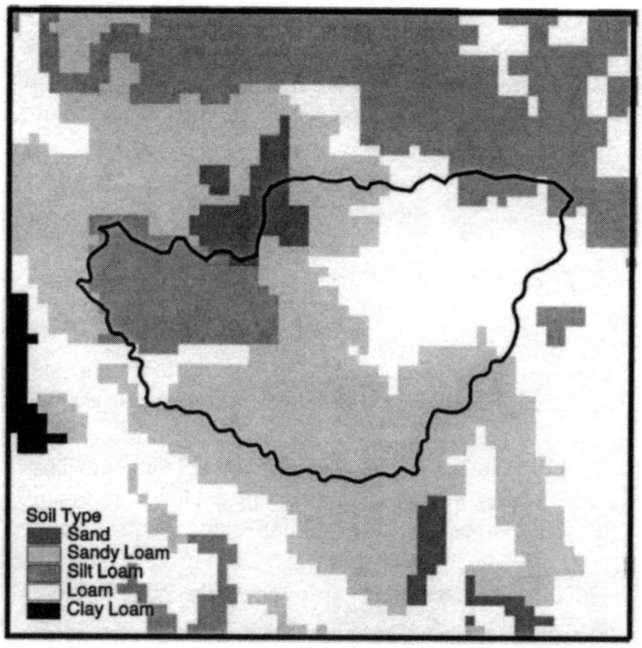

Figure 3. Map of soil textures used in simulations, adapted from U.S. Geological Survey's soils database (STATSGO).

ure 4 and histogram in Figure 5 are drier (wetter) than the model maps and histograms. On July 11, there are clear differences in the shape and height of the model histograms with the HP histogram most closely resembling the ESTAR histogram. Over the course of the dry down, the differences among scenarios diminish, and by July 16 their histograms are largely indistinguishable. From July 11 to July 16, the U histogram changes little, but the HP histogram changes significantly, narrowing and shifting right over time. The HC and HP histograms increasingly resemble a Gaussian curve, reflecting the use of the normal distribution to express sub-grid variability of soil moisture. Why differences among the model histograms decrease during a dry down is addressed in the discussion section.

The EI, EC, and EP scenarios begin on July 11 at 1600 UTC because the first ESTAR map occurs at this time. Because the EC and EP scenarios are updated with ESTAR data, a figure comparing their soil moisture maps is unnecessary. Since the EI scenario is not updated with ESTAR data, it is included in Figure 6a, a time series of study area mean 5-cm soil moisture. In Figure 6a, the means of the U, HC, and HP scenarios are wetter and change more slowly than the ESTAR time series, consistent with Figure 4. Although the U scenario is slightly wetter and the HP scenario slightly drier, there is only a 1–3% of volume difference among the means of the U, HC, and HP scenarios throughout the study period. The evolution of the EI scenario resembles the FULL set of scenarios, diverging quickly from the ESTAR time series after initialization on July 11. The EI means are no more than 3% of volume drier than the U means.

Figure 2 implies that the convective precipitation occurring during the study time period had high spatial variability. The S set of scenarios simplify simulation boundary conditions, applying an average rainfall, soil type, and land cover to the entire study area. Figure 6b compares the S set of scenarios to the ESTAR and U scenario time series, making clear the improvement the higher resolution grid brings to the case study. Throughout the study period, the soil moisture time series of the S set of scenarios are significantly wetter than the ESTAR and U scenario. Compared to the ESTAR, the percent difference between the SP/SC scenarios and the SU scenario is about 20%, implying that the SP/SC scenarios are only a 20% improvement from a very poor result. Similar to the HC and HP scenarios in Figure 6a, the difference between the SC and SP time series is only 0–2% of volume.

4.2. Surface Fluxes

Because there is nothing analogous to the ESTAR soil moisture maps for surface fluxes, Figure 7 compares the spatial variability of the latent heat fluxes from the model scenarios to each other on July 14 2000 UTC. The highest latent heat flux occurs over the reservoir on the western edge of the study area (see U grid), indicating high atmospheric demand during this time. Although the general pattern of the model latent heat fluxes closely follows the pattern of model soil moisture in Figure 4 with areas of wetter (drier) soil moisture corresponding to higher (lower) latent heat flux, the spatial distribution of latent heating reveals larger differences among the model scenarios than the soil moisture maps in Figure 4. The contiguous areas of latent heat flux less than 100 W m^{-2} are much larger in the U grid than in the HC grid. Only a few spots of latent heat flux less than 100 W m^{-2} occur in the HP grid.

The histograms of latent heat flux in Figure 8 further clarify differences among scenarios. Unlike the soil moisture histograms in Figure 5, these differences do not decrease with time. Even when the peaks of the histograms appear to coincide (e.g., July 12), there is noticeable variation in the shapes of the histograms. The EP and EC histograms tend to be the broadest, consistent with the high spatial variability of the ESTAR soil moisture updating their simulations. The scenarios with CV = 0, the EI and U scenarios, have the narrowest histograms. The difference in latent heat flux distribution between the EC and EP scenarios is smaller than the difference between the HC and HP scenarios and noticeably smaller than between the HP and U scenarios. After July 12, the HC and HP scenarios have more grid cells greater than 300 W m^{-2}

Table 2a. Specified Soil Characteristics

	Sand	Sandy Loam	Loam	Clay Loam	Silt Loam
CV (HC scenario)	0.20	0.20	0.25	0.20	0.32
Mean topographic deviation (m)	10	10	30	10	10
Mean hill-to-hill distance (m)	1000	1000	5000	1000	1000
Soil Albedo	0.23	0.22	0.18	0.16	0.20
Fraction Silt	0.00	0.30	0.45	0.55	0.65
Fraction Sand	0.95	0.60	0.35	0.10	0.20
Fraction Clay	0.05	0.10	0.20	0.35	0.15
Study Area %	4	34	37	1	24

The value of the CV for the SC scenario is 0.20.

even though the surface soil moisture of the U scenario (Figures 4, 6) tends to be higher. On July 16, HC and HP scenarios lack the bimodal distribution of the other scenarios resulting from the brief rainfall on July 15. Not only are the HC and HP histograms different from each other, they are significantly different from the other scenarios.

In Figures 7 and 8, there are significant differences in the latent heat flux distributions among scenarios. In Figures 9a-b, differences among the time series of average latent and sensible heat fluxes for the FULL set of scenarios are small, and there is little difference between the EP and EC scenarios. The difference between the latent heat fluxes of the HP and U scenario is less than 25 W m^{-2}, and the difference between EP and EC scenarios is less than 10 W m^{-2}. All scenarios are essentially closed systems in which energy balance must be maintained. Even if the spatial distributions of surface fluxes among scenarios vary, average latent heat flux can be expected to be similar for scenarios with the same boundary conditions.

More conspicuous is the 50–100 W m^{-2} gap between the ESTAR and FULL sets of scenarios. The ESTAR surface soil moisture maps tended to be drier for the sandy loam and loam soils constituting most of the study area. Also, the ESTAR set of scenarios was initiated after the July 10–11 rain event. The generalized initial soil moisture profiles for the bottom layers missed the spatial variability of sub-surface soil moisture that would result from infiltration after this rain event. Because the majority of root mass is specified to occur between 5–25 cm, the flux of water upwards from the deeper soil layers is critical for sustaining evapotranspiration. Hence, drier soil profiles in the ESTAR set of scenarios result in lower mean evapotranspiration and more grid cells (Figure 8) with latent heat fluxes below 200 W m^{-2} than the FULL set of scenarios. Because the evolution of the soil moisture in the EI scenario

Table 2b. Specified Vegetation Characteristics

	Bdlf & Ndlf Trees	Crops (high)	Crops (low)	Shrubs	Surface Water	Grass (high)	Grass (low)
Albedo	0.18	0.15	0.15	0.15	0.10	0.20	0.25
% Green Cover	0.90	0.80	0.10	0.90		0.70	0.30
Leaf Water Potential (m)	-200	-210	-210	-200		-200	-200
Minimum stomatal resistance (s/m)	110	70	70	110		110	110
Maximum stomatal resistance (s/m)	1700	1700	1700	1700		1700	1700
Constant surface roughness length (m)	1.5	0.2	0.0012	0.5	0.0	0.2	0.2
Minimum stomatal operating temperature (K)	270	278	278	283		283	283
Maximum stomatal operating temperature (K)	294	305	305	315		313	313
Cessation of stomatal operating temperature (K)	315	325	325	323		328	328
Leaf Area Index	4.6	5.0	0.7	4.6		2.5	1.0
Study Area %	1.2	1.5	8.2	0.6	0.4	74.5	13.6

The terms "high" and "low" refer to high or low percent green cover. Fields designated "low" have been harvested.

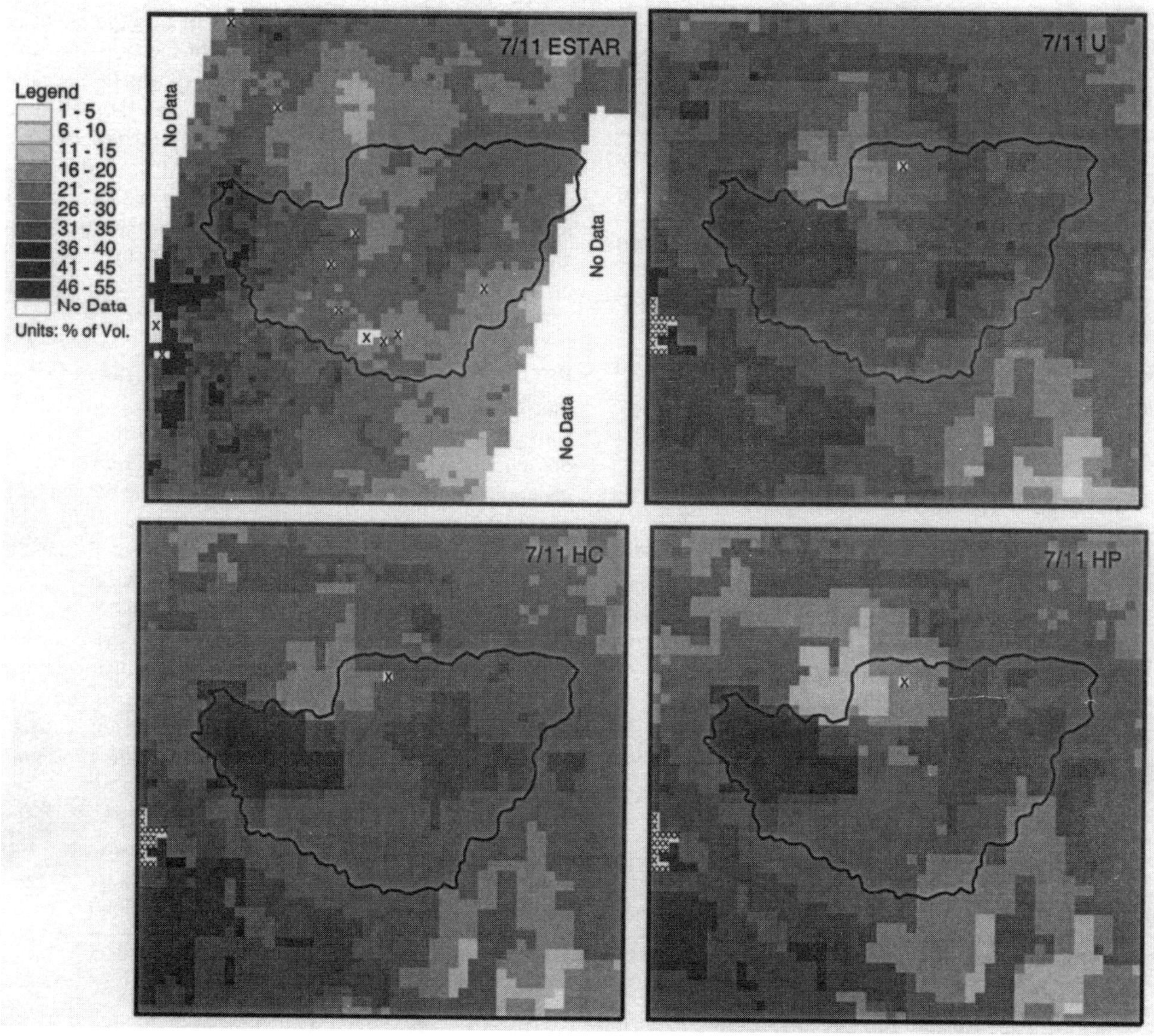

Figure 4a. ESTAR and FULL 5-cm soil moisture maps for 7/11.

is not corrected by assimilation of additional ESTAR data, the EI time series of latent heat flux drifts away from the EC and EP scenarios after July 12.

Figure 9c is a comparison of the time series of the U, EP, and S set of scenarios. As in the time series of soil moisture in Figure 6b, the time series of latent heat fluxes for the S set of scenarios suggests a wetter, cooler Little Washita than the scenarios with the high-resolution grid. There are greater differences among scenarios using a single cell (S) than among scenarios using the high resolution grid (FULL and ESTAR). For the first day after rainfall (i.e., July 12 and July 16), the latent heat flux of the SU scenario is 100 W m^{-2} higher than the SC scenario. During the driest days, there is almost no difference between the SU and SC/SP scenarios. As with the U and HC/HP scenarios, the SU soil moisture (Figure 6b) is higher than the SC/SP soil moisture throughout the study period.

Because there are no maps of observed surface fluxes, observed soil moisture will serve as a proxy for the actual latent heat flux in the study area. There is no suitable proxy covering the entire study area for comparison to model sensible heat flux. For each scenario, the Pearson correlation coefficient (R) between the ESTAR soil moisture and the corresponding model latent heat flux was calculated at each grid cell for 2000 UTC. The correlation between the accumulated rainfall (mm) and model latent heat flux was also calculated.

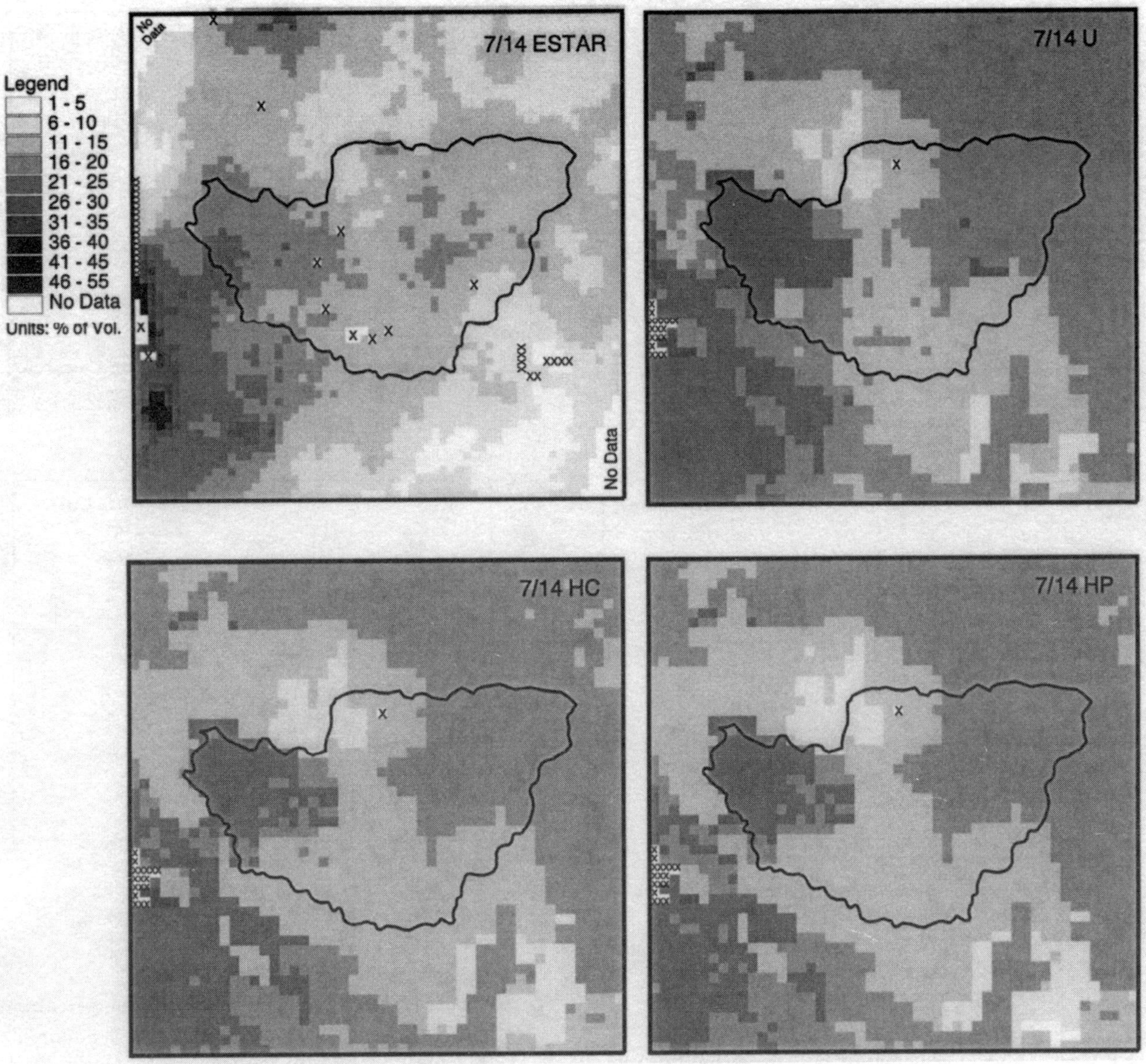

Figure 4b. ESTAR and FULL 5-cm soil moisture maps for 7/14.

The correlations are intended to provide insight into the accuracy of the simulated spatial distribution of surface fluxes in each scenario.

Figure 10a is a time series of the correlation to ESTAR soil moisture, and Figure 10b is the time series of the correlation to accumulated rainfall. Except for the HP scenario, the correlation to accumulated rainfall is higher than the correlation to ESTAR soil moisture. The correlation to accumulated rain is clearly higher for the U scenario but only slightly higher for the HC scenario. Despite the handicap of initialization after the July 10–11 rain event, the latent heat flux of the EC and EP scenarios have higher correlations to ESTAR soil moisture than the U scenario. Since ESTAR soil moisture reflects the rainfall that actually infiltrated, Figure 10a implies that scenarios with sub-grid variability, particularly the HP scenario, produced more realistic spatial distribution of latent heat flux than the scenarios without sub-grid variability. Comparing the three ESTAR scenarios, it appears the EP and EC scenarios benefited from assimilating ESTAR data. The EI scenario is multiply handicapped (simplified sub-surface soil moisture profile, CV = 0, no additional input of ESTAR data), such that the poor results in Figure 10 are not surprising.

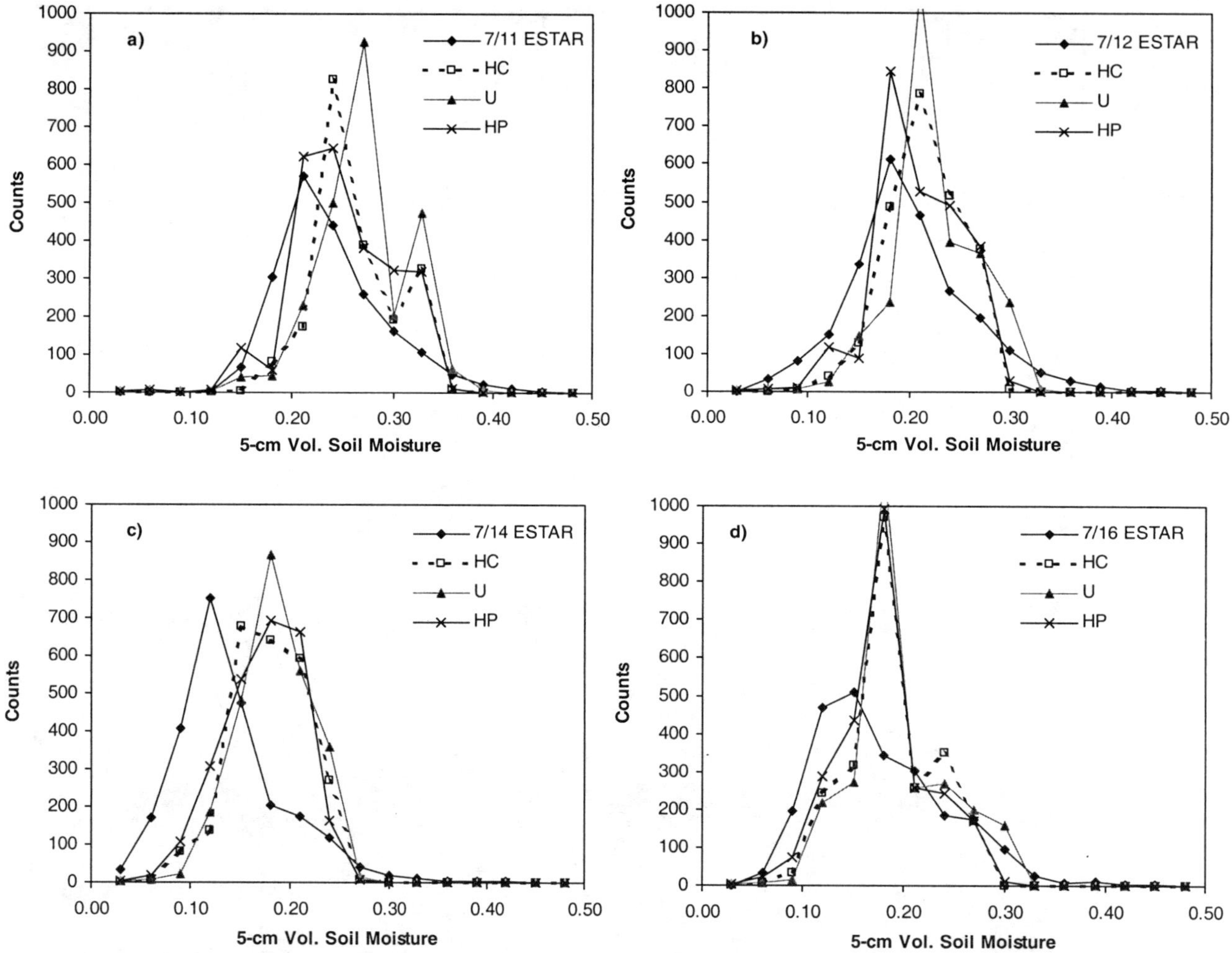

Figure 5. Histograms of surface soil moisture at 1600 UTC for the FULL set of scenarios.

5. DISCUSSION

The CV is involved in calculating both the evapotranspiration and mean soil moisture for scenarios with sub-grid variability (CV > 0). In Figure 5 by July 14, the U, HC, and HP surface soil moisture histograms look similar. Two factors limit differences between simulations with and without sub-grid variability. First, while PLACE accounts for the field-observed relationship that the standard deviation of soil moisture decreases at very low and very high values of mean soil moisture [*Hills and Reynolds*, 1969; *Bell et al.*, 1980; *Famiglietti et al.*, 1998], it forces the normal distribution to converge either as a saturation soil water content or a minimum soil water content (user-specified) is approached. Thus, sub-grid variability mainly affects intermediate values of soil moisture.

The second limiting factor is that PLACE calculates a new, normal soil moisture distribution from the CV at each time step, instead of tracking the changes in the same distribution over time. Consistent with the observations of *Famiglietti et al.* [1999], the ESTAR soil moisture histogram shifts from normal to positively skewed. Since evaporation rates decay exponentially after wetting, patches of wetter and drier soils evaporating at increasingly different rates contribute to the spatial variability of soil moisture and the shift in the soil moisture distribution. Late in a dry down, the simulated soil moisture distribution remains normal rather than becoming positively skewed, resulting in a higher mean and a smaller difference between the highest and lowest values of soil moisture both within and between grid cells (*c.f.* Figure 5).

Despite these constraints, the non-linear relationship built into PLACE between soil moisture and evapotranspiration magnifies small differences in soil moisture among scenarios spatially as long as sufficient deep soil moisture exists to meet atmospheric demand. Throughout the case study, the spatial distribution of latent heat fluxes evolved very differently be-

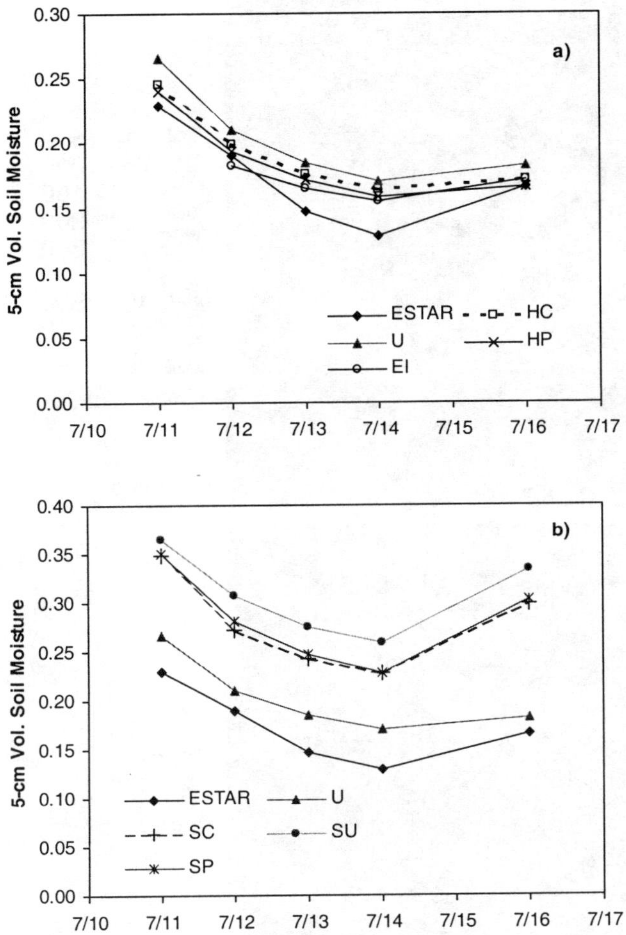

Figure 6. Time series of area-average soil moisture, a) vs. ESTAR, b) S set of scenarios vs. ESTAR.

tween scenarios with and without sub-grid variability. In Figure 8, for July 11 (not shown) and July 12, evapotranspiration is higher in many of the grid cells in the U scenario compared to the HC and HP scenarios. By July 13, this relationship reverses, and more evapotranspiration occurs in the HC and HP scenarios (Figures 7 and 8). Even when the standard deviation of soil moisture decreases significantly, calculation of evapotranspiration in the HC and HP scenarios is still based on a range of values, many of which after July 13 are higher than the single value used to calculate evapotranspiration in the U scenario. Because the EP and EC scenarios began after 10 or more hours of drying and were updated on July 12, there was much less time for these scenarios, compared to the FULL set scenarios, to evolve differently from one another.

The differences between the HC and HP scenarios were much smaller than the differences between either of these scenarios and the U scenario. Equation 1 implies that the CV changes slowly for mean soil moisture above about 15% of volume. For much of the study period in much of the study area, the model soil moisture was near or above this value (Figure 4), so that updating the CV provided limited benefit to the HP scenario results. Incorporating a CV and assimilating additional ESTAR data, clearly benefited the EC and EP scenarios. Even with a simplified sub-surface soil moisture initialization, they had a more realistic pattern of latent heat flux than the U or EI scenarios (*c.f.* Figure 10). The latent heat flux in the EC and EP scenarios had a lower correlation to ESTAR soil moisture than the HC and HP scenarios. It is likely that the poorer simulation of sub-surface soil moisture, an important process control on evapotranspiration [*Mohr et al.*, 2000], in the EC and EP scenarios was responsible for this gap. Unfortunately, the ESTAR set of scenarios could not be initialized on July 9 or even July 10 for a more complete evaluation of soil moisture data assimilation.

For a 2×15 km study area of moderate topography in the Kansas prairie, *Sellers et al.* [1995] and *Sellers et al.* [1997] suggested that process controls such as topographic slope, vegetation parameters, and soil wetness scale linearly from $O(10\text{ km})$ and that simple averages can be used to run SVATs over a wide range of scales. They argued that the reduction of variability during soil drying reduces the impact of non-linear relationships among process controls on area-averaged surface fluxes. In contrast, for an 11.7 km^2 watershed in the same area of Kansas, *Famiglietti and Wood* [1995] and *Wood* [1997] found significant error in simulations using a lumped-model (i.e., spatially averaged process controls). They showed that above a spatial scale of $O(1\text{ km})$ simulations with statistical distribution functions of dominant process controls (e.g., sub-surface soil moisture, vegetation parameters) produced area-averaged evapotranspiration comparable to spatially explicit simulations.

In the S set of scenarios, environmental conditions were averaged over the study area, and the resulting time series (Figures 6b and 9c) differed significantly from the time series of area-average results from the high-resolution scenarios. The evolution of the area-average soil moisture and surface fluxes in the high-resolution scenarios was a far better reflection of the meteorological conditions (i.e., rainfall, atmospheric demand) than the single cell S set of scenarios. *Famiglietti and Wood* [1995] and a study of the Sahel by *Taylor et al.* [1997] concluded that in areas with high spatial variability of rainfall and soil moisture, simulations must account for this spatial variability either explicitly or statistically.

6. CONCLUSION

In this study, PLACE, a general-purpose land surface parameterization, was used to conduct a case study of a 2500 km^2 area around the Little Washita watershed in southwestern

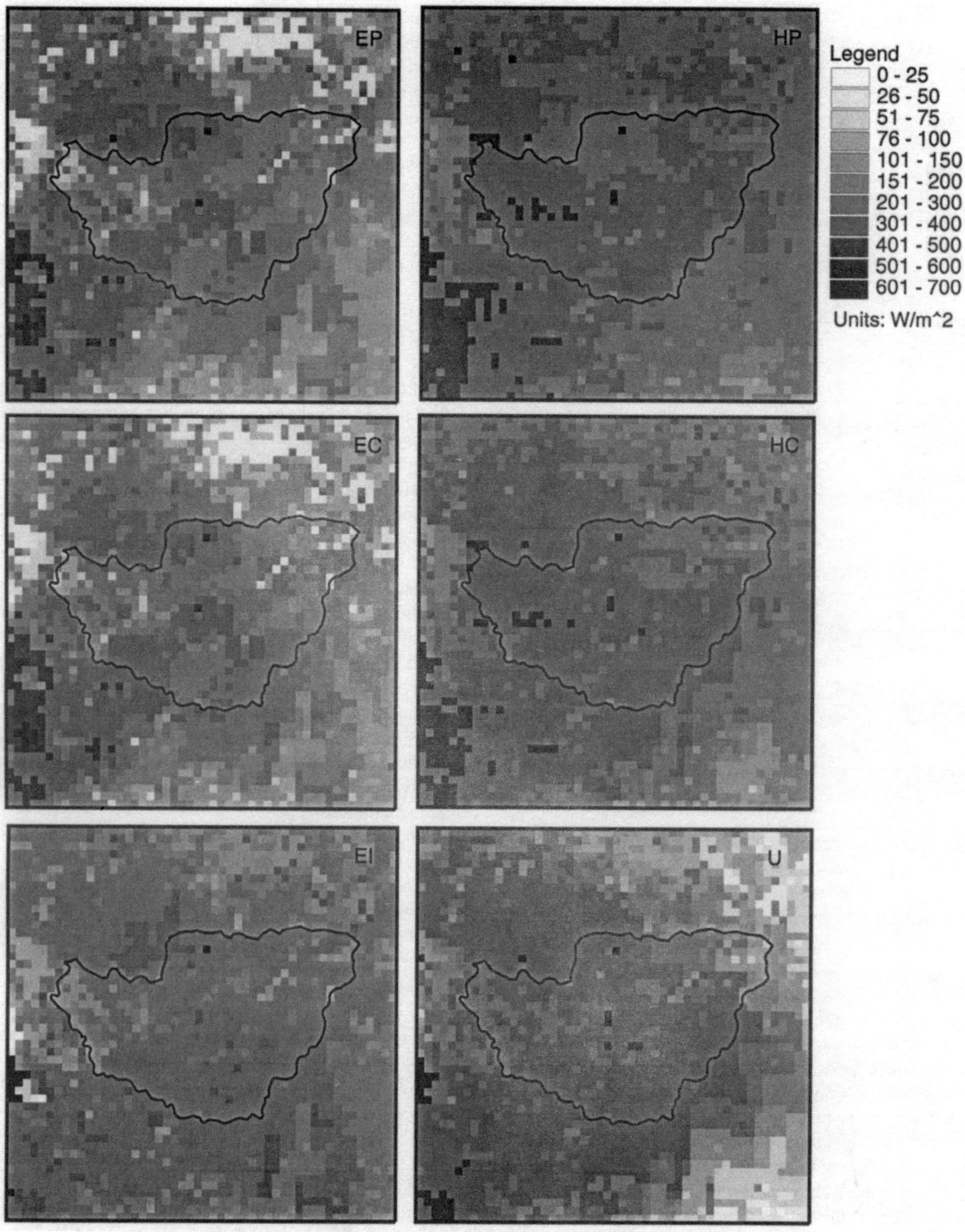

Figure 7. 7/14 2000 UTC model latent heat fluxes.

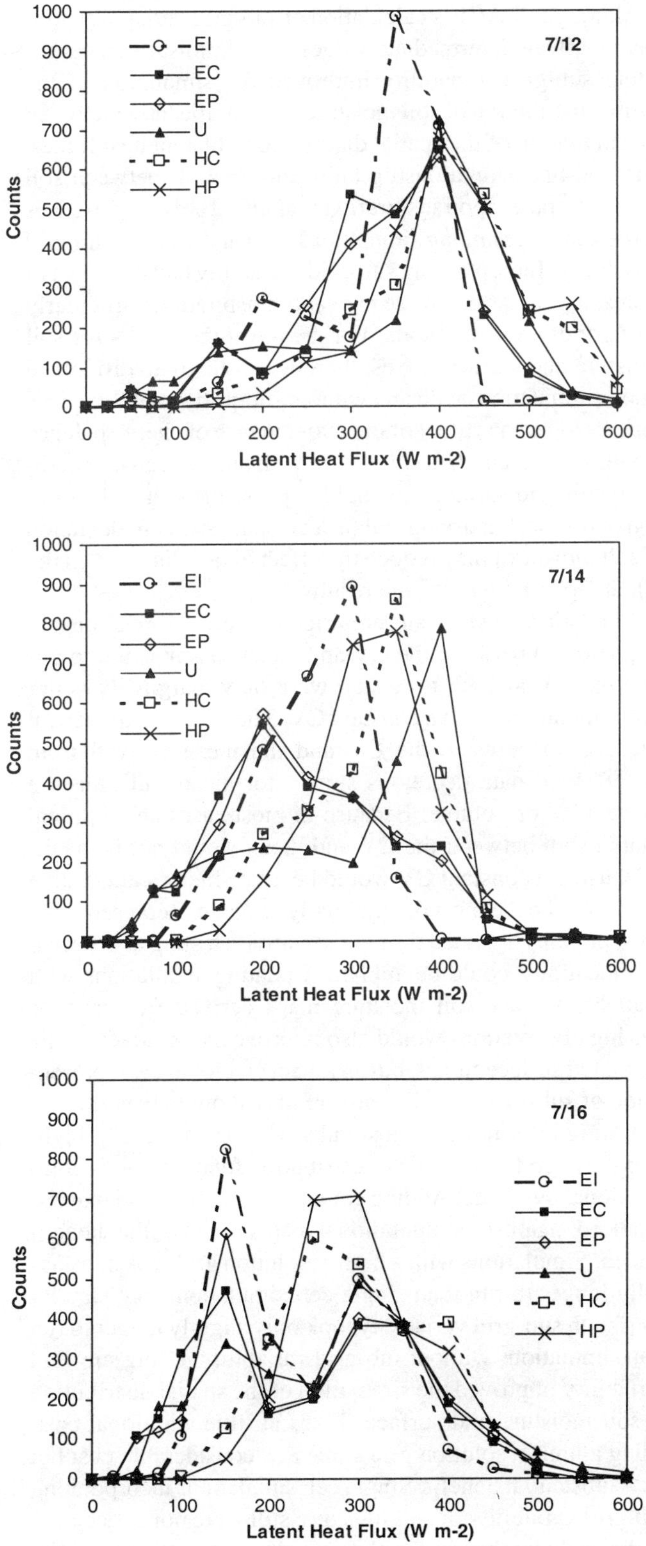

Figure 8. Histograms of latent heat fluxes at 2000 UTC.

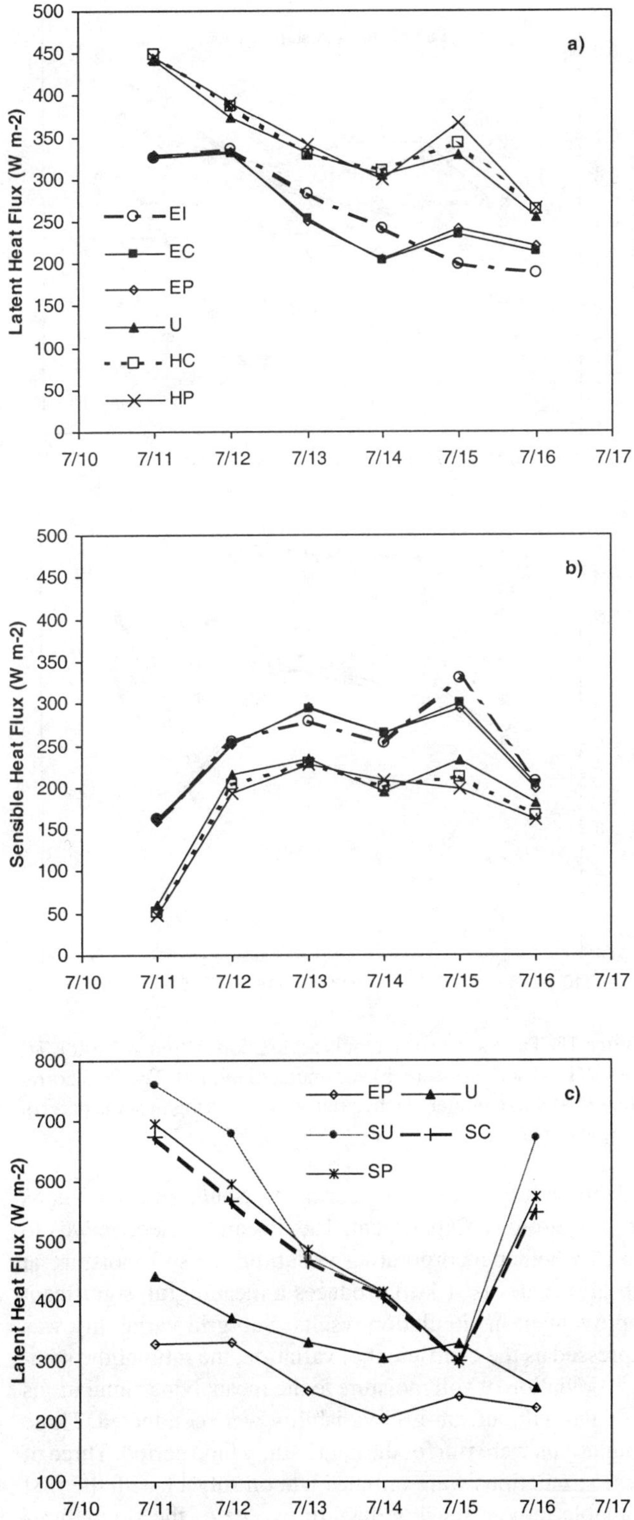

Figure 9. Time series of area-average surface fluxes, a) latent heat flux, b) sensible heat flux, c) latent heat flux for S set of scenarios.

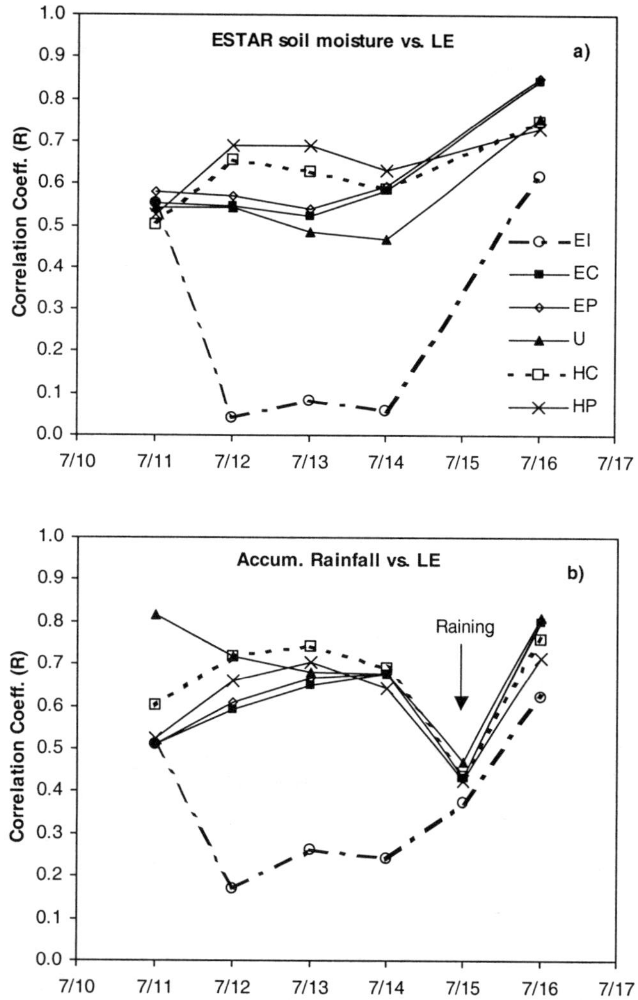

Figure 10. Time series of latent heat flux correlation at 2000 UTC to a) ESTAR soil moisture, b) accumulated rainfall. The 7/15 correlation in b) was calculated when rain was occurring in some parts of the study area.

Oklahoma July 9–16, 1997, during the Southern Great Plains 1997 Hydrology Experiment. The research objective was to assess whether incorporating variability of soil moisture at sub-grid scales (< 1 km) produces a meaningful, significant improvement in simulation results. Sub-grid variability was expressed as the coefficient of variation, the ratio of the standard deviation of soil moisture to the mean. Nine simulations with and without sub-grid variability were conducted. Three simulations were run for the entire study time period. Three of these simulations were initiated late on July 11 with the first available remote sensing data. In two cases the simulations were updated with additional remote sensing data. Three simulations used a single grid cell for comparison to the average results for simulations with a high-resolution grid.

Although PLACE's calculation of sub-grid variability principally affects intermediate values of soil moisture, incorporating sub-grid variability improved the simulation of the spatial distribution of soil moisture and significantly improved the simulation of the spatial distribution of latent heat fluxes. The non-linear relationship built into PLACE between soil moisture and evapotranspiration magnified small differences in soil moisture among simulations with and without sub-grid variability. Incorporating sub-grid variability had the effect of increasing the grid cell mean evapotranspiration particularly for very dry and wet soils. For area-average results for soil moisture and surface fluxes, there was little to no difference among simulations with and without sub-grid variability. This appears to be an artifact of the maintenance of energy balance in a closed system. In a simulation with open boundary conditions, sub-grid variability should improve the spatial distribution of results, but drying soil (a decreasing standard deviation of soil moisture) may reduce the effect of non-linearity in the calculation of area-average results.

Simulations using an empirically derived relationship (Equation 1) between the CV and mean soil moisture to update the CV at each time step were only marginally better than simulations with a constant CV. The inverse exponential relationship between the CV and the mean derived from SGP97 field data decreases slowly for mean soil moisture above 15% of volume. Because in most cases an empirical relationship between the CV and mean would not be available, using a constant CV would be a satisfactory alternative and would be the most operationally realistic. Representative CVs, perhaps obtained through sensitivity testing at observation locations, could be applied. Updating simulations with available surface soil moisture maps derived from remote sensing observations would also improve the accuracy of the simulated surface fluxes, but any inaccuracies in the representation of sub-surface soil moisture may limit its benefit.

Adding information to a simulation in the form of a high-resolution grid or statistical distribution functions adds computational overhead. Adding computational overhead must be balanced against computational capacity and the benefit gained. Simulations with a high-resolution grid took substantially longer to run than single cell simulations, but simulations with sub-grid variability took only slightly longer to run than simulations without sub-grid variability. Using sub-grid variability improved the simulation of the spatial distribution of soil moisture and surface fluxes at little additional cost. Using a high-resolution grid came at a considerable cost but for a substantial benefit. Single cell simulations incorporating sub-grid variability of soil moisture still were not as accurate as the high-resolution simulations. As computing resources evolve and land surface databases improve, higher resolution global and regional simulations can be run, making smaller the scale at which statistical distribution functions should be incorporated.

Although PLACE's (and similar SVATs) method of calculating and applying sub-grid variability constrains the effect of field-scale sub-grid variability, further study is warranted. An obvious application would be studies requiring realistic gradients of surface heating for the generation of mesoscale circulations and cumulus convection [*Clark and Arritt*, 1995; *Hong et al.*, 1995; *Lynn et al.*, 1995; *Avissar and Liu*, 1996; *Lynn et al.*, 1998]. For this application in particular, accurate expression of field-scale sub-grid variability in a demand-limited environment would be beneficial.

Acknowledgments. Necessary data were obtained from the SGP97 field campaign, NWS Arkansas-Red River Forecast Center, Penn State Earth System Science Center, Oklahoma Mesonet, and USDA-ARS LW micronet. Thanks to Tom Jackson (USDA Hydrology Lab), Binayak Mohanty (USDA Salinity Lab), Johanna Devereaux (UTDoGS), Corinna Prietzch (UTDoGS), and Seann Reed (UTCE) for advice and assistance. The manuscript also benefited from our discussions with Dave Baker (NASA-GSFC) and Rolf Reichle (MIT). This research was funded by NSF Grant 9454098, NASA Grant NAG5-6395, and the University of Texas Geology Foundation.

REFERENCES

Anthes, R.A., Enhancement of convective precipitation by mesoscale variations in vegetative covering in semiarid regions, *J. Clim. Appl. Meteorol.*, *23*, 541-554, 1984.

Avissar, R., Conceptual aspects of a statistical-dynamical approach to represent landscape subgrid-scale heterogeneities in atmospheric models, *J. Geophys. Res.*, *97*, 2729-2742, 1992.

Avissar, R., and Y. Liu, Three-dimensional numerical study of shallow convective clouds and precipitation induced by land surface forcing, *J. Geophys. Res.*, *101*, 7499-7518, 1996.

Avissar, R., and R.A. Pielke, A parameterization of heterogeneous land surface for atmospheric numerical models and its impact on regional models, *Mon. Weather Rev.*, *117*, 2113-2136, 1989.

Bell, K.R., B.J. Blanchard, T.J. Schmugge, and M.W. Witczak, Analysis of surface moisture variations within large field sites, *Water Resour. Res.*, *16*, 796-810, 1980.

Charpentier, M.A., and P.M. Groffman, Soil moisture variability within remote sensing pixels, *J. Geophys. Res.*, *97*, 18987-18995, 1992.

Clark, C.A., and R.W. Arritt, Numerical simulations on the effect of soil moisture and vegetation cover on the development of deep convection, *J. Appl. Meteorol.*, *34*, 2029-2045, 1995.

Cosby, B.J., G.M. Hornberger, R.B. Clapp, and T.R. Ginn, A statistical exploration of the relationships of soil moisture characteristics to the physical properties of soils, *Water Resour. Res.*, *20*, 682-690, 1984.

Doraiswamy, P., A.J. Stern, and P.W. Cook, Classification techniques for mapping biophysical parameters in the U.S. Southern Great Plains, *Proceedings of the International Geoscience Remote Sensing Symposium*, IEEE Cat. No. 007803-4403, 862-865, 1998.

Entekhabi, D., and P. Eagleson, Land surface hydrology parameterization for atmospheric general circulation models including subgrid scale spatial variability, *J. Clim.*, *2*, 816-831, 1989.

Entekhabi, D., I. Rodriguez-Iturbe, and F. Castelli, Mutual interaction of soil moisture state and atmospheric processes, *J. Hydrol.*, *184*, 3-17, 1996.

Famiglietti, J.S., J.A. Devereaux, C.A. Laymon, T. Tsegaye, P.R. Houser, T.J. Jackson, S.T. Graham, M. Rodell, and P.J.V. Oevelen, Ground-based investigation of soil moisture variability within remote sensing footprints during the Southern Great Plains 1997 (SGP97) Hydrology Experiment, *Water Resour. Res.*, *35*, 1839-1852, 1999.

Famiglietti, J.S., J.W. Rudnicki, and M. Rodell, Variability in surface moisture content along a hillslope transect: Rattlesnake Hill, Texas, *J. Hydrol.*, *210*, 259-281, 1998.

Famiglietti, J.S., and E.F. Wood, Evapotranspiration and runoff from large land areas: Land surface hydrology for atmospheric general circulation models, *Surv. Geophys.*, *12*, 179-204, 1991.

Famiglietti, J.S., and E.F. Wood, Multiscale modeling of spatially variable water and energy balance processes, *Water Resour. Res.*, *30*, 3061-3078, 1994.

Famiglietti, J.S., and E.F. Wood, Effects of spatial variability and scale on areally averaged evapotranspiration, *Water Resour. Res.*, *31*, 699-712, 1995.

Francis, C.F., J.B. Thornes, A.R. Diaz, F.L. Bermudez, and G.C. Fisher, Topographic control of soil moisture, vegetation cover and land degradation in a moisture stressed Mediterranean environment, *Catena*, *13*, 211-225, 1986.

Ghan, S.J., J.C. Liljegren, W.J. Shaw, J.H. Hubbe, and J.C. Doran, Influence of subgrid variability on surface hydrology, *J. Clim.*, *10*, 3157-3166, 1997.

Gusev, Y.M., O.Y. Busarova, and O.N. Nasonova, Modelling soil water dynamics and evapotranspiration for heterogeneous surfaces of the steppe and forest-steppe zones on a regional scale, *J. Hydrol.*, *206*, 281-297, 1998.

Hawley, M.E., T.J. Jackson, and R.H. McCuen, Surface soil moisture variation on small agricultural watersheds, *J. Hydrol.*, *62*, 179-200, 1983.

Henderson-Sellers, A., A.J. Pitman, P.K. Love, P. Irannejad, and T.H. Chen, The Project for Intercomparison of Land Surface Parameterization Schemes (PILPS): Phases 2 and 3, *Bull. Am. Meteorol. Soc.*, *76*, 489-503, 1995.

Hills, T.C., and S.G. Reynolds, Illustrations of soil moisture variability in selected areas and plots of different sizes, *J. Hydrol.*, *8*, 27-47, 1969.

Hollenbeck, K.J., T.J. Schmugge, G.M. Hornberger, and J.R. Wang, Identifying soil hydraulic heterogeneity by detection of relative change in passive microwave remote sensing observations, *Water Resour. Res.*, *32*, 139-148, 1996.

Hong, X., M.J. Leach, and S. Raman, A sensitivity study of convective cloud formation by vegetation forcing with different atmospheric conditions, *J. Appl. Meteorol.*, *34*, 2008-2028, 1995.

Jackson, T.J., and D.M. LeVine, Mapping soil moisture using an aircraft-based passive microwave instrument: algorithm and example, *J. Hydrol.*, *184*, 85-99, 1996.

Jackson, T.J., D.M. LeVine, A.Y. Hsu, A. Oldak, P.J. Starks, C.T. Swift, J.D. Isham, and M. Haken, Soil moisture mapping at regional scales using microwave radiometry: The Southern Great Plains Hydrology Experiment, *IEEE Trans. Geosci. Rem. Sens.*, *37*, 2136-2151, 1999.

Jackson, T.J., D.M. LeVine, C.T. Swift, T.J. Schmugge, and F.R. Schiebe, Large area mapping of soil moisture using the ESTAR passive microwave radiometer in Washita '92, *Remote Sens. Environ.*, *53*, 27-37, 1995.

Loague, K., Soil water content at R-5. Part 1. Spatial and temporal variability, *J. Hydrol.*, *139*, 233-251, 1992.

Lynn, B.H., D. Rind, and R. Avissar, The importance of mesoscale circulations generated by subgrid-scale landscape heterogeneities in general circulation models, *J. Clim.*, *8*, 191-205, 1995.

Lynn, B.H., W.-K. Tao, and P.J. Wetzel, A study of landscape generated deep moist convection, *Mon. Weather Rev.*, *126*, 928-942, 1998.

Mattikalli, N.M., E.T. Engman, T.J. Jackson, and L.R. Ahuja, Microwave remote sensing of temporal variations of brightness temperature and near-surface soil water content during a watershed-scale field experiment, and its application to the estimation of soil physical properties, *Water Resour. Res.*, *34*, 2289-2299, 1998.

Mohr, K.I., J.S. Famiglietti, A. Boone, and P.J. Starks, Modeling soil moisture and surface flux variability with an untuned land surface scheme: A case study from the Southern Great Plains 1997 Hydrology Experiment, *J. Hydrometeorol.*, *1*, 154-169, 2000.

Noilhan, J., and P. Lacarrere, GCM grid-scale evaporation from mesoscale modeling, *J. Clim.*, *8*, 206-223, 1995.

Nyberg, L., Spatial variability of water content in the covered catchment at Gardsjon, Sweden, *Hydrol. Process.*, *10*, 890-903, 1996.

Owe, M., E.B. Jones, and T.J. Schmugge, Soil moisture variation patterns observed in Hand County, South Dakota, *Water Resour. Bull.*, *18*, 949-954, 1982.

Pan, H.-L., and L. Mahrt, Interaction between soil hydrology and boundary-layer development, *Boundary Layer Meteorol.*, *38*, 185-202, 1987.

Sellers, P.J., M.D. Heiser, F.G. Hall, S.J. Goetz, D.E. Strebel, S.B. Verma, R.L. Desjardins, P.M. Schuepp, and J.I. MacPherson, Effects of spatial variability in topography, vegetation cover and soil moisture on area-averaged surface fluxes: A case study using the FIFE 1989 data, *J. Geophys. Res.*, *100*, 25607-25629, 1995.

Sellers, P.J., M.D. Heiser, F.G. Hall, S.B. Verma, R.L. Desjardins, P.M. Schuepp, and J.I. MacPherson, The impact of using area-averaged land surface properties-topography, vegetation condition, soil wetness- in calculations of intermediate scale (approximately 10 km2) surface-atmosphere heat and moisture fluxes, *J. Hydrol.*, *190*, 269-301, 1997.

Taylor, C.M., R.J. Harding, A.J. Thorpe, and P. Bessemoulin, A mesoscale simulation of land surface heterogeneity from HAPEX-Sahel, *J. Hydrol.*, *189*, 1040-1066, 1997.

Wetzel, P.J., and A. Boone, A Parameterization for Land-Atmosphere-Cloud Exchange (PLACE): Documentation and testing of a detailed process model of the partly cloudy boundary layer over heterogeneous land, *J. Clim.*, *8*, 1810-1837, 1995.

Wetzel, P.J., and J.-T. Chang, Evapotranspiration from nonuniform surfaces: A first approach for short-term numerical weather prediction, *Mon. Weather Rev.*, *116*, 600-621, 1988.

Wood, E.F., Effects of soil moisture aggregation on surface evaporative fluxes, *J. Hydrol.*, *190*, 397-412, 1997.

K.I. Mohr, Department of Earth and Atmospheric Sciences, University at Albany, SUNY, Albany, NY 12222.

J.S. Famiglietti, Department of Geological Sciences, University of Texas at Austin, Austin, TX 78712

A. Boone, Meteo-France,CNRM/GMME/MC2, 42 Ave. Coriolis, 31057 Toulouse Cedex, France

Assimilation of fAPAR and Surface Temperature Into a Land Surface and Vegetation Model

Wolfgang Knorr

Max Planck Institute for Biogeochemistry, Jena, Germany

Venkataraman Lakshmi

University of South Carolina, Columbia, South Carolina

Land surface hydrological processes and vegetation activity form an intricately coupled system that determines the energy and water balance over most of the earth's land surfaces. Therefore, a coupled land surface and vegetation model is particularly suited to simulate hydrological processes at the global scale. This is important since hydrological models at large spatial scales usually suffer from a general lack of data, both for model input and for parameterization. One advantage of such a comprehensive model is, therefore, that it can exploit information that is not directly related to soil moisture, such as vegetation cover or surface skin temperature, information that can rather easily be derived from satellite remote sensing data. In the present study, two assimilation techniques for satellite data are presented with the process based Biosphere Energy-Transfer Hydrology scheme (BETHY): one for vegetation cover, expressed as the fraction of photosynthetically active radiation (fAPAR), and one for surface skin temperature. Both assimilation techniques lead to a significant increase in model-estimated plant-available soil moisture content for most regions, especially in the tropics. It is shown that the fAPAR assimilation generally improves agreement of simulated and observed surface temperatures, as the surface temperature assimilation increases agreement between model and satellite derived vegetation cover. Through this analysis, and through comparison with some field data from the Amazon rainforest, it is concluded that present vegetation and land surface models usually underestimate soil water storage capacity and rooting depth in most tropical and semi-arid environments. The study is intended as a first step towards comprehensive use of satellite data for large-scale hydrological modeling.

1. INTRODUCTION

A critical factor in both regional and global land surface modeling is the availability of spatially resolved input data. These include soil properties, vegetation cover, and variables referring directly to the water and

Land Surface Hydrology, Meteorology, and Climate: Observations and Modeling
Water Science and Application Volume 3, pages 177-200

energy budget, namely temperature and soil moisture. For some time now, various data sets have been available for characterizing land surfaces, such as soils databases (CONUS-SOIL, STATSGO) [*Miller and White,* 1998], and various satellite-based data sets derived from the NOAA/AVHRR sensor, namely the Global Vegetation Index (GVI) [*Gutman et al.,* 1995], the 1-km satellite archive by the International Geosphere Biosphere Program (IGBP) [*Townshend et al.,* 1994], and various data sets from the International Satellite Land Surface Climatology Project (ISLSCP) [*Sellers et al.,* 1995]. Land surface temperature can further be derived from a number of instruments on board polar orbiting satellites, e.g. AVHRR (Advanced Very High Resolution Radiometer) [*Price,* 1984], and TOVS (Tiros Operational Vertical Sounder) [*Susskind et al.,* 1997], and geostationary satellites, e.g. GOES imager (Geostationary Orbiting Earth Satellite) [*Diak,* 1990].

Soil moisture is a particularly important variable in the water and energy budget of land surfaces, such that lack of spatially explicit data of this variable makes the validation of hydrological models very difficult. The widely used approach to compare modeled streamflows at catchment outlets with observations does not offer a satisfactory solution to this problem, because agreement of the two does not insure the correctness of the energy budget. For example, errors in infiltration vs. evapotranspiration could balance each other to produce correct runoff data, while soil moisture and partitioning between sensible and latent heat fluxes could still be incorrect. The inclusion of other data sets to validate individual components of the water and energy budget in a spatially explicit fashion is therefore highly desirable if realism and robustness of hydrological models is to be assured. Both on a regional and global scale, such data can best be provided from satellite measurements.

Two types of satellite data are of particular interest in this context, because they can be directly related to variables routinely calculated in land surface descriptions within hydrological or climate models, at least for those advanced models that include a vegetation description: land surface temperatures and vegetation indices. Satellite observed temperatures provide the necessary spatial and diurnal coverage that make them a good indicator of moisture status. While surface temperature directly influences evapotranspiration through its impact on the saturated vapor pressure, its diurnal range varies significantly with the partitioning between sensible and latent heat, which in turn is connected to soil moisture availability and state of vegetation cover.

The land cover fraction of active, photosynthesizing vegetation, which can be derived from the contrast in reflectance between the visible and the near-infrared solar spectra [*Verstraete,* 1994], can also be used as an indicator of soil moisture status and energy partitioning. Since water is limiting plant growth for most parts of the earth, plants are forced to maximize control over land evapotranspiration through rooting strategies and regulation of transpiration through their stomata. This creates a close coupling between vegetation activity and the energy and water budget.

It is now widely recognized that land surface processes have a significant impact on the global climate [e.g., *Budyko,* 1956; *Geiger,* 1965; *Budyko,* 1974; *Mintz,* 1984; *Garratt,* 1993], acting through evapotranspiration [*Shukla and Mintz,* 1982], water-holding capacity [*Milly and Dunne,* 1994], albedo [*Charney et al.,* 1977] and surface roughness [*Sud et al.,* 1988]. Many of those parameters are influenced by the vegetation cover, which itself is largely determined by the climate [e.g, *Holdridge,* 1947; *Box,* 1981]. Capturing the state and evolution of both terrestrial vegetation and soil moisture are, therefore, not only closely interrelated, but also of primary interest for hydrological modeling and for accurate simulations with global [*Charney et al.,* 1975; *Claussen,* 1997] and regional climate models [*Christensen et al.,* 1997], as well as numerical weather prediction models [*Viterbo and Beljaars,* 1995].

One important application of the satellite data is certainly model validation. However, this is only one part of the task of improving hydrological and surface process modeling. Numerous facts can contribute to errors in the simulated soil moisture, in particular errors in precipitation and solar radiation input, as documented by the Atmospheric Model Intercomparison Project (AMIP) [*Gates et. al.,* 1999]. Therefore, a method for compensating such errors in forcing data is required that makes optimal use of readily available satellite data. This can in principle be achieved through data assimilation, whereby model parameters and state variables are adjusted to optimize agreement between modeled and observed variables, taking into account the relevant error variances.

Data assimilation with the purpose of improving soil moisture estimates has now been pursued for some time, although it is still a rather new field of study [*McLaughlin,* 1995]. Examples are model inversion against microwave satellite data [*Entekhabi et al.,* 1994; *Lakshmi et al.,* 1997], the assimilation of near-surface atmospheric variables into mesoscale models [*Bouttier et *al.*, 1993a,b], and the use of satellite estimates of surface skin temperature to adjust soil moisture [*McNider et al.*, 1994; *Ottle and Vijal-Madjar,* 1994], thus improving estimates of surface fluxes and surface temperature. A nudging method is used by *van den Hurk et al.* [1997] to adjust modeled evaporation rates checked against satellite data in the context of numerical weather prediction. Parameter estimations of hydrological models using microwave satellite data have also been carried out [*Blyth,* 1993].

Assimilation of satellite derived vegetation data is a technique that has been developed in the context of global carbon cycle modeling [*Knorr et al.*, 1995; *Knorr and Heimann,* 1995], and has only very recently been applied to the modeling of land surface – atmosphere interaction [*Knorr and Schulz,* 2001]. In fact, the most widely used approach so far has been to derive various ground-surface properties directly from satellite data [*Dickinson et al.*, 1990; *DeFries and Townshend,* 1994; *Wanner et al.*, 1997] and use them as input to surface process [*Sellers et al.*, 1994] or carbon cycle models [e.g., *Potter et al.*, 1993; *Ruimy et al.*, 1996]. However, such a method, in which remotely sensed surface and vegetation information is used for forcing and not for assimilation, poses two important problems: (1) Data gaps and unfavorable viewing conditions create gaps in simulations, or require corrections based on prior assumptions or other models [e.g., *Los et al.*, 1994]. (2) Often, the satellite derived parameters, such as vegetation cover or leaf area index (LAI), can also be simulated by the model itself [*Knorr,* 2001]; if simulated and satellite derived vegetation cover are different, forcing the model with satellite data is bound to create inconsistencies [*Knorr,* 1998; see below]. In order to make optimal use of all available observations, and at the same time preserve consistency between data and model simulations, it is advisable to treat satellite derived vegetation cover in the same manner as surface temperature in the examples mentioned above: by assimilating the data into a combined surface and vegetation model, adjusting model parameters or state variables until agreement is optimal within present uncertainties.

In the present work, the satellite derived parameter that is assimilated is chosen to be fAPAR, the fraction of absorbed photosynthetically active radiation (PAR) absorbed by vegetation. fAPAR, which is itself a radiative quantity and therefore closely linked to satellite radiances, is also a variable that is computed within the vegetation and land surface model used here. The method is an extension of one presented earlier [*Knorr et al.*, 1995] and is applied to the global scale [see also *Knorr and Schulz,* 2001]. In this study, it is for the first time compared to global-scale assimilation of satellite derived surface temperatures. This second method presented is similar to earlier studies by *Ottle and Vijal-Madjar* [1994] and *McNider et al.* [1994], where the adjusted variable is soil moisture. However, it differs from those earlier studies in the spatial scale, and in the use of a combined vegetation and surface process model. Intercomparison between the two assimilation methods is also intended as an additional test for the assimilation schemes presented.

2. OVERVIEW OF THE BETHY MODEL

The model used in the present study, the Biosphere Energy-Transfer and Hydrology scheme (BETHY), is essentially a simulator of soil-vegetation-atmosphere interaction that uses observed climate data on a global scale. For a full description, the reader is referred to [*Knorr,* 1997] and [*Knorr,* 2000]. The model consists of four parts, as illustrated in Fig. 1: energy and water balance, photosynthesis, carbon balance, and phenology.

2.1. *Energy and Water Balance*

The energy and water balance part, which is the most comprehensive, considers three pools of water, which are soil water, a skin reservoir residing on the vegetation, and snow. Precipitation can fall as rain directly on the ground, or hit the vegetation, filling up the skin reservoir after which throughfall occurs, or it can fall as snow. Rainfall on the ground, throughfall and snow melt all fill up the ground water pool up to a value of w_{max}; if this value is exceeded, runoff occurs. w_{max} is computed from the equations of *Saxton et al.* [1986] as the amount of water stored in the soil between the surface and some assumed rooting depth, and between the water content at the wilting point (defined by 1.8 MPa soil suction) and that at field capacity (defined by 2 mm/day conductivity) [*Federer,* 1979].

Apart from runoff, water leaves the surface as evapotranspiration, thus becoming part of the energy balance. Potential soil evaporation and snow evaporation are assumed to occur at the equilibrium rate as in [*Jarvis and McNaugton,* 1986], until the snow pool is depleted, or, respectively, with actual soil evaporation according to [*Ritchie,* 1972]. Evaporation from the skin reservoir and transpiration, by far the largest flux for most cases, are both computed with the Penman-Monteith formula

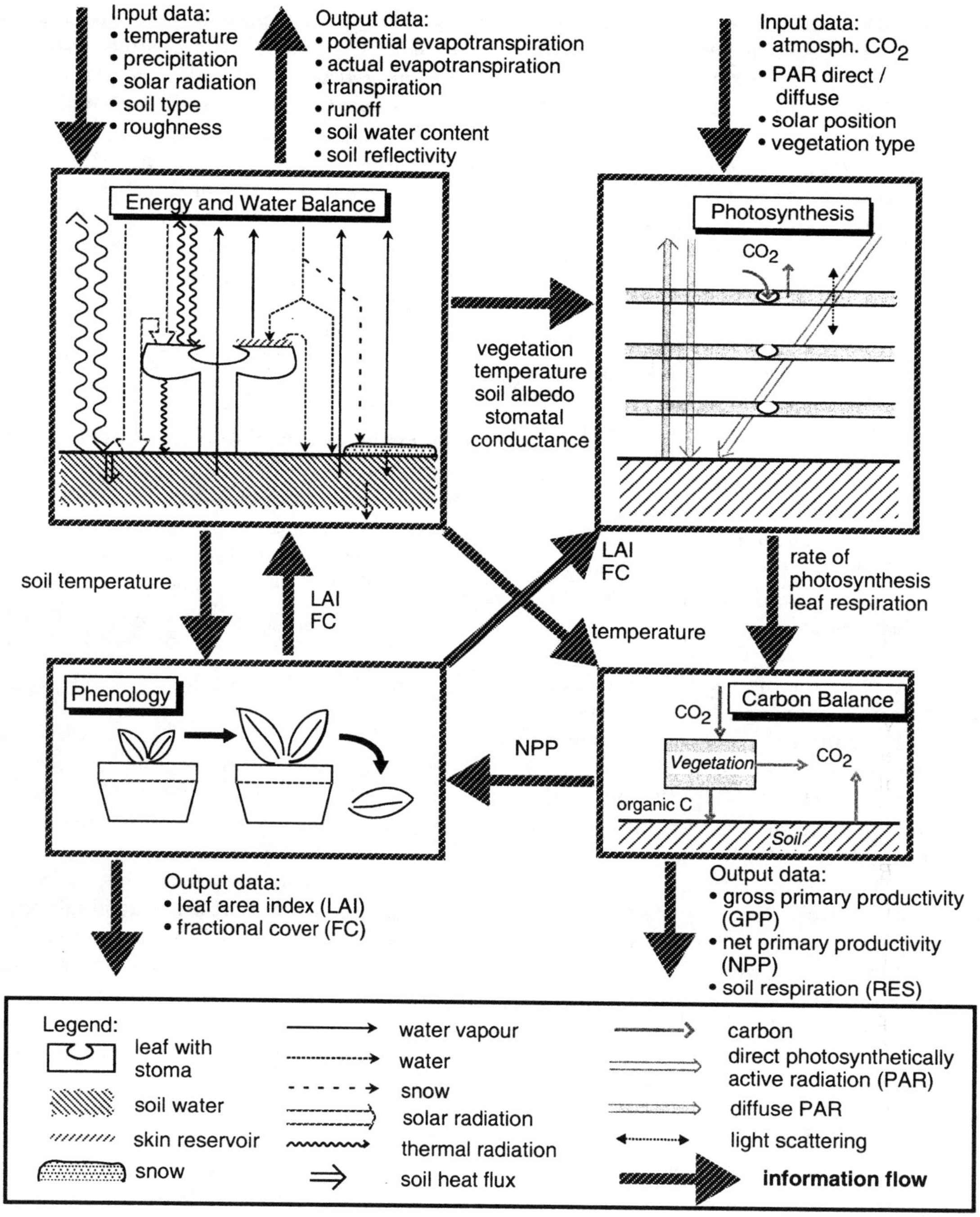

Figure 1. Structure of the BETHY model with input and output as well as information flow between the four model components. Various fluxes of water, light and carbon are also displayed.

[*Monteith,* 1965], the first with only aerodynamic resistance applied to the latent heat flux, the second with an additional canopy resistance (see Section 4).

The radiative balance is computed as incoming solar radiation derived from incoming PAR with a conversion factor depending on cloudiness and solar angle [*Pinker and Laszlo,* 1992], outgoing longwave radiation based on an emissivity of 0.97, and sky radiation depending on air temperature, air vapor pressure, and cloudiness [*Brutsaert,* 1982, p. 139]. Net radiation is computed separately for vegetation and bare soil, the former entering the Penman-Monteith formula, the latter used to compute the equilibrium evaporation rate for the unvegetated surface part. Vegetation albedo is set to 0.15 and soil background albedo as described in Section 5. A soil heat flux, G, is also included, assumed to be a fixed fraction (0.036) of the overall net radiation [*Verma et al.*, 1986] (cf. Section 4).

2.2. Photosynthesis

The photosynthesis part of the BETHY model assumes a horizontally homogeneous vegetation canopy consisting entirely of green vegetation. Optically, vegetation is assumed to behave like randomly orientated scatterers with a single scattering albedo of 0.12 in the photosynthetically active spectrum. To determine the rate of photosynthesis, the Farquhar model [*Farquhar et al.*, 1980] is used for C3 vegetation, and the model of *Collatz et al.* [1992] for C4 plants. Leaf or "dark" respiration, R_d, is calculated as a fraction of the maximum carboxylation rate (0.011 and 0.042 at 25°C, for C3/C4 respectively). Radiation absorption in the foliage is calculated with the two-flux approximation [*Sellers,* 1985] for three layers, using direct and diffuse PAR computed as in [*Weiss and Norman,* 1985]. The fraction of absorbed photosynthetically active radiation (fAPAR) in this scheme is calculated as:

$$f_{APAR} = f_c\{[R(0) + R\downarrow(0) - R\uparrow(0)] \\ -[R(\Lambda) + R\downarrow(\Lambda) - R\uparrow(\Lambda)]\} \\ /(R(0) + R\downarrow(0)) \tag{1}$$

where $R(0)$ and $R\downarrow(0)$ are incoming direct and diffuse PAR at the top of the canopy, $R\uparrow(0)$ the backscattered PAR from the canopy, covering a fraction f_c of the total ground area, and $R(\Lambda)$, $R\downarrow(\Lambda)$ and $R\uparrow(\Lambda)$ the direct and diffuse fluxes at the bottom of the canopy with a leaf area index of Λ (green-leaf surface area per ground area).

2.3. Carbon Balance

Calculation of various respiration terms connects the photosynthesis with the carbon balance part, which computes net primary productivity (NPP) as the difference of gross photosynthesis (also: Gross Primary Productivity, GPP) minus two terms of plant respiration: NPP=GPP$-(R_M + R_G)$. Maintenance respiration, R_M, is computed from leaf respiration (delivered by the photosynthesis part) assuming

$$R_M = R_d/f_{R,l} \tag{2}$$

and growth respiration as

$$R_G = f_G\text{NPP}, \tag{3}$$

where $f_{R,l} = 0.40$ is the leaf fraction of total maintenance respiration, and $f_G = 0.25$ the growth respiration cost [*Ryan,* 1991].

2.4. Phenology

In the context of this study, where the focus is on water and energy balance, the significance of NPP lies in the fact that it limits leaf growth and thus becomes one of the factors controlling the leaf area index (LAI). This has been taken into account when designing the phenology part of the BETHY model, in which the LAI, Λ, is computed as the minimum of a temperature limited value, Λ_T, a water limited value, Λ_W, and a growth limited one, Λ_G. Temperature limitation, which essentially serves the purpose of avoiding frost damage, is prescribed following [*Dickinson et al.* 1993] as

$$\Lambda_T = \begin{cases} 0 & \text{if } T_{0.5} \le T_\phi \\ \hat{\Lambda}\left[1 - \left(\frac{\hat{T}_\phi - T_{0.5}}{\hat{T}_\phi - T_\phi}\right)^2\right] & \text{if } T_\phi < T_{0.5} < \hat{T}_\phi \\ \hat{\Lambda} & \text{if } T_{0.5} \ge \hat{T}_\phi, \end{cases} \tag{4}$$

with the temperature at 0.5 m soil depth, $T_{0.5}$, and standard values of $\hat{\Lambda} = 5$, $T_\phi = 5$°C and $\hat{T}_\phi = 15$°C, with the exception of agriculture, where $T_\phi = 12$°C.

Water limitation of LAI is modeled in the following way: for increasing LAI, Λ_W is set to the value that maximizes NPP; at decreasing LAI, Λ_W is set to the LAI value of the preceding time step unless NPP becomes negative; in this last case, Λ_W is set such that NPP equals zero. The LAI is thus decreased just enough to maintain near-positive NPP values. Eventually, Λ_G accounts for 50% of NPP invested into leaf growth, but cannot be lower than 0.5 to allow leaf growth initiation.

Radiation, energy balance and photosynthesis are all calculated hourly, the water and carbon balances daily, and the phenology scheme is invoked every 10 days.

3. ASSIMILATION OF FAPAR

The fAPAR assimilation technique used here has first been presented by *Knorr* [1998] and was originally developed for improving carbon flux estimates at the global scale [*Knorr,* 1997]. Within this scheme, several parameters are modified until model simulated and satellite derived estimates of fAPAR agree optimally. fAPAR is a suitable quantity indicative of vegetation activity at the land surface, because, as a radiative quan-

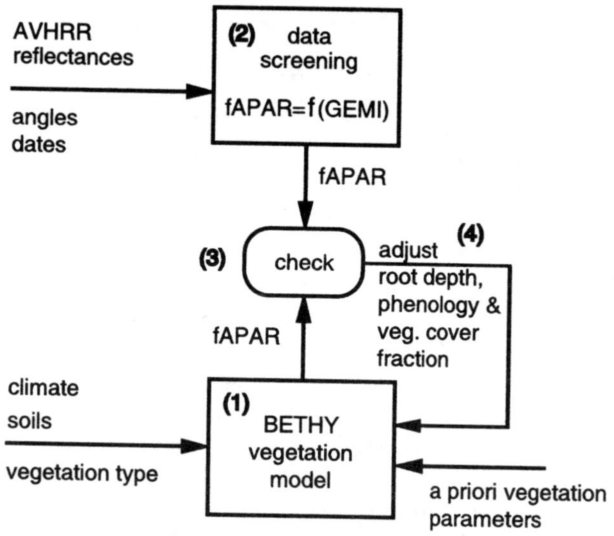

Figure 2. Logical structure of the assimilation procedure for fAPAR. Numbers refer to executed steps as described in the text.

tity, it can be derived more reliably from a vegetation index than e.g. LAI or biomass [*Asrar et al.*, 1992; *Goel and Qin*, 1994; *Gobron et al.*, 1997]. This parameter re-evaluation is done on the basis of the mean state of the model, i.e. the one that results from mean climatic data as input (see Section 5). The method, illustrated in Fig. 2, essentially consists of four steps:

1. Global monthly fields of fAPAR are derived from NOAA-11/AVHRR data at 1 degree resolution after rigorous screening.

2. The Biosphere Energy-Transfer Hydrology model (BETHY) is driven with mean climate data to compute both land-surface carbon and water exchange, and fAPAR for 12 months, also at 1 degree spatial resolution.

3. Satellite derived and model predicted values of fAPAR are compared.

4. Parameters $x_1 \ldots x_n$ of the BETHY model are modified separately for each grid point, and steps 1. to 4. are repeated until a cost function, J, reaches its minimum.

The cost function is defined as

$$J(x_1 \ldots x_n) = \sum_{m=1}^{12} \frac{(f_m - g_m(x_1, x_2, \ldots))^2}{\sigma_f^2} + \sum_{j=1}^{n} \frac{(x_i - y_i)^2}{\sigma_{y,i}^2}, \quad (5)$$

and takes into account the deviation between monthly satellite derived fAPAR, g_m, and model computed fAPAR, f_m, the error variance of fAPAR, σ_f^2, as well as the difference between actual, x_i, and *a priori* model parameters, y_i, with estimated error variances of $\sigma_{y,i}^2$. The actual parameters chosen are shown in Table 1, together with their assumed error ranges and their significance for the assimilation scheme. We assume $\sigma_f^2 = 0.2$ if fAPAR data exist, and $\sigma_f^2 = \infty$ if no data are available after screening. For x_i, we choose the three most important parameters for vegetation activity in this model, representing in turn water limitation ($x_1 \equiv w_{max}$), temperature limitation ($x_2 \equiv T_\phi$), and other, residual limitations that typically have longer times scales than decades ($x_3 \equiv f_c$), such as human land use or nutrient availability. The three parameters are then optimized at each grid cell of 1° latitude by 1° longitude separately in an iterative manner, using the downhill simplex method [*Press et al.*, 1992]. In each iteration, the model is run for 6 years, the first three of which are ignored in order to make the results independent of initial conditions.

4. ASSIMILATION OF SURFACE TEMPERATURE

The second data assimilation technique differs from the first one in two ways: it uses diurnally varying surface skin temperature as a measure of the surface moisture status, and it proceeds by varying the state and not the parameters of the model. A nudging technique, where the state variable soil moisture is varied by small amounts at a daily time step, is chosen because it is more similar to earlier techniques used for the assimilation of either surface temperature or microwave data from satellites (see above). Comparing the results of both strategies will thus be used to check whether using information on vegetation state and cover, through fAPAR assimilation, yields similar results in terms of evapotranspiration estimates and moisture status compared to a more traditional assimilation approach based on surface temperature.

In contrast to fAPAR assimilation, where the model always follows its exact trajectory with only model parameters modified at each new model run, the surface temperature assimilation scheme is not exactly energy and water conserving because the time trajectory of soil moisture is "nudged" at a daily time step in the direction indicated by an optimal combination of satellite derived and modeled diurnal temperature ranges. For these simulations, the BETHY model is driven with precipitation and temperature data from the years 1990 to 1992. The first two years are ignored to insure appro-

Table 1. Measured quantities and model parameters used in the assimilation scheme

Symbol	Model variable	$\sigma_f, \sigma_{y,i}$	Quantity or parameter	Significance
f_m	f_{APAR}	0.2	fAPAR by month	measure of vegetion activity
x_1	w_{max}	$4w_{max}$	bucket size	water limitation
x_2	T_ϕ	2°C	leaf onset/shedding temp.	temp. limitation
x_3	f_c	0.5	vegetation fractional cover	other limitations

priate spin-up of the mopdel, while the last year, 1992, corresponds to the satellite skin temperature observations (see Section 5).

4.1. Computation of Skin Temperatures

The surface skin temperature measured by the satellite, T_{skin}, is a combination of the surface temperature of transpiring plant canopies, T_c, and that of bare soil, T_s. In the model, it is represented by T^m_{skin} in the following way:

$$T^m_{skin} = (1 - t_l)T_c + t_l T_s, \tag{6}$$

where t_l is the long-wave transmissivity of the canopy, computed as

$$t_l = f_c \exp(-\bar{\mu}\Lambda/f_c) + (1 - f_c), \tag{7}$$

with the extinction coefficient $\bar{\mu} = 0.5$, being the average cosine leaf orientation for randomly distributed leaves, the leaf area index Λ, and f_c, the fractional vegetation cover. Thus, the satellite signal depends both on the energy balance of plant canopies and bare soil, and on the amount of vegetation cover.

The two skin temperatures, T_c and T_s, form part of the energy balance, which is computed separately for canopies and bare soil as:

$$R_{nc} - \frac{\rho C_p}{\gamma(r_{ac} + r_c)}(e_s(T_c) - e_a) - \frac{\rho C_p}{r_{ac}}(T_c - T_a) = 0 \tag{8a}$$

and

$$R_{ns} - \frac{\rho C_p}{\gamma(r_{as} + r_s)}(e_s(T_s) - e_a) - \frac{\rho C_p}{r_{as}}(T_s - T_a) - G = 0, \tag{8b}$$

where R_{nc} and R_{ns} are net radiation of canopy and soil, respectively, T_a air temperature, e_s and e_a, in turn, saturated and actual air vapor pressure, r_{ac} and r_{as} aerodynamic resistances for canopy and bare soil, r_c the canopy resistance to transpiration, and r_s the resistance of the soil surface that limits evaporation. ρ, C_p and γ are the density, specific heat and psychrometric constant of air, and G the soil heat flux. Part of the soil heat flux is assumed to originate from the canopy, a quantity that enters Equ. 8a as a component of the net radiation term, R_{nc}, because the canopy does not transfer heat directly to the soil. This term is at the same time added to R_{ns} as an additional heat source:

$$R_{nc} = (1 - t_{l,v})(R_{nL} - G) + (1 - a_c)R_s \tag{9a}$$

and

$$R_{ns} = t_{l,v} R_{nl} + (1 - a_s)R_s + (1 - t_{l,v})G, \tag{9b}$$

where R_{nL} is the net longwave radiation above the canopy level, a_s and a_c soil and canopy shortwave albedo, and R_s the incoming shortwave radiation.

The aerodynamic resistances to vapor and heat are taken as equal to each other and are computed from [*Brutsaert*, 1982],

$$r_a = \frac{1}{k^2 u}(\ln(\frac{z - d}{z_0}))^2 \tag{10}$$

with k, the von Karman constant (0.41), u as the wind speed, z the reference height, z_0 the roughness length, and d the zero plane displacement. For the canopy aerodynamic resistance, r_{ac}, the following parameterization has been fitted to data by *Kelliher et al.* [1993]: $u = 3m/s$, $z_0 = 0.1h_c$, $z = h_c + 2$ m, and $d = 0.56h_c$, where h_c is the canopy height, being 30 cm for short grass, 1 m for shrubs, 2 m for long grass, 15 m for temperate, boreal and drought deciduous trees, and 30 m for tropical evergreen trees. For the bare soil resistance, r_{as}, we use $z_0 = 1$ cm and $d = 0$.

We assume a canopy resistance, r_c, which is the combined resistance of all leaves in the canopy, that is determined by the photosynthetic demand of the canopy, $A_{c,0}$ (the photosynthetic rate limited only by light and photosynthetic capacity), soil water content, w, and the vapor pressure deficit, $e_s(T_a) - e_a$ [*Lindroth and Halldin*, 1986]:

$$r_c = r_{c,0}(1 + b_e(e_s(T_a) - e_a)) \tag{11}$$

where $r_{c,0}$ satisfies

$$A_{c,0} = \frac{C_a - C_{i,0}}{1.6 r_{c,0}} \frac{p}{R(T_a + 273.15)} \tag{12}$$

(with air pressure, p, the universal gas constant, R, and T_a in °C), and b_e is determined such that at 13:00 h (the time of highest evaporative demand), the transpiration rate is equal or less than the root supply rate, denoted S, which can be written as:

$$\frac{\rho C_p}{\gamma(r_{ac}+r_c)}(e_s(T_c)-e_a) \leq S = c_w \frac{w}{w_{max}} \tag{13}$$

The constant c_w is assumed 1 mm/hour, following [*Federer*, 1979]. $C_{i,0}$ is taken as $0.87C_a$ for C3 vegetation and $0.67C_a$ for C4 grasses, with C_a, the free-air concentration of CO_2, at 355μmol/mol.

The soil evaporation, E_s, is assumed to happen in two phases as predicted by the model of *Ritchie* [1972]. During phase 1, it is assumed to proceed at the potential rate with a soil-atmosphere aerodynamic resistance, r_{as}, computed from Equ. 10. After an accumulated soil evaporation of $\Sigma E_1 = 14.29 - 9.23 f_1^{sand}$ (in kg m^{-2}), cumulative evaporation is assumed proportional to the square root of time, with a proportionality constant $\kappa = 5.62 - 2.56 f_1^{sand}$, in kg m^{-2} day$^{-0.5}$, called 'desorptivity'. f_1^{sand} denotes the sand fraction of the uppermost soil layer. The rate of soil evaporation in phase 2 is then given by

$$E_s = \frac{d}{dt}[\kappa\sqrt{t}] \tag{14}$$

where t denotes the time since the last rain event.

Both canopy and bare soil skin temperatures are computed within the Penman-Monteith approximation [*Monteith*, 1965], i.e. by solving Equ. 8a and 8b together with $e_s(T_s) \approx e_s T_a + s(T_a)(T_s - T_a)$ and $e_s(T_c) \approx e_s T_a + s(T_a)(T_c - T_a)$, where s is the derivative of the saturated vapor pressure with respect to temperature.

4.2. The Adjustment Technique

It is now required that model simulations be consistent with the satellite observed temperature via Equ. 6, taking into account the model and observation uncertainties. This is done through minimization of a cost function, J, depending on observed skin temperature, T_{skin}, the *a priori* model variables T_a, T_c and T_s, the adjusted variables T_a', T_c' and T_s', and assumed error variances σ_{skin}^2, σ_a^2, σ_c^2 and σ_s^2:

$$J(T_a', T_c', T_s') = \frac{(T_{skin}'^m - T_{skin})^2}{\sigma_{skin}^2} + \frac{(T_a' - T_a)^2}{\sigma_a^2} + \frac{(\Delta T_c' - \Delta T_c)^2}{\sigma_c^2} + \frac{(\Delta T_s' - \Delta T_s)^2}{\sigma_s^2}, \tag{15}$$

where ΔT_c stands for $T_c - T_a$ etc., and $T_{skin}'^m$ is itself a function of the adjusted temperatures T_c' and T_s', taking the equivalent of Equ. 6. The optimal values of T_a', T_c' and T_s' are found by requiring all three partial derivatives of J to be zero. This leads to a simple three-dimensional matrix equation that is solved for all four observation times per day. Between the observation times, the differences $T_a' - T_a$ and $T_c' - T_c$ are interpolated linearly to yield hourly values of T_a' and T_c'.

The temperature assimilation scheme as described so far builds on previous work by *Lakshmi* [2000] by using a combination of satellite observed and modeled skin temperatures to be fed back into the surface model. For $t_l = 0$ (full vegetation cover) and $\sigma_a = 0$, minimizing J leads to the equation used in the work just cited:

$$T_c' = \frac{\sigma_c^2}{\sigma_c^2 + \sigma_{skin}^2} T_{skin} + \frac{\sigma_{skin}^2}{\sigma_c^2 + \sigma_{skin}^2} T_c$$

with $T_s' = T_s$ and $T_a' = T_a$. The adjusted skin temperatures now imply changes in the water balance because of Equ. 8a (in the case of canopy temperature), e.g. a decrease in $T_c - T_a$ means increased canopy evapotranspiration, and vice versa.

Other than for regional studies at shorter time scales in which $t_l \approx 0$ is mostly fulfilled, applying such an assimilation scheme to the global scale requires some important modifications: (1) For areas with considerable bare-soil fractions, ground surface and atmosphere may be poorly coupled, resulting in large differences $T_a - T_s$, such that differences between observed and modeled skin temperatures would mainly be attributable to uncertainties in r_a, and not in the surface moisture status. Therefore, only T_a' and T_c' are used in the assimilation, the first as a direct input for recomputing the diurnal energy balance, and the second for adjusting canopy transpiration rate and soil moisture, w, using Equ. 13. (2) Vegetation cover, and thus t_l, is not independent of the moisture status and should be adjusted as well. This is done by dividing assimilation of monthly satellite data into steps of $\Delta t = 1$day, using a relaxation time of $\tau = 30$ days, and letting the model adapt its LAI, and thus t_l, at each time step.

Adjustment of soil moisture, w, is done for the time of highest atmospheric demand (13:00 h), consistent with the computation of the stomatal parameter b_e: first, the model-internal transpiration rate, E_c, is computed using the adjusted air temperature, T_a', together with Equ. 13 and 11. Then, the transpiration rate consistent with adjusted skin temperatures is computed as

$$\lambda E_c' = R_{nc} - \frac{\rho C_p}{r_{ac}}(T_c' - T_a') \tag{16}$$

and the adjusted supply rate, S', as

$$S' = S + (E_c' - E_c)\frac{\Delta t}{\tau}. \tag{17}$$

With S', the second part of Equ. 13 can be inverted to yield the adjusted soil water content w', restricted to values between 0 and w_{max}, thus limiting the adjustment to cases where soil water is actually limiting:

$$w' = \frac{w_{max}}{c_w} S' \tag{18}$$

Apart from being computationally inexpensive, this techniques ensures that adjusted fluxes are always consistent with both model physics and vegetation physiology. Only small adjustments of model state variables are allowed, staying within the range of possible values without adjustment. Thus, the complete model is allowed to adapt gradually to observed skin temperatures, which includes adjustment of vegetation amount through changes in soil water content, w.

In the simulations described below, we use fixed error estimates for satellite skin temperature ($\sigma_{skin} = 3°C$) and air temperature ($\sigma_a = 3°C$), and relative error estimates for the temperature differences $\Delta T_c \equiv T_c - T_a$ and $\Delta T_s \equiv T_s - T_a$, which are assumed to result from uncertainties in the aerodynamic resistances, r_{ac} and r_{as}:

$$\begin{aligned} \sigma_c &= \frac{\partial \Delta T_c}{\partial r_{ac}} \sigma_{rac} \\ &= (T_c - T_a)\frac{\sigma_{rac}}{r_{ac}} \end{aligned} \tag{19}$$

and the same for T_s and r_{as}. We assume a relative uncertainty of 100% for the aerodynamic resistances, i.e. $\sigma_{rac}/r_{ac} = 1$ and $\sigma_{ras}/r_{as} = 1$.

5. DATA REQUIREMENTS

All model simulations presented here are done globally and on a monthly basis. Therefore, both climatic input data, satellite observations, and model output are stored as monthly means, with the exception that a monthly average diurnal cycle is used for satellite skin temperatures. The computation of daily fluxes within the BETHY model are done as an internal means of accurately integrating highly non-linear water and carbon balances, including leaf development and shedding.

5.1. Climate, Soils and Vegetation Data

Data requirements of the BETHY model are mean climatological precipitation, temperature, and daily temperature amplitude [*Cramer,* pers. comm.; *Leemans and Cramer,* 1991], number of wet days per month based on station data by Müller [*Friend,* pers. comm.; *Müller,* 1982], and incident photosynthetically active radiation (PAR) taken from the International Satellite Cloud Climatology Project (ISCCP) [*Pinker and Laszlo,* 1992]. Other input data are global maps of soil type [*Dunne and Willmott,* 1996], of soil brightness [*Wilson and Henderson-Sellers,* 1985] and land cover and vegetation type [*Wilson and Henderson-Sellers,* 1985]. The soil types are assigned soil horizons with values of sand, silt and clay content [*Dunne and Willmott,* 1996], from which water content at field capacity and wilting point are computed (cf. Section 2). There are 23 functional plant types with type-specific values of photosynthetic capacity [*Beerling and Quick,* 1995] and pathway, phenological type, height, and leaf area per dry mass [*Schulze et al.,* 1994]. Further, rooting depth is assumed to depend both on vegetation type and on the depth of the soil. It is determined in the following way: if the vegetation are trees and the land cover type savanna, rooting depth is assumed 3 m, otherwise 1 m. If the soil is shallower than this value, then the rooting depth is reduced to the depth of the lowest soil horizon at the grid point concerned. Such values are standard for most vegetation and surface models [e.g., *Raich et al.,* 1991; *Potter et al.,* 1993; *Dickinson et al.,* 1993; *Lüdeke et al.,* 1994; *Henderson-Sellers et al.,* 1996].

For surface temperature assimilation, the monthly mean temperature is corrected by anomalies taken from the operational analyses of National Center for Environmental Prediction (NCEP), and monthly precipitation by anomalies from the Global Precipitation Climatology Center [*GPCC,* 1992; *Rudolf et al.,* 1994].

5.2. Satellite Derived fAPAR Fields

As mentioned earlier, the satellite data used are channels 1 and 2 measured by the AVHRR sensor on board the NOAA series of satellites. We use weekly and atmospherically corrected data from the NOAA GVI archive [*Gutman et al.,* 1995], which were obtained through CESBIO [*Berthelot et al.,* 1994]. The data have a resolution of 1/7° latitude by longitude, comprising the years 1989 and 1990, so that all data were measured on-board the NOAA-11 satellite, which avoids intercalibration problems between different sensors. Preprocessing carried out by CESBIO includes conversion to reflectances [*Kaufman and Holben,* 1993], and atmospheric correction based on the SMAC code [*Rahman and Dedieu,* 1994].

Although much of the cloud contamination has already been removed in the original weekly GVI data, even residual atmospheric contamination can have a serious impact on the retrieval of surface properties [*Gutman,* 1990]. Here, we adopt a rather conservative strategy aimed at avoiding adverse observational conditions and at the same time some of the disadvanteges associated with the widely used maximum compositing approach [*Holben,* 1986]. This latter procedure has been found to lead to temporal biases towards 'greener' periods within a month [*Gutman and Ignatov,* 1996], while it fails to account for angular biases created by the effect of view or sun angles on the observed signal [*Meyer et al.,* 1995]. We also use the vegetation index GEMI [*Pinty and Verstraete,* 1992] instead of the more traditional NDVI, because it has been demonstrated that a number of alternative vegetation indices, among them GEMI, achieve a better signal-to-noise ratio than NDVI if both atmospheric effects and the brightness of the underlying soil are considered noise [*Leprieur et al.,* 1994; *Goel and Qin,* 1994].

For angular screening, data points are required to have sun zenith angles $\theta_s \leq 60°$ and view zenith angles of $\theta \leq 60°$, and to fall outside of the hot-spot range, with the condition

$$\sqrt{\tan^2\theta_s + \tan^2\theta - 2\tan\theta_s\tan\theta\cos\Delta\phi} \leq 0.25, \tag{20}$$

where $\Delta\phi$ is the azimuth angle between sun and viewing directions [*Verstraete et al,* 1990]. Cloud screening is then applied on the basis of GEMI, exploiting its property of assuming low to negative values over clouds [*Flasse and Verstraete,* 1994]. A monthly composite is then computed as the average of the remaining weekly data points, avoiding both 'greenness' and angular biases. The data are then averaged to 1° resolution and areas with residual cloud contamination are screened by requiring a ratio of the time-interpolated maximum composite GEMI to the monthly average (according to the standard procedure) of ≤ 1.1. Finally, averaging remaining data of both years and filling in gaps of only one month by linear interpolation results in a global area-weighted coverage of 74.8% of land areas [*Knorr,* 1997].

As the last processing step, values of GEMI are converted to fraction of absorbed photosynthetically active radiation (fAPAR, Equ. 1) using the following approximation [cf. *Knorr,* 1997]:

$$f_{APAR} = \begin{cases} 0 & \text{if GEMI} \leq 0.380, \\ 1 & \text{if GEMI} \geq 0.720, \\ -0.122 - 1.061\text{GEMI} & \\ \quad +3.639\text{GEMI}^2 & \text{else.} \end{cases} \tag{21}$$

This relationship has been tested with a radiative transfer model [*Gobron et al.,* 1997] and accounts for increasing residual atmospheric contamination over humid, densely vegetated areas. It was found to be rather insensitive to typical uncertainties in leaf reflectance, or leaf angle distribution [*Knorr,* 1997]. The error in fAPAR usual lies below 0.1, although an uncertainty of 0.2 is used in the assimilation procedure (see above).

5.3. Satellite Observed Surface Temperatures

The satellite-observed surface skin temperatures used here are monthly averages of routinely processed daily surface temperatures from the Tiros Operational Vertical Sounder (TOVS), flown on the NOAA spacecraft since 1978. They are computed as radiative temperature from the thermal channels 8, 18 and 19 of the High Resolution InfraRed Sounder (HIRS2), which forms part of TOVS [*Susskind et al.,* 1997]. The data are available as 1°×1° gridded fields together with local time of observation, and are matched to the grid used by the surface model BETHY. Missing values are assigned to areas of excessive cloudiness (> 80%) before monthly averaging is carried out.

Here, we use 12 monthly averages of four different daily overpass times from 1992. Two times come from the NOAA-11 satellite with an equatorial local overpass time of approximately 3:15 a.m. and p.m., respectively, and two from NOAA-12 with overpass times at the equator close to 7:30 a.m. and p.m. For observations away from the equator, or away form nadir view, local observation times can differ by up to 3 hours in polar regions. Therefore, local observation times are used in conjunction with surface temperatures.

It is likely that some bias exists between the daily average surface temperatures computed by the vegetation and surface model and the satellite observations, for example through errors in the spatial interpolation of the station data used to construct the temperature climatology of Cramer and Leemans, or within surface temperature retrieval. We therefore restrict the use of satellite observed temperatures essentially to the diurnal temperature range by carrying out a base line correction of the four diurnal values. This correction is done iteratively using the following steps:

1. A temperature offset is computed that is the difference between the average model predicted skin temperature, T^m_{skin} (Equ. 6) at the observation times t_i, and the mean of the satellite skin temperatures, T_{skin}, at the same times. This yields the first corrected skin temperatures, $T^1_{skin}(t) = T_{skin}(t) + T^1_{offset}$, with:

$$T^1_{offset} = \frac{1}{n_{obs}} \sum_{i=1}^{n_{obs}} [T^m_{skin}(t_i) - T_{skin}(t_i)]$$

2. The corrected skin temperatures, T^1_{skin}, are then restricted to the range allowed by the energy balance, i.e. to between T_a and $T_a + R_{nc}r_{ac}/(\rho C_p) + R_{ns}r_{as}/(\rho C_p)$, yielding skin temperatures T'^1_{skin}.

3. The correction resulting from the previous step is then averaged to adjust the temperature offset to the next iteration step

$$T^2_{offset} = T^1_{offset} + \frac{1}{n_{obs}} \sum_{i=1}^{n_{obs}} \left[T'^1_{skin}(t_i) - T^1_{skin}(t_i)\right]$$

4. Steps 2 and 3 are carried out until T^n_{offset} converges, yielding the final corrected satellite skin temperatures

$$T^{\infty}_{skin}(t) = T_{skin}(t) + T^n_{offset}$$

It is assumed that no inversion occurs between the surface and the reference height z, 2 m above either bare ground or canopy, in accordance with the assumptions made in the BETHY surface scheme. In the case of actual inversion, as it might be possible in the winter over snow cover, the scheme yields $T^{\infty}_{skin}(t) = T_a(t)$ and thus $E'_c = 0$ according to Equ. 16, which is consistent with the assumption that no canopy transpiration occurs under such conditions. Further, if none of the adjusted satellite skin temperatures lie outside the possible range allowed by T_a and the energy balance, the procedure simply assumes the same average for satellite and model skin temperatures and no adjustment of soil water content, w, takes place (cf. Equ. 16 to 18). In most cases, however, relatively more weight is given to observation times where temperatures are better constrained, effectively adjusting the minimum temperatures and always leaving the diurnal cycle unchanged.

6. RESULTS

Before the impact of fAPAR and surface temperature assimilation on simulated soil water and evapotranspiration is assessed, both schemes are cross-checked against the global satellite observation used by the other as an independent data source. This is to check if and for what regions fAPAR assimilation also increases consistency with satellite derived surface temperature, and vice versa.

To further validate conclusions drawn from the global simulations, we also compare simulated and observed soil moisture for a particular site in the Amazon rainforest where both schemes have a particularly large impact. Simultaneous consideration of carbon and water fluxes for this site is used to explain the mechanisms underlying the results of the fAPAR assimilation technique. A further sensitivity analysis of the model is also used to assess the robustness of the results against various model parameter uncertainties.

6.1. Validation With Global Observations

The effect of the fAPAR assimilation scheme on the simulated surface temperature range is shown on Fig. 3 and compared to the corresponding satellite data. Here, we use the difference between the early morning and early afternoon overpass times as an indication of diurnal temperature range, showing the temperature difference averaged over July 1992 measurements with the TOVS instrument (top of figure), and compare it to simulations using the same times at each pixel as given by the satellite. The *a priori* simulation, i.e. without parameter adjustment through fAPAR assimilation (center of figure) tends to overestimate the diurnal range of surface temperature in most regions, both of the tropics and the temperate zone. The extent of dried out, low soil moisture areas also seems to be too large, which indicates that soil moisture might be underestimated in the standard model run (see below). After fAPAR assimilation (bottom of figure), however, the model obtains a rather good agreement with satellite data for many regions. Notably the Amazon basin, which was simulated dry for large south-eastern portions, now turns out to be simulated rather well, as

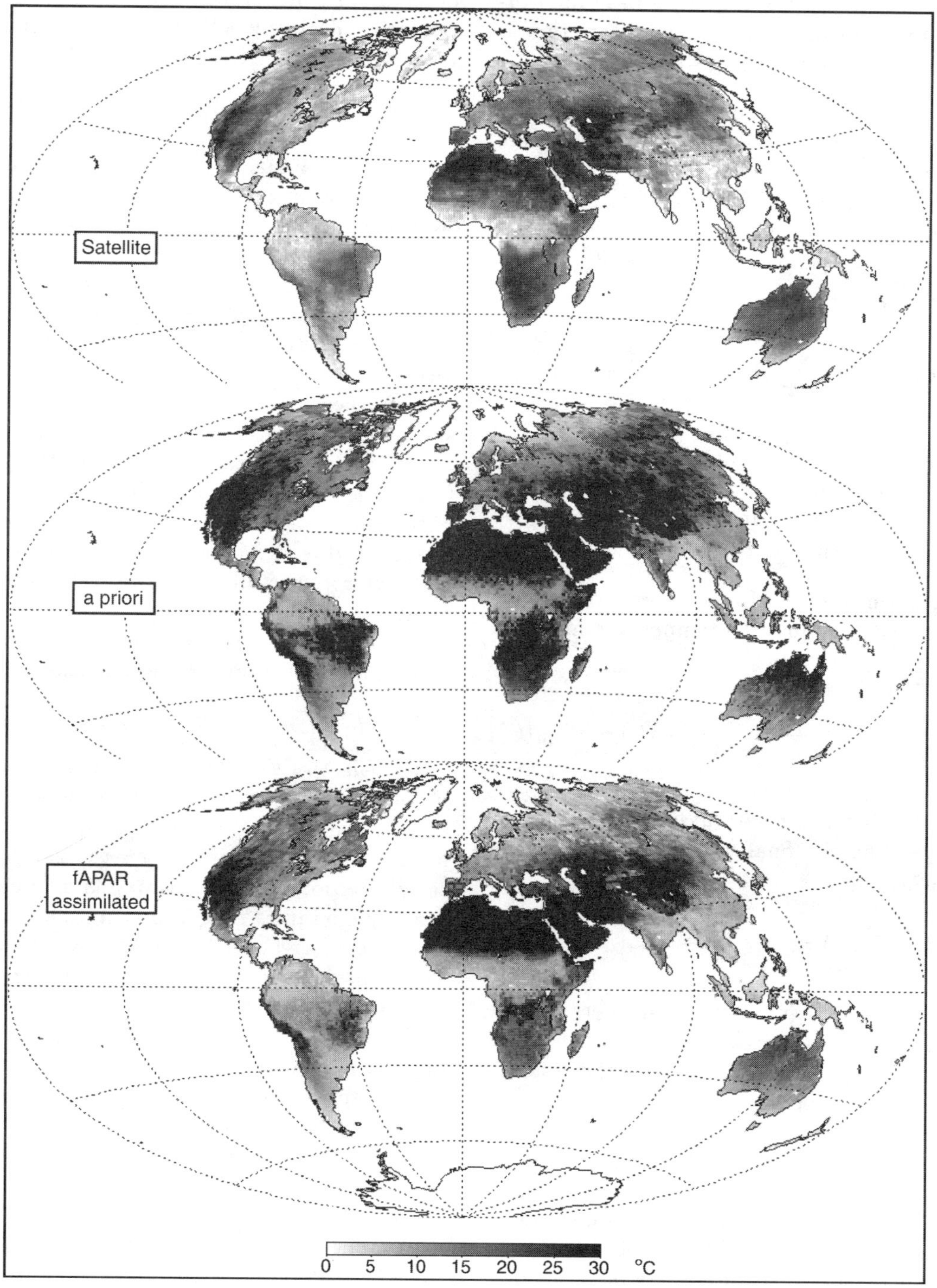

Figure 3. Temperature difference between the early afternoon overpass time (around 3:15 p.m. at the equator) and the early morning overpass time (3:15 a.m.) measurered from satellite with the TOVS instrument for July 1992 (top), simulated by the BETHY model without data assimilation (center), and after assimilation of fAPAR (bottom).

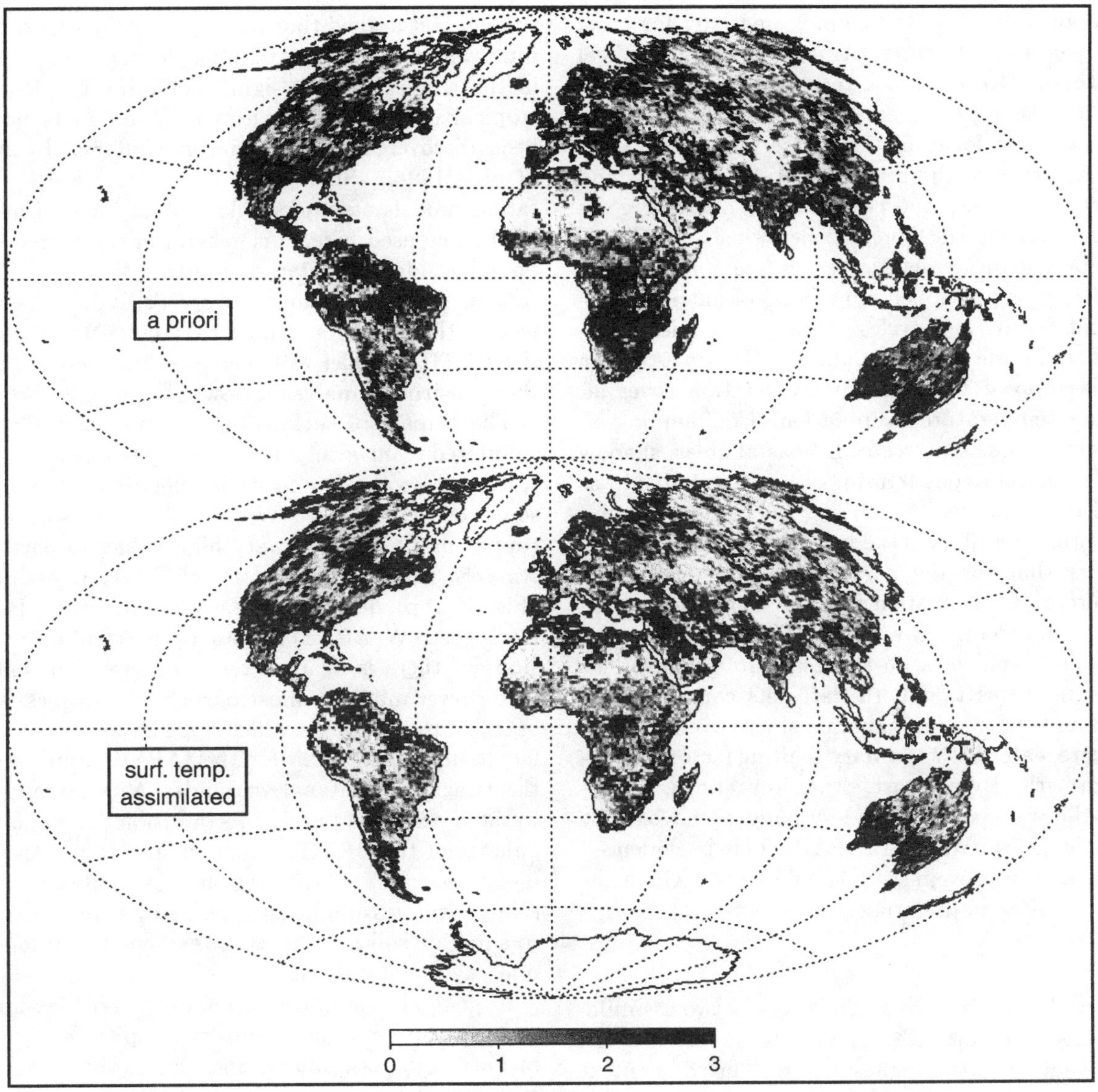

Figure 4. χ^2 statistics for monthly fAPAR fit to satellite observations for simulations using meteorological data of 1992 (top), and the same after assimilating satellite observed skin temperature of 1992 (bottom).

are eastern North America, Europe, southeast Asia and Australia. Only for large mountain areas (Tibetan or Andean Plateaus, central Rocky Mountains), and some desert areas (e.g. in the Sahara and the Arabian desert), the temperature range seems to be generally overestimated. Possibly, the soil brightness given by Wilson and Henderson-Sellers sometimes underestimates soil albedo, leading to too much net radiation and thus too high a diurnal temperature difference.

An independent test of the surface temperature assimilation is shown on Fig. 4, using a χ^2 statistics to measure the goodness of fit between satellite and model derived fAPAR:

$$\chi^2 = \frac{1}{n} \sum_{m=1}^{n} \frac{(f_m - g_m)^2}{\sigma_f^2} \qquad (22)$$

where f_m is satellite-derived fAPAR, g_m modeled fAPAR $\sigma_f = 0.2$, and n the number of valid measurements, f_m. Since simulations are for 1992 and observed data a mean of 1989 and 1990, some uncertainty resulting from interannual climate variations should be allowed for. It appears that in the tropic, especially for most of the areas around central Amazonia and in central and southern Africa, surface temperature assimilation leads to a clear improvement of the simulations. There is a similar effect for some temperate regions of rather dry climate, such as in parts of the U.S., southern and southeastern Europe, and central Asia. No change in χ^2 is achieved for the boreal zone, where the model tends to overpredict vegetation cover [see *Knorr,* 1997], because there is no explicit simulation of nutrient limitation and disturbance cycles. Finally, some desert areas with small amounts of rainfall (N Sahara, Arabian Peninsula) appear to overpredict vegetation cover after surface temperature assimilation. Confounding effects might include sea breezes in coastal areas, shadow casting in mountainous terrain, or lower soil albedo as assumed in the model.

Comparing the effect of the two assimilation schemes, it appears that for the particular model used here, both sources of information indicate too little vegetation cover and too much water stress for large sections of the tropical and the semi-arid temperate zones. This generally indicates that both data sets contain useful information for detecting moisture status over large regions where water is the primary limiting factor for vegetation growth. By contrast, other limitations, not related to the water cycle or energy budget, can only be detected by information on fAPAR. This is evidenced by the lack of improvement of simulated fAPAR in the boreal zone after assimilating surface temperature.

6.2. Impact on Simulated Soil Moisture

After comparing and cross-checking the two assimilation schemes, we now assess the results in terms of the variable that is of the highest interest in the context of the present study, i.e. with simulated plant-available soil moisture. It turns out that both schemes have a rather large impact on this variable, at least in tropical regions, while this impact somewhat differs from region to region. Fig. 5 and 6 both show the simulated plant-available soil moisture for July before and after assimilation, Fig. 5 for the mean climate and fAPAR assimilation, and Fig. 6 for 1992 and surface temperature assimilation.

As Fig. 5 shows for July, there is a general trend for increasing soil moisture content after assimilation of fAPAR. A similar picture appears for all other months (see below). The reason for this tendency appears to be a general underestimate of rooting depth, assuming values of not more than one to three meters for the *a priori* model simulations. *Kleidon and Heimann* [1998] have in fact argued that rooting strategies tend to optimize water use, leading to much deeper roots, especially in the tropics, with a significant impact on simulated tropical climate. Such deep reaching roots have also been discovered in the Amazon rainforest by *Nepstad et al.* [1994]. Since most current surface and vegetation models assume similar values for rooting depth as the one used here, this general underestimate might be of considerable interested within the area of surface process modeling [*Knorr and Schulz,* 2001]. The robustness of those results against parameter uncertainties of the BETHY model will therefore be assessed in a further sensitivity analysis below.

The impact of surface temperature assimilation on simulated plant-available soil moisture is shown on Fig. 6, restricted to the low latitudes between 30° north and south of the equator. While assimilated soil moisture (upper map) lacks the very high values in parts of the Amazon rainforest found for the fAPAR assimilation (Fig. 5, top), for some areas (N Amazonia, Brazilian Northeast, W Africa, SE Asia) the results are rather similar: there is again a general increase in soil moisture (lower map) for most parts of the tropics.

One important area which appears wetter for the surface temperature than for the fAPAR assimilation are the Brazilian Pantanal wetlands. As a large river fed wetland area, the purely precipitation driven soil water balance of the BETHY model is unable to capture the moisture status of this region. By contrast, the skin temperature assimilation scheme, by nudging w, is able to simulate soil moisture that exceeds the amount supplied by precipitation.

In general, the patterns found for both assimilation schemes are remarkably similar, despite the differences in both data assimilated and the method used (parameter re-estimate vs. nudging of status variable).

6.3. Comparison With Soil Moisture Measurements From Amazonia

One of the main results so far has been that soil moisture availability for plant use is much higher in large parts of the tropics than previously assumed in most models. As mentioned before, *Nepstad et al.* [1994] have discovered that water can be supplied to tropical rainforests from rather deep soil layers during pronounced dry seasons. Tropical evergreen forest generally tends to establish itself where annual precipitation exceeds annual potential evapotranspiration [*Box,* 1981], even if prolonged dry seasons exist, because they are able to rely on a rather large soil water storage capacity [*Kleidon and Heimann,* 1998]. This is exactly

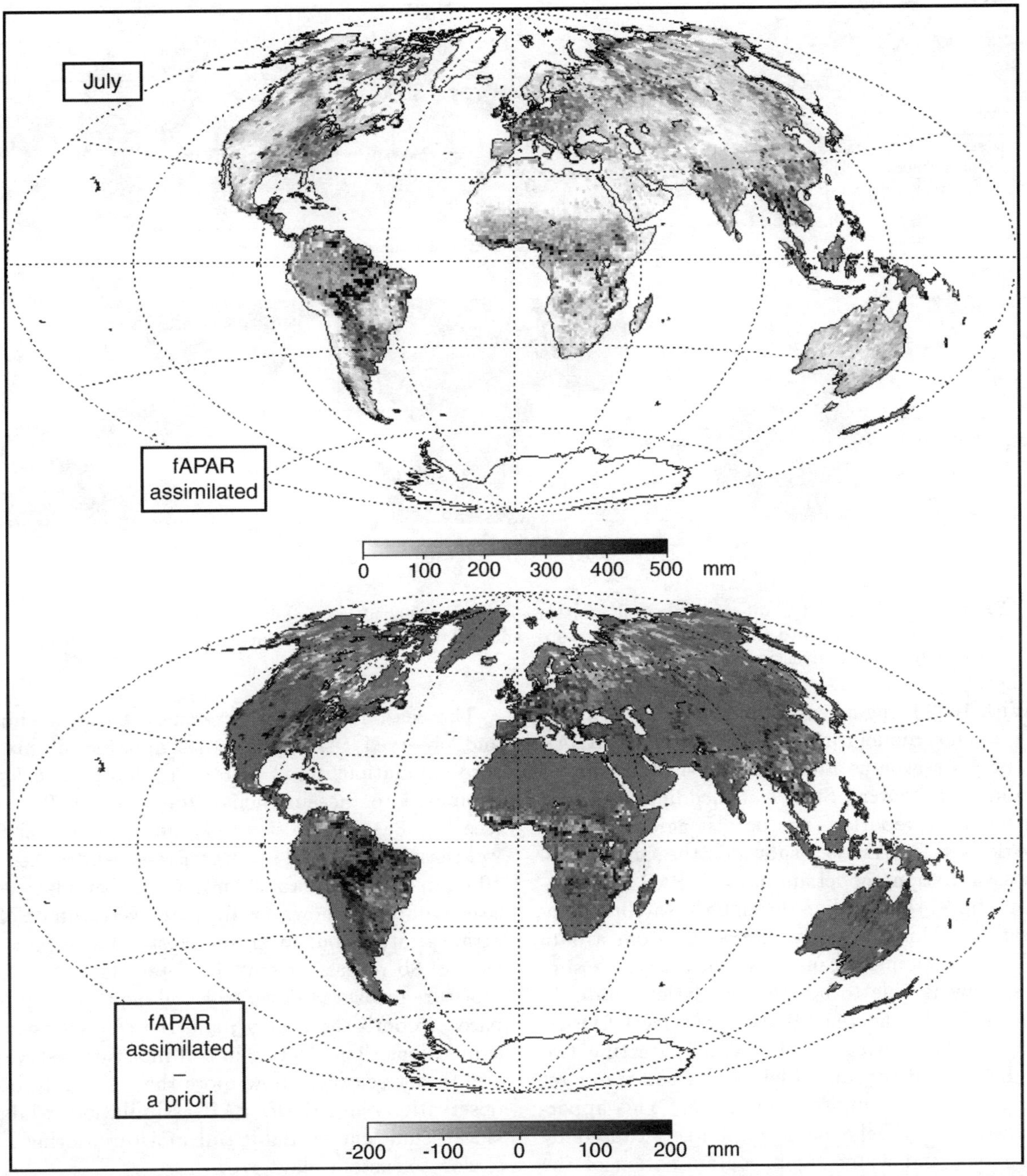

Figure 5. Plant-available soil moisture for July simulated with the BETHY model after fAPAR assimilation (top), and the difference to the model run without satellite data assimilation (bottom).

what the fAPAR assimilation scheme achieves, as will be demonstrated next.

A comparison of simulated and satellite measured fAPAR for the site investigated by Nepstad and coworkers is shown on Fig. 7a. The simulation was run using daily precipitation data from *Nepstad et al.* [1994] for Paragominas (3°S, 48°W) for the year 1992, with vegetation parameters set to tropical rainforest, except that *a priori* rooting depth was set to 2 m, instead of 1 m. 2 m were chosen for better comparison with published soil moisture measurements, which, given the results by Nepstad et al., can still be considered "shal-

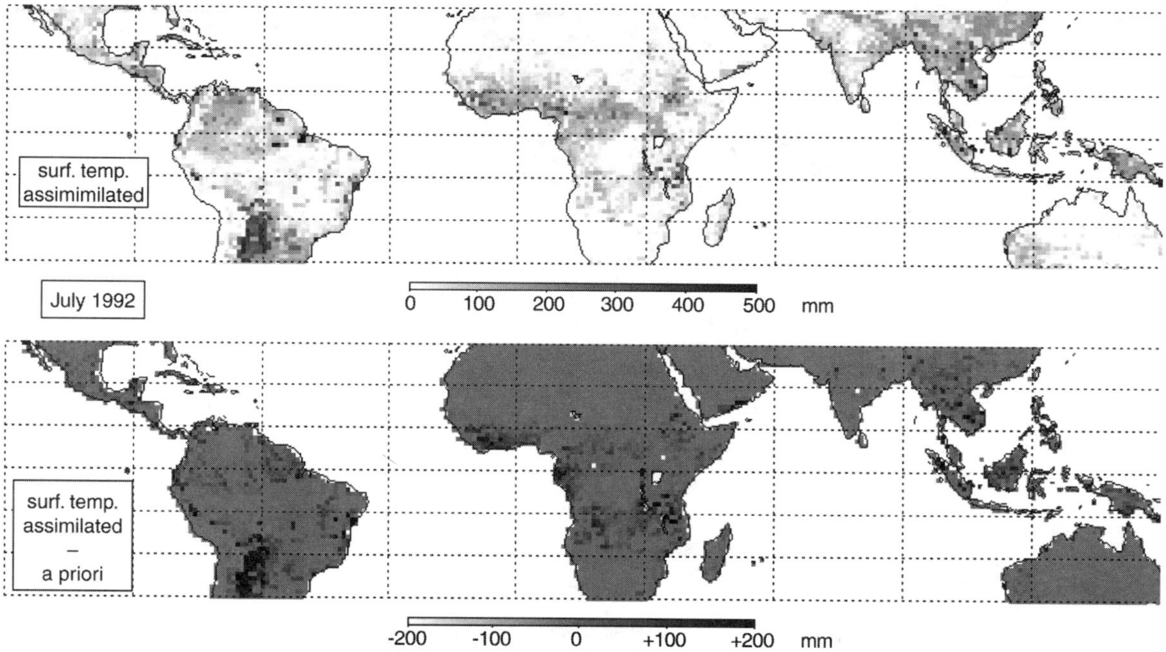

Figure 6. Simulated plant-available soil moisture for July 1992 after assimilation of satellite derived surface temperature into the BETHY model (top), and difference of this simulation minus the resulting soil moisture without assimilation (bottom).

low". With 1992 being an exceptionally dry year at the site, the *a priori* model run shows a decrease in fAPAR during the dry season as a result of leaf shedding forced by declining soil water. Such leaf shedding, however, has neither been reported, nor does it agree with the satellite derived fAPAR data shown on the same graph.

After assimilation, model simulated fAPAR values result in an almost straight line through the satellite measured values. This is already interesting from a data processing standpoint, because it shows that the assimilation scheme is able to use the vegetation model to filter out noise in the fAPAR data. In fact, fAPAR values are higher during the dry season, starting two months before leaf shedding (delayed by soil water reserves), and lower during the wet season. This apparently contradictory behavior is most likely caused by residual cloud and water vapor contamination of the satellite data during the wet season.

The effect of the assimilation procedure on simulated evapotranspiration is shown on Fig. 7b. Because the forest is now truly evergreen, high evapotranspiration rates persist through the dry season, although somewhat reduced due to declining soil water reserves. If used within a global or regional climate model, such drastic changes in seasonal evapotranspiration can be expected to have a strong impact on the simulated surface climate [*Knorr and Schulz,* 2001].

The reason for the discrepancy between simulated and observed fAPAR becomes apparent if soil moisture simulations for 2 m rooting depth (*a priori*) are compared to measurements from the top 2 m of soil (see Fig. 8a). While there is a good agreement for the wet season, measured soil water reserves decline by only 100 mm as opposed to 200 mm for the simulations. The reason does not, however, lie in an overestimate of evapotranspiration, but in an underestimate of root depth, as Fig. 8b reveals. Here, the total plant-available soil moisture measured down to a soil depth of 8 m is compared to both the *a priori* and the fAPAR assimilated simulations. The decline in soil moisture reserves from wet to dry season is now more than 400 mm, both for observations and the fAPAR assimilation, while there is still sufficient available soil moisture during the dry season to sustain plant growth.

As the fAPAR assimilation scheme seems to be able to capture some of the tropical-rainforest hydrology as bserved for this site, it may be instructive to further investigate the modeling assumptions that lead to this rather important result. Some insight into those mechanisms is offered by Fig. 9. It shows net primary productivity (NPP), i.e. the carbon uptake of vegetation after allowing for respiration losses, for three cases: *a priori*, assimilated, and with fAPAR prescribed by fixing the leaf area index (LAI) at five, the value reported

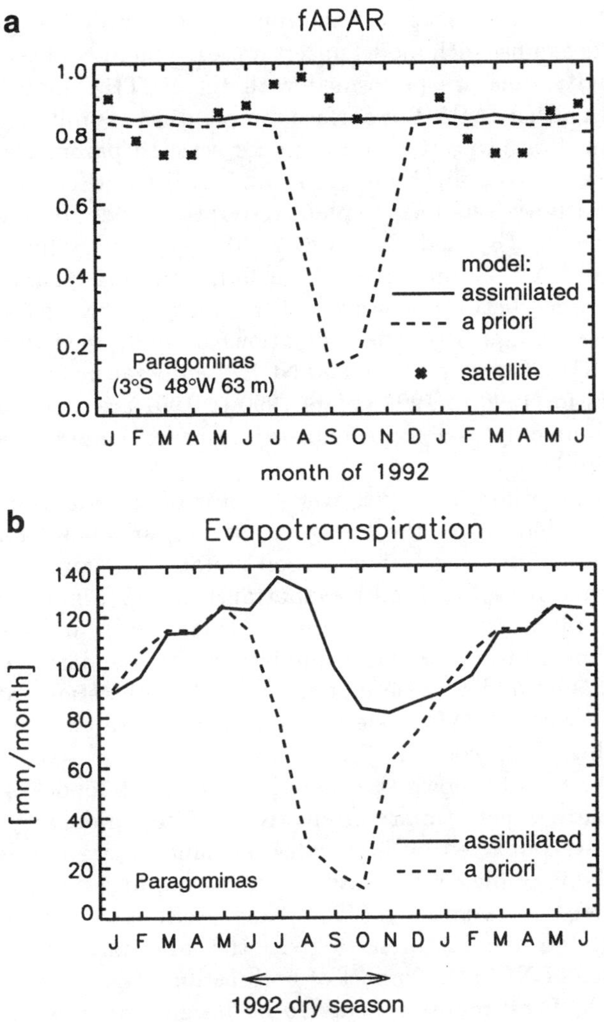

Figure 7. Satellite derived fAPAR and simulations of fAPAR (a) and evapotranspiration (b) for a site in the eastern Amazonian rainforest. Simulations use the BETHY vegetation model with and without assimilation of the satellite derived fAPAR values. Daily precipitation values were taken from *Nepstad et al.* [1994].

by the authors of the field study. In the *a priori* case, NPP declines to zero already in August. As the phenology scheme is designed to protect the vegetation from negative values of NPP, leaf shedding starts and stabilizes NPP around zero, even though the LAI, and with it fAPAR, does not fall to zero entirely (cf. Fig. 7a). By contrast, if fAPAR is prescribed to the model, a strategy followed by some global modeling groups (see introduction), the additional respiration costs required to maintain a high LAI let NPP values drop heavily below zero. It is thus the carbon balance that forces leaf shedding in the model, and thus leads to a close link between fAPAR and the surface water balance, a mechanism exploited in the assimilation scheme presented here.

One conclusion from this simulation is that photosynthesis and gross primary productivity (GPP) of evergreen tropical rainforests should decline only moderately even during prolonged dry seasons, to sustain NPP significantly higher than zero. In fact, for an evergreen forest site with similar climatic conditions in the Brazilian province of Rondônia, no decline in net carbon fluxes to the forest could be found [*Grace et al.*,

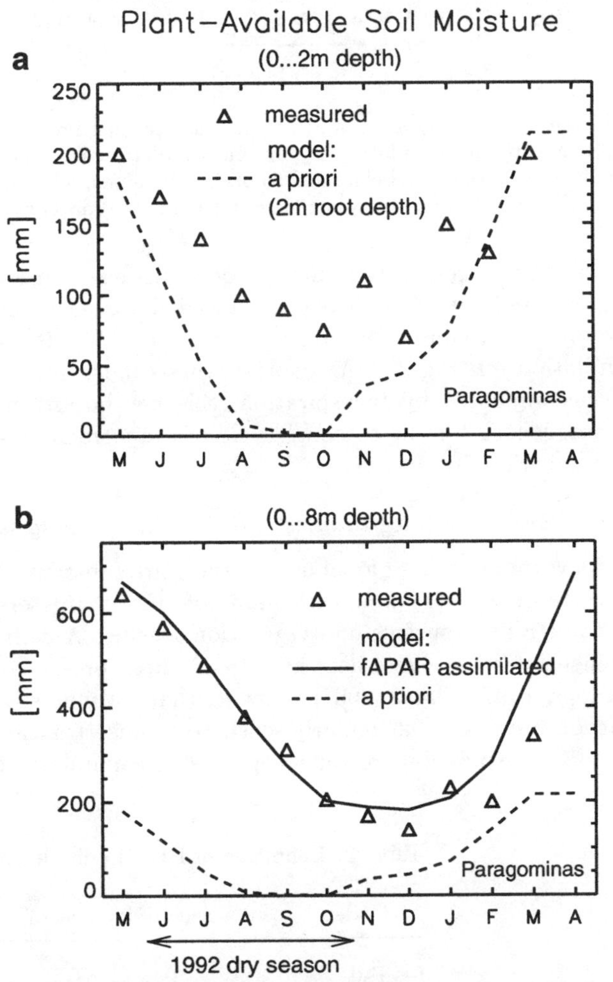

Figure 8. Comparison of measured plant-available soil moisture by *Nepstad et al.* [1994] with simulations. (a) Measurements from the top 2 m of soil and simulations with a rooting depth set also to 2 m. (b) Measured plant-available soil moisture down to a soil depth of 8 m, simulations either for 2 m root dept and without assimilation, or after assimilating satellite-derived fAPAR, thus modifying rooting depth and maximum plant-available soil moisture content.

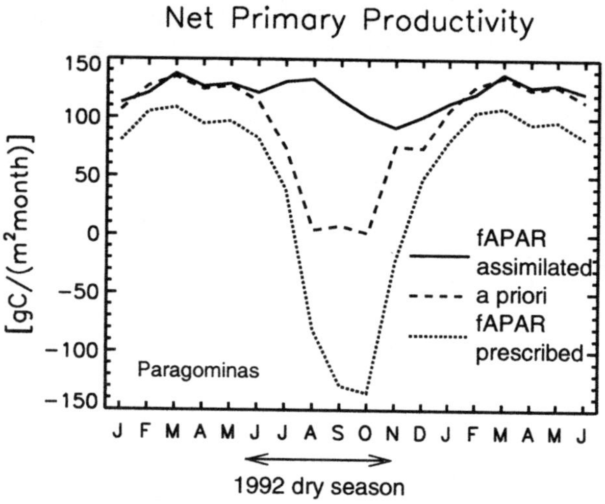

Figure 9. Monthly simulated net primary productivity for the same model simulations, plus one simulation in which the leaf are index has been fixed to observed values, without changing maximum plant-available soil moisture content.

1995]. The forest appeared to photosynthesize almost unrestricted all through a pronounced dry season, as simulated here for the Paragominas site after fAPAR assimilation (cf. Fig. 9). As photosynthesis must always be accompanied by transpiration, this has important consequences for the seasonal cycle of evapotranspiration (cf. Fig. 7b).

6.4. Parameter Uncertainties and Sensitivity Analysis

As with any other global model, there are substantial uncertainties associated with many of the parameters of the BETHY surface and vegetation scheme. A comprehensive error assessment has already been presented [*Knorr*, 1997; *Knorr*, 2001], showing that estimates of carbon fluxes are particularly sensitive to uncertainties in plant respiration, rooting depth, and a number of other factors. To assess the robustness of previous results against such model uncertainties, a number of sensitivity runs are performed with the BETHY model, with either fAPAR or surface temperature assimilation turned on. Uncertainty ranges for selected parameters are adopted from [*Knorr*, 2001] and are listed in Table 2. Sensitivity tests refer to plant respiration (code 'R', parameters $f_{R,l}$ and f_G, Equ. 2, 3), vegetation albedo (code 'A', parameter a_c, Equ. 9a), and aerodynamic exchange between canopy and free air (code 'X', parameter u, Equ. 10). The simulations are performed with the GPCC precipitation and NCEP temperature anomalies for 1990 to 1992 (years 1990 to 1991 are used for spin-up only, as described above for surface temperature assimilation).

The results, averaged over a sector of eastern Amazonia that covers the Paragominas site, are shown for monthly net vegetation carbon uptake (net primary productivity, Fig. 10a,b), evapotranspiration (Fig. 10c,d), and plant-available soil moisture (Fig. 10e,f), and are compared to the *a priori* simulation with standard parameters and no assimilation, as well as the standard runs with fAPAR (code 'f0') and surface temperature assimilation (code 'Ts0').

As already shown elsewhere [*Knorr*, 1997], model estimates of net primary productivity (NPP) in the tropics are impacted by high rates of plant respiration as a result of high temperatures, increasing uncertainties in NPP substantially. The same result is shown on Fig. 10a and b. In fact, given the large uncertainty range of NPP, the impact of both assimilation schemes on NPP estimates appears to be insignificant, except possibly for the beginning of the dry season (May and June).

Comparing the impact of the two assimilation schemes on the simulated seasonal cycle of evapotranspiration (Fig. 10c and d), we find that fAPAR assimilation

Table 2. Definition of BETHY model versions for the sensitivity analysis

Code	Sensitivity test of	Description	Standard
f0, Ts0	(standard)		
fR+, TsR+	plant respiration	$f_{R,l} = 0.3$, $f_G = 0.3$	$f_{R,l} = 0.4$
fR-, TsR-	"	$f_{R,l} = 0.5$, $f_G = 0.2$	$f_G = 0.25$
fA+, TsA+	canopy albedo	$a_c = 0.20$	$a_c = 0.15$
fA-, TsA-	"	$a_c = 0.10$	
fX+, TsX+	aerodynamic exchange	$u = 6.0$m/s	$u = 3.0$m/s
fX-, TsX-	"	$u = 1.5$m/s	

Cf. Equ. 2 ($f_{R,l}$), Equ. 3 (f_G), Equ. 9a (a_c) and Equ. 10 (u).

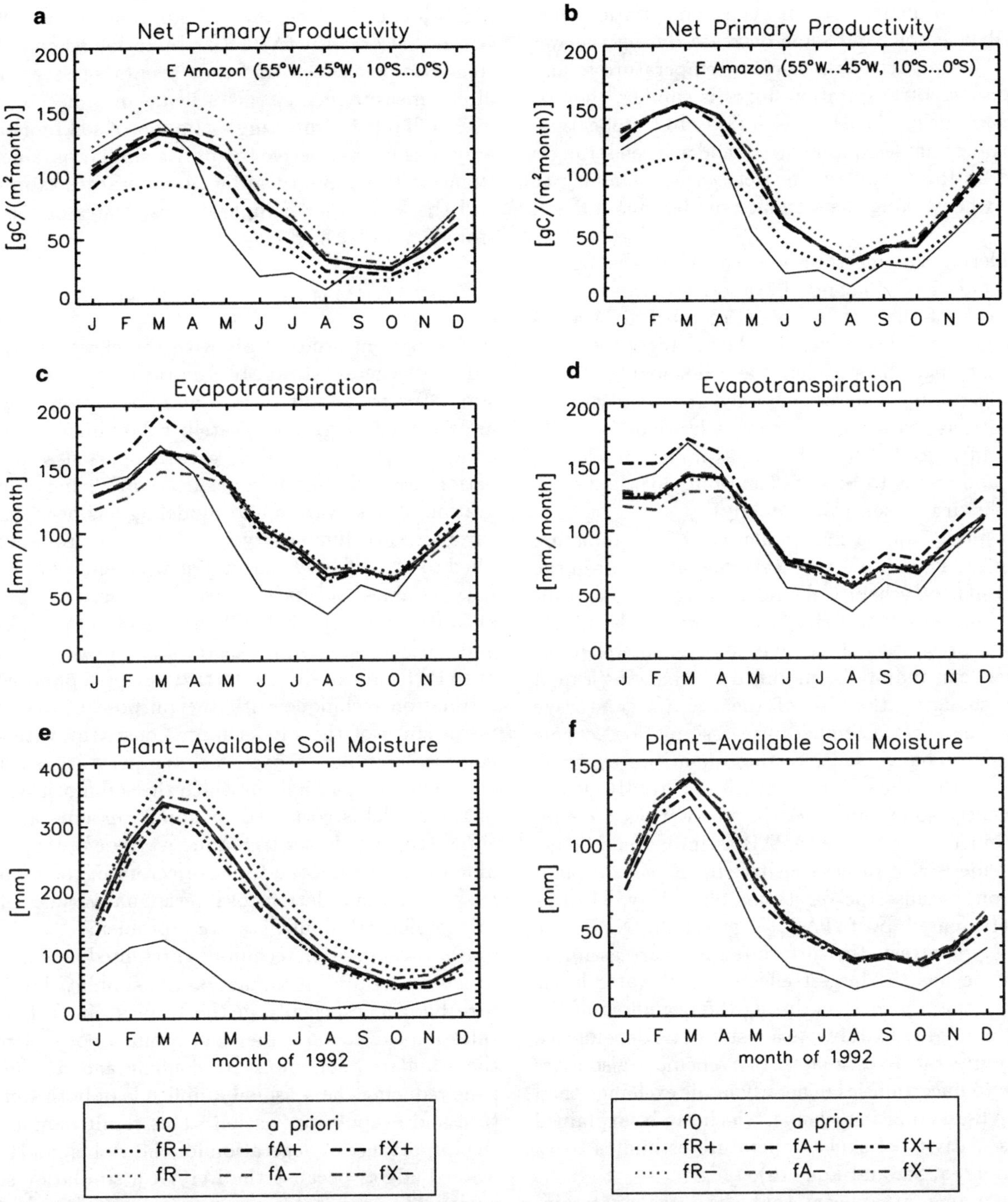

Figure 10. Seasonal cycle of simulated net primary productivity (a, b), evapotranspiration (c, d), and plant-available soil moisture (e, f) for a section of the eastern Amazonian rainforest, using climate data for 1992. Simulations are without satellite data assimilation in the standard model configuration of the BETHY model (*a priori*), and for a range of model versions both with fAPAR assimilation (a, c, d), marked 'f', and with surface temparature assimilation (b, d, f), marked 'Ts'. '0' refers to the standard configuration, and 'R+' to 'X-' to sensitivity runs reflecting estimates of model parameter uncertainties, as described in the text.

again leads to an increase in evapotranspiration during the dry season, and some decrease during the wet season. The impact of the surface temperature assimilation on evapotranspiration fluxes is similar: there is an increase during the dry season (May to October), although somewhat less pronounced, and an even stronger decrease at the height of the wet season (January to March), when taking the standard model runs 'f0' and 'Ts0'.

Considering the uncertainty ranges produced by the model simulations, changes of evapotranspiration rates at the peak of the wet season (December to March) can be considered insignificant, while changes for other months appear to be significant and lie outside of estimated uncertainty ranges. Generally, uncertainty ranges for evapotranspiration are much smaller than for NPP, and the difference between the two assimilation schemes is small and seems to be significant only for the beginning of the dry season (May to July).

When finally comparing simulations of soil moisture (Fig. 10e and f), the fundamental difference between the two assimilation schemes should be taken into consideration: while fAPAR assimilation works with strictly prescribed precipitation, so that dry-season soil moisture levels can mainly be modified by changing runoff through changing the size of the soil water storage capacity, the surface temperature assimilation scheme is allowed to change soil moisture quantities by small amounts, so that the water balance is not strictly closed. Consequently, soil moisture during the wet season is increased much more by the fAPAR assimilation scheme. This scheme is also rather sensitive to changes in plant respiration, because the vegetation phenology, which is ultimately sensed by fAPAR, is driven by NPP (see above). By contrast, the surface temperature assimilation scheme has the largest effect on soil water levels relative to the *a priori* run during the beginning of the dry season in May to July. As a result of its dependence on skin temperature estimates, this scheme is also more sensitive to uncertainties in aerodynamic exchange than the fAPAR assimilation scheme, which can be explained by the sensitivity of skin temperature estimates to r_a (cf. Equ. 6, 8a, 8a, and Equ. 10).

Summarizing, it appears that the large uncertainties in NPP estimates do not affect the main conclusion drawn above, that dry-season evapotranspiration is much larger in many tropical environments than usually simulated with low values of rooting depth. Both assimilation schemes lead to the same changes in simulated evapotranspiration rates within given model uncertainties. Soil moisture estimates, however, turn out to be highly dependent on the assimilation method chosen. Given that only the fAPAR assimilation scheme strictly conserves soil water, and that results agree with some direct measurements, this scheme may be considered more reliable for inferring estimates of soil moisture. It may thus be instructive to apply a similar parameter estimation technique to surface temperature assimilation and check whether comparable estimates of soil water storage are obtained.

7. SUMMARY AND CONCLUSION

The present study deals with the effect of land surface and vegetation variables on the global water budget simulated by a land surface and vegetation model. It aims at the integration of satellite data into such a modeling framework, given the general sparseness of observations for global-scale land surface modeling. The integration of observation and modeling has been achieved through two different types of data assimilation, applied to two different sets of global, satellite-derived surface variables: vegetation cover, expressed as fraction of absorbed photosynthetically active radiation (fAPAR), and surface skin temperature. While the assimilation of fAPAR information is carried out by a parameter reestimation technique with the purpose of making the mean state of the model more consistent with observations, assimilation of surface temperature is achieved with a nudging scheme in which the soil moisture status of the model is corrected by small amounts at a daily time step. The latter technique is, therefore, more suitable for correcting the time trajectory of the model, for example for simulations of interannual variations when meteorological input data are a problem.

Both assimilation techniques presented here are shown to have a significant impact on the simulated soil moisture budget, especially in the tropics. In fact, the results appear to be somewhat similar, both increasing the simulated soil moisture content and at the same time reducing the seasonal amplitude of both soil moisture and evapotranspiration. This result demonstrates the importance of vegetation for the global land surface water budget, because the fAPAR assimilation scheme achieves to use vegetation cover as an effective indicator of soil moisture and evapotranspiration.

Hydrological models have much evolved since the use of simple bucket schemes in the late 1960s, often by incorporating many effects related to vegetation activity. The issue is now to make better use of existing large-scale observations to improve model simulations. For example, for large parts of the tropics, where most of the water and energy turnover takes place, coverage of traditional meteorological data is often highly insufficient. Future large-scale modeling schemes should, therefore, try to account for the information contained in various satellite data. While this study is certainly only a first attempt into that direction, the significance and usefulness of global satellite information has clearly been demonstrated. In addition, a methodology has been presented that may later allow the integration of both types of satellite data into one common assimilation scheme. To further validate the results gained here, systematic comparison with routinely observed soil moisture would also be desirable. Since detailed observations will always be carried out only for a very limited number of sites, there is a clear need for further progress in routine satellite observation of soil moisture. The quality of large-scale hydrological simulations could then greatly benefit from the integration of such data with the modeling and data assimilation techniques presented here.

Acknowldegments. The first author would like to express his gratitude to Michel Verstraete and Bernard Pinty for their help with satellite data interpretation and processing, and to Garik Gutman for his comments and suggestions for the original manuscript. Support by Wolfgang Cramer and Andrew Friend, who have provided climate data fields, is gratefully acknowledged.

REFERENCES

Asrar, G., R. B. Myneni, and B. J. Choudhury, Spatial heterogeneity in vegetation canopies and remote sensing of absorbed photosynthetically active radiation: a modeling study. *Remote Sens. Environ., 41*, 85-1031, 1992.

Beerling, D. J., and W. P. Quick, A new technique for estimating rates of carboxylation and electron transport in leaves of C3 plants for use in dynamic global vegetation models. *Global Change Biol., 1*, 289-294, 1995.

Berthelot, B., Dedieu, G., Cabot, F. and Adam, S. Estimation of surface reflectance and vegetation index using NOAA/AVHRR: Methods and results at global scale. Communication for the 6th international symposium on physical measurements and signatures in remote sensing, ISPRS, Val d'Isère, France, Jan. 17 to 21, 1994.

Blyth, K., The use of microwave remote sensing to improve spatial parameterization of hydrological models, *J. Hydrol., 152*, 103-129, 1993.

Bouttier, F., J.-F.Mahfouf and J. Noilhan, Sequential assimilation of soil moisture from atmospheric low-level parameters. Part I: Sensitivity and calibration studies, *J. Climate, 32*, 1335-1351, 1993a.

Bouttier, F., J.-F. Mahfouf and J. Noilhan, Sequential assimilation of soil moisture from atmospheric low-level parameters. Part II: Implementation in a mesoscale model, *J. Climate, 32*, 1352-1364, 1993b.

Box, E. O. *Macroclimate and plant forms: An introduction to predictive modeling in phytogeography,* 174pp., Junk, The Hague, Netherlands, 1981.

Brutsaert, W. *Evaporation into the atmosphere,* 299 pp., Reidel, Dordrecht, Netherlands, 1982.

Budyko, M. I., *Heat balance of the Earth's surface* (in Russian), Gidrometeoizdat, 255 pp., 1956.

Budyko, M.I., *Climate and Life,* Academic Press, New York, 508 pp, 1974.

Charney, J. G., P. H. Stone, and W. J. Quirk, Drought in the Sahara: A biogeophysical feedback mechanism, *Science, 187,* 434-435, 1975.

Charney, J. G., W. J. Quirk, S.-H. Chow, and J. Kornfield, A comparative study of the effects of albedo change on drought in semi-arid regions, *J. Atmos. Sci., 34,* 1366-1385, 1977.

Christensen, J. H., B. Machenhauer, R. G. Jones, C. Schar, P. M. Ruti, M. Castro, and G. Visconti, Validation of present-day regional climate simulations over Europe - LAM simulations with observed boundary conditions, *Climate Dyn., 13,* 489-5061, 1997.

Claussen, M., The greening of the Sahara during the mid-holocene - results of an interactive atmosphere-biome model, *Global Ecol. Biogeogr. Lett., 6,* 369-377, 1997.

Collatz, G. J, M. Ribas-Carbo, and J. A. Berry, Coupled photosynthesis-stomatal conductance model for leaves of C_4 plants, *Aust. J. Plant Physiol., 19,* 519-538, 1992.

DeFries, R. S., and J. R. G. Townshend, NDVI-derived land cover classifications at a global scale, *Int. J. Remote Sensing, 15,* 3567-3586, 1994.

Dickinson, R. E., B. Pinty, and M. M. Verstraete, Relating surface albedos in GCM to remotely sensed data, *Agric. For. Meteor., 52,* 109-131, 1990.

Dickinson, R. E., A. Henderson-Sellers, A., and P. J. Kennedy, Biosphere-atmosphere transfer scheme (BATS) version 1e as coupled to the NCAR community climate model, *NCAR Technical Note NCAR/TN-387+STR,* National Center for Atmospheric Research, Boulder, Colorado, 1993.

Entekhabi, D., H. Nakamura, and E. G. Njoku, Solving the inverse problem for soil moisture and temperature profiles by sequential assimilation of multifrequency remotely sensed observations, *IEEE Trans. Geosc. Remote Sens., 32,* 438-448, 1994.

Garratt, J. R., Sensitivity of climate simulations to land-surface and atmospheric boundary-layer treatments - a review, *J. Climate, 6,* 419-449, 1993.

Geiger, R., *The Climate near the Ground,* Harvard University Press, Cambridge, Mass., 1965.

Diak, G. R., Evaluation of heat flux, moisture flux and aerodynamic roughness at the land surface from knowledge of PBL height and satellite derived skin temperatures, *Agric. For. Meteorol., 52,* 181-198, 1990.

Dunne, K. A., and C. J. Willmott, Global distribution of plant-extractable water capacity of soil, *Int. J. Climat., 16,* 841-859, 1996.

Farquhar, G. D., S. von Caemmerer, and J. A. Berry, A biochemical model of photosynthesis in leaves of C_3 species, *Planta, 149,* 78-90, 1980.

Federer, C. A., A soil-plant-atmosphere model for transpiration and availability of soil water, *Water Resour. Res., 15,* 555-562, 1979.

Flasse, S., and M. M. Verstraete, Monitoring the environment with vegetation indices: comparison of NDVI and GEMI using AVHRR data over Africa, in *Vegetation, Modelling and Climate Change Effects,* edited by F. Veroustraete and R. Ceulemans, pp. 107-135, Academic, The Hague, Netherlands, 1994.

Gates, W. L., J. S. Boyle, C. Covey, C. G. Dease, C. M. Doutriaux, R. S. Drach, M. Fiorino, P. J. Gleckler, J. J. Hnilo, S. M. Marlias, T. J. Phillips, G. L. Potter, B. D. Santer, K. R Sperber, K. E. Taylor, and D. N. Williams, An overview of the results of the Atmospheric Model Intercomparison Project (AMIP I), *Bulletin of the American Meteorological Society, 80,* 29-55, 1999.

Gobron, N., B. Pinty, M. M. Verstraete, and Y. Govaerts, A semi-discrete model for the scattering of light by vegetation, *J. Geophys. Res., 102,* 9431-9446, 1997.

Goel, N. S., and W. Qin, Influences of canopy architecture on relationships between various vegetation indices and LAI and FPAR: A computer simulation, *Remote Sens. Rev., 10,* 309-347, 1994.

GPCC, Monthly Precipitation Estimates Based on Gauge Measurements on the Continents for the Year 1987 (Preliminary Results) and Future Requirements, *Report No. DWD/K7/WZN-1992/08-1,* Global Precipitation Climatology Centre, Deutscher Wetterdienst, Offenbach, Germany, 1992.

Grace, J., J. Lloyd, J. Mcintyre, A. Miranda, P. Meir, H. Miranda, C. Nobre, J. Moncrieff, J. Massheder, Y. Mahli, I. Wright, and J. Gash, Carbon dioxide uptake by an undisturbed tropical rain forest in South-West Amazonia, 1992-1993, *Science, 270,* 778-780, 1995.

Gutman, G., Review of the workshop on the use of satellite-derived vegetation indices in weather and climate prediction models, *Bull. Am. Meteorol. Soc., 71,* 1458-1463, 1990.

Gutman, G., and A. Ignatov, The relative merit of cloud/clear identification in the NOAA/NASA Pathfinder AVHRR 10-day composites, *Int. J. Remote Sensing, 17,* 3295-3304, 1996.

Gutman, G., D. Tarpley, A. Ignatov, and S. Olson, The enhanced NOAA global land dataset from the Advanced Very High Resolution Radiometer, *Bull. Am. Meteorol. Soc., 76,* 1141-1156, 1995.

Henderson-Sellers, A., K. McGuffie, and A. J. Pitman, The Project for Intercomparison of Land-Surface Parameterization Schemes (PILPS): 1992 to 1995. Climate Dyn., 12, 849-859, 1996.

Holben, B. N., Characteristics of maximum-value composite images from temporal AVHRR data, *Int. J. Remote Sens., 7,* 1417-1434, 1986.

Holdridge, L. R., Determination of world formations from simple climatic data, *Science, 105,* 193-215, 1947.

Jarvis, P. G., and K. G. McNaughton, Stomatal control of transpiration: scaling up from leaf to region. *Adv. Ecol. Res.* 15, 1-49, 1986.

Kaufman, Y. J., and B. N. Holben, Calibration of the AVHRR visible and near infrared bands by atmospheric scattering, ocean glint, and desert reflection, *Int. J. Remote Sensing, 14,* 21-52, 1993.

van den Hurk, B. J., W. Bastiaanssen, H. Pelgrum and E. Meijgaard, A new methodology for assimilation of initial soil moisture fields in weather prediction models using Meteosat and NOAA data, *J. Appl. Meteorol., 36,* 1271-1283, 1997.

Kelliher, F. M., R. Leuning, and E.-D. Schulze, Evaporation and canopy characteristics of coniferous forests and grasslands. *Oecologia, 95,* 152-163, 1993.

Kleidon, A. and M. Heimann, Optimised rooting depth and its impacts on the simulated climate of an atmospheric general circulation model, *Geophys. Res. Lett., 25,* 345-348, 1998.

Knorr, W., and M. Heimann, Impact of drought stress and other factors on seasonal land biosphere CO_2 exchange studied through an atmospheric tracer transport model, *Tellus, 47B,* 471-489, 1995.

Knorr, W., N. Gobron, Ph. Martin, B. Pinty, M. M. Verstraete and G. Dedieu, Constraining a climate driven vegetation model with satellite data, *Proceedings of the Photosynthesis and Remote Sensing Symposium,* EARSeL, Paris, pp. 269-279, 1995.

Knorr, W., Satellitengestützte Fernerkundung und Modellierung des globalen CO_2-Austauschs der Landvegetation: Eine Synthese, Examensarbeit Nr. 49, Max-Planck-Institut für Meteorologie, Hamburg, Germany. Available in English through
http://www.bgc-jena.mpg.de/ wolfgang.knorr, 1997.

Knorr W., Constraining a mechanistic vegetation model with satellite data,, IEEE International Geoscience and Remote Sensing Symposium Proceedings, Seattle, July 6 to 10, 1998.

Knorr, W., Annual and interannual CO2 exchanges of the terrestrial biosphere: process-based simulations and uncertainties, *Global Ecol. Biogeogr., 9,* 225-252, 2000.

Knorr, W., and J.-P. Schulz, Using Satellite Data Assimilation to Infer Global Soil Moisture Status and Vegetation Feedback to Climate, in *Remote Sensing and Climate Modeling: Synergies and Limitations,* edited by M. Beniston and M. M. Verstraete, Advances in Global Change Research series, Kluwer, Dordrecht, Netherlands and Boston, Mass., 2001.

Lakshmi, V., E. F. Wood and B. J. Choudhury, Evaluation of Special Sensor Microwave Imager Satellite data for regional soil moisture estimation over Red River Basin, *J. Appl. Meteorol., 36,* 1309-1328, 1997.

Lakshmi, V., A Simple Surface Temperature Assimilation Scheme for Use in Land Surface Models, *Water Resour. Res.*, 2000, in press.

Leemans, R., and W. Cramer, The IIASA climate database for mean monthly values of temperature, precipitation and cloudiness on a terrestrial grid, *RR-91-18,* Institute of Applied Systems Analysis, Laxenburg, Austria, 1991.

Leprieur, C., M. M. Verstraete, and B. Pinty, Evaluation of the performance of various vegetation indices to retrieve vegetation cover from AVHRR data, *Remote Sens. Rev., 10,* 265-284, 1994.

Lindroth, A., and S. Halldin, Numerical analysis of pine forest evaporation and surface resistance, *Agric. For. Meteorol., 38,* 59-79, 1986.

Los, S. O., C. O. Justice, and C. J. Tucker, A global 1° by 1° NDVI data set for climate studies derived from the GIMMS continental NDVI data, *Int. J. Remote Sensing, 15,* 3493-3518, 1994.

Lüdeke, M. K. B., F.-W. Badeck, R. D. Otto, C. Hger, S.

Dnges, J. Kindermann, G. Wrth, T. Lang, U. Jkel, A. Klaudus, P. Ramge, S. Habermehl, and G. H. Kohlmaier, The Frankfurt Biosphere Model: a global process oriented model of seasonal and long-term CO_2 exchange between terrestrial ecosystems and the atmosphere, I. Model description and illustrative results for cold deciduous and boreal forests, *Clim. Res., 4*, 143-166, 1994.

McLaughlin, D., Recent developments in hydrologic data assimilation, *Reviews of Geophysics*, Supplement, U.S. National Report to International Union of Geodesy and Geophysics 1991-1994, pp. 977-985, 1995.

Meyer, D., M. M. Verstraete, and B. Pinty, The effect of surface anisotropy and viewing geometry on the estimation of NDVI from AVHRR, *Remote Sensing Rev., 12,* 3-27, 1995.

Miller, D. A., and R. A. White, A conterminous United States multi-layer soil characteristics data set for regional climate and hydrology, *Earth Interactions, 2,* 1998.

Müller, M. J. *Selected climatic data for a global set of standard stations for vegetation science,* Junk, The Hague, Netherlands, 1982.

McNider, R. T., A. J. Song, D. M. Casey, P. J. Wetzel, W. L. Crosson and R. M. Rabin, Towards a dynamic-thermodynamic assimilation of satellite surface temperature in numerical atmospheric models, *Mon. Wea. Rev., 122,* 2784-2803, 1994.

Milly, P. C. D., and K. A. Dunne, Sensitivity of the global water cycle to the water-holding capacity of land, *J. Climate, 7,* 506-526, 1994.

Mintz, Y., The sensitivity of numerically simulated climates to land-surface boundary conditions, in *The Global Climate,* edited by J.T. Houghton, pp. 79-105, Cambridge University Press, Cambridge, U.K., 1984.

Monteith, J.L., Evaporation and environment, *Symp. Soc. Exp. Biol., 19,* 205-234, 1965.

Nepstad, D. C., C. R. de Carvalho, E. A. Davidson, P. H. Jipp, P. A. Lefebvre, G. H. Negeiros, E. D. da Silva, T. A. Stone, S. E. Trumbore, and S. Vieira, The role of deep roots in the hydrological and carbon cycles of Amazonian forests and pastures, *Nature, 372,* 666-669, 1994.

Ottle, C. and D. Vijal-Madjar, Assimilation of soil moisture inferred from infrared remote sensing in a hydrological model over the HAPEX-MOBILHY region, *J. Hydrol., 158,* 241-264, 1994.

Pinker, R. T. and Laszlo, I., Global distribution of photosynthetically active radiaton as observed from satellites, *J. Clim., 5,* 56-65, 1992.

Pinty, B., and M. M. Verstraete, GEMI: A non-linear index to monitor global vegetation from satellites, *Vegetatio, 101,* 1335-1372, 1992.

Potter, S. C., J. T. Randerson, C. B. Field, P. A. Matson, P. M. Vitousek, H. A. Mooney, and S. A. Klooster, Terrestrial ecosystem production: a process model based on global satellite and surface data, *Global Biogeochem. Cycles, 7,* 811-841, 1993.

Press, W. H., S. A. Teukolsky, W. T. Vetterling, and B. P. Flannery, *Numerical Recipes in Fortran,* pp. 402-406, 2nd edition, Cambridge University Press, Cambridge, U.K., 1992.

Price, J. C., Land surface temperature measurements from the split window channels of the NOAA-7 Advanced Very High Resolution Radiometer, *J. Geophys. Res., 89,* 7231-7237, 1984.

Raich, J. W., E. B. Rastetter, J. M. Melillo, D. W. Kicklighter, P. A. Steudler, B. J. Peterson, A. L. Grace, B. Moore III, and C. J. Vörösmarty, Potential net primary productivity in South America: application of a global model, *Ecol. Appl., 1,* 399-429, 1991.

Ritchie, J. T., Model for predicting evaporation from a row crop with incomplete cover, *Water Resour. Res., 8,* 1204-1213, 1972.

Rudolf, B., H. Hauschild, W. Rueth, and U. Schneider, Terrestrial precipitation analysis: operational method and required density of point measuremenets, in *Global Precipitations and Climate Change,* edited by M. Desbois and F. Desalmond, NATO ASI Series I, Vol. 26, pp. 173-186, Springer-Verlag, Berlin, 1994.

Rahman, H., and G. Dedieu, SMAC: A simplified method for the atmospheric correction of satellite measurements in the solar spectrum, *Int. J. Remote Sensing 15,* 123-143, 1994.

Ruimy, A., G. Dedieu, and B. Saugier, TURC: A diagnostic model of continental gross primary productivity and net primary productivity, *Global Biogeochem. Cycles, 10,* 269-285, 1996.

Ryan, M. G., Effects of climate change on plant respiration, *Ecol. Appl., 1,* 157-167, 1991.

Saxton, K. E., W. J. Rawls, J. S. Romberger, and R. I. Papendic, Estimating generalized soil-water characteristics, *Soil Sci. Soc. Am. J., 50,* 1031-1036, 1986.

Schulze, E.-D., F. M. Kelliher, Ch. Krner, J. Lloyd, and R. Leuning, Relationships among maximum stomatal conductance, ecosystem surface conductance, carbon assimilation rate, and plant nitrogen nutrition: a global ecology scaling exercise, *Ann. Rev. Ecol. Syst., 25,* 629-660, 1994.

Sellers, P.J., Canopy reflectance, photosynthesis and transpiration, *Int. J. Remote Sensing, 6,* 1335-1372, 1985.

Sellers, P. J., C. J. Tucker, G. J. Collatz, S. O. Los, C. O.Justice, D. A. Dazlich and D. A.Randall, A global 1° by 1° NDVI data set for climate studies. Part 2: The generation of global fields of terrestrial biophysical parameters from the NDVI, *Int. J. Remote Sensing, 15,* 3519-3545, 1994.

Sellers, P. J., B. W. Meeson, F. G. Hall, G. Asrar, R. E. Murphy, R. A. Schiffer, F. P. Bretherton, R. E. Dickinson, R. G. Ellingson, C. B. Field, K. F. Huemmrich, C. O. Justice, J. M. Melack, N. T. Roulet, D. S. Schimel, and P. D. Try, Remote sensing of the land surface for studies of global change: Models - algorithms - experiments, *Remote Sens. Environ., 51,* 3-26, 1995.

Shukla, J., and Y. Mintz, Influence of land-surface evapotranspiration on the earth's climate., *Science* **215,** 1498-1501, 1982.

Sud, Y. C., J. Shukla and Y. Mintz, Influence of land surface roughness on atmospheric circulation and precipitation: A sensitivity study with a general circulation model, *J. Appl. Meteorol., 27,* 1036-1054, 1988.

Susskind, J., P. Piraino, L. Rokke, L. Iredell, and A. Mehta, Characteristics of the TOVS Pathfinder Path A Data Set, *Bull. Am. Meteorol. Soc., 78,* 1449-1472, 1997.

Townshend, J. R. G., C. O. Justice, D. Skole, J.-P. Malingreau, J. Cihlar, P. Teillet, F. Sadowski, and S. Ruttenberg, The 1 km resolution global data set: Needs of the International Geosphere Biosphere Programme, *Int. J. Remote Sens., 15,* 3417-3441, 1994.

Verma, S. B., D. D. Baldocchi, D. E. Anderson, D. R. Matt, and R. J. Clement, Eddy fluxes of CO_2, water vapor and sensible heat over a deciduous forest, *Boundary Layer Meteorol., 36,* 71-91, 1986.

Verstraete, M. M., Retrieving canopy properties from remote sensing measurements, in *Imaging Spectrometry - a tool for Environmental Observations,* edited by J. Hill and J. Mégier, pp. 109-123, ECSC, EEC, EAEC, Brussels and Luxemburg, 1994.

Verstraete, M. M., B. Pinty, and R. E. Dickinson, A physical model of the bidirectional reflectance of vegetation canopies. 1. Theory, *J. Geophys. Res., 95,* 11775-11765, 1990.

Viterbo, P., and A. C. M. Beljaars, An improved land surface parameterization scheme in the ECMWF model and its validation, *J. Climate, 8,* 2716-2748, 1995.

Wanner W., A. H. Strahler, B. Hu, P. Lewis, J. P. Müller, X. Li, C. L. B. Schaaf, and M. J. Barnsley, Global retrieval of bidirectional reflectance and albedo over land from EOS MODIS and MISR data - Theory and algorithm, *J. Geophys. Res., 102(D14),* 17143-17161, 1997.

Weiss, A. and J. A. Norman, Partitioning solar radiation into direct and diffuse, visible and near-infrared components, *Agric. For. Meteorol., 34,* 205-213, 1985.

Wilson, M. F., and A. Henderson-Sellers, A global archive of land cover and soils data for use in general circulation models, *J. Clim., 5,* 119-143, 1985.

W. Knorr, Max Planck Institute for Biogeochemistry, Carl-Zeiss-Promenade 10, 07745 Jena, Germany. (e-mail: wknorr@bgc-jena.mpg.de)

V. Lakshmi, Department of Geological Sciences, University of South Carolina, Columbia, SC 29208, U.S.A.

Experimental Design and Initial Results From the Mahurangi River Variability Experiment: MARVEX

Ross Woods[1], Rodger Grayson[2], Andrew Western[2], Maurice Duncan[1], David Wilson[2], Rodger Young[2], Richard Ibbitt[1], Roddy Henderson[1], Tom McMahon[2]

Spatial variability is at the heart of many of today's challenges in hydrology, but few of our hydrological theories take account of the richness in structure and diversity that comprises a real catchment. A key constraint on our ability to more comprehensively represent variability is the lack of appropriate field data. The MAhurangi River Variability EXperiment (MARVEX) is specifically designed to answer the question "what are the most important sources of variability in streamflow and hydrological response" for a temperate area of New Zealand and to provide data for the testing and development of spatial modelling methods. This chapter describes the experimental set up and rationale for the MARVEX data collection and presents some of the preliminary data and analysis related to rainfall-runoff response and soil moisture behaviour. At the time of writing, the MARVEX project is ongoing and we are actively collecting data, both on a continuous basis, and via intensive field campaigns. Coordinated, extensive field studies such as MARVEX are very expensive to undertake, yet we firmly believe that to make progress on some of the key outstanding questions in hydrology, information of this sort is vital. Without this type of information, we have no way of testing the models or theoretical concepts that embody our understanding of hydrological phenomena and the science of hydrology will remain severely constrained.

1. INTRODUCTION

Spatial variability is at the heart of many of today's challenges in hydrology. Much has been discovered about the structure and characteristic features of idealised hydrological systems, assuming either constant temporal forcing, uniform spatial forcing, or uniform media and geometry. Well-established theories exist for the motion of water in prismatic channels, as sheet flow over a plane, and in uniform porous media. However, few of our hydrological theories take account of the richness in structure and diversity that comprises a real catchment, without taking the reductionist, information-hungry step of applying the theory to spatial elements small enough to have effectively uniform inputs, uniform geometry and uniform flow resistance. At the very best, uniform assumptions have been relaxed by using theoretical distributions but these have been too constrained to represent the variability apparent in nature. As a result, most attempts to apply hydrological theory require either (i) the subdivision of the space-time domain into regions that are assumed to have uniform properties or (ii) the assumption that a heterogeneous system can be modelled by an equivalent uniform system.

[1]National Institute for Water and Atmospheric Research, New Zealand

[2]Centre for Environmental Applied Hydrology, Department of Civil and Environmental Engineering, University of Melbourne, Australia

Land Surface Hydrology, Meteorology, and Climate: Observations and Modeling
Water Science and Application Volume 3, pages 201-213

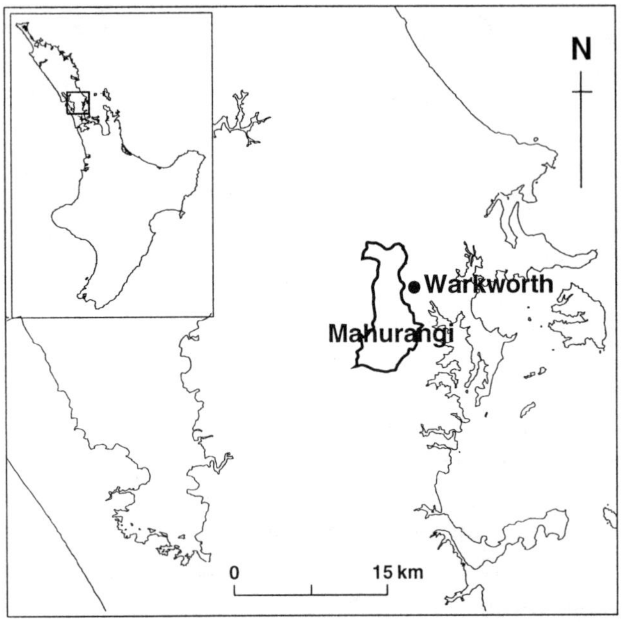

Figure 1. Location map for Mahurangi catchment.

The temporal discretisation approach has served hydrology adequately, because of the ready availability of daily and sub-daily data to force and calibrate hydrology models, and the fact that the catchments themselves usually change slowly relative to the temporal variability of the forcing. However, the difficulty in collecting spatially variable data for atmospheric forcing, catchment media, and hydrological responses has created severe practical and theoretical difficulties. It is unclear to what extent the many potential sources of spatial variability need to be incorporated in hydrological theory, models, and decision-making. At present, it is not realistic to include all possible sources of variability in any of these contexts. As a result, the selection of spatially variable phenomena to resolve in a hydrology model is usually made using intuition and data availability, rather than a quantitative assessment of the importance of the phenomena to the problem at hand.

A key constraint on our ability to more comprehensively represent variability is the lack of appropriate field data. There are only a small number of studies where spatial and temporal data of hydrological response have been collected to a level of detail that enables sources of variability to be properly defined. The USDA sites at Walnut Gulch *[Houser et al.,* 2000]; Goodwin Creek [*DeCoursey*, 1982; *Blackmarr*, 1995] and Reynolds Creek [*Seyfried and Wilcox*, 1995] amongst others, are exceptions where data collection and analysis for a range of environments is ongoing. There are also a small number of shorter term, targeted studies centred on particular processes such as the shallow subsurface flow work at Minifelt [*Lamb et al,* 2000]; the soil moisture work at Tarrawarra [*Western and Grayson*, 1998, 2000]; the flow pathways work at Panola [*Freer et al.*, 1998] and the saturated source area work of *Merz and Plate* [1997] at Weiherbach, to name a few. These studies have all shown that detailed spatial measurements can provide insights that are not possible with more conventional monitoring, and have provided data that for the first time, are capable of properly testing spatial simulations from distributed parameter hydrological models [*Grayson and Blöschl*, 2000].

The MAhurangi River Variability EXperiment (MARVEX) is specifically designed to answer the question "what are the most important sources of variability in streamflow and hydrological response" for a temperate area of New Zealand and to provide data for the testing and development of spatial modelling methods. The size of the project was too big for a single research group to undertake. Therefore a consortium of researchers, each with a particular interest in complimentary aspects of spatial and temporal variability, was formed to work together on the collection and analysis of a comprehensive set of field measurements described in the following section. These were designed to answer questions related to the effects on hydrological response of variations in soils, vegetation and precipitation and changes in scale. Ultimately we seek to provide a quantitative, predictive framework which explains, for a given climatic regime, the "transitions" in dominant sources of hydrological variability with changing time and space scales. We also seek to develop methods of representing the way in which small-scale variability is manifested at larger scales.

This chapter describes the experimental set up and rationale for the MARVEX data collection and presents some of the preliminary data and analysis related to rainfall-runoff response and soil moisture behaviour. The more detailed analyses to be undertaken in the near future are described and the specific hypotheses being tested are elucidated.

2. SITE DESCRIPTION

2.1. General

The Mahurangi River catchment drains 50 km^2 of steep hills and gently rolling lowlands located 70 km north of Auckland, New Zealand (Figure 1). This part of New Zealand experiences a warm, humid climate (Figure 2), with typical annual rainfalls of 1600 mm (maximum rainfall is usually in July, the middle of the austral winter) and annual pan evaporation of approximately 1300mm (maximum monthly temperature and pan evaporation occurring in January or February). Frosts are rare, and snow and ice are unknown. In late summer (February and March), the rem-

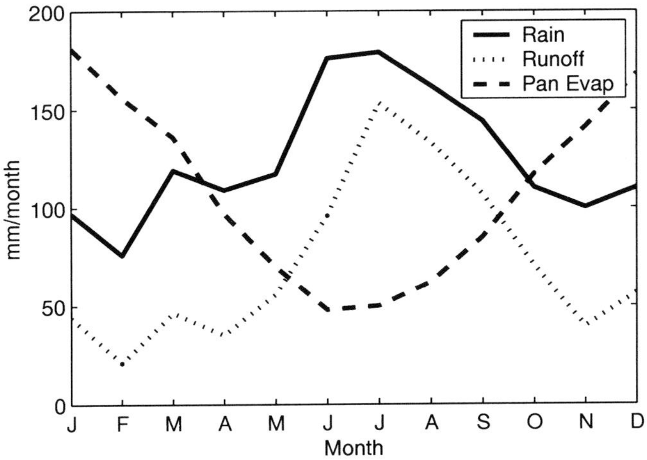

Figure 2. Mean monthly rainfall, runoff, and pan evaporation for Mahurangi.

nants of a tropical cyclone occasionally pass over northern New Zealand, producing intense bursts of rain. Convective activity is significant over the summer, whereas the majority of the winter rain comes from frontal systems.

The drainage network of the Mahurangi River divides the catchment into northern and southern subcatchments (Figure 3), whose junction is 5 km west of Warkworth, a town of 12,000 inhabitants at the mouth of the Mahurangi River, where it discharges into the Mahurangi estuary. Catchment elevations range from 250m above sea level on the northern and southern boundaries, to near sea level at Warkworth on the east coast (Plate 1a). The catchment's soils have developed over Waitemata sandstones, which typically display alternating layers of sandstone and siltstone, deposited during the periodic descents of New Zealand beneath sea level in the recent geological past. Most soils in the catchment are clay loams (Plate 1b), no more than a metre deep (see Table 1 for a typical profile description). Before the beginning of the 20th century, the dominant vegetation in the area was a mixture of native hardwood and podocarp forests. Sheep, cattle and deer now graze half of the catchment (the central lowlands), which is planted predominantly in pasture. Another quarter of the catchment is now in plantation forestry (predominantly *Pinus radiata* in the southern hill country), while the northern quarter of the catchment is a mixture of native forests and pasture (Plate 1c).

2.2. *Hydrometric and Soil Moisture Measurements*

A network of 29 stream gauges and 13 raingauges throughout the catchment form the backbone of the MARVEX network (Figure 3). This spatially intensive network was installed in 1997-98, and is planned to operate for 3-4 years. At most of the stream gauges, water levels are measured every two minutes using float and counterweight with compound v-notch weirs (a few sites use natural cross-sections). Streamflows are estimated with weir formulae, checked against current meter gaugings for each site. Rainfall depths are measured every two minutes using standard 200 mm collectors and 0.2mm tipping buckets. Monthly accumulations are corrected using bulk collecting raingauges adjacent to every recording gauge. Data for water level and rainfall are stored on site, and downloaded during monthly site visits.

Two weather radars are also available to augment the rainfall observations (Figure 3). The Physics Department of the University of Auckland deploys a mobile X-band radar for intensive campaigns of 1-2 months duration. This radar is sited in the southwest corner of the catchment and has a range of 15 km. It is configured to resolve rainfall on a 150m grid over the Mahurangi catchment, and provide an image every 5 seconds (typically amalgamated to two-minute average values for hydrological analyses). In addition, New Zealand's MetService operates a C-band radar located 10km northwest of the catchment at an elevation of 440m asl. It has a range of 250 km. Data from this radar is recorded once every 15 minutes, and values are interpolated onto a 1km grid.

The soil moisture component of MARVEX uses two sampling strategies: a network of continuously recording

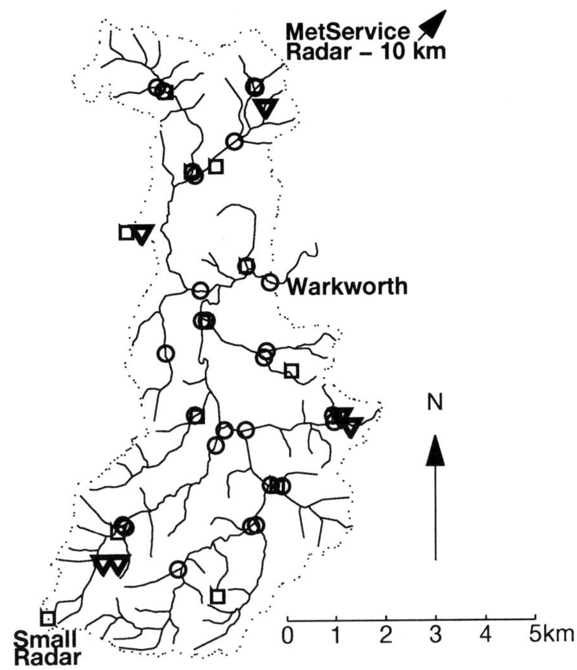

Figure 3. The MARVEX stream gauges (circles), raingauges (squares), soil moisture instrumentation (triangles) and weather radar sites.

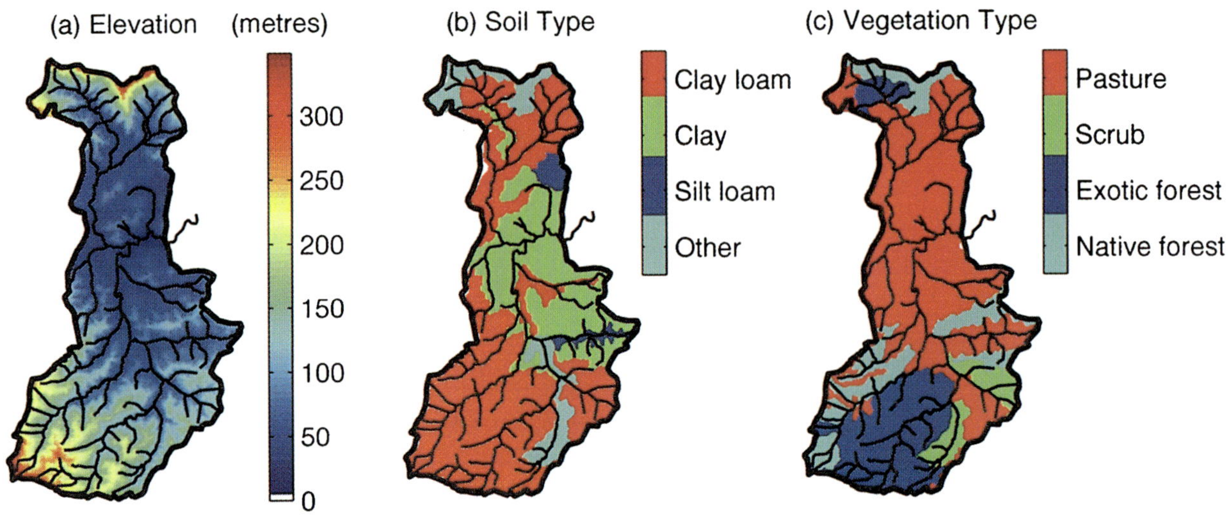

Plate 1. Maps of (a) elevation, (b) soil type and (c) vegetation type, each overlaid by the river network and catchment boundary.

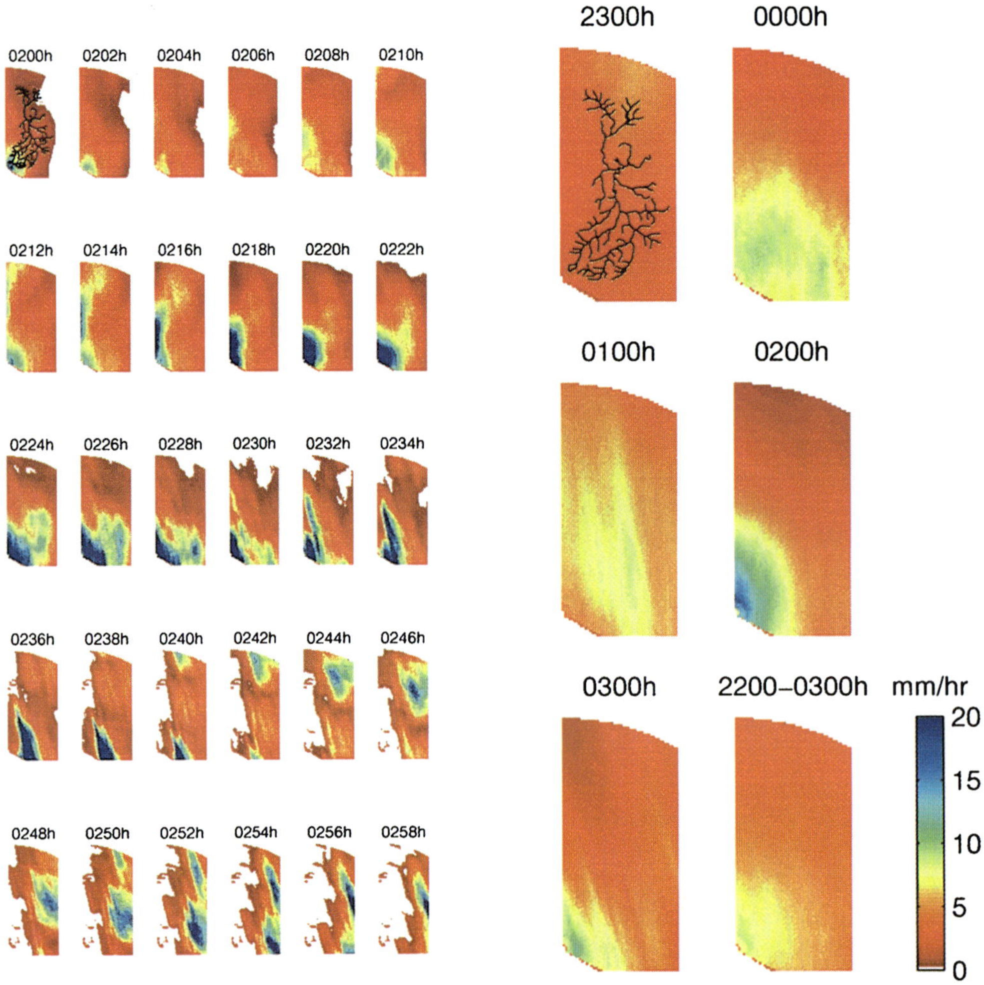

Plate 2. Radar rainfall data over the Mahurangi catchment (a) rainfall every two minutes from 0200h to 0300h (local time) on 11 August, 1998; (b) five 1-hour rainfall averages and a 5-hour average map, from 2200h August 10 to 0300h August 11. The river network is over-plotted on the first image for scale. Images are labelled with the time at the end of the averaging period.

Table 1. Typical soil profile at Mahurangi (Whangaripo clay soil: 16% sand, 41% silt, 43% clay, saturated hydraulic conductivity 0.4 mm/h).

Horizon	Depth (cm)	Description	Porosity (%)
A1	0 – 8	Very dark gray clay; friable; medium nutty structure: few pores; numerous roots.	45.7
AB	8 – 20	Intermingled yellow and brown clay; red mottles; firm; fine nutty structure; some pores; many worms.	58.6
B1	20 46	Pale yellow brown clay; slightly sticky; red mottles; firm; medium nutty structure; some pores; numerous roots.	60.9
B2	46-90	Intermingled yellow and brown clay; firm; coarse and fine nutty structure; few pores; many roots.	60.3
B3	90-120	Matrix of clay and strongly weathered sandstone; massive and firm; medium nutty structure; moderate porosity.	N/A
C	120-	Red weathered sandstone.	N/A

sensors, and seasonal mapping campaigns in selected sub-catchments. Three types of instrumentation are used: manually read access tubes using a Neutron Moisture Meter (NMM), continuously recording in-situ Campbell Scientific CS615s and spatial mapping using Time Domain Reflectometry (TDR) mounted on a specially designed all terrain vehicle. The data from the seasonal mapping campaigns are used to interpret the limited point data from the recorded locations in a spatial context, providing spatially and temporally variable estimates of soil water storage across the catchment. In total there are 36 continuously recorded sensors at 18 sites, 12 in pasture and 6 in forested country. These measure soil moisture and soil temperature (to enable corrections to the CS615 calibrations) in the top 300 mm and over a 250 mm depth at the base of the solum (generally from around 500-800 mm) at 30 minute intervals. Each site has a NMM access tube to a depth of 1.5 m that is measured monthly at eight depths. These sites were chosen to provide a "best estimate" of catchment-wide temporal response of soil moisture over the depth of soil that is likely to be a major influence on runoff response. Obviously it would have been ideal to have more instruments but this was not possible, so a key challenge was to choose locations across the catchment that would provide the most information with the available resources.

The detailed location of the 18 installations was based on the following rationale. Preliminary work by *Grayson and Western* [1998], based on several soil moisture data sets from Australia and the U.S.A., indicated that there were likely to be certain parts of landscapes where the local soil moisture content represented the mean of a large area, irrespective of whether the general soil water contents were high or low. These were termed Catchment Average Soil Moisture Monitoring sites (CASMM sites). Such sites are ideal locations for the insitu devices. Unfortunately the preliminary nature of that work, and the relatively small areas covered by the data used in the *Grayson and Western* [1998] study, precluded any definitive methodology for choosing these sites a priori. Nevertheless, consideration of the processes that dominate the spatial distribution of soil moisture indicate that CASMM sites should occur at locations with the following qualities:

- neither strongly divergent or convergent areas;
- on the hillslope proper rather than top of a ridge or bottom of a gully;
- on an area with an average slope;
- a topographic aspect which is likely to receive average solar radiation inputs; and
- soils and vegetation which are representative of a larger area.

In addition to knowing something about mean behaviour, it is necessary for this study to have some measure of the extremes and therefore the overall range of the soil moisture distributions. For this purpose, sites with more extreme topographic positions (e.g. on ridges and close to gullies) are also needed. Three pasture areas and two forest areas were chosen as being generally representative of the broader Mahurangi catchment. Three instrument nests were installed at each area (upper slope, mid slope and lower slope locations), with an additional three sites at the focus sub-catchment known as "Satellite". The final choice of sites is shown in Figure 3.

In the three pasture areas, the mapping of soil moisture is done seasonally using the Green Machine [*Tyndale-Biscoe et al.*, 1998; *Western and Grayson*, 1998], an all terrain vehicle with hydraulic insertion of TDR and precise geolocation via differential GPS linked to an on-board recording facility. Sampling consists of a grid covering each sub-catchment area, strategically placed transects to define small-scale (sub-grid) variability and mapping of saturated source areas. Dates of visits and average soil moisture values for the pasture sub-catchments are shown in Table 2.

Table 2. Dates of the intensive field campaigns and average soil moisture at the three pasture sites.

Campaign dates	Average soil moisture at Satellite (%v/v)	Average soil moisture at Claydons (%v/v)	Average soil moisture at Carrans (%v/v)
Apr. 1998	30.9	36.0	38.3
Aug. 1998	51.9	51.6	55.9
Nov. 1998	31.4	41.1	53.6
Feb. 1999	22.7	26.4	24.0
May 1999	41.8	38.8	40.1

Table 3 shows the sampling regimes at each site. At the forest sites, where the Green Machine could not operate, an alternative method for obtaining some measure of spatial variability was needed. Here we installed a series of NMM access tubes that could be monitored on the occasions of the intensive field campaigns. The forest topography consists of large-scale hillslopes (slope lengths approximately 300 m) which are heavily dissected creating smaller hillslopes, approximately 30 m long. The 30 NMM sites in the forest sample divergent, planar and convergent locations at these two obvious scales.

2.3. Other MARVEX Components

As mentioned, a consortium of researchers are involved in MARVEX, the core of which are the NIWA and University of Melbourne teams as well as the radar meteorology group from University of Auckland's Physics Department. In addition to these groups, teams from several other institutions are involved, although their results are not discussed in this chapter:

- Landcare Research New Zealand and the Institute of Geological and Nuclear Sciences (GNS) are undertaking a detailed study of flow pathways using isotopic and solute tracers on an area within the Satellite catchment. The Institute of Geological and Nuclear Sciences is also studying between-catchment variability in sources of streamflow, using isotopic techniques.
- HortResearch and the University of Waikato are using scintillometer, eddy correlation and microlysimeter methods for the measurement of evaporation during the spring and summer of 1999.
- The University of Texas at Austin is studying shallow surface soil moisture and remote sensing data.

Table 3. Details of soil moisture instrumentation at MARVEX sites.

Site	Sampling regime
Satellite (pasture)	3 CS615 and soil temperature sites every 30 min for depths of 0-300 mm & 400-600 mm; 6 NMM access tubes measured monthly; 40m x 40m TDR grid measured over 300 mm (approx. 400 points); 1000m of transect at 4m spacing using TDR during intensive campaigns; Mapping of saturated areas during intensive campaigns.
Carrans (pasture)	3 CS615 and soil temperature sites every 30 min for depths of 0-300 mm & 400-600 mm; 3 NMM access tubes measured monthly; 10m x 10m TDR grid measured over 300 mm (approx. 500 points); 300m of transect at 4m spacing; 50 m by 100 m area sampled on 4 m by 4 m grid using TDR during intensive campaigns.
Claydons (pasture)	3 CS615 and soil temperature sites every 30 min for depths of 0-300 mm & 400-600 mm; 3 NMM access tubes measured monthly; 20m x 20m TDR grid measured over 300 mm (approx. 300 points); 500m of transect at 4m spacing using TDR during intensive campaigns.
Marine Rd. West (forest)	3 CS615 and soil temperature sites every 30 min for depths of 0-300 mm & 600-800 mm; 3 NMM access tubes measured monthly; 15 NMM access tubes measured during intensive campaigns.
Marine Rd. East (forest)	3 CS615 and soil temperature sites every 30 min for depths of 0-300 mm & 600-800 mm; 3 NMM access tubes measured monthly; 15 NMM access tubes measured during intensive campaigns.

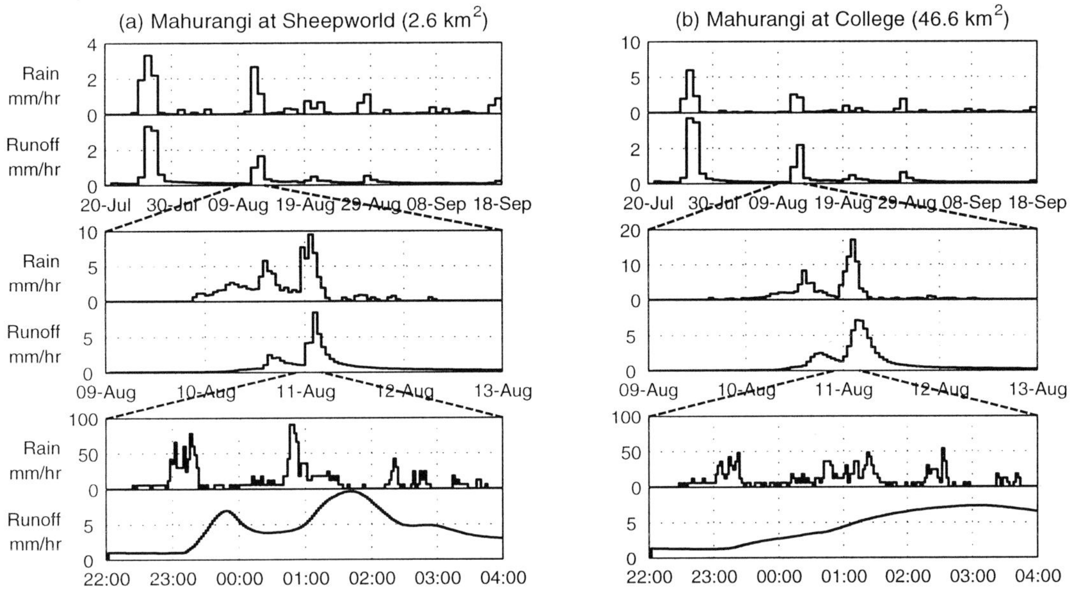

Figure 4. Rainfall-runoff data for two Mahurangi catchments at three temporal resolutions: 1 day, 1 hour, and 2 minutes: (a) Mahurangi at Sheepworld (2.6 km^2); (b) Mahurangi at College (46.6 km^2).

- The University of Western Australia has begun hydrological modelling of process controls on the spatial variability of MARVEX water balance.

3. INITIAL RESULTS AND DISCUSSION

At the time of writing, the MARVEX project is ongoing and we are actively collecting data, both on a continuous basis, and via intensive field campaigns. Here we present some data and initial analyses to give an indication of the richness of the information being collected. However, much of this work is of a preliminary nature, and is still to be expanded and investigated in more detail. In a later section we discuss some of the proposed analyses and expand on the specific questions being addressed in the data analysis.

3.1. Effects of Time-scales on Rainfall-Runoff Relationships

We begin with an investigation of catchment response times to rainfall events. Figure 4a shows rainfall and runoff data for a 2.6 km^2 catchment (rain and flow both measured at Mahurangi at Sheepworld, in the northwest of the catchment), averaged with three timescales: daily, hourly and 2-minute. Shorter time series are shown for the finer time intervals: the two-month period from which the data are drawn is in winter 1998. At daily and hourly timesteps there is a clear correspondence between rainfall peaks and flow peaks, with no delay at the daily scale, and a 1-hour lag at the hourly timescale. From the 2-minute data it appears that flow is smoothly varying at that scale, whereas rainfall is quite intermittent.

To check this for another catchment, Figure 4b shows rainfall and runoff data in the same format for the entire 46.6 km^2 Mahurangi catchment (the raingauge used for Figure 4b is near the centre of the catchment, at Toovey Rd). Similar comments apply for the daily and hourly data, although the hourly data for this larger catchment shows marginally more smoothing. The difference at the 2-minute scale is clear, with only a single peak for the larger catchment.

It is well known that hydrographs from larger catchments are usually smoother; however, the relative roles of rainfall, hillslope processes, and channel network processes in providing this smoothing are not well established. This filtering can take place via at least two mechanisms:

- the temporal pattern of point rainfall varies within a catchment, so the catchment-average rainfall time series will be smoother than the point rainfall records;
- travel times to a river basin outlet vary within a catchment, so that rain falling near the outlet may exit the catchment at the same time as rain which fell earlier, but further from the outlet.

Before addressing this question in more detail, we first consider the additional complexity driven by the spatial variability of rainfall.

3.2. Combined Effects of Space and Time-scales on Rainfall-Runoff Relationships

Observations of detailed space-time rainfall are not widely used in hydrology, but where available, the results are very interesting. Plate 2a shows a rainfall image every two minutes for an hour early on August 11, 1998. The sequence of images shows the development and movement of rainfall cells across the catchment. These cells are approximately 1-5km across, and move at speeds of the order 40 kmh^{-1} ($\sim$10 ms^{-1}). They can cross the entire Mahurangi catchment (5-10km wide) in perhaps 15 minutes. Since we have already seen that the observed catchment response times for storms are 1 hour or longer, it is possible that the two-minute detail is superfluous to an understanding of storm response for the monitored MARVEX catchments. However, for smaller catchments, or for runoff processes that depend strongly on 'instantaneous' rain intensity, this detail could still be significant. For MARVEX, what is probably more important is that an accurate estimate is made for the spatial pattern of time-averaged rainfall for periods of the order 1 hour. Plate 2b shows five one-hour rainfall maps, and a five-hour map, calculated by averaging the radar images. The visual impression of spatial rainfall at this timescale is that it varies rather unpredictably from hour to hour; this is a significant contrast to the two-minute data in Plate 2a. Further analysis is required to place these hourly accumulations in the larger space and time context of the storm structure; the coarser C-band radar data will be used for this purpose.

What are the consequences of this for storm response? Plate 3 shows maps of hourly runoff totals for five hours, starting at 2300h on August 10 (that is, starting 1 hour after the rainfall maps). As a first approximation, rainfall-runoff relationships can be inferred by comparing runoff from each hour with the preceding hour's rainfall (see Plate 2b). Some clear correspondences of rainfall and runoff patterns can be seen, although one needs to make allowance for upstream inflows when interpreting the lower catchment recorders. Heavy bursts of rain in the southwest corner of the catchment (e.g., Plate 2b, time 0200h) show clearly in the runoff data (Plate 3, time 0300h). There may also be more subtle signatures of the rainfall pattern, such as in Plate 2b, time 0100h, where a narrow north-south line of intense rainfall may be associated with localised high flows in the northeast of the catchment at time 0200h. A considerable amount of additional checking and analysis is needed to draw conclusions, but the richness and complexity of the data is clear now.

3.3. Soil Moisture Patterns

Installation of the soil moisture instrumentation was completed in April 1998 and five intensive field campaigns have been conducted since (see Table 2) with another planned for October/November 1999. As is apparent from Table 2, the sampling campaigns have covered a range of soil moisture conditions from very wet (immediately following an extremely wet period in June-July 1998) to relatively dry (at least for this environment). Plate 4 shows the TDR measurements for August 1998 and February 1999 for the pasture site known as "Satellite". Figure 5 shows elevation and moisture content along the southern transect at the Satellite site for February 1999, and Figure 6 shows the relationship between soil moisture and the topographic wetness index of *Beven and Kirkby* [1978] for the data in Figure 5.

It is clear from these figures that there is little correlation between topographic position and soil moisture in the top 300 mm at the Satellite site for these two sampling occasions. Regression analysis was conducted on all surveys to date using the topographic variables of aspect, specific upslope area, slope, curvature and wetness index. None of the correlation coefficients were above 11%, confirming the visual assessment that topography has little influence over soil moisture content in the top 300 mm. Data from the deeper instruments appear to confirm this observation. This apparent lack of topographic influence, particularly in the wetter periods, is yet to be fully explored but points to the hydrological response being controlled by deeper groundwater processes or to lateral processes on the hill slopes being strongly influenced by preferred flow pathways. Analysis of the mapped saturated areas and the drilling of deep piezometers, as well as the flow pathway work underway (see above) should assist in explaining this observed behaviour, particularly for wetter periods. During drier times, the influence of topography is expected to be limited [*Western et al.*, 1999; *Grayson et al.*, 1997; *Barling et al.*, 1994].

3.4. Influence of Soil Moisture on Rainfall-Runoff Relationships

Detailed analysis of the effects of soil moisture on runoff relationships is yet to be conducted; however, some preliminary comments can be made. Figure 7 shows a time series of soil moisture, precipitation and runoff for a 34 ha sub-catchment at the Satellite site. Figure 8 is a plot of av-

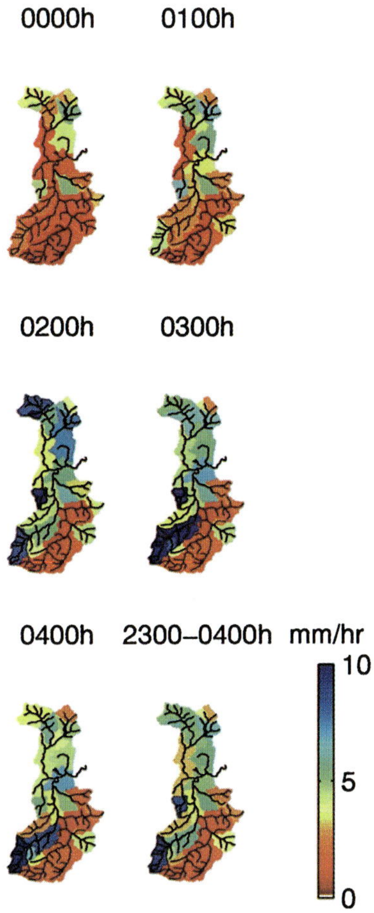

Plate 3. Runoff data for the Mahurangi catchment: five 1-hour average maps and a 5-hour average map, from 2300h August 10 to 0400h August 11. Images are labelled with the time at the end of the averaging period.

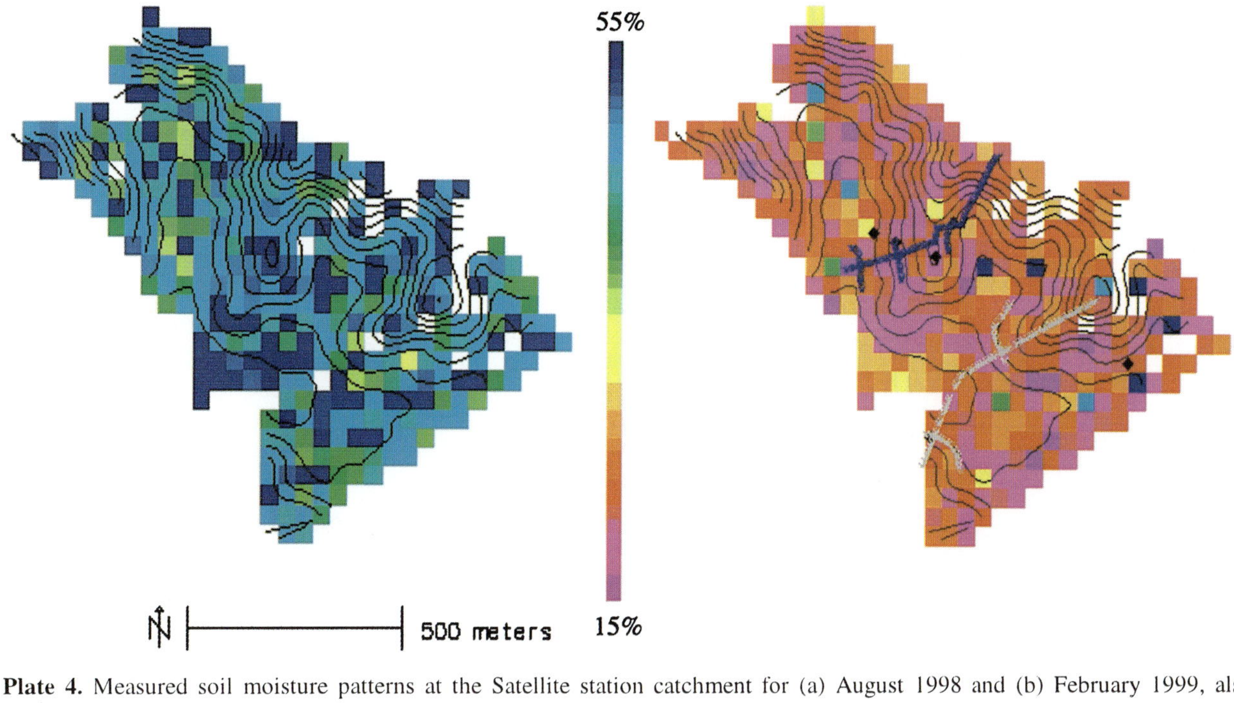

Plate 4. Measured soil moisture patterns at the Satellite station catchment for (a) August 1998 and (b) February 1999, also showing the location of CS615s and transects.

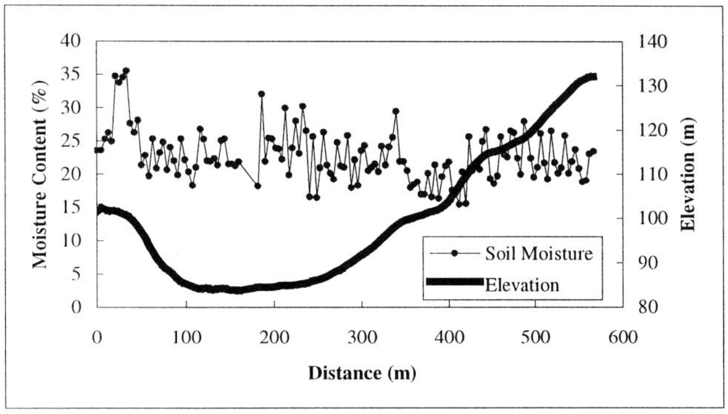

Figure 5. Transect data at 4 m spacings from Satellite, Feb. 1999.

erage soil moisture versus runoff ratio (runoff/precipitation) for this catchment and the 10 ha Tarrawarra catchment [*Western and Grayson*, 1998] for daily events larger than 5 mm. The threshold nature of the runoff process is readily apparent with significant runoff being generated only for moisture contents above about 42%. It would appear that in these circumstances, accurate measurement of soil moisture should be useful for runoff prediction. It is also likely that at these high average moisture contents, the spatial distribution of soil water will be critical in the prediction of runoff behaviour and will therefore need to be well predicted, if the data from insitu recorded points are to be useful.

4. WHERE TO FROM HERE?

Intensive MARVEX data collection is in progress and will continue throughout 1999. The rainfall and streamflow components of the network are planned to continue until at least the end of 2000, in collaboration with other MARVEX components (described above).

The long-term research plan for analysis of catchment-scale data proceeds in three steps: first, we will infer from the data what are the dominant controls on variability; the answer to this question is expected to vary among hydrological phenomena. For example, from the data presented above we might expect to find that spatial patterns of hourly rainfall will determine the observed spatial patterns of peak runoff. In contrast, observed spatial patterns of mean annual runoff may instead be controlled by spatial patterns of vegetation. The second step is to understand which physical processes lead to changes in the dominant controls on variability for storm runoff, low flows, etc. This step will include detailed sensitivity analysis using fully-distributed hydrology models, which will first be tested using MARVEX data (see below). To expand on our simplistic example, we speculate that rainfall would be more important than vegetation in controlling spatial storm runoff patterns because hourly catchment rainfall values are 'more variable' in space than is the interception capacity of vegetation (these statements are necessarily vague at this stage).

The final step is to develop appropriate ways to quantify this understanding of process control on variability. At present, detailed simulation using a time-stepping, spatially-distributed hydrology model is one of the few quantitative approaches available. Such studies require more data than is commonly available, and probably more data than is really necessary to answer the question. We suggest this approach is analogous to solving the complete equations of fluid motion in order to determine whether flow is turbulent or laminar. Fluid mechanics has the Reynolds number to answer

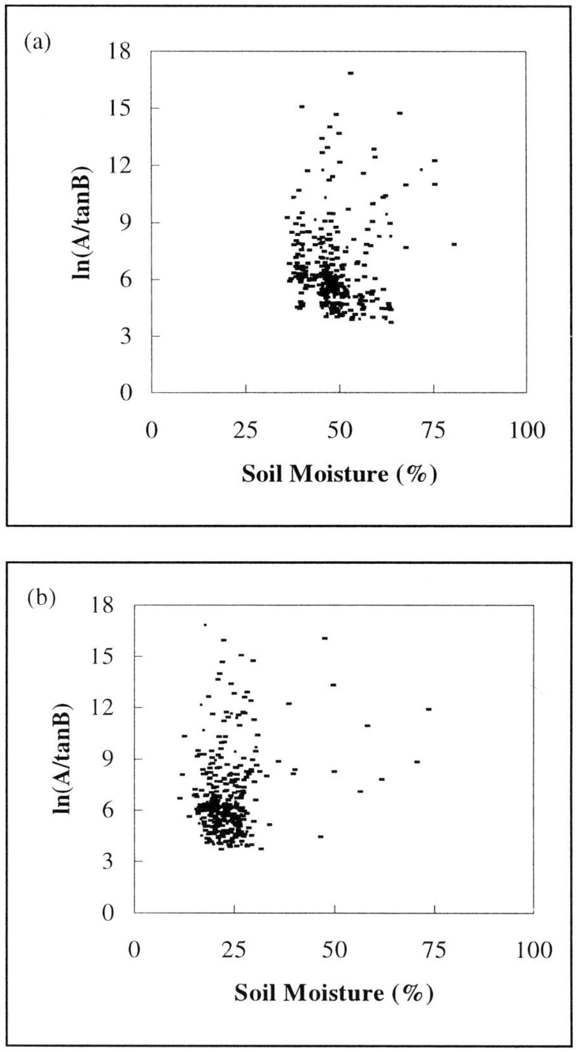

Figure 6. Soil moisture versus topographic wetness index for (a) August 1998 (winter) and (b) February 1999 (summer).

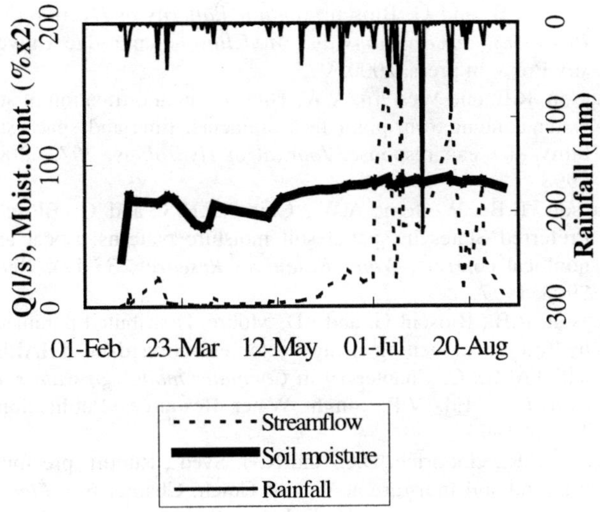

Figure 7. Soil moisture, precipitation and runoff for a 34ha sub-catchment at the Satellite site for 7 months in 1998.

this question: is there a correspondingly direct way to answer questions about variability in hydrology? Our goal here is to develop dimensionless numbers for the magnitude of sources of hydrological variability. These would permit hydrologists to make order-of-magnitude estimates of the roles played by variability in climate, soil, vegetation, geology and topography, and thus to identify the dominant sources of variability in the hydrological response of a river basin.

In the short term, we will expand the storm runoff analysis to use the full database; the next step is to evaluate a new theory which predicts the relative importance of multiple sources of space-time variability in storm runoff [*Woods and Sivapalan*, 1999]. Further research will focus on investigating and testing theories for multiple sources of variability in seasonal water balance and summer low flows, using the dimensionless water balance theory of *Woods* [2000]. A number of smaller 'spin-off' studies will also be generated, from issues that arise during the data analysis.

Collection of soil moisture data will continue throughout 1999, as will laboratory based analyses designed to check/revise standard calibrations for the TDR, NMM and CS615 instruments. The soil moisture related analyses will focus on developing predictors of spatial patterns for extrapolation across landscapes using both the MARVEX data and similar data being collected from sites in Australia. Essentially the question to be answered is "how best do we interpret limited point measurements in a spatial context?" This work will involve defining and characterising the temporal dynamics of spatial soil moisture patterns and the development of methods for representing "sub-grid" variability in soil moisture across different modelling scales. A range of dynamic simulation models will be applied to the MARVEX sites, including Thales [*Grayson et al.*, 1995; *Western et al.*, 1999], VCaST [*Woods*, 1997] and ALSIS [*Shao et al.*, 1997]. These will be used in conjunction with the soil moisture, precipitation, runoff, soils and vegetation data to illuminate the process controls on the variability of hydrological response across the MARVEX catchment.

The notion of CASMM sites [*Grayson and Western*, 1998] will be further explored and it is hoped to develop methods for predicting CASMM locations a priori using the extensive MARVEX data base along with an expanding data base from Australian sites. Now that the technology is available for relatively cheap, continuous monitoring of soil moisture, the utility of such data for real-time runoff prediction will be explored, along with the definition of the siting and operational requirements of such instrumentation for use as part of a real-time forecasting system.

5. CONCLUDING REMARKS

We envisage that the database being developed at MARVEX will become available to the wider scientific community early in 2001. In the meantime, groups who wish to become involved in the project are welcome to do so, provided they bring expertise that adds to the overall thrust of MARVEX and provides information that is useful to existing collaborators.

Coordinated, extensive field studies such as MARVEX are very expensive to undertake, yet we firmly believe that

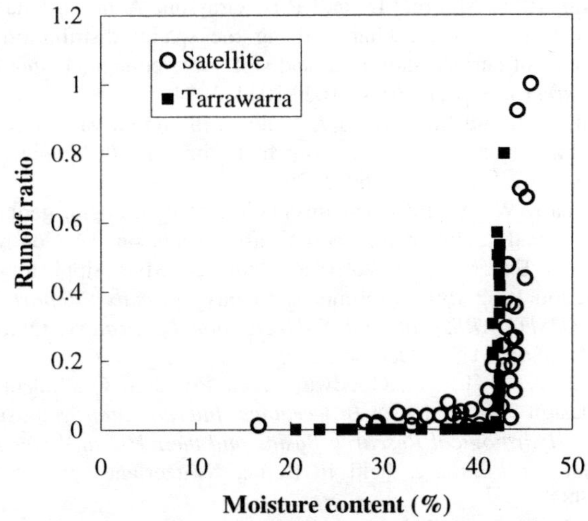

Figure 8. Catchment average soil moisture versus runoff ratio for a 34ha sub-catchment at the Satellite site and the 10 ha Tarrawarra catchment for precipitation events greater than 5mm.

to make progress on some of the key outstanding questions in hydrology, information of this sort is vital. Without this type of information, we have no way of testing the models or theoretical concepts that embody our understanding of hydrological phenomena and the science of hydrology will remain severely constrained [*Grayson and Blöschl*, 2000]. Countries like Australia and New Zealand do not have the institutions that are able to maintain long-term research stations. Instead, funds must be sought on a 3 or 5 year basis for specific projects which makes the execution of large exercises like MARVEX problematic. Nevertheless, MARVEX has shown that with a coordinated effort by a group of researchers (and some "luck" / foresight from research funding agencies) it is possible. It is also clear that once a core of work begins, expansion via the involvement of new groups is relatively simple. We suspect that, with the pressures on research funding around the world, and certainly in countries with a small population base, collaborative studies such as MARVEX will provide the primary avenue for comprehensive, field-based studies.

Acknowledgments. The authors wish to acknowledge the financial support of the Australian Research Council (grants C39813076, A39801842, C39804872) and the New Zealand Public Good Science Fund (contracts CO1635, CO1817). We also wish to express our gratitude to the Auckland Regional Council for supplying hydrological data, to Pete Pattinson, Graeme Mackay, and Kathy Walter (all of NIWA), who have contributed considerable technical expertise, and to the Mahurangi residents who have generously given us access to their land.

REFERENCES

Barling, R.D., Moore I.D. and R.B. Grayson, A quasi-dynamic wetness index for characterising the spatial distribution of zones of surface saturation and soil water content, *Water Resources Research, 30*(4), 1029 1044, 1994.

Beven, K. J. and M. J. Kirkby, A physically-based variable contributing area model of basin hydrology, *Hydrological Sciences Bulletin, 24*, 43-69, 1979.

Blackmarr, W. A. (Ed.), Documentation of hydrologic, geomorphic and sediment transport measurements on the Goodwin Creek Experimental Watershed, Northern Mississippi, for the period 1982-1993, Preliminary Release, *Research Report No. 3, USDA-ARS, National Sedimentation Laboratory*, Oxford, MS 38655, USA, October, 1995.

DeCoursey, D.G., The Goodwin Creek Research Catchment. I. Design Philosophy, in *Proceedings, International Symposium on Hydrological Research Basins and their Use in Water Resources Planning,* held in Berne, Switzerland, pp. 65-74, 1982.

Freer, J., McDonnell, J., Beven, K.J., Brammer, D., Burns, D.A., Hooper, R.P., and C. Kendall, Topographic controls on subsurface storm flow at the hillslope-scale for two hydrologically distinct small catchments, *Hydrological Processes, 11*(9), 1347-1352, 1997.

Grayson, R.B. and G. Blöschl, *Spatial Patterns in Hydrological Processes: Observations and Modelling,* Cambridge University Press, in press, 2000.

Grayson, R.B. and Western, A.W. Towards areal estimation of soil water content from point measurements: time and space stability of mean response, *Journal of Hydrology, 207*, 68-82, 1998.

Grayson, R.B., Western, A.W., Chiew F.H.S. and G. Blöschl, Preferred States in spatial soil moisture patterns: Local and nonlocal controls, *Water Resources Research, 33*(12), 2897-2908, 1997.

Grayson, R.B., Blöschl G. and I.D. Moore, Distributed parameter hydrologic modelling using vector elevation data: THALES and TAPES-C, Chapter 19 in *Computer models of watershed hydrology*. Ed. V.P. Singh. Water Resources Publications. 1130p, 1995.

Houser, P.R., Goodrich, D.C. and K.H. Syed , Runoff, precipitation and soil moisture at Walnut Gulch, Chapter 6 in *Spatial Patterns in Catchment Hydrology: Observations and Modelling.* Cambridge University Press, in press, 2000.

Lamb, R, Beven K and S. Myrabø. Shallow groundwater response at Minifelt . Chapter 11 in *Spatial Patterns in Catchment Hydrology: Observations and Modelling.* Cambridge University Press, in press, 2000.

Merz, B. and E. J. Plate, An analysis of the effects of spatial variability of soil and soil moisture on runoff, *Water Resources Research, 33*(12), 2909-2922, 1997.

Seyfried, M. S. and B. P. Wilcox, Scale and the nature of spatial variability: Field examples having implications for hydrologic modeling, *Water Resources Research 31*(1), 173-184, 1995.

Shao, Y., L. M. Leslie, R. K. Munro, P. Irannejad, W. F. Lyons, R. Morison, D. Short, and M. S. Wood, Soil moisture prediction over the Australian Continent, *Meteorology and Atmospheric Physics, 63*(3-4), 195-215, 1997.

Tyndale-Biscoe, J. P., G. A. Moore, and A. W. Western, A system for collecting spatially variable terrain data, *Computers and Electronics in Agriculture 19*(2), 113-128, 1998.

Western, A.W. and R.B. Grayson, Soil moisture and runoff processes at Tarrawarra, Chapter 9 in *Spatial Patterns in Catchment Hydrology: Observations and Modelling.* Cambridge University Press. In press, 2000.

Western, A.W. and R.B. Grayson, The Tarrawarra data set: Soil moisture patterns, soil characteristics and hydrological flux measurements, *Water Resources Research 34*(10), 2765-2768, 1998.

Western, A.W., Grayson, R.B., Blöschl, G., Willgoose, G. and T.A. McMahon, Observed spatial organisation of soil moisture and relation to terrain indices, *Water Resources Research, 35*(3), 797-810, 1999.

Western, A.W., Grayson, R.B., and T. R. Green, The Tarrawarra project: High resolution spatial measurement, modelling and analysis of soil moisture and hydrological response, *Hydrological Processes, 13*, 633-652 1999.

Woods, R. A. MARVEX: Identifying the causes of spatially-variable hydrology, *Proceedings of the 24th Hydrology and Water Resources Symposium*, Auckland, November 1997, 551, 1997.

Woods, R.A., and M. Sivapalan, A synthesis of space-time variability in storm response: rainfall, runoff generation and routing, *Water Resources Research, 35*(8), 2469-2485, 1999.

Woods, R.A., The relative roles of climate, soil, vegetation and topography in determining seasonal and long-term catchment water balance, submitted to *Advances in Water Resources*, 2000.

Ross Woods, National Institute for Water and Atmospheric Research, P.O. Box 8602, Christchurch, New Zealand

Rodger Grayson, Department of Civil and Environmental Engineering, University of Melbourne, Victoria 3010, Australia

Andrew Western, Department of Civil and Environmental Engineering, University of Melbourne, Victoria 3010, Australia

Maurice Duncan, National Institute for Water and Atmospheric Research, P.O. Box 8602, Christchurch, New Zealand

David Wilson, Department of Civil and Environmental Engineering, University of Melbourne, Victoria 3010, Australia

Rodger Young, Department of Civil and Environmental Engineering, University of Melbourne, Victoria 3010, Australia

Richard Ibbitt, National Institute for Water and Atmospheric Research, P.O. Box 8602, Christchurch, New Zealand

Roddy Henderson, National Institute for Water and Atmospheric Research, P.O. Box 8602, Christchurch, New Zealand

Tom McMahon, Department of Civil and Environmental Engineering, University of Melbourne, Victoria 3010, Australia

Integration of Land Observations and Modeling: Experiences and Strategies of a Large Scale Experiment

Richard G. Lawford

NOAA Office of Global Programs, Silver Spring, Maryland

Significant advances in understanding and predicting the behavior of the coupled land-atmosphere system on large space and time scales are most frequently achieved through regional experiments that marshal substantial resources and entrain a large range of multi-disciplinary expertise. These regional experiments inevitably involve integrated observational and modeling strategies. This paper reviews the advances that have been made through extensive field campaigns carried out under the International Satellite Land Surface Climatology Project (ISLSCP) and more recently under the World Climate Research Programme's (WCRP) Global Energy and Water Cycle Experiment (GEWEX) and the GEWEX Continental-scale International Project (GCIP). The extended description of the GCIP initiative outlines its strategy for integrating observations and models and using them in understanding the mechanisms and relative importance of various hydrometeorological processes and phenomena, in closing water and energy budgets for the Mississippi River Basin, for improving data assimilation systems and products, in model evaluation studies, and in their application for improved water resources management. The interplay of observations and models has led to improved capabilities that not only have increased the ability of the research community to address complex problems but also have benefited operational forecast services.

1. INTRODUCTION

Understanding and modeling land surface processes over all relevant space and time scales are complex challenges requiring a comprehensive, integrated effort combining observational initiatives, model development and applications. In addition to addressing the diversity of climate regimes and land surface types that should be represented in comprehensive land surface and hydrology models, studies are needed to provide an understanding of a large number of atmospheric and terrestrial processes. Expertise in a range of disciplines including meteorology, hydrology, soil science and ecology must be enlisted for these studies. Given the complexity and diversity of land-atmosphere interactions, it is not surprising that land surface modeling has advanced most rapidly during the past decade through large scale experiments and intense field campaigns. These large initiatives have provided measurements of a wide range of variables and parameters simultaneously because they marshaled comprehensive instrumentation, infrastructure, and expertise in the same research area, thereby reducing the number of unknowns in descriptions of the complex earth-atmosphere system.

Land Surface Hydrology, Meteorology, and Climate: Observations and Modeling
Water Science and Application Volume 3, pages 215-230

Single discipline large-scale experiments have occurred over the last 40-50 years. For example, in the 1960's and

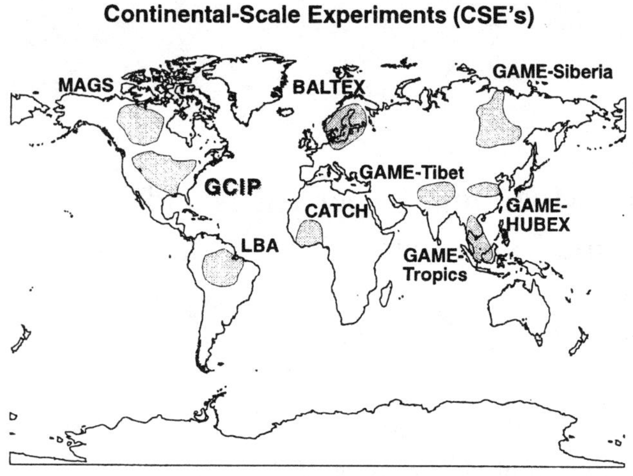

Figure 1. Location of the Five Continental-Scale Experiments (CSEs).

1970's a large percentage of the hydrological research community was involved in research in well instrumented watersheds as part of the International Hydrological Decade. Intensive land surface experiments conducted under the International Satellite Land Surface Climatology Project (ISLSCP) in the mid-1980s to early 1990s extended over the area of a model grid square (approx. 100 km X 100 km) and brought together large interdisciplinary teams. Subsequently, in the 1990s, studies were carried out under the auspices of the World Climate Research Programme (WCRP) on the scale of large river basins where regional water budgets could be studied in their entirety.

These larger scale studies required extensive planning and coordination. This chapter outlines some lessons from large-scale land based projects with particular emphasis on the GEWEX (Global Energy and Water Cycle Experiment) Continental-scale International Project (GCIP). This project was designed initially to quantify and describe water and energy budgets, including the land surface water budget, on a continental scale. However, it has expanded to deal with other issues including predictability and applications.

2. PROGRAMMATIC CONTEXT FOR GCIP AND OTHER LARGE-SCALE LAND ATMOSPHERE EXPERIMENTS

Many large-scale experiments involving the land and climate are undertaken under the auspices of the World Climate Research Program (WCRP) and International Geosphere Biosphere Program (IGBP). Projects within the IGBP tend to focus on process understanding and interdisciplinary linkages, while those in WCRP document climate processes and the climate system. Within the WCRP program, studies of regional water and energy budgets are part of the GEWEX program.

Historically, the synthesis of observations and models has proceeded most rapidly where there has been an integrating paradigm and some criteria to help guide the endeavor towards the ultimate strategic solution. GEWEX consists of two principal elements: the development of global data sets designed around specific water and energy budget parameters, and studies in different regions aimed at closing the regional water and energy budgets over different time and space scales. The closure of regional water balances over specific land surface areas has proven to be a very useful integrating theme for GEWEX. A total of eight large land areas form the five Continental-scale experiments being coordinated by the GEWEX Hydrometeorology Panel (GHP). Figure 1 shows the locations of the five principal GEWEX Continental Scale Experiments (CSEs). GCIP, the initial CSE, is being carried out in the Mississippi River Basin under the leadership of the USA. Other continental experiments include the MacKenzie GEWEX Study (MAGS) led by Canada, the Baltic Sea Experiment (BALTEX) led by Germany, the GEWEX Asian Monsoon Experiment (GAME) led by Japan, and the Large-Scale Biosphere Atmospheric Experiment in Amazonia (LBA) led by Brazil, the European Community, and the USA. The CATCH initiative is an emerging collaborative regional study being coordinated by France and Benin under the auspices of GEWEX.

The GEWEX Hydrometeorology Panel also integrates the activities of the International Satellite Land Surface Climatology Project (ISLSCP), the Global Runoff Data Center (GRDC), and, to some extent, the GEWEX Precipitation Climatology Project (GPCP) with those of the CSEs. ISLSCP has implemented three intensive observational campaigns continuing for a number of consecutive weeks in different seasons and over areas of approximately 100 km x 100 km in different climate regimes. Projects that fall under this initiative have provided many important data sets for model development including:

1. Tropical data from the HAPEX-Sahel in Africa.
2. Prairie grassland data for Kansas from FIFE (First ISLSCP Field Experiment).
3. Boreal forest data from the northern forests of Saskatchewan and Manitoba (BOREAS).

These projects have provided a large number of independent data sets that are being used for model validation [e.g., *Chen et al.*, 1996]. GCIP has relied on the integration of observations and models as a strategy at both the programmatic and scientific levels. The ultimate goal of GCIP is "to demonstrate a capability to predict changes in water resources on time scales up to seasonal, annual, and inter-

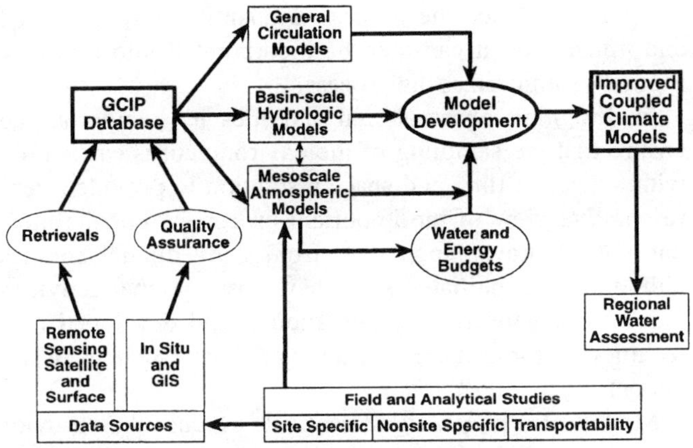

Figure 2. Schematic outlining the modeling paradigm implemented in GCIP [*IGPO*, 1994].

annual as an integral part of a climate prediction system" [*NRC*, 1998].

A long-term strategy for GCIP involves the development of comprehensive land surface models that can become an integral part of a global climate model (GCM). These models are needed for climate prediction, and for the development of scenarios that project the climatic consequences of greenhouse gas increases and land use change. A first step in demonstrating a predictive capability involves quantifying regional water and energy budgets on seasonal to annual time scales as a basis for model validation. In addition, observations have been used to quantify processes, estimate parameters, and facilitate model development and testing. Furthermore where a clear science goal has existed, the contributions of process studies, observations and models have been highly integrated around the modeling paradigm shown in Figure 2.

The Mississippi River Basin was chosen by an international GEWEX panel for research on regional water and energy budgets because of the extensive data collection networks that exist within the basin. However, as research planning progressed, it became clear that some land surface processes, important for the global climate such as permafrost and tropical forests could not be modeled using only data from the Mississippi River Basin. Subsequently, other continental scale experiments (CSEs) were initiated to address processes that did not occur in the Mississippi River Basin. These experiments include: land-atmosphere-ocean interactions in BALTEX; the role of land processes in the Asian monsoon in GAME; the role of tropical forests in LBA; and the role of cold region processes in producing runoff for the Arctic Ocean in MAGS.

In order to address a broad range of issues using the limited resources available, GCIP undertook its observational program in four phases. Each phase was associated with a different sub-basin and involved extensive data set development for that sub-basin. Each component of the Mississippi River Basin was identified as a Large Scale Area (LSA) [*IGPO*, 1994]. The periods for which extensive data sets were or are being acquired are shown in Figure 3.

To achieve its mission GCIP has established five working objectives, namely:

1. Determine and explain the annual, interannual and spatial variability of water and energy cycles in the Mississippi River Basin.
2. Develop and evaluate coupled hydrologic-atmospheric models at resolutions appropriate to large-scale continental basins.
3. Develop and evaluate atmospheric, land and coupled data assimilation schemes that incorporate both remote and in-situ observations.
4. Improve the utility of hydrologic predictions for water resource management up to seasonal and interannual time scales.
5. Provide access to GCIP in-situ, remote sensing and model data sets for use in GCIP and as benchmarks for model evaluation.

Science programs must be directed at central hypotheses to contribute to the advancement of science. In addition, the techniques developed in these studies must have demonstrable benefit for the supporting agencies in order to obtain continuing financial support. For example, the space agencies have supported GEWEX because they believe that the techniques being produced by GEWEX are improving the value of remote sensing data for environmental monitoring. The National Oceanic and Atmospheric Administration (NOAA) has supported GCIP be-

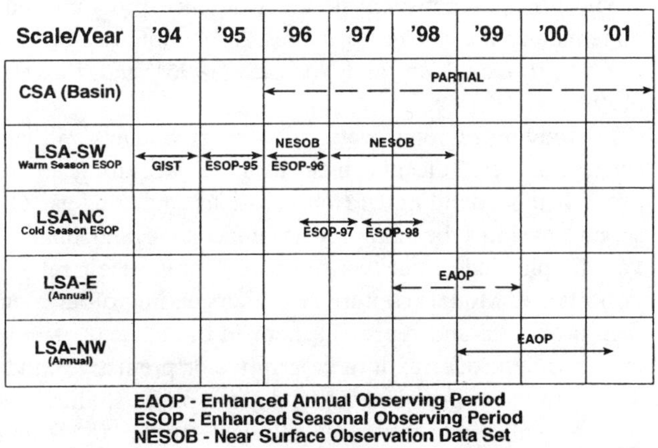

Figure 3. Periods for which special GCIP large-scale data sets are available [*IGPO*, 1999].

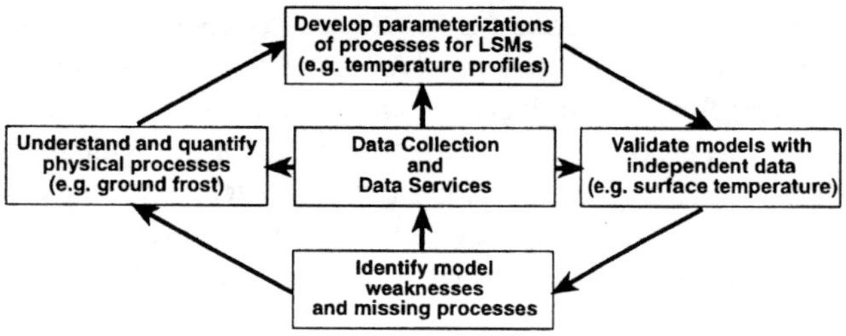

Figure 4. Example of the linkages in GCIP between observations and models.

cause of the improvements it provides for monitoring and predicting important weather and climate variables such as precipitation.

To address the unique features of each LSA the following hypotheses were adopted:

1. Land surface processes (soil moisture and vegetation) and the low level jet control the summer precipitation patterns and regional hydrology of the semiarid area of the southwestern Mississippi River Basin (LSA-SW).
2. The atmosphere is relatively decoupled from the land surface during winter but begins to be significantly coupled when surface changes occur (e.g., snowmelt, thaw) in response to the onset of spring (LSA-NC).
3. Precipitation and runoff variations and extremes are predictable at climate time scales in complex topography (LSA-E).
4. Coupled land-atmosphere interactions and terrain modulate large-scale circulation patterns over the entire annual cycle. These effects can be simulated on daily to seasonal time scales (LSA-NW).

The linkages between the GCIP science program components and the above objectives and hypotheses are discussed in Coughlan and Avissar (1996) and Lawford (1999).

In studying regional water and energy budgets and land surface process, the two main tools for acquiring an improved understanding are observations and models. Observations must be analyzed to understand and quantify various physical processes and to determine their relative importance. Models capture the understanding of land surface processes and represent them in the context of other larger scale processes in a descriptive or predictive mode. The normal progression for land-atmosphere studies is to rely on data and statistical/analytical studies to demonstrate the existence of a relationship. After showing the existence of a relationship, specific hypotheses are formulated and targeted data collection activities and analyses are pursued. Once the process is quantitatively described and understood it can then be incorporated into a model. Figure 4 summarizes the process.

In designing observational activities it is important to ensure that the sampling of the environment is carried out with sufficient time and space resolution to provide a representative view of the important processes. For the most part GCIP draws upon data from conventional sources although it has initiated some new observational activities (such as soil moisture observations), and developed new techniques for sensing certain variables (such as water vapor).

Models also play a significant role because they provide a synthesis of the information known at any point in time and allow that knowledge to be used to examine a range of initial conditions. They also facilitate sensitivity studies, and provide a capability to predict future conditions or states based on different forcing functions.

In the case of large-scale experiments such as GCIP, models have also become useful frameworks within which new process understanding can be codified and utilized. In this role they allow one to interpolate or extrapolate from areas with extensive observations into ungauged areas as needed. Although not frequently used in GCIP for this purpose, models can also be used to simulate processes and thereby facilitate the in-depth examination of the nature and range of their operation. For example, the South American Low Level Jet that was first described by the wind and moisture fields produced by the Eta Model [*Berbery and Collini,* 1999] and has since been substantiated by observations in Bolivia [*Douglas,* personal communication].

3. CLOSING THE WATER BUDGET

The GCIP focus on regional water and energy budgets forces realism upon a program that would like to measure all components of the water and energy budgets and reduce the error in these budgets to near-zero. Given the errors and inadequacies of models and observational systems, the best estimates of the water and energy budget components appear to result from optimizing the mix of conventional observations, remote sensing products, special observations and models. Furthermore, closure of the water and energy budgets demands that the uncertainty associated with measurements and model estimates also be quantified. Consequently, a component of observational activities involves determining: 1) to what level of accuracy a particular sensor can observe a specific variable and how these measurements should be calibrated if they are incorrect; 2) where the sensors should be placed to obtain representative measurements; 3) how much sampling or averaging the sensor should do before a reading is recorded (and to assess whether averages or maximum values of the sampled values will be recorded as measurements), and 4) how spatially distributed fields can be generated from point measurements or spatially limited data.

In order to characterize and understand the processes that govern water budgets it is necessary to assemble and analyze observational data sets. To quantify water and energy budgets it is also essential to undertake modeling studies, because there are some variables such as areal evapotranspiration that are needed for closing budgets but cannot be reliably obtained from existing observations. Preliminary work in closing the atmospheric water budget was undertaken by Ropelewski and Yarosh (1998) using radiosonde data from stations around the perimeter of the basin. This exercise involved solving the following equation for the atmospheric water budget components:

$$dq_a/dt = C_q - (P-E) \qquad (1)$$

where q_a - total atmospheric water mass per unit horizontal area,
C_q- net inflow of atmospheric water into the air above the basin,
P - precipitation in the basin, and
E - evaporation from the basin.

The result of this work indicated that there was a large difference in the moisture flux into the Mississippi River Basin between 0000Z (1800 LST) and 1200Z (0600 LST) during the summer months. The runoff estimated by substituting the values of C_q, q_a and P into equation (1) is 0.2 mm/day or approximately one-third of the observed runoff (0.6 mm/day) based on streamflow measurements at Vicksburg. This discrepancy can be attributed to the fact that the radiosondes are launched at 12-hour intervals and at times when moisture transport processes may be near their minimum [*Yarosh et al.*, 1999] leading to underestimates of the moisture convergence into the basin. For example, profiler data demonstrate that the low level jet reaches peak intensity at 0300Z (2100 LST) to 0500Z (2300 LST). It is possible to use the monthly runoff values to provide adjusted flux estimates, although this approach is even less satisfactory at shorter time scales because of time lags in the runoff generation process.

To derive basin scale estimates of moisture convergence during those times when radiosondes are not being launched it is necessary to use a high-resolution data assimilation model to infer the moisture convergence or analyze profiler measurements. At present the availability of profiler data is insufficient to circle the basin with measurements and to give high resolution estimates of moisture convergence into the basin. To assess the degree of uncertainty in atmospheric convergence fields estimated from 12-hour radiosondes, computations have been done using the data assimilation product from the Eta model. Yarosh et al. (1999) showed that the moisture flux convergence increases significantly if modeled fluxes are computed every three hours rather than every twelve hours. This increase provides strong evidence that the time resolution of upper air soundings is the main source of error in closing the budget.

Water budget studies have also uncovered other problems, particularly during the winter months. The computation of monthly evaporation based on atmospheric moisture budgets indicates that negative values are not uncommon during these months [*Ropelewski and Yarosh*, 1998]. These negative values may be the result of errors in computing flux measurements, or problems in estimating cold season precipitation, or both. Independent studies confirm that the measurement of solid precipitation using conventional gages is a problem. Typically, unshielded Balfour gauges undercatch snow by 8 to 33% for wind speeds of 6 mph. Of the many procedures available for correcting solid precipitation measurements, two are under consideration in GCIP. The first procedure, developed by the World Meteorological Organization (WMO) Intercomparison Study, makes corrections for wind speed, wetting losses and evaporation. In order to provide more realistic estimates of snowfall GCIP also has developed correction factors for the northern states based on the exposures of the gauges and their locations relative to surrounding topography and vegetation [*Peck*, personal communication]. Daily average wind speeds from the nearest synoptic station are transferred to the site and corrections are made based on that estimate and gauge exposure.

Modeling studies are being carried out in GCIP to understand the interplay between energy budgets, atmospheric dynamics, and regional water budgets. Roads et al. (1998) have analyzed water and energy budgets for NCEP's global reanalysis over the Mississippi River Basin and have shown that budget residuals are an important component of both water and energy budgets. Sources of these residuals include cumulus convection, diffusive transports in the boundary layer, and adjustments in large-scale transports of moisture and temperature. The role of larger scale forcing is also important to document if the nature of moisture fluxes into the Mississippi River Basin are to be fully understand. For example, Dirmeyer and Brubaker (1999) have shown that the climate conditions over the Mississippi River Basin change from drought to flood depending on the intensity of the moisture influx on the southern edge of the Mississippi River Basin.

From these studies it is clear that data and models must be used together in data assimilation mode to define the water budget over the Mississippi River Basin. However, this also means that not only the uncertainties in the data but also the data assimilation model may limit the accuracy of the moisture flux estimates. Recognition of this fact has caused GCIP to place considerable emphasis on improving models, particularly over land. The other CSEs have also encountered similar problems with the temporal resolution of water budgets estimates. In MAGS, Intensive Observing Periods (IOPs) were held with enhanced soundings to provide data at six-hour intervals to close the budget. Although these data are not fully analyzed they are expected to demonstrate the benefits of higher time resolution observations for that area. Other CSEs are also undertaking research to improve model capabilities for data assimilation purposes. Within GCIP, model development efforts have focused on the land component of regional models in the belief that the correct representation of soil, vegetation, and runoff processes will improve the ability of these models to simulate atmospheric conditions and, in particular, boundary layer dynamics and precipitation processes.

4. IMPROVING DATA ASSIMILATION PRODUCTS

Within GCIP three operational models are being used to produce data assimilation products that are then stored in an archive at the National Center for Atmospheric Research (NCAR). GCIP has relied on operational Numerical Weather Prediction (NWP) centers to provide its data assimilation products because these centers have access to the full suite of real time observations. The three models are the Eta model run at the National Centers for Environmental Prediction (NCEP), the Mesoscale Analysis and Prediction System (MAPS) model being developed at the Forecast System Laboratory (FSL), and the Global Environmental Multiscale (GEM) Model used at the Canadian Meteorological Center. GCIP investigators have assessed the sensitivity of outputs from these models to changes in the representation of soil moisture, snow, ground frost, snow melt, vegetation, and boundary layer processes, and have identified and implemented a number of improvements.

The Eta model is one of the main tools used by the National Weather Service in providing routine forecasts. It operates on a latitude-longitude grid with Eta vertical coordinates. GCIP uses a developmental version for the data assimilation function of the model where new parameterization schemes can be tested. The Mesoscale Analysis and Prediction System (MAPS) model is unique in its use of isentropic surfaces whereby, arguably, it produces more realistic water vapor transports than other models.

Initially, in 1995, a bucket scheme was used in the Eta model. Predictably, this model did not do well at estimating surface temperatures in situations where soil moisture values varied significantly. However, the improved representation of soil moisture processes in a more sophisticated land surface scheme resulted in significant improvements in the temperature and specific humidity simulations [*Yucel et al.*, 1998]. The majority of these improvements have been the direct result of research carried out by GCIP. During this period the Office of Hydrology worked closely with NCEP to improve the land component of the numerical model by incorporating the Oregon State University's (OSU) soil moisture model. This model provided additional layers of soil and improved the representation of moisture movement in the soil. The benefits of this approach were most evident in improvements to the level of accuracy in surface air temperature fields. Given that the operational centers provide products for the science community to analyze and evaluate and, through GCIP, they have full access to the new modeling techniques developed in the academic community, these centers have both the incentive and capability to improve the land surface components of their models and thereby increase the quality of their data assimilation and prediction products.

The ISLSCP experiments have been very useful in documenting processes over small to intermediate size areas and have been remarkably successful in contributing to the development of improved land surface models. For example, Chen et al. (1996) demonstrated how comparisons of snow simulations with different land surface schemes and one ISLSCP data set aided in the selection of a land surface scheme for an Eta model upgrade. On the larger scale the European Centre for Medium-Range Weather Forecasting (ECMWF) has used albedo measurements from BOREAS to improve the representation of boreal forest effects over Euroasia during winter months. This change has led to improvements in winter air temperature forecasts over that region.

5. THE USE OF DATA IN MODEL DEVELOPMENT AND SIMULATIONS

Given the need for data assimilation products for water and energy budget studies and the lack of good data for a number of water budget parameters such as soil moisture and evaporation, GCIP has fostered the development of a number of land surface schemes that contribute to both mesoscale modeling and global climate modeling. The complexity of land-atmosphere interactions, including vegetation effects, make them difficult to model. Uncertainties in outputs from these models arise from errors in the data inputs, from the non-representativeness of parameters in the model and from the limitations in the model framework [*Gupta et al.*, 1999].

The development of regional scale coupled land-atmosphere models in GCIP has progressed on a number of fronts. Academic studies have been examining the role of various processes using a range of models. Within this community the following models and land surface schemes are being developed and applied: RAMS, MM5, Simple Biospheric Model (SSiB), Biosphere-Atmosphere Transfer Scheme (BATS), Eta, MAPS, GEM, OSU Land Surface Model and Surface Water Budget (SWB) model. Through the GCIP NOAA Core Project and other related initiatives, Koren et al. (1999) and Benjamin (personal communication) have incorporated a number of techniques developed through academic research into their developmental and operational models.

In general, the development of Land Surface Models (LSMs) in GCIP has involved the following approaches:

1. Improving representation of physical processes in models through process studies.
2. Carrying out special studies and analyses to estimate parameters for use in the models
3. Comparing model outputs and observations for the same period and location to determine where the model needs improvement.
4. Undertaking model intercomparisons to assess the sensitivity of various land-surface schemes to their process representations and parameter values.

5.1 Initialization

GCIP's longer-term mission of monthly to seasonal predictions of water cycle variables is its most challenging model development issue because seasonal forecasting entails both initial conditions and boundary conditions. If the initial conditions are not correctly specified then the feedback from day 1 to day 2 will be wrong and the boundary conditions forcing the atmosphere on day 2 will be unrealistic. These errors accumulate and become compounded over time. In a climate model with a 10 year run these errors may average out so that the general influences of the boundary conditions will prevail. At the seasonal time scale, however, the memory in variables such as soil moisture will persist to the extent that erroneous initial soil moisture fields could cause errors to propagate for many months unless the conditions of the land surface are reinitialized by a climate extreme (such as heavy rains or floods) resulting from major external forcing such as an El Niño event.

One of the most critical initialization problems involves establishing correct soil moisture fields when observations of this variable are very limited. The importance of initial soil moisture fields has been stressed in a number of studies including Betts et al. (1996), Viterbo and Betts (1999), and Koster and Suarez (1999). In the absence of observations this variable is frequently initialized with a model derived value. Snow cover and snow pack are also important initial fields for models. GCIP has developed a hemispheric snow cover product based on satellite data that is used routinely to initialize the Eta model. Carroll et al. (1998) have developed procedures for initializing the water equivalent of the snow pack.

5.2 Model Structure

A second major challenge in model development involves model formulation. As has been noted, it is very difficult to capture the full complexity of land-atmosphere processes. Often in regional models this is done by tuning the model to a particular region so it accurately simulates processes important in that region (e.g. snow or tropical rainforest). Many of the efforts directed to building Soil Vegetation Atmosphere Transfer Schemes (SVATS) have followed the practice of incorporating many processes defined in one dimension from site studies into a single column version of the model and then applying this representation with different tuning of the parameters over the entire globe. It has been argued that this approach fails to capture the complexity of processes like runoff, that take place in the horizontal plane and involve interactions between the grid squares. GCIP has emphasized the testing of model modifications in a mesoscale model framework because it is possible to see the spatial and temporal influences of the change without having to wait for the results of a multi-year model run as would be the case if a full GCM model was used.

GCIP has supported the Project for the Intercomparison of Land Surface Parameterization Schemes (PILPS) for a number of years as its primary model intercomparison activity. Some of the intercomparison studies have led to significant improvements in GCIP models. In recent years the focus of these studies has been one-dimensional model components aimed at improving LSMs [*Shao and Henderson-Sellers*, 1996]. Initially, the focus of these model intercomparisons was to assess their ability to reproduce surface latent and sensible heat fluxes. However, with the intercomparison in the Red-Arkansas basin in PILPS-2c, the focus shifted to determining errors in simulated runoff and infiltration. The PILPS 2c intercomparison [*Lohmann et al.*, 1998; *Liang et al.*, 1998; *Wood et al.*, 1998] was carried out over the GCIP LSA-SW, demonstrated that there is a very clear distinction between models that have a large amount of the evaporation coming from the soil surface (50-80% with lower values in the summer months) and those that have only 10-20% of the evaporation coming from the soil surface.

Hydrologic models can produce different results due to their formulation. The structure and calibration requirements for distributed models are different from those of lumped parameter models. Lumped parameter models are one-dimensional for a catchment with a single average value for the precipitation input, a single equation to describe the rainfall runoff relationship and no representation of the internal dynamics of the basin. Furthermore, the vegetation, hydraulic conductivity and topography are all represented by one value per parameter in spite of the heteorgeneity of these parameters across the basin.

Distributed models, such as the VIC model, are divided either into Cartesian grid squares or small hydrologic units, and different parameter values are assigned to each unit or grid square. The water balance is then solved for a grid square, and the runoff is routed into an adjacent downstream grid square, where it is added to the runoff generated in that grid square before being routed to the next downstream grid square, and so forth, until the basin streamflow is computed.

The development of macroscale hydrological models that incorporate more process understanding is a goal for GCIP. The VIC model provides a framework for working towards this goal. Although the early VIC model contained fewer parameters to be calibrated, it operated at such coarse resolution (1 x 1) it did not simulate streamflow accurately [*Abdulla et al.*, 1996]. However, recent efforts to increase the resolution to 1/8 x1/8 grid squares have resulted in improved performance [*Lettenmaier*, personal communication]. Distributed hydrologic models are also able to make use of the high resolution distributed data fields available from NEXRAD radar and satellite data systems.

Development work on coupled models has focused on both the land surface component of atmospheric and hydrologic models, and on the interchange of land surface schemes between the two types of models. Abdulla et. al. (1996) have shown that the VIC model can produce good flux estimates suggesting that this model could be modified and used as an alternative to a SVATS in a climate model. Land surface schemes of hydrologic origin generally have a greater ability to capture runoff and streamflow processes and hence provide results that are more relevant to water managers. On the other hand SVATS generally represent vegetation processes more effectively than do hydrologic models.

Within GCIP, significant effort is being directed to the development of a system that will provide a full suite of initial land conditions for an operational prediction system. This Land Data Assimilation System (LDAS) initiative involves the use of distributed hydrologic models as well as SVATS. Through the collective research of the NOAA Core GCIP project, NASA's Goddard Space Flight Center (GSFC) and several universities an LDAS is being developed which, when completed, will be able to assimilate satellite data, rainfall measurements and soil moisture measurements. This system uses the Eta model as its basic framework and then allows land surface schemes and hydrologic models such as MOSAIC, Surface Water Balance (SWB) and VIC models to be interfaced with it interchangeably. When run off-line this LDAS will produce highly reliable surface fields for climate analysis and diagnostic studies.

5.3 Parameter Specification

The third major challenge in model development involves the estimation of parameters that represent certain physical processes in models. In some cases the parameters (and even the variables in the models) have no counterparts in the physical world. This is seen as an acceptable approach where neural networks are used to develop statistical representations of reality [*Cotton*, personal communication]. In other cases a surrogate for a physical parameter is used in a model [*Entin et al.*, 1999]. For example, soil wetness is a modeled parameter that is used in place of soil moisture in a number of models. Another problem arises from the need to specify a single value for a model grid square for a land surface parameter to represent a large heterogeneous area. Issues related to the effects of spatial heterogeneity and the consequence of using a single variable to represent large spatial variability must be addressed.

Observations have played a major role in model development in GCIP. Initially modelers used experiments like FIFE (First ISLSCP Field Experiment) in Kansas to develop parameter estimates for models [*Sellers, et al.*, 1988]. As GCIP data became available they were used to assess the accuracy of model outputs and data assimilation products. In other applications, particularly in the LSA-North Central, process studies and related data sets have been useful in characterizing and quantifying certain cold season hydrometeorological processes (e.g., ground frost) so they could be included in models. According to Baker (personal communication), specific observational data sets that have proved useful are:

1. Snow measurements used in the validation, calibration, and testing of snow accumulation and ablation models.
2. Frost depth measurements used in the validation, calibration and testing of new algorithms for simulating frozen soil extent.
3. Soil moisture measurements used for calibration and testing of model water balance and validation, and calibration of frozen soil algorithms
4. Spatial frost measurements used to develop statistical distributions to represent the spatial distribution of frozen soil.

The issue of parameter specification is important for both atmospheric and hydrologic models. However, the approach often differs according to the model being used. Since most atmospheric models are derived from a basic system of equations for momentum, energy and mass budgets and a thermodynamic equation, they have constants that are tied to basic atmospheric processes. Traditionally, it has been possible to document the processes involved and use physical constants in the equations to provide physically based process representations. Generally, these physical constants are not constrained by the model format so the equations can be transferred from one model to another. Scale effects are often dealt with by setting terms that are negligible at the model's operating resolution to zero. However, things become more complex when land surface heterogeneity is added to these models because there are several ways to handle this heterogeneity. Frequently, single values are used to characterize the net effect of these processes on a grid square by grid square basis. This is particularly true when the processes are measured at a point but they are not known or are only poorly estimated for an area.

In the case of lumped hydrologic models and some distributed hydrologic models the parameterization process poses another challenge. In many cases the model relates input to output without a clear physical process being defined in the relationship. As a result it is not known whether there is a significant statistical relationship between the constants in the model and a measurable parameter in the real world. The high degree of calibration in lumped parameter models makes it difficult to transfer them from one sub-basin to another without the time-consuming task of assembling large historical data bases and recalibrating the parameters.

When modifications are made to models, parameters are often changed one at a time. However, model performance depends on the relationship between all the parameters. Consequently, attention must be given to the interaction between parameters when making a change to even one parameter. A model that has been tuned on the basis of a given parameter having a certain value may produce unexpected results when that parameter is changed. GCIP research on parameter estimation by Gupta et al. (1999) includes the development of techniques for optimizing model calibrations by determining how all the parameters should be changed when one parameter is modified. These techniques are being applied to SVATS (specifically BATS and the LSM used in the Eta model).

A study to develop techniques for regionalizing lumped hydrologic model parameterizations from small highly instrumented watersheds to ungauged catchments using an automated scheme for specifying a priori parameter values is being undertaken through the MOPEX (Model Parameter Estimation Experiment) Project [*Schaake*, personal communication]. Some parameters such as Normalized Difference Vegetation Index (NDVI) and the ratio of precipitation to potential evapotranspiration are correlated. Consequently, it may be possible to simplify the parameterizations in some models using this relationship. In addition to acquiring more historical data for model calibration, GCIP has also developed high resolution data sets for soil parameters such as water holding capacity [*Miller and White*, 1999] that are used for specifying parameters in LSMs.

6. CRITICAL LAND SURFACE PROCESSES

The following section discusses the broad range of observations within GCIP needed to support model development. Based on the experience of the PILPS 2b experiments [*Shao and Henderson-Sellers*, 1996] and through the recognition of model limitations identified from other studies, GCIP has supported research to more effectively represent physical processes related to vegetation, soil moisture, surface heterogeneity, cold season and runoff processes.

6.1 Vegetation

Vegetation contributes to the complexity of land-atmosphere interactions due to its spatial heteorogeneities, and its strong diurnal and seasonal cycles. Plant-atmosphere interactions are very complex because plants change from being photosynthetically active during the day to being inactive at night. Their effects influence the water budget through transpiration and the energy budget through albedo effects. Models must take into account the albedo effects of seasonal changes in vegetation as it is

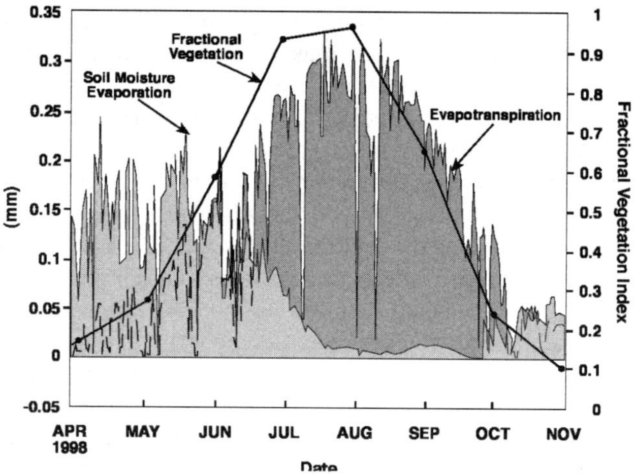

Figure 5. Magnitudes of daily soil evaporation and evapotranspiration over a field in the Little Washita (Oklahoma) basin during the growing season of 1997 (*Meyers*, 1998).

covered either fully or partially by winter snow. North American coniferous trees, however, retain their needles in winter and the albedo remains low unless they accumulate a large amount of snow on their branches. Rapid changes can occur in the albedo as the snow melts in the spring and the albedo of the underlying soil and vegetation changes, particularly in areas with deciduous trees. Ensuring that all of these processes are adequately understood and represented at appropriate time and space scales in mesoscale models and coarser climate models is a central challenge for land surface modelers.

To fully represent vegetation forcing in models a large range of variability must be included in the representation of transpiration. The period of active growth (and hence the seasonal cycle of transpiration) is dependent on the plant species. Baldocchi and Meyers (1998) have studied of transpiration rates in broad leaf forests and other vegetative regimes using a mix of local scale techniques including micrometeorological measurements (eddy covariance), physiology (sap-flow) and hydrological (watershed) measurements. According to their results transpiration rates can range from a midday average of 100-200 W/m^2 for jack pine forests to an average of 300-500 W/m^2 for wheat growing at midlatitudes.

Given this range of transpiration rates, it is evident that a knowledge of the spatial distribution of vegetation type is critical for estimating areal evapotranspiration. Vegetation type has large spatial variability at the scale of a watershed. In the case of the Mississippi River Basin the dominant vegetation covers are croplands (21.36%), grasslands (21.04%), cropland/ natural vegetation mix (20.6%) and deciduous broadleaf forest (16.37%) [*Gallo*, personal communication]. This vegetation distribution was derived from the IGBP land cover classes stored at the Earth Resources Observation Satellite (EROS) data center. Other measures of vegetation cover of interest to modelers include the Normalized Difference Vegetation Index (NDVI) and the fractional vegetation cover. When vegetation is inactive, moisture from the soil evaporates directly to the atmosphere. When the vegetation is growing, the roots uptake moisture from the soil and transfer it to the atmosphere. This process results in a net drying of the soil in the summer. Figure 5 shows the relative importance of soil evaporation and transpiration during the growing season. Evapotranspiration is intimately tied to the carbon cycle. Decreases in evapotranspiration due to soil moisture deficits are associated with plant water stresses that decrease photosynthesis rates and the amount of carbon fixation [*Meyers*, 1998]. The role of root distribution on the subsurface and atmospheric moisture budgets is not fully known. Furthermore, roots tend not to be adequately represented in models. This topic represents an opportunity area for new research.

For deciduous tress seasonal effects on albedo and moisture fluxes at mid and high latitudes are dramatic. However, the effects are generally less dramatic for conifer stands. However, over the Boreal forest, it was found that the Bowen ratios remained very large (sensible heat flux high, latent heat flux low) until the ground frost thawed and the trees began to green [*Sellers et al.*, 1998]. The sensitivity of atmospheric processes to vegetation effects in coupled models has been demonstrated by Xue et al. (1996). They found that a more realistic representation of the phenology of the vegetation cover during the summer months leads to more negative values of the lifted index (a commonly used index of atmospheric instability) and hence, more convective precipitation.

Baldocchi and Meyers (1998) have documented the land-atmosphere interactions over different vegetation types by combining boundary layer data at these sites with satellite data to define the mesoscale characteristics of surface heat fluxes. Atmospheric stability exerts an important control on these fluxes. GCIP research will assess the adequacy of current techniques for parameterizing boundary layer processes, particularly during the spring. Baker et al., (1999) have shown that, in the spring, the development of the boundary layer during the day is dependent on the snow cover on the ground and the large scale radiation forcing.

6.2 Soil Moisture

Soil moisture is a critical control on the feedback between the land and the atmosphere. Under sunny conditions dry soil is characterized by warm temperatures and large sensible heat fluxes while wet soil is characterized by cooler temperatures and larger latent heat fluxes. GCIP has had success in reproducing vertical profiles of soil moisture with one-dimensional water balance models. However, the large heterogeneity of soil moisture, even on small spatial scales, has been a major obstacle in moving beyond a one-dimensional representation of soil-atmosphere interactions. Using data sets from the Former Soviet Union, Vinnikov et al. (1996) have shown that soil moisture departures from average scale linearly on both hydrologic and atmospheric space scales with decorrelation lengths of 400 km and 1 to 2 months. This characteristic scaling at climatological time and space scales offers promise that anomalies can be used rather than absolute values of soil moisture in climate modeling. With some adjustments a number of mesoscale models could be capable of using relative soil moisture values for validation purposes.

Soil moisture is a critical variable in the definition of the initial conditions. In NWP and climate forecasts initial conditions are very critical for determining whether soil moisture will enhance precipitation on a daily basis and over a season.

Numerous studies have shown the implications of soil moisture for the prediction of precipitation. During the flood of 1993, models with enhanced soil moisture schemes properly accounted for precipitation. Koster and Suarez (1999) have shown that the inclusion of actual estimates of soil moisture over the USA in place of climatological soil moisture has a greater impact on seasonal predictions of summer precipitation than the use of measured Sea Surface Temperatures (SSTs) in place of climatological SSTs. However, for seasonal prediction of precipitation, both initial and boundary conditions must be considered.

More comprehensive assessments of soil moisture effects will be undertaken now that high frequency soil moisture fields from Oklahoma have become available for the GCIP/CART/ARM soil moisture sites [*Schneider*, personal communication], the Agricultural Research Service (ARS) and the Oklahoma mesonet soil moisture networks. These networks rely on the TDR sensor and a great deal of calibration work is needed to standardize the measurements and make them more useful for model validation. However, calibrating soil moisture measurements is a challenge. The Time Delay Reflectometry (TDR) sensor measures the dielectric constant which is correlated with hydraulic conditions of the soil and hence the moisture in the soil. Different soil types respond differently, consequently it becomes important to know where the sensors are located and the soil profiles at each site. The calibration is further complicated because of the complex calibration curve of the TDR sensor, which behaves differently on wetting, than it does when drying out. High frequency observations in Oklahoma are being used to provide soil moisture fields that will be useful in assessing the ability of models to simulate the drying out of moist landscapes. The results of NASA's SGP97 and SGP99 (Southern Great Plains 1997 Study and Southern Great Plains 1999 Study) are also contributing to characterizing the two dimensional soil moisture fields.

In considering the interactions of models and observations, two (of the many) research questions that arise are:

1. To what extent can existing remote sensing data be used to characterize the values of soil moisture and its scaling properties?
2. To what extent does the scaling of soil moisture in models match the scaling of soil moisture in nature?

6.3 Surface Heteorogeneity

For GCIP to be successful, those processes important at the mesoscale must be identified, quantified, understood and incorporated into models. The procedures for taking the understanding of a specific physical process and representing it in a form that is useful in models requires the ability to document the physical process at representative locations and to upscale it so the measurements taken at a site can be made applicable for a much wider area. This approach involves studies to characterize heterogeneity effects, to understand their relative importance and find ways of characterizing it in models.

Heterogeneity is present in every hydrometeorological field from precipitation to vegetation. The consequences of heterogeneity are more significant for some parameters and variables than others. For example, the heterogeneity of precipitation seriously affects the ability of many hydrological models to simulate runoff when using spatially averaged precipitation as input. [*Leung et al.*, 1996; *Schaake et al.*, 1996]. Methods for characterizing the heteorogeneity of precipitation involve the development of downscaling algorithms [*Venugopal et al.*, 1999] and statistical formulations of the spatial variability of precipitation.

The Regional Atmospheric Modeling System (RAMS) is being used to demonstrate the effects of shape and size of sub-grid surface heterogeneities on mesoscale circulations [*Avissar and Liu,* 1996]. According to their analysis these mesoscale heterogeneities can be responsible for creating mesoscale circulations with their own localized precipitation patterns. This effect could reduce the confidence often placed in simple linear procedures used to average properties for heterogeneous fields. The results of this modeling have been clarified in the LBA experiment in the Amazon where observations have shown that sensible heat fluxes are very strong over clear cut areas. Different methods of representing surface heteorogeneity using aggregation techniques are also being assessed by Arian et al. (1997). According to their analysis, the aggregation method provides reasonable surface flux estimates for most seasons over most of the continental USA.

6.4 Cold Season Processes

Studies in the North Central area of the Mississippi River Basin have been directed at quantitatively describing the snow covered land surface and its interactions with the atmosphere. During winter, the surface is relatively decoupled from the atmosphere. With the onset of spring, the snow melts, the ground thaws, the surface warms, and the processes of sensible heating and evaporation once again become important to the climate. Observational studies by Kunkel et al. (1999) in Illinois have documented the heterogeneity of the snow cover during melt. Data on ground heat fluxes, snow melt rates, meltwater ponding on the surface, and ground melting are being provided by Baker et al. (1999). Several hydrologic models are being tested in order to better interpret these data. Baker (personal communication) has found that the SOIL model is more effective than the SHAW model in representing heat fluxes due to difficulties in the long wave radiation component in the SHAW model. GCIP is supporting the improved representation of snow processes in the RAMS [*Liston et al.*, 1999], SSiB [*Sun et. al.*, 1999] and BATS [*Yang, et al.*, 1999] models.

Large-scale land-based climate anomalies can influence larger regions during subsequent seasons. One of the most important Euroasian telecommunications involves the correlation between the European snow cover and the intensity of the Indian monsoon the following summer. Based on some analysis of snow cover in North America, Gutzler (personal communication) is studying possible connections between snow cover and the extent and the onset of the monsoon in North America.

6.5 Runoff/Discharge

Stream discharge is an important variable to use in model validation. Modelers have become increasingly aware that stream and river channels transport significant amounts of water from an area and that the modeling of runoff represents a robust test of the extent to which a given land surface scheme can represent the land surface processes. The PILPS 2c model intercomparison, which included both SVATS models and distributed hydrologic models, assessed the ability of these models to simulate runoff [*Lohmann et al.*, 1998]. Results from PILPS 2c showed that many land surface schemes produce reasonable runoff estimates but do not adequately represent the process whereby the runoff is generated because the amount of water coming from base flow is either dominant or virtually non-existent. However, models that represent hydrologic processes such as the Variable Infiltration Capacity model seemed to reproduce a reasonable balance between surface runoff and base flow.

There are serious limitations in representing runoff processes for the Mississippi River Basin. One such limitation is the status of our understanding of hydrologic processes. The traditional Hortonian flow paradigm of runoff flowing over the land surface, has been shown to occur less frequently than initially thought, particularly in drier areas such as the western part of the Mississippi River Basin. As noted earlier, many hydrological models have used a "lumped-parameter" paradigm where single value parameters and inputs are used for catchments. This results rameters and inputs are used for catchments. This results in the development of relationships that are very specific to the area. Accordingly, processes such as bank storage, hydraulic properties of the streambed, recharge and discharge, etc. are all implicitly included in these calibrated parameters and constants. A new paradigm for large-scale experiments is physical understanding, and improved process representation through the development of distributed models. However, the linear nature of the routing processes, the complex dynamic nature of local stream hydraulics, the varying travel times and the presence of lakes and reservoirs have made it difficult to integrate streamflow into coarse resolution distributed models. In spite of these challenges new parameterizations and formulations for cold season processes including freeze/thaw cycles are being incorporated into the VIC model and the domain of the model has been adapted to the North Central and Northwest areas of the Mississippi River Basin through the development of new river routing schemes [*Cherkauer and Lettenmaier*, 1999].

7. APPLICATIONS OF KNOWLEDGE

The benefits of improved prediction techniques depend on the degree to which the prediction products meet the user's needs. The utility of the forecasts depends on the degree to which the user community accepts and applies the model outputs. Within the operational NWP centers the development of operational prediction systems has benefited from model research and model development streams that vigorously interact with each other. However, in areas such as hydrology, where there is a larger diversity in the formulation of the model platforms, the opportunities to influence operations have not been as significant. Furthermore, users must be educated on the ways of taking advantage of new forecast products if they are going to apply them. A number of obstacles have limited the penetration of new GCIP technologies in this area. These include:

1. The difficulty of demonstrating the performance of GCIP models in comparison to operational models when GCIP investigators don't have access to real-time data flows.
2. The reluctance of operational offices to change familiar procedures.
3. The educational effort needed to get users to accept new forecast products.

In spite of these limitations, GCIP has taken steps to encourage the transfer of its knowledge to operational agencies and water resource decision-makers. A number of demonstration projects have been established to assess and demonstrate the value of improved predictions for water resource managers. In most of these projects climate forecasts or historical conditions are used along with a downscaling model to produce predictions at a scale of relevance to a watershed. The most commonly used downscaling techniques are embedded high-resolution mesoscale models and statistical techniques. These later techniques include the representation of the spatial variability of precipitation as a means of accounting for the heterogeneity in a model. When using downscaling approaches it is necessary to ensure that the larger scale model can provide reliable boundary conditions to the smaller scale model. This approach is in an experimental stage and issues remain related to the degree to which errors in values provided by the global model at the boundaries of the mesoscale model propagate into the central part of the domain, thereby influencing the reliability of results.

Four research areas in water resource applications are being addressed within GCIP to more effectively meet the needs of water resource managers [*Lawford*, 1999]. These areas include:

1. Determining the needs of water managers for climate information and predictions.
2. Assessing the need for methods for adjusting climate models outputs for model biases.
3. Representing the uncertainties in climate predictions and formulations of probalistic quantitative precipitation forecasts in hydrologic models.
4. Evaluating climate forecasts using skill measures relevant to water resource applications.

Climate precipitation forecasts are particularly relevant for water supply problems. One potential use of seasonal forecasts is decision-making for reservoir operations. Depending on the location of these reservoir operations, there is a need to meet the social and economic objectives without losing water through spillage and without putting populations at risk from flooding. Often the decisions made by dam operators are constrained by government regulations. In cases such as the Saylorville reservoir where distinct criteria for management decisions have been defined, Georgakakos et al. (1998), found that providing a forecast is not enough, one must formulate the forecast in probability terms and improve the decision making model to enable it to accept stochastic (or probabilistic) information and to allow it to adapt or "learn" as it proceeds. Georgakakos et al. (1998) used the European Climate Model - Hamburg Version (ECHAM) climate prediction and the Office of Hydrology's Sacramento model as input for decision making. The results of the Georgakakos study indicated that the benefits for an Iowa reservoir were greatest when the forecast did include a range of values rather than a specific value. Furthermore, a system that could adjust to differences between the expected and actual conditions on a daily basis provided the best results in terms of the overall management of the water.

8. SUMMARY

The integration of data and models in addressing regional water and energy budgets is an important activity in GCIP. Advances in model development in support of these budgets have been facilitated by establishing data sets that are comprehensive and relevant for the needs of model development; data delivery systems that are affordable and readily accessible; and model development work that has clear relevance to, and a strong influence on, operational NWP models. Where data are missing for model initialization and parameter estimation, specialized models have been able to provide the necessary inputs. In addition to assisting in closing regional water and energy budgets, models have been successfully used within GCIP to quantify processes, to provide predictions for water resource studies, and to provide data assimilation products for climate studies. These developments have also benefited NCEP's operational weather services because it has adopted a number of model innovations developed by the academic community.

APPENDIX A
Acronyms

ARM Atmospheric Radiation Measurement
ARS Agricultural Research Service
BALTEX Baltic Sea Experiment
BATS Biosphere-Atmosphere Transfer Scheme

BOREAS	Boreal Ecosystem Atmosphere Study
CART	Clouds and Radiation Testbed
CATCH	Coupling of the Tropical Atmosphere and Hydrological Cycle (English Translation)
CSE	Continental Scale Experiment
ECHAM	European Climate Model-Hamburg Version
ECMWF	European Centre for Medium-Range Weather Forecasting
EROS	Earth Resources Observation Satellite
FIFE	First ISLSCP Field Experiment
FSL	Forecast Systems Laboratory
GAME	GEWEX Asian Monsoon Experiment
GCIP	GEWEX Continental-scale International Project
GCM	Global Climate Model
GEM	Global Environmental Multiscale (Model)
GEWEX	Global Energy and Water Cycle Experiment
GHP	GEWEX Hydrometeorology Panel
GPCP	Global Precipitation Climatology Project
GRDC	Global Runoff Data Centre
GSFC	Goddard Space Flight Center (NASA)
IGBP	International Geosphere Biosphere Project
IGPO	International GEWEX Project Office
IOP	Intensive Observing Period
ISLSCP	International Satellite Cloud Climatology Project
LBA	Large Scale Biosphere-Atmosphere
LDAS	Land Data Assimilation System
LSA	Large-Scale Area
LSA-E	Large-Scale Area-East
LSA-NC	Large-Scale Area -North Central
LSA-NW	Large-Scale Area - Northwest
LSA-SW	Large-Scale Area Southwest
LSM	Land Surface Model
MAGS	Mackenzie GEWEX Study
MAPS	Mesoscale Analysis and Prediction System
MM5	Mesoscale Model (NCAR)
MOPEX	Model Parameter Estimation Experiment
NASA	National Aeronautics and Space Administration
NCAR	National Center for Atmospheric Research
NCEP	National Centers for Environmental Prediction
NDVI	Normalized Difference Vegetation Index
NEXRAD	Next Generation Radar
NOAA	National Oceanic and Atmospheric Administration
NRC	National Research Council
NWP	Numerical Weather Prediction
OSU	Oregon State University
PILPS	Project for the Intercomparison of Land Surface Parameterization Schemes
RAMS	Regional Area Modeling System (Colorado State University)
SAST	Snow-Atmosphere Soil Transfer
SGP97	Southern Great Plains 1997 Study
SGP99	Southern Great Plains 1999 Study
SHAW	Simultaneous Heat and Water (model)
SSiB	Simple Biosphere Model
SST	Sea Surface Temperature
SVATS	Soil Vegetation Atmospheric Transfer Scheme
SWB	Surface Water Budget
TDR	Time Delay Reflectometry
VIC	Variable Infiltration Capacity (Hydrologic Model)
WCRP	World Climate Research Programme
WMO	World Meteorological Organization

9. REFERENCES:

Abdulla, F. A., D. P. Lettenmaier, E. F. Wood, and J. A. Smith, Application of a macroscale hydrologic model to estimate the water balanace of the Arkansas-Red River Basin, *J. Geophys. Res.*, Vol. 101, No. D3, 7449-7459, 1996.

Arian, A.M., W.J. Shuttleworth, Z.-L. Yang, J. Michaud, and J. Dolman, Mapping surface cover parameters using aggregation rules and remotely sensed land cover classes, *Q.J.R. Meteor. Soc.*, 123, 2325-2348, 1997.

Avissar, R., and Y. Liu, A three-dimensional numerical study of shallow convective clouds and precipitation induced by land-surface forcing, *J. Geophys. Res.*, 101, 7499-7518, 1996.

Baker, J. M., K. J. Davis, and G. C. Liknes, Surface energy balance and boundary layer development during snowmelt, *J. Geophys. Res.*, Vol. 104, No. D16, 19,611-19, 621, 1999.

Baldocchi, D. and T. Meyers, On using eco-physiological, micrometeorological and biogeochemical theory to evaluate carbon dioxide, water vapor and trace gas fluxes over vegetation: a perspective. *Agricultural and Forest Meteorology* , 90. 1-25, 1998.

Berbery, E., and E. A. Collini, Springtime precipitation and water vapor flux over Southeastern South America. Submitted to *Mo. Weather Rev.*, Vol. 128, 1328-1346, 1999.

Carroll, S. S., T. R. Carroll and R. W. Poston, Spatial modeling and prediction of snow water equivalent using ground-based, airborne, and satellite snow data. *J. Geophys. Res.*, Vol. 104, No. D16, 19, 623- 19, 629, 1998.

Chen, F., K. Mitchell, J. Schaake, Y. Xue, H.-L Pan. V. Koren, Q. Y. Duan, M. Ek and A. Betts, Modeling of land surface evaporation by four schemes and comparison with FIFE observations. *J. Geophys. Res.*, 101, No. D3, 7251-7268, 1996.

Cherkauer, K. A., and D. P. Lettenmaier, Hydrologic effects of frozen soils in the upper Mississippi River Basin, *J. Geophys. Res.*, Vol. 104, No. D16, 19,599-19,610, 1999.

Coughlan, M., and R. Avissar, The Global Energy and Water Cycle Experiment (GEWEX) Continental Scale International Project (GCIP): An Overview, *J. Geophys. Res.*, Vol., 101, No. D3, 7139 – 7147, 1996.

Dirmeyer, P.A., and K. L. Brubaker, Contrasting evaporative moisture sources during the drought of 1988 and flood of 1993, *J. Geophys. Res.*, Vol. 104, No. D16, 19,383-19,397, 1999.

Entin, J. K., A. Robock, K. Y. Vinnikov, V. Zabelin, S. Liu, A.

Namshai, and T. Adysaures, Evaluation of Global Soil Wetness Project Soil Moisture Simulations. *J. of Meteor Soc. of Japan*, Vol., 77, No. 1B, 183-199, 1999.

Georgakakos, A. P., K. P. Georgakakos, and N.E. Graham, Assessment of benefits of climate forecasts for reservoir management in the GCIP region, *GEWEX News*, 8(3), 5-7, 1998.

Gupta, H. V., L. A. Bastidas, S. Sorooshian, W.J. Shuttleworth, and Z.-L. Yang, Parameter estimates of a land surface scheme using multi-criteria methods, J. *Geophys Res.*, Vol. 104, No. D16, 19,491-19,503, 1999.

International GEWEX Project Office, GCIP Major Activities Plan for the Large Scale Area-Northwest (Missouri River Basin), IGPO Publication Series No. 33, International GEWEX Project Office, Washington, DC, 1999.

International GEWEX Project Office, Implementation Plan for the GEWEX Continental-scale International Project (GCIP); Volume II, Research. International GEWEX Project Office, Washington, DC, 1999.

Koren, V., J. Schaake, K. Mitchell, Q.-Y. Duan, F. Chen, and J. M. Baker, A parameterization of snowpack and frozen ground intended for NCEP and climate weather models, *J. Geophys. Res.*, Vol. 104, No. D16, 19,569-19,585, 1999.

Koster, R.D. and M.J. Suarez, A Simple Framework for examining the interannual variability of land surface mositure fluxes, *J. of Climate*, 12:1911-1917, 1999.

Kunkel, K. KE., S. A. Isard, S. E. Hollinger, B. Gleason, and M. Belding, Spatial heteorogeneity of albedo over a snow-covered agricultural landscape, *J. Geophys. Res.*, Vol. 104, No. D16, 19,425-19,434, 1999.

Lawford, R. G., A midterm report on the GEWEX Continental-scale International Project (GCIP). J. *Geophys. Res.*, Vol. 104, No. D16, 19,279-19,292, 1999.

Leung, *L.R*, and S.J. Ghan, Parameterizing subgrid orographic precipitation and surface cover in claimte models, *Mo. Weath. Rev.*, 126, 3271-3291, 1998.

Liang, X., E.F. Wood, D.P. Lettenmaier, D. Lohmann, A. Boone, S. Chang, F. Chen, Y. Dai, C. Desborough, R. E. Dickinson, Q. Duan, M. Ek, Y.M. Gusev, P. Irannejad, R. Koster, K.E. Mitchell, O.N. Nasonova, J. Noilhan, J. Schaake, A. Schlosser, Y. Shao, B. Shmakin, D. Verseghy, K. Warrach, P. Wetzel, Y. Xue, Z.-L. Yang and Q.-C. Zeng., The Project for the Intercomparison of Land-Surface Parameterization Schemes (PILPS): Phase 2c, Red-Arkansas River basin experiment, 2, Spatial and temporal analysis of energy fluxes, *Global Planetary Change*, 19, 137-159, 1998.

Liston, G. E., R. A. Pielke, Sr., and E. M. Greene, Improving first-order snow-related deficiencies in a regional climate model, *J. Geophys Res.*, Vol. 104, No. D16, 19, 559 - 19, 567, 1999.

Lohmann, D., D.P. Lettenmaier, X. Liang, E.F. Wood, A, Boone, S. Chang, F. Chen, Y. Dai, C. Desborough, R. E. Dickinson, Q. Duan, M. Ek, Y.M. Gusev, P. Irannejad, R. Koster, K.E. Mitchell, O.N. Nasonova, J. Noilhan, J. Schaake, A. Schlosser, Y. Shao, A.B. Shmakin, D. Verseghy, K. Warrach, P. Wetzel, Y. Xue, Z.-L. Yang and Q.-C. Zeng, The Project for Intercomparison of Land-Surface Parameterization Schemes (PILPS): Phase 2c, Red-Arkansas River basin experiment, 3, Spatial and temporal analysis of water fluxes, *Global Planetary Change*, 19, 161-176, 1998.

Meyers, T. P, Inter-annual Variations in Summertime Water and CO_2 fluxes over Rangeland. (For submission to *Agricultural and Forest Meteorology)*, 1998.

Miller, D. A. and R.A. White, A conterminous United States multi-layer soil characteristics data set for regional climate and hydrology modeling, Earth Interactions 2, 1999. (Available at http://EarthInteractions.org), 1999.

National Research Council, GCIP: A Review of Progress and Opportunities, National Academy Press, Washington, 93 pp, 1998.

Roads, J. O., S.–C. Chen, M. Kanamitsu, and H. Juang, Vertical structure of humidity and temperature budget residuals over the Mississippi River basin, *J. Geophys. Res.*, 103, 3741-3759, 1998.

Ropelewski, C. F., and E. S. Yarosh, The observed mean annual cycle of moisture budgets over the central United States (1973-92), *J. Clim.*, 11, 2180-2190, 1998.

Schaake, J.C. Jr., V.I. Koren, Q. Duan, K. Mitchell and F. Chen, Simple Water Balance Model for estimating runoff at different spatial and temporal scales. *J. Geophys. Res.* 101(D3), 7461-7475, 1996.

Sellers, P. J., F. G. Hall, G. Asrar, D.E. Strebel, and R. E. Murphy, The First ISLSCP Field Experiment (FIFE*), Bull. Am. Meteor. Soc.*, 69(1), 22-27, 1998.

Shao, Y., and A. Henderson-Sellers, Modelling Soil moisture: A Project for the Intercomparison of Land-Surface Parameterization Schemes, Phase 2(b*), J. Geophys. Res.*, 101, 7461-7475, 1996.

Sun, S., J. Jin, and Y. Xue, A simple snow-atmosphere soil transfer (SAST) model, *J. Geophys. Res.*, Vol., 104, No. D16, 19,587-19,598, 1999.

Venugopal, V., E. Foufoula-Georgiou, and V. Zapozhnikov, A space-time downscaling model for rainfall, *J. Geophys. Res.*, Vol. 104, No. D16, 19,705-19,721, 1999.

Viterbo, P. and A.K. Betts, Impact of the ECMWF reanalysis soil water on forecasts of the July 1993 Mississippi flood, *J. Geophys. Res.*, Vol. 104, No. D16, 19,361-19,366, 1999.

Vinnikov, K. Y., A. Robock, N. A. Sperankaya and C. A. Schlosser, Scales of temporal and spatial variability of midlatitude soil moisture, *J. Geophys. Res.*, Vol. 101, No. D3, 7163-7174, 1996.

Wood, E.F., D.P. Lettenmaier, X. Liang, D. Lohmann, A. Boone, S. Chang, F. Chen, Y. Dai, R. E. Dickinson, Q. Duan, M. Ek, Y.M. Gusev, F. Habets, P. Irannejad, R. Koster, K.E. Mitchell, O.N. Nasonova, J. Noilhan, J. Schaake, A. Schlosser, Y. Shao, A.B. Shmakin, D. Verseghy, K. Warrach, P. Wetzel, Y. Xue, Z.-L. Yang and Q.-C. Zeng, The Project for the Intercomparison of Land-Surface Parameterization Schemes (PILPS): Phase 2c, Red-Arkansas River basin experiment, 1. Experiment description and summary intercomparisons, *Global Planetary Change*, 19, 115-135, 1998.

Xue, Y., M. Fenessey, and P. J. Sellers, Impact of vegetative properties on U.S. summer weather prediction. *J. Geophys. Res.*, 101, 7419-7430, 1996.

Yang, Z. L., R. E. Dickinson, A. Robock, and K. Y. Vinnikov,

Validation of the snow-sub-model of the Biosphere Atmosphere Transfer Scheme with Russian snow cover and meteorological observational data, *J. Clim.*, 10 353-373, 1997.

Yarosh, E., C. Ropelewski, and E. H. Berbery, Biases of the observed atmospheric water budgets over the central United States, *J. Geophys. Res.*, Vol. 104, No. D16, 19, 349- 19, 360, 1999.

Yucel, I., W. J. Shuttleworth, J. Washburne, and F. Chen, Evaluating NCEP Eta model derived data against observations, *Mon. Weath. Rev.*, 126, 1977-1991, 1998.

Richard G. Lawford, NOAA Office of Global Programs, Suite 1210, 1100 Wayne Ave., Silver Spring, MD 20904

Hydrological Implications of the El Niño-Southern Oscillation (ENSO): Observations and Hydrologic Forecasting

Thomas C. Piechota

Department of Civil and Environmental Engieering, University of Nevada, Las Vegas, Nevada

Francis H. S. Chiew

Cooperative Research Centre for Catchment Hydrology, University of Melbourne, Australia

John A. Dracup

Department of Civin and Environmental Engineering, University of California, Berkeley, California

This study investigates the hydrologic implications of the El Niño-Southern Oscillation (ENSO) and potential for making long-range streamflow forecasts in eastern Australia and the western United States. These two regions are particularly interesting due to the strong effect of ENSO on their hydrology. In eastern and south-eastern Australia, there is a strong tendency for below normal hydrologic conditions following an El Niño event. The relationship is strong enough to develop a seasonal streamflow forecast. In the western U.S., the potential of making a six-month lead time forecast of spring-summer runoff is demonstrated for the Columbia River basin. The skill associated with these forecasts is better than the baseline "climatology" condition and there are potential benefits of using these forecasts in the management of water resource systems. The potential benefits of using this forecast are demonstrated for an urban water supply system and for a rural irrigation system in eastern Australia.

INTRODUCTION

In the past decade, there has been major progress in documenting and modeling large-scale ocean/atmosphere interactions, such as the El Niño-Southern Oscillation (ENSO), and in understanding the role that these large-scale processes play in long-range climate and hydrologic prediction. ENSO refers to the interaction of El Niño, defined as the periodic large scale warming of the central-eastern equatorial Pacific Ocean, with the Southern Oscillation, the large scale ocean and atmosphere interaction that exist in the tropical Pacific. The ENSO phenomenon causes, simultaneously, droughts in Australia, New Zealand, and Southern Africa and devastating floods in North America, Peru, and Ecuador [*Ropelewski and Halpert*, 1987; *Rasmusson*, 1985]. The warm phase of ENSO is called "El Niño," while the cold phase is called La Niña [*Philander*, 1990].

Due to advances in the atmospheric science and oceanographic communities, models have been developed that can predict up to one year in advance the onset of an ENSO event [e.g., *Zebiak and Cane*, 1987; *Barnett et al.*, 1993; *Ji et al.*, 1996; *Penland and Magorian*, 1993; *Jiang et al.* 1995; *Chen et al.* 1995]. The forecasting of ENSO events

Land Surface Hydrology, Meteorology, and Climate: Observations and Modeling
Water Science and Application Volume 3, pages 231-246

is invaluable for protecting land and water resources against drought and floods, for determining current and future water allocation for agricultural irrigation and urban uses, and for directing the reservoir operations that produce hydroelectric power .

The impact of ENSO on the hydrologic parameters of streamflow, precipitation, and snow water equivalent has been analyzed extensively [*Ropelewski and Halpert*, 1986; *Yarnal and Diaz*, 1986; *Ropelewski and Halpert*, 1987; *Cayan and Peterson*, 1989; *Ropelewski and Halpert*, 1989; *Schoner and Nicholson*, 1989; *Kuhnel et al.*, 1990; *Redmond and Koch*, 1991; *Halpert and Ropelewski*, 1992; *Mechoso and Iribarren*, 1992; *Cayan and Webb*, 1992; *Kahya and Dracup*, 1993; *Woolhiser et al.*, 1993; *Chiew et al.*, 1994; *Dracup and Kahya*, 1994; *Kahya and Dracup*, 1994; *Gingras and Adamowski*, 1995; *Cayan*, 1996; *Guetter and Georgakakos*, 1996; *Piechota and Dracup*, 1996; *Piechota et al.*, 1997; *Shabbar et al.*, 1997; *Smith and Ropelewski*, 1997; *Sun and Furbish*, 1997; *Chiew et al.*, 1998, *Gershunov*, 1998, *Gershunov and Barnett*, 1998, *Dettinger et al.*, 2000]. A partial listing of ENSO-hydrology studies is also included in *Liu et al.* [1998].

An understanding of the impacts of ENSO events has additional benefits. Research has shown that the ENSO is an important factor when forecasting seasonal streamflow [e.g., *Simpson et al.*, 1993; *McKerchar and Pearson*, 1994; *Moss et al.*, 1994; *Liu et al.*, 1998; *Piechota et al.*, 1998; *Piechota and Dracup*, 1999; *Wang and Eltahir*, 1999; *Dettinger et al.*, 1999]. Unfortunately, the strong association between ENSO and streamflow has not inspired researchers to thoroughly develop methods for using ENSO information in water resources management. Yet a seasonal streamflow forecast is invaluable to water resource planners and managers, who need such information if they are to allocate water supplies fairly and meet equitably the demands of competing water uses (e.g., agricultural, domestic, industrial, commercial, environmental). This is particularly relevant in Australia and the western United States where the evapotranspirative demand is high and where the inter-annual variability of streamflow is higher than most parts of the world [see *McMahon et al.*, 1992].

The study presented here investigates the effects of ENSO on the regional hydrology of eastern Australia and the western United States. In eastern Australia, anomalous precipitation is produced due to the presence of high pressure and cooler than normal sea surface temperatures in the Australian region. In the western U.S., changes in storm tracks coupled with an enhanced subtropical jet stream produce wet conditions in the southwest and dry conditions in the northwest. Two examples of making long-range streamflow forecasts in Australia and the western U.S. are presented and the potential benefits of using ENSO information are explored. Finally, directions of future research

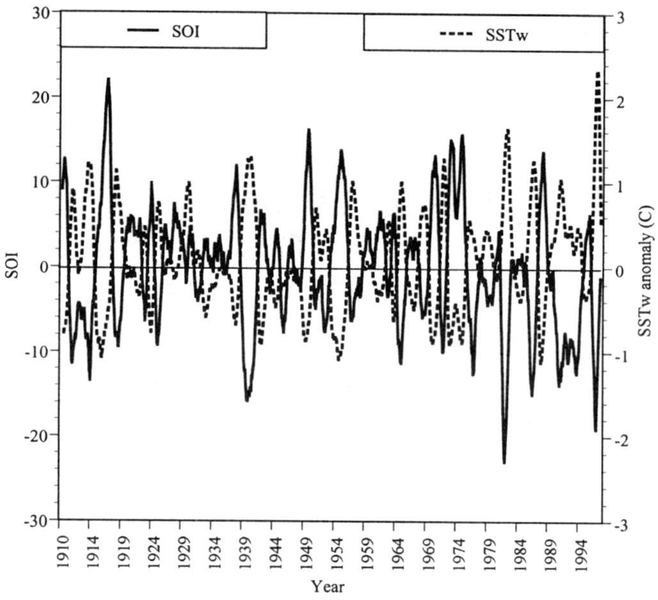

Figure 1. The 12 month running average of the Southern Oscillation Index (SOI) and the Wright Sea Surface Temperature (SST_w) time series (1910-1998). The SOI is based on the definition of *Troup* [1965] and the SST_w is based on the definition of *Wright* [1989].

and the use of ENSO information in hydrology is discussed.

ENSO AND STREAMFLOW DATA

ENSO activity is typically monitored by observing the sea level pressures and sea surface temperatures in the equatorial Pacific. Figure 1 presents the 12 month running average of two ENSO indicators, the Southern Oscillation Index (SOI) and the Wright SST (SST_w). The SOI is the most common indicator of ENSO activity that measures the difference in sea level pressures between Tahiti and Darwin, Australia. The Troup SOI [*Troup, 1965*] is defined as the standardized difference of the sea level pressures at Tahiti minus Darwin multiplied by a factor of 10. The Troup SOI time series described in *Allan et al.* [1996] is available from 1876 to the present. Other measures of sea level pressure indices describe anomalous circulation patterns in other parts of the world such as the Pacific North American (PNA), the North Atlantic Oscillation (NAO), and the Pacific Decadal Oscillation (PDO).

The other commonly used indicator of ENSO activity is equatorial Pacific sea surface temperatures (SSTs). The longest available time series of published SST anomalies for the equatorial Pacific includes the years 1872 to 1986 [*Wright, 1989*] and is commonly referred to as the Wright

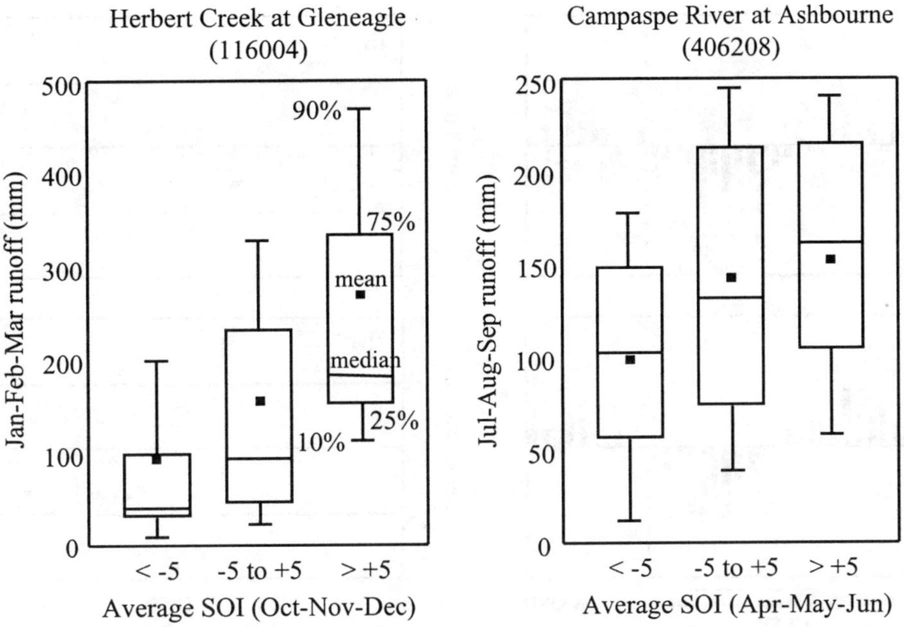

Figure 2. Whisker plots showing means and runoff percentiles over three months in two Australian catchments for three SOI categories. The SOI is a three-month average and leads streamflow by one season as noted. The top (bottom) of the whisker represent the 90th (10th) percentiles and the top (bottom) of the box represent the 75th (25th) percentiles.

SST (SST_w). The SST_w covers a large region in the tropical Pacific from approximately 180^0 W to 90^0W and 6^0S to 6^0N. Other SST measures of ENSO activity are available for smaller regions such as the NINO12, NINO3, NINO4 SST series. Researchers have also developed their own SST series that describe different modes of variability. For instance, *Drosdowsky and Chambers* [1998], developed time series of the major modes of variability in the Pacific and Indian Oceans based on empirical orthogonal functions (EOFs) of the gridded SST values.

Streamflow data are taken from six unimpaired Australian catchments (see Figure 7), which represent different regions of Australia with an uninterrupted 47-year record of monthly streamflow covering the period 1950 to 1996.

Streamflow records for the western U.S. and the Columbia River Basin are available from the U.S. Geological Survey Hydroclimatic Data Network (HCDN) [*Slack and Landwehr*, 1992]. Monthly data from 79 stations in the western U.S., seven within the Columbia River Basin, were extracted for a 56-year span, 1933 to 1988, and updated to 1992, using the individual state data bases available from the U.S. Geological Survey on the Internet. These stations have an uninterrupted 60-year record from 1933 to 1992 and represent medium to large drainage basins. In addition, streamflow data from the Columbia River station at The Dalles (station 14105700) was used for the period 1911 to 1992 (82 years).

ENSO-STREAMFLOW TELECONNECTION

Australia

There is a clear relationship between ENSO and hydrologic variations in Australia. Two early studies [*McBride and Nicholls*, 1983; and *Nicholls*, 1989] observed the influence of ENSO on Australian precipitation. The influence of ENSO on Australian streamflow is seen in Figure 2 which presents whisker plots of seasonal runoff in two catchments. In Figure 2, the runoff is generally higher when the SOI has a high positive value (i.e., La Niña) and vice versa.

The streamflow-ENSO signal is consistent throughout eastern and southeastern Australia, where important agricultural regions and more than three quarters of the population are located. In these regions, a below normal rainfall and streamflow season typically begins around the middle of an El Niño year and extends for about nine months until early the following year. This is illustrated by the 36-month aggregate El Niño composites of rainfall and streamflow in Figure 3a. The composites show the log-normal percentiles of rainfall and streamflow averaged over 14 and seven El Niño years respectively, and over many locations in eastern Australia, analyzed using the harmonic, season detection and index time series methods

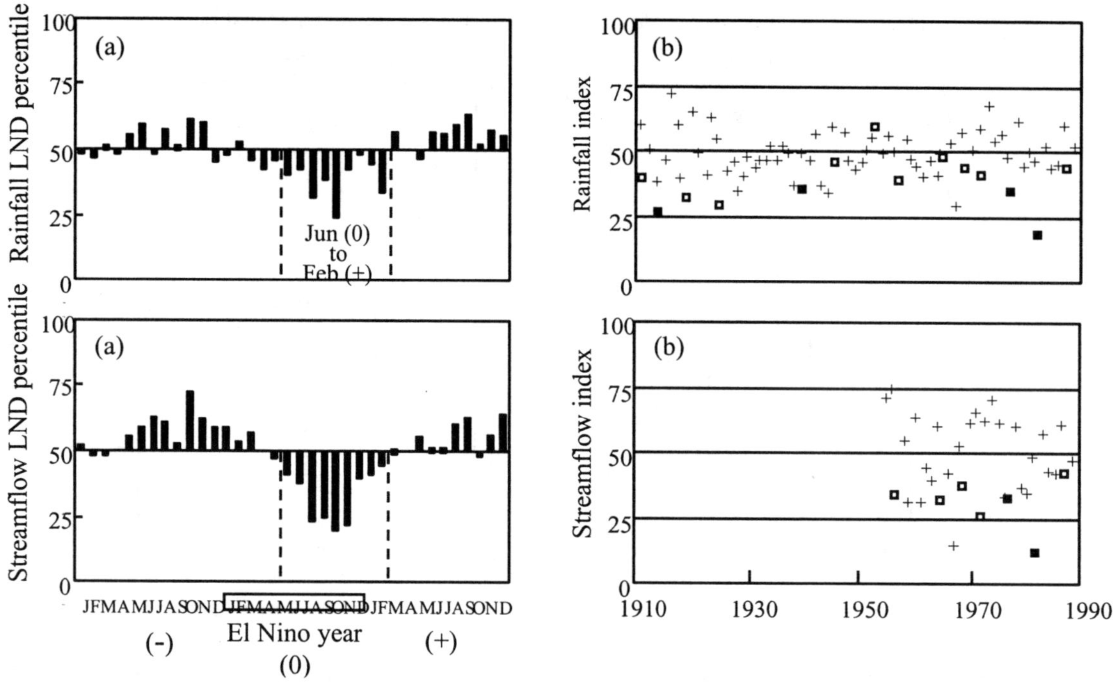

Figure 3. Thirty-six month aggregate El Niño composites of rainfall (119 stations) and streamflow (22 catchments) in eastern Australia and index time series for June to February rainfall and streamflow (in the index time series plots, the moderate (SOI < -5) and strong (SOI < -10) El Niño years are indicated by open and darkened squares respectively).

of *Ropelewski and Halpert* [1986]. The hydroclimate-ENSO teleconnection is also highlighted in the index time series plots (Figure 3b) of percentiles averaged over the signal season in Figure 3a, which show that the below normal rainfall and streamflow generally occur during El Nino years (see *Chiew et al.* [1998] for a detailed description of the analyses). Figure 3 also reflects the strong correlation that exists between rainfall and streamflow in eastern Australia. This is further investigated in this study when developing long-range streamflow forecast predictors.

Figure 4 shows the typical correlations between seasonal rainfall and ENSO, streamflow and ENSO and the serial correlations of rainfall and streamflow, for various lags, in north-east, south-east and south-west Australia [see *Chiew et al.*, 1998, 1999a, for more details]. There is little serial correlation in rainfall while the serial correlation in streamflow in south-east and south-west Australia is statistically significant for most parts of the year. The streamflow-ENSO lag correlations are also generally greater than the rainfall-ENSO correlations. The statistically significant lag correlations suggest that the ENSO indicators and the serial correlation in streamflow can be used to forecast streamflow several months in advance. The reason for these observations is the delayed response in the rainfall-runoff process due to soil and groundwater storage, giving the streamflow data a memory of conditions over several months. In other regions of the world, such as the western U.S., there is even more of a lag response between anomalous precipitation and streamflow due to snowmelt runoff. In Australia, however, there is not a significant amount of snowpack that produces large volumes of runoff.

Pacific Northwest United States

The spatial characteristics between ENSO and streamflow in the western U.S. are also well documented. In general, the northwest region experiences below normal streamflow during El Niño years while the southwest experiences above normal streamflow. This north-south opposition is seen in Figure 5 which presents the correlation between 79 western U.S. streamflow stations and two ENSO indicators (SOI and SST_w) at various lag periods. The streamflow is for the spring-summer period (April-September) and the ENSO indicators are from prior three month periods (i.e., Lag 1=January-March; Lag 3=October-December from the previous year; Lag 7=July-September from the previous year).

In Figure 5, the correlation patterns for both SOI and SST_w are similar; however, the signs are opposite (i.e.,, positive correlation in the northwest using SOI and negative correlation using SST_w). The highest correlation coef-

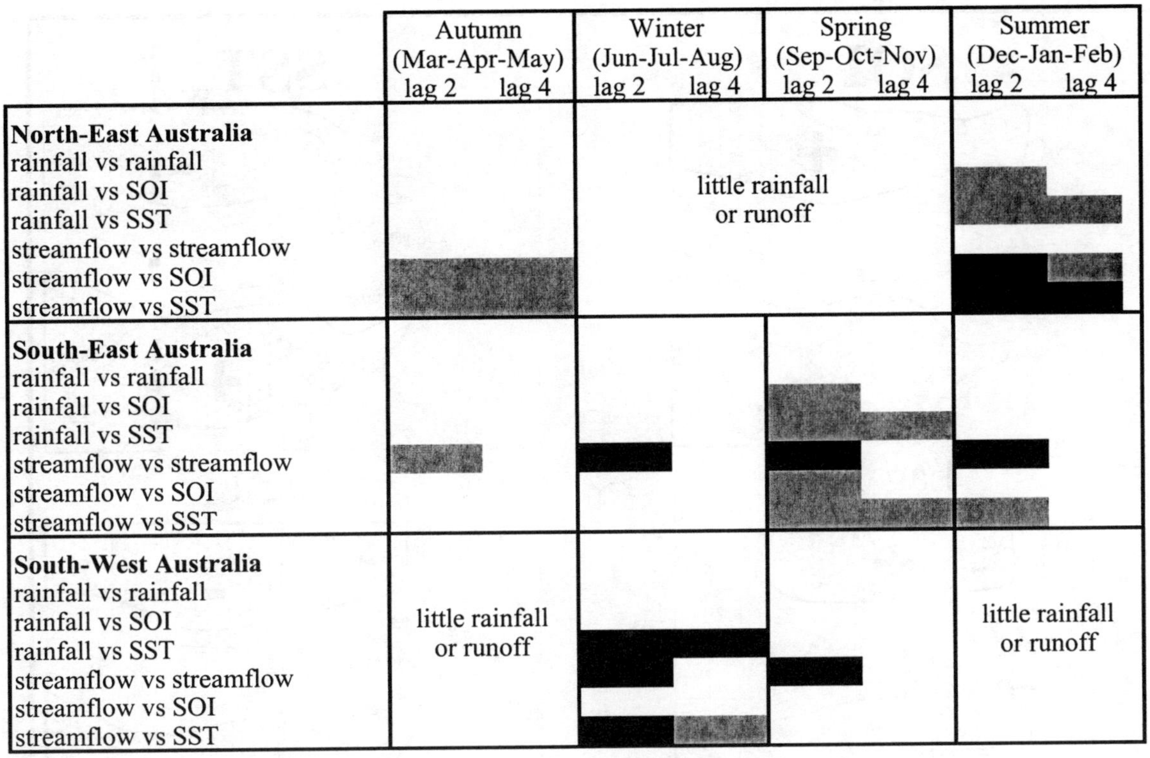

Figure 4. Typical lag correlations of the linear regression between rainfall, streamflow and ENSO in north-east, south-east and south-west Australia (the dark shades indicate that the correlation is above 0.5 in the analyses of over 50 years of time series data, and the lighter shades indicate that the correlation is above 0.4 - in the analyses of 50 years of data, a correlation of 0.3 is statistically significant at the 5% level).

ficients are in the northwest (Montana, Eastern Washington, Northern Idaho) and in the southwest (New Mexico, Arizona, Southern California). These correlation patterns (significant at the 99% confidence level) extend back to the Lag 7 period and are noted by the shaded regions in Figure 5.

The lag correlation in the Pacific northwest is further demonstrated in Figure 6a and 6b which present the 24 month composite of streamflow from the Columbia River at the Dalles and the SOI for El Niño and La Niña years; the first 12 months represent the El Niño (La Niña) year, and the next 12 months are year following the El Niño (La Niña). The composite represents the average of 18 (12) El Niño (La Niña) events from 1911 to 1992. In general, the negative peak in the SOI occurs between September and December of the El Niño year, and the peak streamflow anomaly (departure from the mean) occurs in the summer of the year after El Niño. It is this lag relationship that could make long-range streamflow forecasting possible in the Pacific northwest.

Currently, water supply forecasts for spring and summer runoff are made by the U.S. National Weather Service in conjunction with the National Resources Conservation Service of the U.S. Department of Agriculture [*USDA, 1999*]. These streamflow forecasts are based on current snowpack data drawn from field surveys and remote sensing; a discussion of this regression-based approach is provided by *Garen* [1992]. Because snow accumulation does not begin until December or January, a long-range forecast with a three- to seven- month lead time of spring-summer runoff is not possible. However, working with the ENSO-hydrology relationship could allow such long term forecasts. Following is a methodology for using ENSO indicators to make a long term seasonal streamflow forecast.

SEASONAL STREAMFLOW FORECASTING

Much of the past research on the development of a seasonal streamflow forecast using ENSO indicators has focused on making a categorical streamflow forecast (i.e., below normal, normal, above normal). This is consistent with many of the climate forecasts issued by such agencies as the U.S. Climate Prediction Center and the Australian Bureau of Meteorology. In the management of a water resources system, however, an authority would ideally prefer a forecast with continuous exceedance probabilities. An

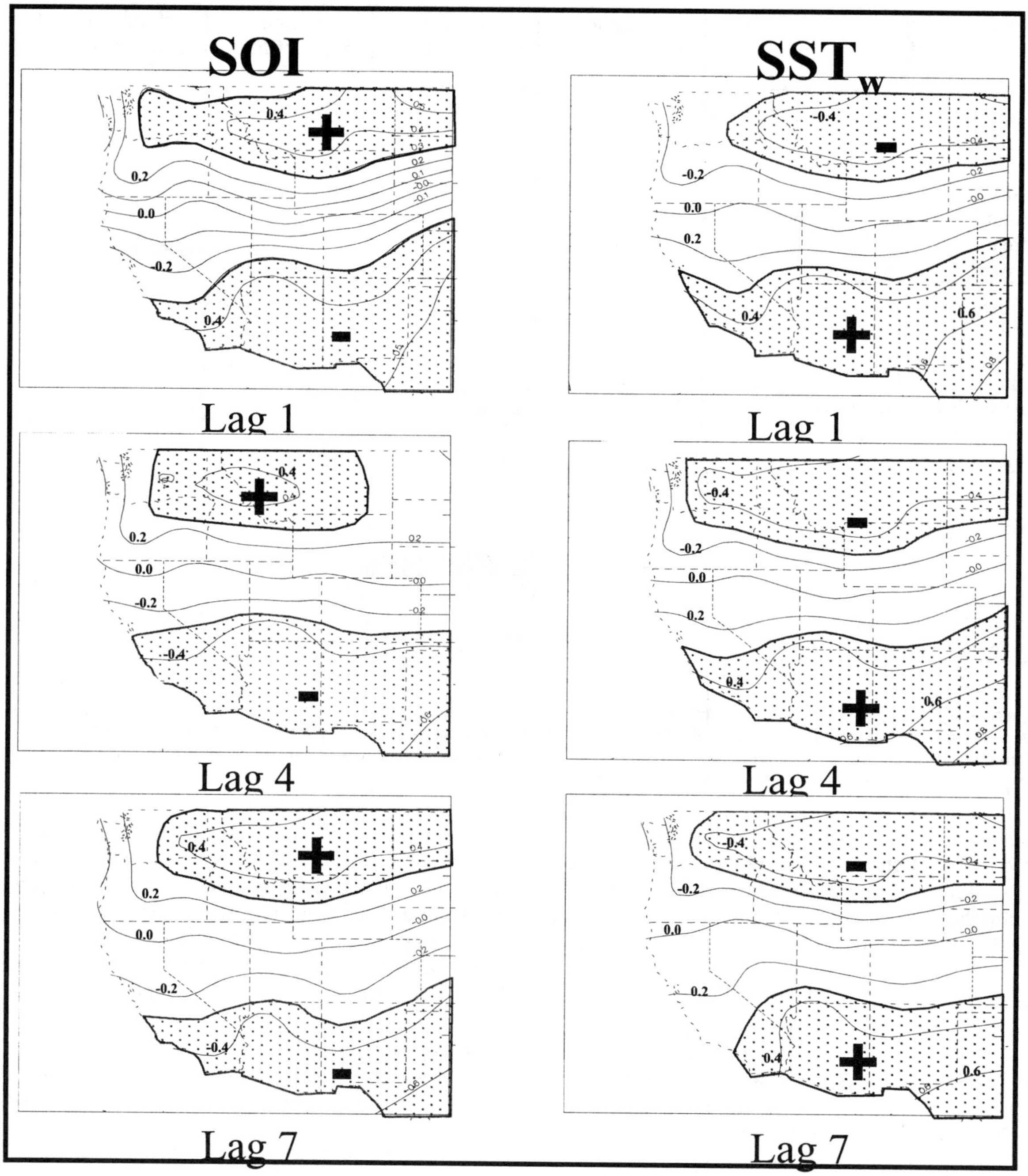

Figure 5. Correlation coefficients of April-September streamflow in the western U.S. with the SOI and SST$_w$ at three different lag periods. Lag 1 = January-March, Lag 3 = October-December from the previous year, and Lag 7 = July-September from the previous year. The shaded regions are significant at the 99% confidence level.

exceedance probability is defined as the probability that the specified value, or the streamflow amount, will be equal to or exceeded during a given time period. An exceedance probability forecast can be used depending on an assumed level of risk. For example, a water authority may choose to take a 10% risk, which would correspond to a streamflow value that has a 90% probability of exceedance. A "climatology" forecast with no skill can be generated by dividing the rank of each historical value by the total number of years in the record. The streamflow forecast developed

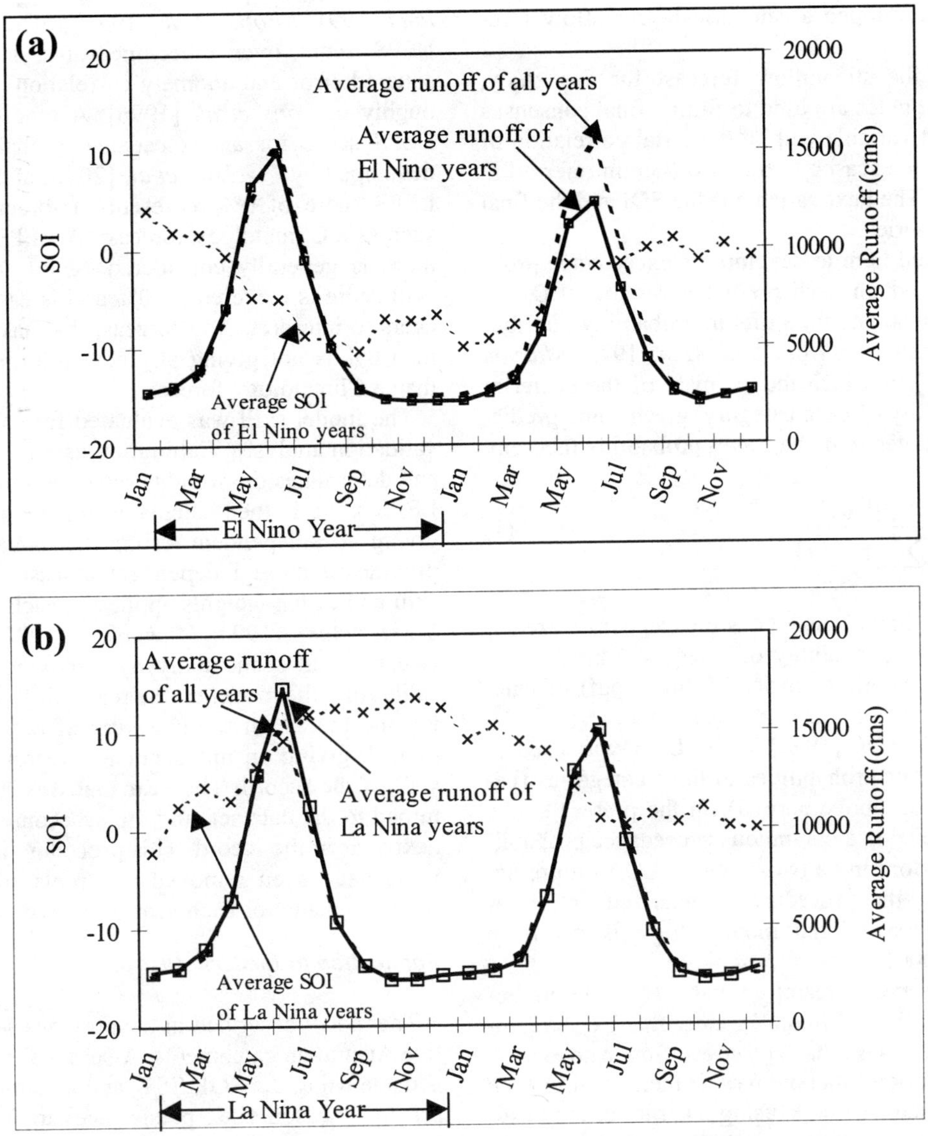

Figure 6. The 24 month composite of runoff from the Columbia River at the Dalles and the Southern Oscillation Index (SOI) for (a) El Niño and (b) La Niña years where the first 12 months represent the El Niño (La Niña) years and the other 12 months are the year after El Niño (La Niña). The composite represents the average of 18 (12) El Niño (La Niña) events from 1911 to 1992.

here is a continuous exceedance probability curve that can be used for any assumed risk level.

A continuous exceedance probability forecast can be made by several methods. In the first method, the predictand values (e.g., streamflow) are divided into groups, or clusters, depending on the magnitude of the predictor variable (e.g., an ENSO indicator). The probability distribution of each group of predictands then forms the exceedance probability curves. The second method is a regression between the predictand and the predictor variable(s). The quantification of the errors in the regression provides a direct estimate of the distribution and, thus, of the exceedance probability forecast. A third alternative is presented here that uses linear discriminant analysis (LDA) to empirically fit data and forecasts probability of exceedance of streamflow amounts. Two advantages are found with this method: it considers the continuous relationship between the predictand and the predictor, and it does not assume a particular model structure. It suffers, however, from its semi-empiricism: fitting the model to the data points assumes that the historical data represents the entire population. This research builds on the study of *Piechota et al.* [1998] who developed a categorical streamflow forecast.

In developing the streamflow forecast for this study, three potential variables are used to form a final consensus forecast. The first variable will be the serial correlation of streamflow from one season to the next (herein referred to as "persistence"). The next variable is the SOI and the final variable is a SST series.

LDA is presented here to develop the exceedance probability curves based on each predictor variable. LDA is typically used to evaluate the shifts in probability distributions of categorized data [*Afifi and Azen*, 1979; *Maryon and Storey*, 1985] in which the estimate of the posterior (forecast) probability of each category, given some predictor values, is estimated with the Bayes probability theorem:

$$\mathrm{Prob}\left(Q_i / X\right) = \frac{p_i f_i(x)}{\sum_{i=1}^{k} p_i f_i(x)} \tag{1}$$

where X is the predictor value, Q_i is the category i streamflow, p_i is the prior probability of category i streamflow, and $f_i(x)$ is the probability density function (pdf) of category i based on the prior season X value.

In the *Piechota et al.* [1998] study, LDA was used to estimate the posterior probabilities of three categories (below normal, normal, above normal). In the present study, LDA is used to develop a continuous exceedance probability forecast by performing a two-category LDA on predictor variable using the procedure summarized below. A complete description of the methodology is given in *Piechota et al.* [2000].

Given some observed streamflow value (Q_i), the predictor variable (X) is divided into a "greater than" Q_i streamflow category and a "less than" Q_i streamflow category. A probability distribution function, $f(x_i)$, is then fitted to each subset of the predictor variable using a kernel density estimator [*Piechota et al.*, 1998]. Next, the posterior probability of streamflow occurring in the "greater than" category (i.e., exceedance probability) is found at some given X using equation 1. The procedure is repeated for all Q_i values to create the exceedance probability curve. The last step is to fit a curve through the points. This is done by first creating an upper and lower envelope of the forecast points and the fitted curve is taken as the vertical average of the upper and lower envelopes.

A measure of skill is necessary to assess the reliability of a forecast. The Linear Error in Probability Space (LEPS) score is one measure of skill that was developed originally to assess the position of the forecast and the position of the observed values in the cumulative probability distribution (non-exceedance probability); the LEPS score can be used for continuous and categorical variables [*Ward and Folland*, 1991; *Potts et al.*, 1996]. The advantages of the LEPS score over other measures, such as root-mean-squared error and anomaly correlation, are discussed thoroughly in *Potts et al.* [1996]. A modified version of the LEPS score for an exceedance probability forecast was developed by *Piechota et al.* [2000] and is applied here. A LEPS score of 0% represents a forecast with "no skill" such as a Climatology forecast. A LEPS score of +10% or more is generally considered "good skill." This type of skill score is centered on 0% and is easier for water managers to interpret. Any forecast that has a LEPS score less than 0% is not giving any new information and is worst than a Climatology forecast.

The model skill was evaluated for calibration and cross validation analyses. Calibration uses all years of the record and determines the weights separately for each station. The LEPS score is then calculated for the same years of data, giving no independent testing. Cross-validation, however, provides a more independent assessment of the forecast skill and of the weights applied to each model [*Elsner and Schmertmann*, 1994; *Michaelsen*, 1987]. Cross-validation is performed sequentially by removing one year of data, calibrating the model on the remaining years, and then testing the model on the one year of data that has been removed, giving an independent forecast for that particular year of the record. The year that was removed is then returned to the data set, and the procedure is repeated on the next year in the record. This procedure is repeated until all years have been removed and replaced and independent forecasts made for each year of the record.

Application to Eastern Australia

The model described in the previous section is applied to five Australian catchments. A seasonal streamflow forecast is made using the SOI, SST, and streamflow data from the previous season (i.e., persistence) to forecast streamflow for the next season. For example, the spring (Sep-Nov) streamflow is forecasted using the winter (Jun-Aug) SOI, SST, and streamflow data. The 12 SST series identified by *Drosdowsky and Chambers* [1998] were tested and the series that had the highest correlation with streamflow in the next season was used in the model. The seasons in this section are for the southern hemisphere and are defined as summer (Dec-Feb), autumn (Mar-May), winter (Jun-Aug), and spring (Sep-Nov). The streamflow data is taken from five unimpaired Australian catchments (see Figure 7), which represent five different regions of Australia with an uninterrupted 47-year record of monthly streamflow covering the period 1950 to 1996.

A summary of the results is presented in Tables 1 and 2. Table 1 presents the optimized weights when all the data are used to calibrate the model. Table 2 presents the LEPS

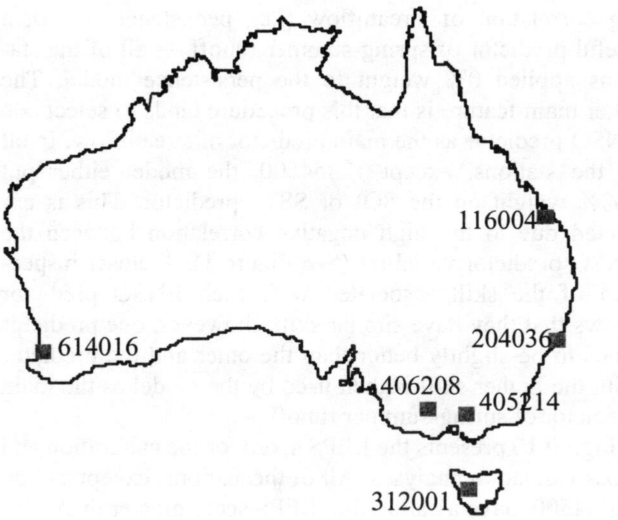

Figure 7. Location of the Australian streamflow stations used in this study.

score for the calibration and cross-validation studies. In both tables, the LEPS scores for the high flow seasons are marked with a * symbol.

The weights in Table 1 suggest, that except for station 312001, persistence is the best predictor of winter and spring streamflow. The results are more variable in summer and autumn; and in several stations, the SOI or SST are combined with persistence to give the best forecasts. The results are consistent with previous studies indicating that the streamflow-ENSO link is strongest in late spring and summer months (*Chiew et al.*, 1998).

Table 2 indicates that for the calibration analyses the LEPS scores are always greater than 10%, which is generally considered a LEPS score with good skill. The LEPS score in Table 2 for the cross-validation analyses drops considerably when compared to the LEPS score for the calibration analysis. Nevertheless, it is encouraging to note that the LEPS score for many stations are still above 10%, as in 11 of the 20 analyses. The LEPS score for practically all the analyses is also above zero, indicating that the forecast model has better skill than does a model forecasting with basic climatology. The exception is perhaps the Tasmanian station (312001), which has been shown in previous studies to have a relatively weak streamflow-ENSO link (*Chiew et al.*, 1999a).

Figure 8 presents examples of a poor and good exceedance probability forecast curves for station 405214 during the spring (Sep-Nov) streamflow forecast. At this station, the forecast is entirely based on persistence (Jun-Aug streamflow). The 1952 forecast is an example of good forecast (LEPS score of 32.4%). If a 90% probability of exceedance forecast were used for managing the water

Table 1 Predictor variables used for each season and the corresponding optimized weights. The SST series are from *Drosdowsky and Chamber* [1998]. Per: persistence; SOI: Southern Oscillation Index.

Station	Summer (DJF)	Autumn (MAM)	Winter (JJA)	Spring (SON)
405214	Per (94%) SST2 (6%)	SOI (30%) SST1 (58%) SST6 (12%)	Per (100%)*	Per (100%)*
614016	Per (100%)	Per (100%)	Per (100%)*	Per (100%)*
312001	SST6 (100%)	Per (100%)*	SST10 (100%*)	SST1 (100%)*
204036	Per (62%)* SOI (38%)*	Per (62%)* SOI (38%)*	Per (100%)*	Per (100%)
116004	SST9 (100%*)	Per (100%)*	Per (100%)	Per (100%)

* Season where the flow is at least 15% of the total annual flow

Table 2 Summary of LEPS scores for calibration and cross validation (in parentheses) analyses.

Station	Summer (DJF)	Autumn (MAM)	Winter (JJA)	Spring (SON)
405214	24.8% (14.4%)	14.7% (0.1%)	16.6% (9.6%)*	26.4% (17.4%)*
614016	30.8% (24.1%)	30.0% (23.2%)	21.1 (14.0%)*	28.4% (21.2%)*
312001	12.6% (0.2%)	12.6% (3.0%)*	10.0% (-6.9%)*	17.7% (5.9%)*
204036	23.4% (8.4%)*	28.6% (12.1%)*	30.6% (25.7%)*	28.0% (18.5%)
116004	18.2% (1.0%)*	18.1% (-0.9%)*	35.9% (22.1%)	22.6% (11.5%)

* Season where the flow is at least 15% of the total annual flow

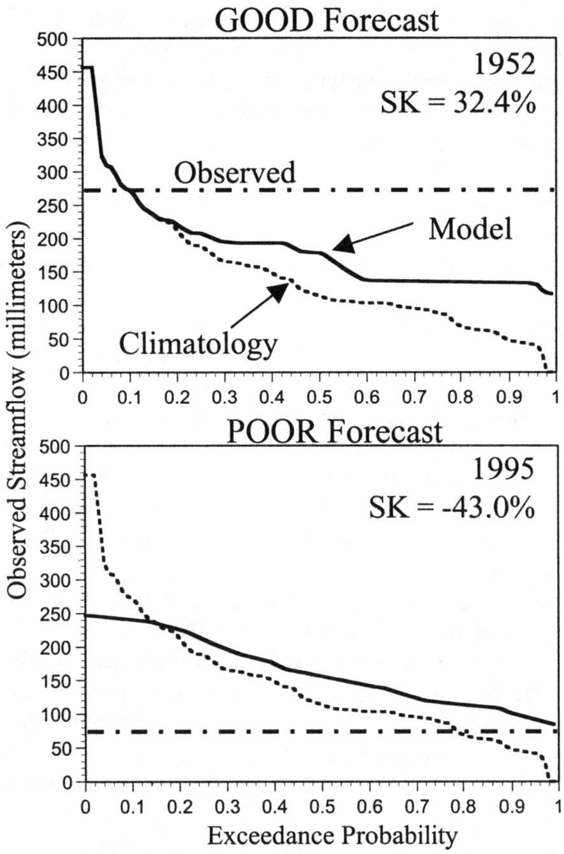

Figure 8. Examples of Poor and Good forecasts developed for the Spring (SON) season at station 405214 in Australia. The year of the forecast and LEPS score for that year are given in the top right corner of each plot. The horizontal line in each plot represents the observed streamflow value. The dashed curve is the climatology forecast and solid curve is the model forecast.

resources system, the user would have correctly anticipated an inflow of 135 mm, a figure higher than that arrived at the 90% level using only climatological information, 50 mm. The actual flow was 275 mm. Risks are associated with the use of poor forecasts, however. For example, if the 90% probability of exceedance forecast for 1995 were depended upon, the user would have anticipated an inflow of 110 mm, but the actual inflow was only 75 mm.

Application to the Columbia River Basin

The streamflow forecast described earlier is also applied to eight catchments in the Columbia River basin located in the Pacific northwest United States (See Figure 9). This application of the streamflow forecast will focus on a long lead time forecast (6 months in advance) of spring-summer runoff.

Table 3 presents the optimized weights for each model from the calibration analysis. It is clear that the six-month lag correlation of streamflow (i.e., persistence) is not a useful predictor of spring-summer runoff — all of the stations applied 0% weight to the persistence model. The other main feature is that this procedure tends to select one ENSO predictor as the main predictor of streamflow. In all of the stations, except 12354500, the model either put 100% weight on the SOI or SST_w predictor. This is expected due to the high negative correlation between the ENSO predictor variables (See Figure 1). A closer inspection of the skill associated with each ENSO predictor shows that they have similar skill; however, one predictor tends to be slightly better than the other and the predictor with the higher skill is then used by the model as the main predictor of spring-summer runoff.

Figure 10 presents the LEPS score for the calibration and cross validation analyses. All of the stations, except station 12404500, have a calibration LEPS score greater than 10% which is generally considered a LEPS score with good skill. The LEPS scores for the cross validation analysis drop considerably; however, it is encouraging that the LEPS scores are all better than Climatology, except at station 12404500. It is noteworthy, that the cross validation analysis at station 12454900 has a LEPS score slightly greater than 10%.

Examples of the exceedance probability curves developed using this procedure and the climatology forecast are shown for good and poor forecasts at station 12354500 in Figure 11. The 1973 forecast is an example of a good forecast that has a LEPS score of 29.4%. The observed April-September runoff was 226 x 10^3 hectare-meters which is below normal — the normal runoff is approximately 493 x 10^3 hectare-meters. If the 75% exceedance probability forecast was used, the model would have given a forecast of approximately 265 x 10^3 hectare-meters. A higher value (400 x 10^3 hectare-meters) would have been used if only the Climatology forecast was used. Risks are associated with the use of poor forecasts, however. For example, if the 75% probability of exceedance forecast for 1989 were depended upon, the user would have anticipated a April-September runoff of 583 x 10^3 hectare-meters, but the actual runoff was 434 x 10^3 hectare-meters. Nevertheless, the positive skill scores in many of the analyses suggest that there are more good forecasts than poor forecasts.

USE OF SEASONAL STREAMFLOW FORECASTS FOR WATER RESOURCES MANAGEMENT

The use of seasonal climate and hydrologic forecasts in the management of water resource systems is an emerging area of research. This is particularly relevant due to the fact that water authorities worldwide take a conservative approach to management of their water supply systems. Water authorities, with a reliable streamflow forecast, can allocate water supplies optimally for competing water users

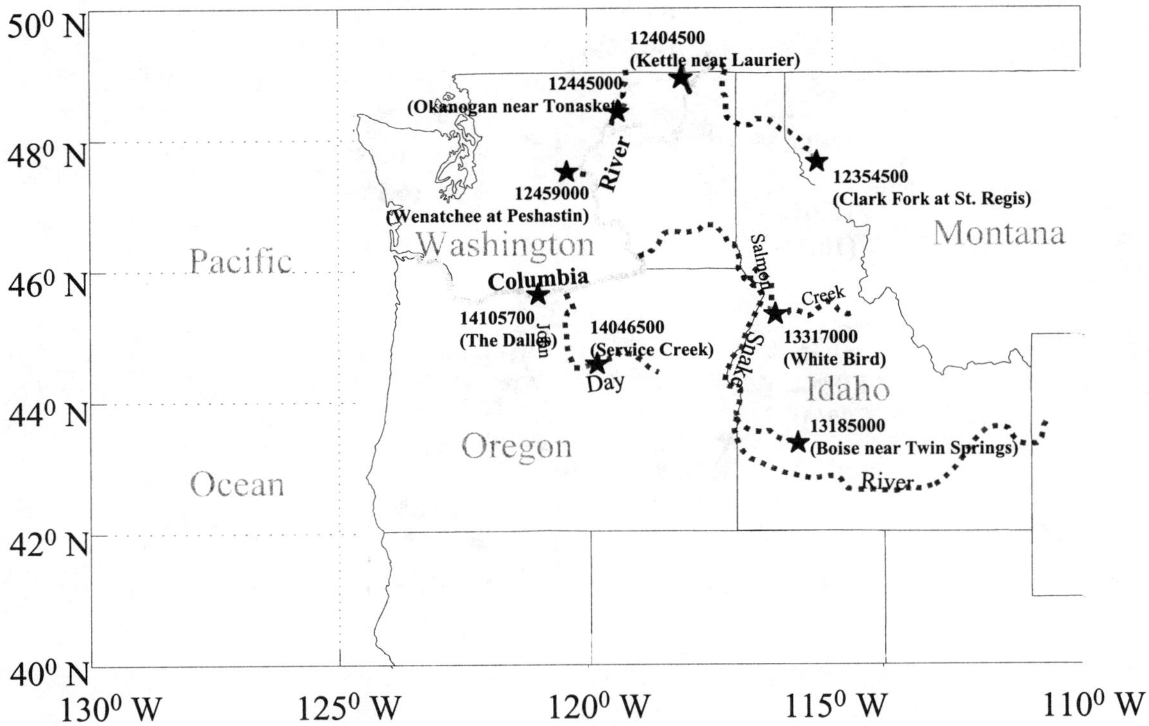

Figure 9. Location of the eight stations used to represent the Columbia River Basin with unimpaired streamflow. The length of record at Station 14105700 is 1911 to 1992, and the remaining stations have data from 1933 to 1992.

Table 3 Optimized weights for each model in calibration analysis.

Station Number	Persistence	SOI	SST_w
12354500	0.0	0.8	0.2
12404500	0.0	1.0	0.0
12445000	0.0	0.0	1.0
12459000	0.0	0.0	1.0
13185000	0.0	0.0	1.0
13317000	0.0	0.0	1.0
14046500	0.0	1.0	0.0
14105700	0.0	1.0	0.0

(e.g., hydropower generation, agricultural, domestic) and for the maintenance of environmental flows.

The value of seasonal streamflow forecasts is demonstrated by *Chiew et al.* [1999b] for an urban water supply system and a rural irrigation system in south-east Australia. For the urban water supply system, a water supply model for a township of 10,000 people was used to investigate the benefits of using ENSO information to derive water restriction rules. The study showed that incorporating SOI in establishing water restriction rules for the urban water supply system reduced the overall loss impact to the community by almost 30%. The loss impact was defined as the total loss associated with a certain restriction stage times a duration factor that reflects the duration of the water restriction.

For a rural irrigation system, simulations were made of four alternative water management and allocation options for a 84,600 km^2 catchment using almost 100 years of data are reproduced in Figure 12. In the "no risk" option, the cropping area is determined based on the water allocation announced by the water agency at the start of the planting season, and this allocation is determined based only on the available water in the reservoir. In the "farmer risk" option, the cropping area is determined based on the farmers' anticipated final allocation, which is estimated as a linear function of the announced allocation and antecedent flow conditions. In the "agency risk" option, the allocation announced by the water agency is determined based on the available storage (or carryover storage) plus the 90% probability of exceedance inflow for the coming months forecasted using SOI. The "combined risks" option considers both the farmers and the water agency taking the above risks.

Figure 12(a) indicates that in about 50% of the years, the summer cropping areas in the four options are the same because there is sufficient water in the reservoir to announce a maximum allocation. In other years, the cropping area, and therefore the amount of irrigation water used, increases with the risks. Figure 12(b) shows that with greater risks, there is a higher chance of the crops failing, with insufficient water to sustain all the planted

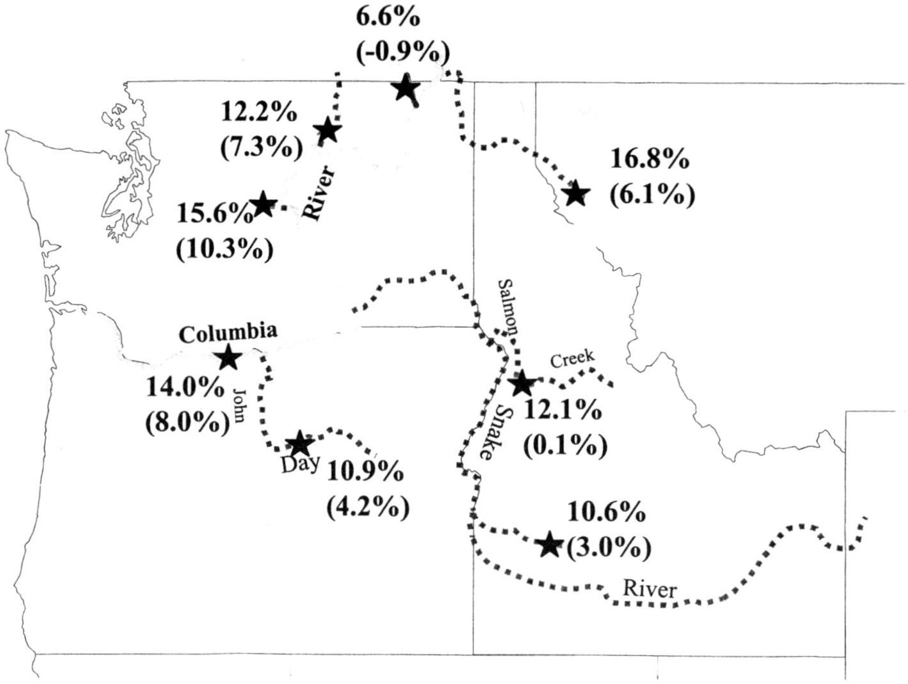

Figure 10. Summary of the LEPS scores for calibration and cross validation analyses. The LEPS score for the cross validation analysis is shown in parentheses.

crops in 30% and 40% of the years in the "farmer risk" and "combined risks" simulations, respectively. There is no crop failure in the "agency risk" option because the allocation announced by the water agency cannot be reduced, and the conservative carry-over storage used by the agency prevents the reservoir from being emptied in this simulation. However, there are potential risks involved as shown in Figure 12(c) where the reservoir is drawn to much lower levels in the "agency risk" option compared to the "farmer risk" option.

The above study is limited by the constraints of the model, the many assumptions in the simulations, and simplification of the system. Nevertheless, the results suggest that although there are risks involved, there is a net benefit in using streamflow forecasts derived from ENSO and the serial correlation in streamflow to help manage water resources systems.

Despite the potential benefits, water agencies still continue to take a very conservative approach in managing water resources systems because of both political reasons and concerns about water shortfalls [*Long and McMahon*, 1996]. Nevertheless, there are increasing numbers of studies investigating the use of seasonal streamflow forecasts that also take into account other factors in managing land and water resources systems. It is likely that these studies will lead to the use of seasonal streamflow forecast and the decisions on irrigation water allocation and environmental flow requirements will be more realistically based.

DISCUSSION AND CONCLUSIONS

There are interesting comparisons that can be made of the ENSO influence on the hydrology of eastern Australia and the western United States. The ENSO influence on eastern Australia hydrology is stronger than the impacts experienced in the western United States. There is also a difference in the timing of the response in streamflow to ENSO forcing. In parts of eastern Australia, the prior season ENSO indicator along with persistence may be good predictors of the next season streamflow. In the western U.S., however, there is a longer lag in the ENSO response and persistence is not a useful predictor for a six month lag period. The influence in both of these regions is strong enough to warrant long-range probabilistic forecasts of streamflow that are currently not formally made for these regions.

In Australia, the generation of streamflow is not snowmelt driven, like the western U.S., and water authorities do not issue a seasonal streamflow forecast. Thus, water authorities take a very conservative approach to managing water resource systems. For example, they will assume that the resources available for a given water year consist only of the water that has been stored by the beginning of the year, less evaporative and distribution losses, plus the minimum historically observed inflows (or seasonal inflows of high probability of exceedance) over the year. The results of this study and previous studies demonstrate that

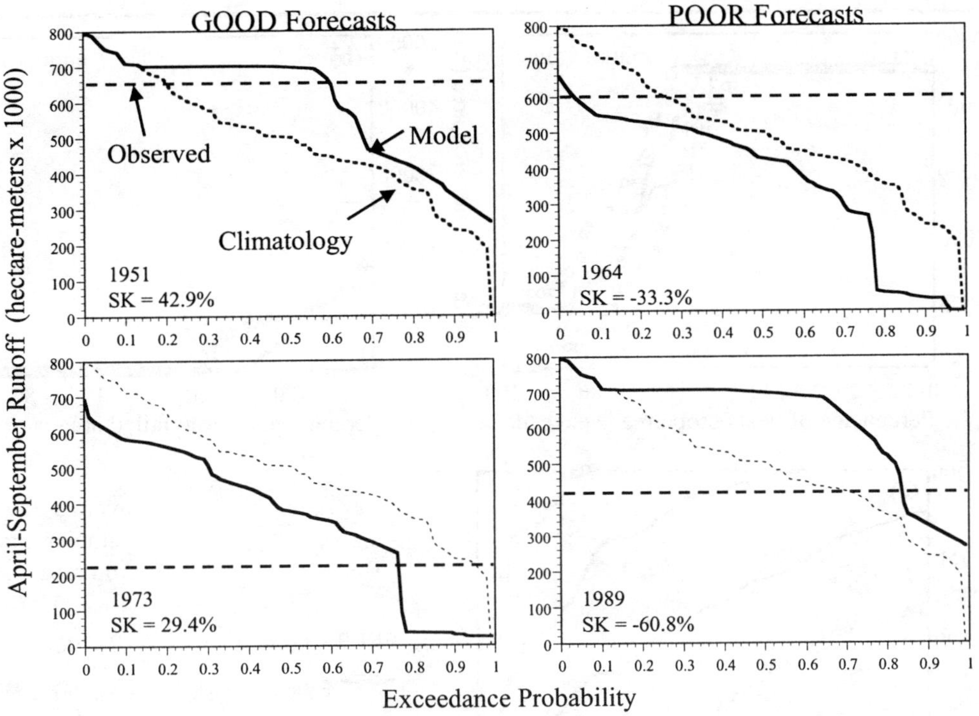

Figure 11. Examples of Good and Poor forecasts for the six month lead time forecast of April-September runoff for station 12354500. The horizontal line in each plot represents the observed streamflow, the dashed curve is the Climatology forecast, and the solid curve is the model forecast.

ENSO information can be used to better manage water resource systems in Australia by making a seasonal streamflow forecast (i.e., forecasting of streamflow for the next season).

In the western U.S., the long lag time between the peak of an ENSO event and the hydrologic anomaly make it possible to make a long-range streamflow forecast (i.e., six month lead time). This delayed response is partially due to the dominance of snowmelt runoff in this region. *Garen* [1993] has studied the use of the SOI in streamflow forecasts for the Flathead River in Montana and found it possible to make long range forecasts of spring-summer runoff. The work presented here extends *Garen's* [1993] preliminary investigation by making an exceedance probability forecasts for a large river basin based on ENSO indicators.

An alternative to the statistical approaches presented here is the use of physically-based hydrologic models to make long range streamflow forecasts. The general approach is to initialize macro-scale hydrologic models with current hydrologic conditions (e.g., soil moisture) and then run the models out using historical traces of hydrologic data. This produces a series of forecasts that can be used to develop a final exceedance probability forecast. This approach is currently being used by the U.S. National Weather Service in the Advance Hydrologic Prediction System (AHPS) which can produce forecasts with lead times of a few days to several months. Currently, the AHPS provides 90-day exceedance probability flows for the Des Moines River basin. A similar approach has been implemented by *Hamlet et al.* [1998] for the Columbia River basin; however, they incorporated ENSO information by conditioning the forecast on the current ENSO phase.

Finally, it should be noted that the work presented here focuses on the seasonal impact of ENSO on regional hydrology. The ability to forecast seasonal streamflow is important for the management of water supply systems. The influence of ENSO on extreme hydrologic events that occur on a daily time scale, however, is also very important for management of flood control systems. Recent research [*Gershunov*, 1998; *Gershunov and Barnett*, 1998; *Cayan et al.*, 1999] has investigated the accentuation of ENSO effects on extreme hydrologic events in the U.S. and shown that the ENSO phase affects the frequency of daily precipitation events and the amount of precipitation on a wet day in many regions of the United States.

The preliminary results of this study reveal some of the practical uses of ENSO information for managing water supply and flood control systems. However, there is a need to better understand the impact of ENSO on hydrologic processes at different spatial and temporal scales, and the other mechanisms that cause extreme hydrologic events.

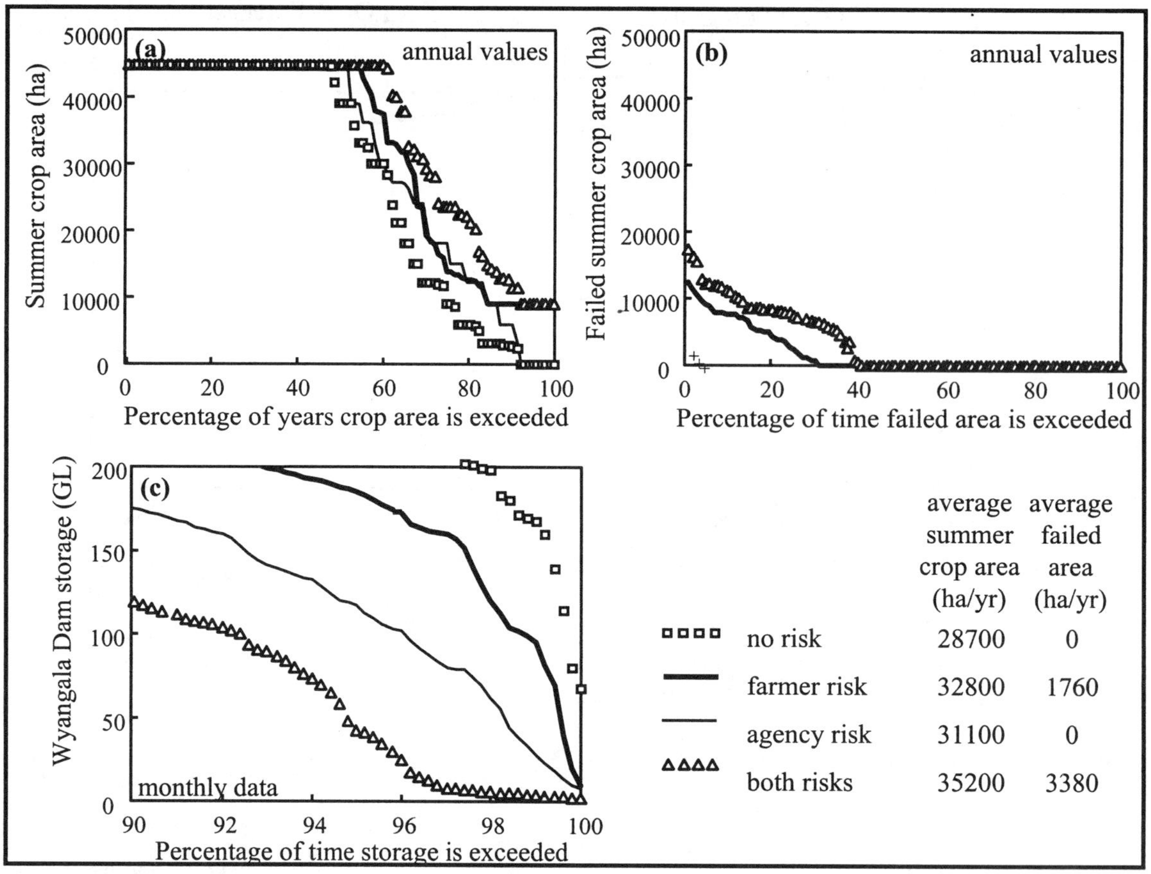

Figure 12. Results of model simulations of alternative water management and allocation options in the Lachlan River catchment in south-east Australia. Exceedance probability curves are shown for (a) total summer crop area (hectares, ha), (b) total failed summer crop area (hectares, ha), and (c) Wyangala Dam storage (gigaliters, GL) using four different management strategies.

The increased knowledge of these linkages will provide water authorities with the evidence, and confidence, to incorporate climate information into the management of water resource systems.

Acknowledgements. The authors especially wish to thank Trevor Casey, Australian BOM, for providing the SST time series. The work at the University of Nevada and the University of California is supported in part by the University of California Pacific Rim Research Program and the National Science Foundation under award EAR 9421030. The work at the University of Melbourne is supported by the Department of Civil and Environmental Engineering and the Australian Land and Water Resources Research and Development Corporation.

REFERENCES

Allan, R.J., G.S. Beard, A. Close, A.L. Herczeg, P.D. Jones, and H.J. Simpson, Mean sea level pressure indices of the El Niño-Southern Oscillation: relevance to discharge in South-Eastern Australia, *CSIRO Division of Water Resources, Canberra, Divisional Report 96/1*, 23 pp., 1996.

Afifi, A.A. and S.P. Azen, *Statistical Analysis - a Computer Oriented Approach,* Academic Press, New York, xx + 442 pp., 1979.

Barnett, T.P., M. Latif, N. Graham, M. Flugel, S. Pazan, and W. White, ENSO and ENSO-related predictability: Part 1- prediction of equatorial Pacific sea surface temperatures with a hybrid coupled ocean-atmosphere model, *J. Clim., 6,* 1545-1566, 1993.

Cayan, D.R., Interannual climate variability and snowpack in the western United States, *J. Clim., 9(5)*, 928-948, 1996.

Cayan, D.R., and D.H. Peterson, The influence of north Pacific atmospheric circulation on streamflow in the west, in *Aspects of Climate Variability in the Pacific and the Western Americas,* Geophysical Monograph Series, Vol. 55, pp. 375-397, AGU, Washington DC, 1989.

Cayan, D.R. and R.H. Webb, El Niño/Southern Oscillation and streamflow in the United States, in *El Niño: Historical and Paleoclimatic Aspects of the Southern Oscillation,* edited by H.F. Diaz and V. Markgraf, Cambridge University Press, pp. 29-68, 1992

Cayan, D.R., K.T. Redmond, L. Riddle, Accentuation of ENSO effects on extreme hydrologic events over the western United States. *J. Clim.*, in press, 1999.

Chen, D., Zebiak, S.E., Busalacchi, A.J., and Mark A. Cane, An improved procedure for El Niño forecasting: implications for predictability, *Science 269(5231)*, 1699-1702, 1995.

Chiew, F.H.S, T.A. McMahon, J.A. Dracup, and T. Piechota, El-Niño/Southern Oscillation and streamflow patterns in south-east Australia, *Australian Civil Engineering Transactions, Inst. Engr. Aust., CE 36(4)*, 285-291, 1994.

Chiew, F.H.S., T.C. Piechota, J.A. Dracup, and T.A. McMahon, El Niño/Southern Oscillation and Australian rainfall, streamflow and drought: links and potential for forecasting, *J. Hydrol., 204 (1-4)*, 138-149, 1998.

Chiew, F.H.S., T.A. McMahon, S.L. Zhou, and T.C. Piechota, Streamflow variability, seasonal forecasting and water resources systems, in *Applications of Seasonal Climate Forecasting in Agricultural and Natural Ecosystems - The Australian Experience,* edited by G. Hammer, N. Nicholls and C. Mitchell, Kluwer Academic, in press, 1999a.

Chiew, F.H.S., S.L. Zhou, K.R. Panta, P.D. Erlanger, T.A. McMahon, and N.M. Clarkson, Use of seasonal streamflow forecasts for water supply management, *Proceedings of the 25th International Hydrology and Water Resources Symposium,* Brisbane, July 1999, Institution of Engineers, Australia, in press, 1999b.

Dettinger, M.D., D.R. Cayan, and K.T. Redmond, United States streamflow probabilities based on forecasted La Niña, winter-spring 2000, *Experimental Long Lead Forecast Bulletin*, 8(4), 1999.

Dettinger, M.D., D.R. Cayan, G.M. McCabe, and J.A. Marengo, Multiscale streamflow variability associated with El Niño/Southern Oscillation, in *El Niño and the Southern Oscillation – Multiscale Variability and Global and Regional Impacts*, Cambridge University Press, 113-146, 2000.

Dracup, J.A., and E. Kahya, The relationships between U.S. streamflow and La Niña events, *Water Resour. Res., 30(7)*, 2133-2141, 1994.

Drosdowsky, W., and L. Chambers, Near global sea surface temperature anomalies as predictors of Australian seasonal rainfall, Bureau of Meteorology Research Centre, Australia, *Research Report 65*, 39 pp, 1998.

Elsner, J.B., and C. P. Schmertmann, Assessing forecast skill through cross validation, *Weather and Forecasting, 9*, 619-624, 1994.

Garen, D.C., Improved techniques in regression-based streamflow volume forecasting, *J. Water Resour. Planning and Management, 118(6)*, 654-670, 1992.

Garen, D.C., Revised surface-water supply index for western United States, *J. Water Resour. Planning and Management, 119(4)*, 437-454, 1993.

Gershunov, A., ENSO influence on intraseasonal extreme rainfall and temperature frequencies in the contiguous United States: implications for long-range predictability, *J. Clim., 11(12)*, 3192-3203, 1998.

Gershunov, A., and T.P. Barnett, ENSO influence on intraseasonal extreme rainfall and temperature frequencies in the contiguous United States: observations and model results, J. Clim., 11 (7), 1575-1586, 1998.

Gingras, D. and K. Adamowski, The impact of El Niño-Southern Oscillation on central Canadian floods and droughts, *Canadian Journal of Civil Engineering, 22(4)*, 834-837, 1995.

Guetter, A.K., and K.P. Georgakakos, Are the El-Niño and La-Niña predictors of Iowa River seasonal flow?, *J. Appl. Meteorol., 35(5)*, 690-705, 1996.

Halpert, M.S., and C.F. Ropelewski, Surface temperatures patterns associated with the Southern Oscillation, *J. Clim., 5(6)*, 577-593, 1992.

Hamlet, A.F., D.P. Lettenmaier, and B. Nijssen, Columbia River basin streamflow forecasting based on ENSO climate signal using a macro scale hydrology model and resampling meteorological data, *EOS, Transaction, American Geophysical Union, 1998 Spring Meeting, Vol. 79(17)*, S117, 1998.

Ji, M., A. Leetmaa and V.E. Kousky, Coupled model forecasts of ENSO during the 1980s and 1990s at the National Meteorological Center, *J. Clim., 9(12)*, 3105-3120, 1996.

Jiang, N., D. Neelin and M. Ghil, Quasi-quadrennial and quasi-biennial variability in the equatorial Pacific, *Climate Dynamics, 12*, 101-112, 1995.

Kahya, E., and J.A. Dracup, U.S. streamflow patterns in relation to the El Niño/Southern Oscillation, *Water Resour. Res., 29(8)*, 2491-2503, 1993.

Kahya, E., and J.A. Dracup, The influences of Type 1 El Niño and La Niña events on streamflows in the southwestern United States, *J. Clim., 7(6)*, 965-976, 1994.

Kuhnel, I., T.A. McMahon, B.L. Finlayson, A. Haines, P.H. Whetton, and T.T. Gibson, Climatic influences on streamflow variability: A comparison between Southeastern Australia and Southeastern United States of America, *Water Resour. Res., 26(19)*, 2483-2496, 1990.

Liu, Z., J.B. Valdés, and D. Entekhabi, Merging and error analysis of regional hydrometeoroloigc anomaly forecasts conditioned on climate precursors, *Water Resour. Res., 34(8)*, 1959-1969, 1998.

Long, A.B. and T.A. McMahon, Review of research and development opportunities for using seasonal climate forecasts in the Australian water industry, *Land and Water Resources Research and Development Corporation, Occasional Paper CV02/96*, 46 pp, 1996.

Maryon R.H. and A.M. Storey, A multivariate statistical model for forecasting anomalies of half-monthly mean surface pressure, *J. Climatol. 5*, 561-578, 1985.

McBride, J.L., and N. Nicholls, Seasonal relationships between Australian rainfall and the Southern Oscillation, *Mon. Wea. Rev., 111*, 1998-2004, 1983.

McKerchar, A.I., and C.P. Pearson, Forecasts of seasonal river flows using Southern Oscillation Index, *J. of Hydrol. (NZ), 32(2)*, 16-29, 1994.

McMahon, T.A., B.L. Finlayson, A.T. Haines, and R. Srikanthan, *Global Runoff - Continental Comparisons of Annual Flows and Peak Discharges*, Catena Paperback, Cremlingen-Destedt, 166 pp, 1992.

Mechoso, C.R., and G.P. Iribarren, Streamflow in southeastern South America and the Southern Oscillation, *J. Clim., 5(12)*, 1535-1539, 1992.

Michaelsen, J., Cross-validation in statistical climate forecast models, *J. Clim. Appl. Meteor., 26*, 1589-1600, 1987.

Moss, M.E., Pearson, C.P., and A.I. McKerchar, The Southern Oscillation index as a predictor of the probability of low streamflows in New Zealand, *Water Resour. Res., 30(10)*, 2717-2723, 1994.

Nicholls, N., Sea surface temperatures and Australian winter rainfall, *J. Clim., 2*, 965-973, 1989.

Penland, C., and T. Magorian, Prediction of Niño 3 sea-surface temperatures using linear inverse-modeling, *J. Clim., 6*, 1067-1076, 1993.

Philander, S.G., *El Niño, La Niña, and the Southern Oscillation,* Academic Press, San Diego, Calif., 293 pp., 1990.

Piechota, T.C., and J.A. Dracup, Drought and regional hydrologic variations in the United States: associations with the El Niño/Southern Oscillation, *Water Resour. Res., 32(5),* 1359-1373, 1996.

Piechota, T.C., J.A. Dracup, and R.G. Fovell, Western U.S. streamflow and atmospheric circulation patterns during El Niño-Southern Oscillation (ENSO), *J. Hydrol., 201,* 249-271, 1997.

Piechota T.C., F.H.S. Chiew, J.A. Dracup, and T.A. McMahon, Seasonal streamflow forecasting in Eastern Australia and the El Niño-Southern Oscillation, *Water Resour. Res., 34(11),* 3035-3044, 1998.

Piechota, T.C., F.H.S. Chiew, J.A. Dracup, and T.A. McMahon, Development of an exceedance probability streamflow forecast using the El Niño-Southern Oscillation and sea surface temperatures, J. *Hydrologic Engr,* in press, 2000.

Piechota, T.C., and J.A. Dracup, Long range streamflow forecasting using El Niño-Southern Oscillation Indicators, *J, Hydrologic Engr., 4(2),* 144-151, 1999.

Potts, J.M., C.K. Folland, I.T. Jolliffe, and D. Sexton, Revised "LEPS" scores for assessing climate model simulations and long-range forecasts, *J. Clim., 9, 34-53,* 1996.

Rasmusson, E.M., El Niño and variations in climate, *Am. Sci.,73,* 168-177, 1985.

Redmond, K.T., and R.W. Koch, Surface climate and streamflow variability in the western United States and their relationship to large-scale circulation indices, *Water Resour. Res., 27(9),* 2381-2399, 1991.

Ropelewski, C.F., and M.S. Halpert, North American precipitation and temperature patterns associated with the El Niño/Southern Oscillation (ENSO), *Mon. Wea. Rev., 114,* 2352-2362, 1986.

Ropelewski, C.F., and M.S. Halpert, Global and regional scale precipitation patterns associated with the El Niño/Southern Oscillation, *Mon. Wea. Rev., 115,* 1606-1626, 1987.

Ropelewski, C.F., and M.S. Halpert, Precipitation patterns associated with the high index phase of the Southern Oscillation, *J. Clim., 2,* 268-284, 1989.

Schoner T., and S.E. Nicholson, The relationship between California rainfall and ENSO events, *J. Clim., 2,* 1258-1269, 1989.

Shabbar, A., B. Bonsal, and M. Khandekar, Canadian precipitation patterns associated with the Southern Oscillation, *J. Clim., 10(12),* 3016-3027, 1997.

Simpson, H.J., M.A. Cane, S.K. Lin, and S.E. Zebiak, Forecasting annual discharge of River Murray, Australia, from a geophysical model of ENSO, *J. Clim., 6,* 386-390, 1993.

Slack, J.R., and J.M. Landwehr, Hydro-Climatic Data Network (HCDN): A U.S. Geological Survey streamflow data set for the United States for the study of climate variations, 1874-1988, *U.S. Geological Survey Open-File Report 92-129,* 1992.

Smith, T.M. and C.F. Ropelewski, Quantifying Southern Oscillation-precipitation relationships from an atmospheric GCM, *J. Clim., 10(9),* 2277-2284, 1997.

Sun, H.B. and D.J. Furbish, Annual precipitation and river discharges in Florida in response to El Nino and La Nina sea surface temperature anomalies, *J. Hydrol., 199(1-2),* 74-87, 1997.

Troup, A.J., The Southern Oscillation, *Quarterly Journal of the Royal Meteorological Society, 91,* 490-506, 1965.

U.S. Department of Agriculture (USDA), Natural Resources Conservation Service, Water Supply Outlook for the Western United States. Portland, OR., 1999.

Wang, G. and E.A.B. Eltahir, Use of ENSO information in medium- and long-range forecasting of the Nile floods. *J. Clim.,* in press, 1999.

Ward, M.N. and C.K. Folland, Prediction of seasonal rainfall in the North Nordeste Brazil using eigenvectors of sea-surface temperature, *Int. J. Climatol., 11,* 711-743, 1991.

Woolhiser, D.A., and T.O. Keefer, Southern Oscillation effects on daily precipitation in the Southwestern United States, *Water Resour. Res., 29 (4),* 1287-1295, 1993.

Wright, P.B., Homogenized long-period Southern Oscillation indices, *Int. J. Climatol., 9,* 33-54, 1989.

Yarnal, B., and H.F. Diaz, Relationships between extremes of the Southern Oscillation and the winter climate of the Anglo-American Pacific Coast, *J. Climatol., 6,* 197-219, 1986.

Zebiak, S.E. and M.A. Cane, A model of the El Niño-Southern Oscillation, *Mon. Wea. Rev., 115,* 2262-2278, 1987.

Department of Civil and Environmental Engineering, University of Nevada, Las Vegas, 4505 Maryland Parkway, Box 454015, Las Vegas, NV 89154-4015, piechota@ce.unlv.edu.

Cooperative Research Centre for Catchment Hydrology, University of Melbourne, Parkville, Victoria 3052, Australia.

Department of Civil and Environmental Engineering, University of California, 760 Davis Hall, Berkeley, CA 94720-1710.